THE CONFERENCE ON

COMPUTERS IN PHYSICS INSTRUCTION

PROCEEDINGS

August 1–5, 1988
Raleigh, North Carolina

THE CONFERENCE ON

COMPUTERS
IN PHYSICS
INSTRUCTION

PROCEEDINGS

Editors
Edward F. Redish and John S. Risley

Managing Editor
Nancy Margolis

ADDISON-WESLEY PUBLISHING COMPANY, INC.
THE ADVANCED BOOK PROGRAM
Redwood City, California • Menlo Park, California • Reading, Massachusetts
New York • Amsterdam • Don Mills, Ontario • Sydney • Bonn • Madrid
Singapore • Tokyo • San Juan • Wokingham, United Kingdom

Publisher: *Allan M. Wylde*
Production Manager: *Jan V. Benes*
Electronic Production Consultant: *Mona Zeftel*
Promotions Manager: *Laura Likely*
Cover Design: *Michael A. Rogondino*

Library of Congress Cataloging-in-Publication Data
Conference on Computers in Physics Instruction (1988 : North Carolina State
  University)
     The Conference on Computers in Physics Instruction : proceedings / [edited
  by] Edward F. Redish, John S. Risley.
          p. cm.
     Conference held at North Carolina State University, Aug. 1988.
     Includes index.
     1. Physics--Study and teaching--Data processing--Congresses. 2. Physics--
  Computer-assisted instruction--Congresses. I. Redish, Edward F. II. Risley, John
  S. III. Title.
  QC30.C636 1990   530'.07'8--dc19   90-364
  ISBN 0-201-16306-3

A complete list of trademarks appears on page 542.

This book was typeset in Quark Xpress™ using a Macintosh computer. Camera-
ready output is from an Apple LaserWriter II NTX Printer.

ABCDEFGHIJ-MA-89

A000005942237

Contents

Section 3. Computers in the Physics Laboratory

Invited Papers

Contributed Papers

Section 4. Physics Education Research and Computers

Section 5. Computational Physics and Spreadsheets

Section 6. Computer Tutorials in Physics

Section 7. Physics Lecture Demonstrations Using Computers

Section 8. Authoring Tools and Programming Languages

Section 9. Computer Utilities for Teaching Physics

Preface

The computer is the dynamo of the information age. The raw computing power currently available in a desktop computer far outstrips what was commonly available on mainframes twenty-five years ago. The software now available to us is perhaps even more revolutionary than the raw power; it has led to fundamentally new ways to interact with the computer. Computers are revolutionizing activities ranging from accounting to composing rock music. Physics researchers, accustomed to being at the forefront of technology, have been deeply affected by the computer revolution. This effect has serious implications for both how we teach physics and what we teach.

The question is not whether the computer can be used to teach physics. It *is* being used to teach physics. Rather, the question is how the computer can be used to teach physics most effectively. The potential contributions of the computer to the teaching of physics—and to the understanding of physics—are immense. We must take this opportunity to think hard about what we really want our students to learn. The computer can let us rearrange the curriculum into units compelled by the logical progression of the material instead of by the computational limitations of our students or the physical limitations of our laboratory facilities.

But if we are to use the opportunity of the computer revolution to rethink the physics curriculum, we must be realistic and thorough. We must begin the task of curriculum reform by rethinking both content and delivery at all levels. Standard physics texts link courses from the high school to the graduate level, and these links are interdependent. The innovations in how we teach are now happening at all levels—from the elementary school to the graduate school. Thus our reform efforts will require the interaction of state-of-the-art researchers, college teachers interested in curriculum reform, high school physics teachers, education specialists, and experts in high technology.

There are many possibilities for using the computer to teach physics that are obvious and many more that we haven't yet thought of. A computer in the physics classroom can supplement lecture demonstrations. It can help students solve homework problems and provide drill and practice when they don't understand. It can help build their intuition either directly or in conjunction with lab work. It can help students take and analyze data in the lab, and it can let them begin independent research at a very early stage. Some of these computer uses will have strong positive impacts on student learning; some will have negative impacts. Our task as teachers is to classify uses according to effectiveness. We must purge uses that don't work, and test, refine, and retain uses that do.

The Conference on Computers in Physics Instruction

To allow physics teachers and software developers in physics education to come together and see the state of the art in using computers to teach physics, we organized a conference on Computers in Physics Instruction, held August 1-5, 1988, at the McKimmon Center on the campus of North Carolina State University in Raleigh, North Carolina. Nearly four 400 attended from all over the United States and a dozen foreign countries.

The idea for the conference evolved from conversations that took place among members of the physics community, particularly at the biannual meetings of the American Association of Physics Teachers and at a workshop of sixty of the country's leading practitioners in computer teaching held at Dickinson College in February 1986. That group specifically recommended that a large national meeting be held.

The conference on Computers in Physics Instruction included 39 invited lectures and 122 contributed presentations. The conference introduced a number of innovations in the hope of increasing interactions and stimulating future contacts. Besides the oral presentations, there were poster-demonstration sessions, one-hour mini-workshops, and "open" computers that were used to display, discuss, and exchange software. Upon arrival, all conference participants received a package of twenty-one diskettes with thirty-five software programs for five different kinds of computers. (To obtain a set of these diskettes, use the order form in the back of this book.)

The Proceedings

The invited talks and contributions in these proceedings reflect the great diversity of ways physicists use computers to teach physics.

One possibility is, of course, to use the computer in the laboratory to take data and to analyze and display the results quickly and easily. This can simplify some of the technical details of the laboratory and expand its power. The use of the computer for data acquisition and analysis is having a powerful impact on researchers and by implication on our training of graduate students. At the introductory, level the computer can help clarify the meaning of the data and lead to substantial improvement in the student's understanding of difficult physical laws and concepts.

A second major use of the computer is to build models and simulate experiments. Model building is the fundamental component of the theoretical side of physics, and having the computer available vastly expands both the class of problems we can consider and the kinds of models we can build. In some cases we can simulate real-world phenomena that we are prevented from studying in the laboratory by constraints of time, expense, danger, or feasibility. For example, planetary motion is too slow to permit extended observation in a single semester. But many orbits can be simulated and studied conveniently and quickly on the computer screen. Furthermore, our simulations can extend to "what if's." We can try models that do not occur in the real world to see what the implication would be. What would happen if we changed the gravitational-force law a little?

One of the most valuable uses of simulation is to allow students to visualize, through display and animation, concepts that they may find difficult to understand from looking at equations and a single static picture in a text. A number of papers in this volume present a variety of innovative examples of how to use simulations in this way.

In some areas of contemporary physics, computers are more than helpful; they are essential. Much of the work in modern-physics research involves extensive computation. For example, the fields of nonlinear dynamics and chaos. A number of papers in this volume discuss how one might approach introducing the concepts of chaos into the curriculum using the computer. The subject of computational physics is discussed extensively. The implications of modern computer developments for the research environment are discussed in articles on supercomputers and on *Mathematica*, a formidable program that combines calculation, symbolic manipulation, graphing, and text processing.

Papers in this volume also discuss a diverse variety of important issues, including cognitive issues, the role of contemporary physics, and the use of the laserdisc. Other papers deal with the use of the computer as an instructional assistant—or evaluation and recording student performance, and for keeping track of the day-by-day details of teaching, not only in physics, but in any discipline.

Another focus is the practical matter of producing physics courseware. Several papers deal with authoring tools and programming languages, focusing on their special applicability to physics. Specific computer utilities that allow physics applications are also evaluated and described.

For those who wish to explore the possibilities for incorporating computers into physics instruction and to explore programs already in place, there are descriptions of the many instructional physics projects now underway in the United States. These projects range from incorporating a physics curriculum into a fully coordinated local area network, through well-funded microcomputer-based laboratories, to individual efforts to incorporate a single simulation into a tradition curriculum.

Finally, some of the papers deal with the central problem of communicating with one another on the role of the computer in physics teaching. The conference on Computers in Physics Instruction is one such forum, but there are others. Papers in this volume discuss the preparation of educational software for publication, the criteria that will be used to evaluate it, and various publishing possibilities. There is, for example a discussion of *Physics Academic Software*, a joint project of the American Institute of Physics, the American Physical Society, and the American Association of Physics Teachers, which will serve as a new mode of simplifying access to high-quality academic software.

Communicating Teaching Innovations

The conference on Computers in Physics Instruction brought some of the best and most creative workers in many diverse groups together, both to further their interaction and to produce a volume that would have a broad impact.

Educational developments seem to have a much smaller "diffusion coefficient" than corresponding innovations in research. Researchers regularly publish their work, report on it at conferences, read the literature, and travel from university to university giving seminars on their latest work.

There is a great deal of activity in computer use in the classroom, but much of it fails to get published or disseminated beyond the local environment. The American Association of Physics Teachers publishes journals and resource volumes, and holds workshops that reach a significant number of the faithful, but the large majority of college physics teachers remain almost completely unaware of such efforts.

Even if they do become interested, the literature is highly diffuse. Important papers are reported in such journals as the *Journal of Research in Science Teaching* and *Computers in Education*, journals that may not be as familiar to the research physicist as *The Physical Review* or the *Review of Scientific Instruments*. Part of what we are trying to do with this volume is to give an overview of what is now happening and to provide a link (through references in the papers) to the rapidly growing relevant literature. We have assembled a smorgasbord of papers on a wide variety of topics relating to the role of the computer in physics education.

The computer has wrought great changes in the practice of the profession of physics. These changes have serious implications for physics teaching. We offer this proceedings to those who would accept both the changes and their implications and decide to involve themselves in the challenging and exciting subject of teaching physics with the computer.

Edward F. Redish and John S. Risley, Scientific Editors
Nancy H. Margolis, Managing Editor

Organization

Chairmen

John S. Risley, Department of Physics, North Carolina State University, Raleigh, NC 27695-8202, Tel. (919) 737-2524
BITNET: Risley@NCSUPHYS

Edward F. Redish, Department of Physics and Astronomy, University of Maryland, College Park, MD 20742, Tel. (301) 454-7383
BITNET: Redish@UMCINCOM

Steering Committee

Gordon J. Aubrecht, Ohio State University
Edward H. Carlson, Michigan State University
J. Richard Christman, U.S. Coast Guard Academy
Dewey I. Dykstra, Boise State University
Robert G. Fuller, University of Nebraska
Karen L. Johnston, North Carolina State University
Eric Lane, University of Tennessee
Priscilla W. Laws, Dickinson College
John W. Layman, University of Maryland
William M. McDonald, University of Maryland
John C. Park, North Carolina State University
Richard E. Swanson, U.S. Air Force Academy
Edwin F. Taylor, Massachusetts Institute of Technology
Jack M. Wilson, American Association of Physics Teachers

Advisory Committee

Bill G. Aldridge, National Science Teachers Association
Arnold B. Arons, University of Washington
Alfred Bork, University of California–Irvine
Blas Cabrera, Stanford University
Arthur Eisenkraft, Fox Lane High School, Bedford, NY
Robert H. Good, California State University–Hayward
C. Frank Griffin, University of Akron
Don F. Holcomb, Cornell University
Elisha R. Huggins, Dartmouth College
David P. Koch, Hoover High School, North Canton, OH

Steven E. Koonin, California Institute of Technology
Jill H. Larkin, Carnegie Mellon University
Lillian C. McDermott, University of Washington
Charles W. Misner, University of Maryland
Jim Nelson, Harriton High School, Rosemont, PA
Joseph R. Priest, Miami University
F. Reif, University of California–Berkeley
John S. Rigden, University of Missouri
Judah L. Schwartz, Harvard University
Bruce A. Sherwood, Carnegie Mellon University
William J. Thompson, University of North Carolina
Robert F. Tinker, Technical Education Research Centers, Cambridge, MA
Carol-Ann W. Tripp, Providence Country Day School, E. Providence, RI
David Vernier, Portland, OR
Dean A. Zollman, Kansas State University

Local Arrangements Committee

Catherine Cavness
Dewey I. Dykstra
Robert A. Egler
Margaret H. Gjertsen
Hugh Haskell
Karen Johnston
Nancy H. Margolis
John T. May
J. Richard Mowat
John C. Park
George W. Parker
Richard R. Patty
Prabha Ramakrishnan
E. F. Redish
John S. Risley
L. W. Seagondollar
Michelle Stone
Jack M. Wilson
Bruce Winston

Conference Sponsors

Associations

American Institute of Physics
American Physical Society
American Association of Physics Teachers
National Science Teachers Association

Foundations

National Science Foundation
Research Triangle Foundation of North Carolina
The Annenberg/CPB Project
The Pew Charitable Trusts

Corporate Sponsors

Apple Computer
Becton Dickinson Research Center
Digital Equipment Corporation
E. I. du Pont de Nemours & Company
IBM*
SAS Institute, Inc.
Sun Microsystems, Inc.
Vernier Software

*Major funding for this conference has been provided by IBM.

The Computer's Impact on the Physics Curriculum

What Works: Settings for Computer Innovators in Physics Education

Edwin F. Taylor

Department of Physics, Massachusetts Institute of Technology, Cambridge, MA 02139

"What works?" no longer means, "What programs run without crashing?" High-quality uses of computers in physics education now exist in sufficient quantity that we can begin to study settings in which they grow and are effective.

The treasure is people. As always, the creator works with a cloudy but compelling vision, in the chaos and disorganization of the never-before-realized. As always, the innovator applies the resulting product to a specific situation, finding a unique local loophole through which novelty can be slipped between the forces of nurturance and antagonism. We have no definition of creativity or "innovativeness," much less a way to identify them in advance. Instead we ask a question that may have an answer: Given creators and innovators, how can they be encouraged? We seek a measure of generality by viewing half a dozen creators and innovators in their settings, asking retrospectively how those nearby helped them, or at least got out of their way.

The people presented here are engaged in a variety of projects—projects that will be thoroughly described during this conference. No doubt some of these efforts will survive and grow, and others will inform us before we pass on to our next preoccupation. But this is not a study of projects; rather it is a study—a few vignettes, really—of the settings in which projects originate and grow to fruition.

In each of the following vignettes the description of a project and questions about a particular institution are epitomized in a single person. On some of these projects several or many other people are full participants. I apologize in advance to these colleagues for over-personalizing the analysis in the interest of brevity.

My research methodology in preparing this paper would curl the hair of a professional sociologist. I chatted with a few people and summarized my own knowledge and perceptions. Everyone described had a chance to correct errors in the paper, but the result is not an objective study. These are my friends and colleagues. I like them and admire the work they do. But I will try to tell no lies and to be as balanced and complete as time and space allow.

Before the professional observers go to work on us, let us talk among ourselves about what nurtures our own creativity, what nests to make for ourselves and our projects, what kinds of help to seek out, and how to show others that what we have done has merit and provides a payoff for students.

Joe Redish: Project M.U.P.P.E.T.

Professor Edward F. ("Joe") Redish runs M.U.P.P.E.T., which stands for the Maryland University Physics Project in Educational Technology. The setting is that of a major research center, the flagship university of the Maryland system of higher education. Like Steve Koonin, the subject of a later vignette, Joe Redish is a theoretical nuclear research physicist. Our task is to understand how Project M.U.P.P.E.T. is nurtured in the setting of the University of Maryland, how Joe Redish juggles competing demands of research and educational innovation, and what other aspects of a major university influence the survival of educational change in physics.

Project M.U.P.P.E.T. has several components, and Joe Redish, cochairman of this conference, is here to describe them in a talk later today. A major part of Project M.U.P.P.E.T. is a small introductory class that he teaches for prospective physics majors wherein students do many of their homework exercises by programming solutions in Pascal. Each student also carries out an independent project on the computer, solving some problem of physical importance. Using computers in this way helps students to analyze realistic systems, such as projectiles with air resistance or the large-angle pendulum, and systems whose behavior cannot be described in analytic form, such as three mutually gravitating bodies. It also allows entirely new topics into the curriculum, such as chaotic dynamics.

In addition to driving M.U.P.P.E.T., Joe Redish carries a full load of research in theoretical nuclear physics, being principal investigator of research grants totaling about $500,000 per year. The advantage for introductory students of having a teacher who is a research worker is obvious: Joe provides a professional role model and introduces current topics into the course as appropriate. When he gives a colloquium in nuclear physics at another university, Redish often asks to give a second talk to the local faculty about his new class.

The University of Maryland has a large physics department—seventy-nine at my count in the directory—and no simple generalization describes their reaction to Project M.U.P.P.E.T. Some oppose research physicists' diverting their energies in this way. Others line up to help teach the course. At least one fellow faculty member, who finds Joe Redish "an exceptional teacher at all levels," wonders how successful he will be in "passing on the course to someone less joyous, with a moderate and decent interest in computers"—the implication being that Joe's interest in computers is "indecent." This colleague reports that, by and large, "the faculty is embracing the idea" of Redish's new course.

Grants are understood in a research university, and on one level the grant that supports M.U.P.P.E.T., from the Fund for the Improvement of Post-Secondary Education (FIPSE), is just one more grant, helping to make educational innovation legitimate. On another level it is an exception, confusing those who find it hard to believe that one person can be heavily involved in education as well as specialty research. Recently Redish was given the Leo Schubert Award for Teaching by the Washington Academy of Sciences. He routinely sent a copy of the award letter to the physics department salary committee. Later the committee chairman stopped him in the hall and asked, "Have you then given up research entirely?"

Many listeners will recall that the American Association of Physics Teachers (AAPT) has its offices near the University of Maryland, and know that the department provides part-time faculty positions to some in that organization. You will not be surprised to learn that AAPT moved to College Park during the time that Joe

Redish was chairman of the Maryland Department of Physics and Astronomy. In a graceful phrase, Redish says that AAPT "curves the local metric" toward education.

Neither the AAPT nor the activities of Project M.U.P.P.E.T. are much noticed by many on the physics research faculty, but both operations are strongly encouraged by the current chairman and others in the higher administration, in part because of the good publicity they bring to the University. The University of Maryland is a state school, and a state school must be responsive to the educational needs of the public. In my superficial investigation there was conflicting testimony on how important this administrative perception is in the considerable support given to M.U.P.P.E.T.

Joe Redish seems exuberant about his involvement in both educational innovation and research. He says he is "learning a tremendous amount" from running the computer-intensive class, which provides the occasion for "thinking through fundamental physics subjects in detail" and stimulates strategies of computer analysis that have direct application in his research. For him it is "intellectually creative," and he does not know when he will give up the exhausting double life.

Once in a while you run into someone who actually fits an ideal about which we have become cynical. Such an ideal is the professor who combines education and research into a mutually reinforcing enterprise. Joe Redish seems to have achieved this dynamic golden balance. And Maryland is the research university with the human face that provides for him the nourishing setting. The productivity of the result is inspiring, while its rarity, thank goodness, permits the rest of us to go on being cynical.

Priscilla Laws: Workshop Physics

Priscilla Laws is for me the archetypal educational entrepreneur in a small liberal arts college. Open, energetic, comfortable in many settings, dedicated to physics, she is also aware of how the world works and is skilled in building a local nest in which her project is nourished and thrives.

Priscilla Laws and colleagues at Dickinson College in Carlisle, Pennsylvania, run an introductory physics course called Workshop Physics. Students in Workshop Physics spend their time in the laboratory carrying out assigned projects. Laboratory work makes heavy use of the microcomputer-based experiments developed by Bob Tinker, described in a later vignette. Increasingly data summaries and report writing are also being done on the laboratory's personal computers. Students read the textbook and do homework exercises as usual. The instructor may interrupt the students in the laboratory and talk for a while, but Workshop Physics has no regular lectures of the kind we are used to. For details, listen to Priscilla Laws' talk later today and attend the workshops she gives during this conference.

General support comes from a FIPSE grant. But watch how Priscilla Laws multiplies this resource. First, FIPSE does not pay for equipment, so Priscilla obtained $50,000 from her institution to purchase computers. That takes chutzpah. But what takes my breath away is that she talked the college into using the overhead payment from the FIPSE grant to renovate the physics laboratories for her classes. Nor can she be said to have exploited Dickinson in this regard: the success of her FIPSE grant was a positive factor in the later award of a large grant to the college from another source for disciplines other than physics. In brief, Priscilla Laws shows us how to follow the Biblical injunction to "be wise as serpents and innocent as doves."

As always, when we look a little deeper, things are not so neat and become a lot more ambiguous. The Dickinson College Department of Physics and Astronomy has five and one-half full-time-equivalent positions divided among seven people. Three of these people do research in the conventional mode and two do less conventional research—one of these two being Priscilla's husband Ken, who is well known for his studies in the physics of ballet. The written policy of the college in examining candidates for tenure and salary advancement, is to consider teaching first, research/scholarship second, and commnity service third. The academic dean and the personnel committee say that "'scholarship' means activity that culminates in evaluatable results, and that the evaluation is done by peers. Articles on the pedagogy one is implementing is one form that we have lots of instances of; for us that's just as good as original mainstream research."[1]

Thus official Dickinson policy. On the other hand, outside evaluators for the physics department insist that faculty should be more involved in mainstream research, and this is also the opinion of some influential college faculty in other departments. Even a full professor like Priscilla Laws feels the influence of the unofficial presumption that "real research" would be preferable to her present activities of developing curriculum and creating student materials. She largely ignores this viewpoint, but just as it nags at her, it must nag at others considering educational innovation, at Dickinson and at other small colleges.

The record shows, and Priscilla Laws feels, that her project is well supported by departmental colleagues and the administration at Dickinson, as well as by outside funding. We have seen that she herself has been active in ensuring that this support is forthcoming. And if the immaculate ideal is not quite matched by the rumpled reality—in Malcolm Parlett's phrase—well, that is life.

Steven Koonin: Computational Tools for the Theoretical Physicist

It is ironic that the California Institute of Technology (Caltech), the most research-oriented of the institutions on our list, is the easiest place for faculty to carry out educational innovation. Steve Koonin had only to ask for an academic quarter off to develop a new course in order to be granted his wish and enough graduate assistants get him started.

Steve Koonin is a theoretical nuclear research physicist. Computers are important in his work, and he has spent a lot of time teaching programming to graduate and undergraduate students. In the early 80s, out of the blue, IBM gave him one of the early PCs, and he became intrigued with its possibilities for teaching.

After he had played with this machine for about a year and a half, IBM gave Caltech 500 personal computers, of which 20 IBM XTs ended up in a physics laboratory. The university called for faculty suggestions, and Koonin proposed that he teach a course in using computers for theoretical physics. The institution gave him the fall quarter off and provided two graduate teaching assistants for the summer as well as that fall to help him with the programming. This was done with university money—and informally, without written proposals or competition among faculty. As Steve Koonin puts it, "Caltech is very lean on administration. It is assumed that if a faculty member wants to try out something, he or she should be supported in doing so."

So Koonin and his assistants prepared programs and class notes during the summer and fall quarters of 1983, and he taught the class for the first time in the spring quarter of 1984. The following summer he was given one teaching assistant. During that summer and fall he wrote a book, now published under the title *Computational Physics*.[2] By the end of the second year he was done with the project and now has returned to research full time, declining to participate in subsequent educational innovations.

How was Koonin's involvement in educational innovation viewed by his colleagues? We are talking here of a small school (800 undergraduates, 800 graduates) with a relatively big physics faculty of 42 members. Koonin feels that other physics faculty did not think of his activity as deviation from research. At the end of the two years, he gave a regular physics department colloquium on his computer teaching project. At that time (1985) most faculty did not appreciate the capabilities of small machines, so there was a great deal of interest.

In fact, during the two years of his project, Koonin continued a full program of research. He says it "almost killed" him to continue both activities—"a real drain"—and admits that there was "something of a notch" in his usual average of seven research publications per year. That is why he has refused to undertake further educational innovations. His plenary lecture at this conference may be our last chance to hear him on this subject.

I am impressed with Caltech's easy and efficient ways of encouraging its faculty in educational innovation. In this case at least, informality, trust, and open-handedness have paid off handsomely for faculty and students and for the rest of us in the physics teaching community at large.

Perhaps more research physicists, at least those with tenure, could undertake intense but limited episodes of educational innovation, especially those innovations, like Steve Koonin's, that leave the rest of us with a residue of gold.

Robert Tinker: TERC and Microcomputer-Based Laboratory

"TERC should not exist", declares Dr. Robert F. Tinker, speaking of the organization that pays his salary. TERC is the Technical Education Research Centers, Inc., an independent nonprofit educational development group in Cambridge, Massachusetts.

Bob Tinker directs one of three parts of TERC, a part called the Technology Center, which develops curricula, hardware, and software for primary through college education. They originated the *Microcomputer-Based Laboratory*, which transforms student laboratory experience by providing powerful experiment inputs, analysis, and display routines. Best known is the motion detector that immediately displays the position, velocity, and acceleration of an object, such as a walking student. The *Microcomputer-Based Laboratory* has been used successfully from grade school through introductory college physics. It received an award as one of the ten best software titles by *Classroom Computer Learning* magazine. Priscilla Laws, the subject of an earlier vignette, has incorporated it into her Workshop Physics course at Dickinson College. Tinker will describe the *Microcomputer-Based Laboratory* in a plenary lecture later in this conference.

By saying "TERC should not exist," Bob means that the tasks TERC does—innovation in technical and scientific education—should really be done at colleges and universities. Why *does* TERC exist? "Every flourishing organization has a secret weapon," says Bob Tinker, "usually a person." TERC's secret weapon is its president, Arthur H. Nelson. For 20 years Arthur Nelson has spent mornings in profit-making activities (for example, renting space to technical enterprises along Boston's Route 128) and afternoons in the community of nonprofit enterprises, one of which is TERC. No simple description fits Arthur Nelson, so let his history introduce him.

In 1943 Arthur H. Nelson graduated as a physics major from the University of Kansas. This was wartime, and a recruiter from the Radiation Laboratory brought him to the Massachusetts Institute of Technology (MIT), where he worked on radar and on the AWACS plane that would have been the airborne command center if the invasion of Japan had taken place. After the war he consulted on missile guidance with a major electronic manufacturing firm while getting his degree at Harvard Law School ("studying the laws of man as well as the laws of nature") and taking classes at Harvard Business School. He founded a company that did electronics work for the government, selling the company ten years later.

In 1966 Arthur Nelson and several presidents and past presidents of the American Technical Education Association, founded TERC to develop curricula for technicians, paramedics, and other support personnel for professionals in emerging technologies. With the help of many faculty from MIT and other universities, as well as scientists from business, TERC produced widely adopted curricula in biomedical technology, electromechanical technology, nuclear medicine, and nuclear technology.

In the mid-70s Bob Tinker, a fresh Ph.D. from MIT, taught physics at Amherst College, worked for the Commission on College Physics, and then came to TERC to develop a curriculum in electronics for scientists. By the late 70s he was insisting that TERC act on the promise of computers in technical and science education. By the early 80s they developed motion-detector equipment and the *Microcomputer-Based Laboratory*, which in 1983 attracted one of the first big secondary education grants from the newly refunded National Science Foundation.

TERC has not always thrived. Three times its existence has been threatened by cutbacks in federal funding. Three times a combination of staff inventiveness,

TERC's experience with over 200 projects, and Arthur Nelson's generosity and business sense has kept it alive. An outsider's impression is that month-to-month success rests on a collegial guiding structure consisting of Nelson and a senior staff that has been with TERC for over a decade. As a member of this guiding staff, Bob Tinker has strongly influenced the setting that nourishes his projects.

The past is prologue. We ask again: Why is TERC not displaced by competing organizations in universities? Nelson thinks it is because universities find it difficult to collaborate with a wide variety of groups: other universities, technical schools and community colleges and public school systems, professional societies, or with a mix of profit and nonprofit groups. In brief, universities have turf problems. TERC is a nonprofit group that threatens no one. It has a single focus: to identify national needs in technical, scientific, and special needs education and to bring together the resources to satisfy these needs. Without turf problems, TERC can use as resources what Nelson calls "the best, wherever they are located."

I have two additional guesses as to why TERC is not displaced by university projects. First, it has little institutional inertia. TERC can—and to survive TERC must—identify areas with unserved populations, formulate a plan to move on their needs, and apply to fund the plan before the rest of us have even rumbled up to the problem.

Second, TERC has taken as its playing field the very turf where success can be objectively verified. An incompetent technician is more quickly identified than an incompetent professor of literature. Nature herself helps you sort out physics students who know their way around a laboratory from students who do not. I believe it is no accident that one of the three centers at TERC concerns evaluation and a second concerns the handicapped, another field where success is demonstrable.

In thinking about TERC I find myself murmuring, "only in America." Where else could you find a private-enterprise entrepreneur nurturing a nonprofit center that uses government funds and collaborates with a spectrum of public, private, profit, and nonprofit organizations to improve education in technology and science, and for those with special needs?

MIT Project Athena: Graphics Utilities in Special Relativity

A strong motivation for giving this talk has been to work through my own experience of developing physics educational software with help from Project Athena at MIT and to compare it with the experiences of other developers in different settings. You recognize immediately that objectivity in reporting my own story is even more suspect than in reporting the stories of friends and colleagues in other schools. Happily, help has arrived: Professors Donald Schon and Sherry Turkle and coworkers at MIT, have carried out a careful series of studies of MIT's Project Athena. What follows has benefited from their report.[3]

Project Athena at MIT has been a major effort to explore the uses of computers in education across the span of academic disciplines. For its first five years (1983–88), Project Athena was heavily funded by Digital Equipment Corporation and the International Business Machines corporation. MIT itself raised $20 million for individual faculty educational development projects.

My student programmers and I are substantial beneficiaries of Project Athena, receiving a total of approximately $160,000 over three years in awards based on competitive internal proposals. This money has gone largely for student wages and part of my salary, but we received much more. To this total must be added the Athena-provided computers on which the programs were developed and used by students, and the administrative services of the Department of Physics. I am extremely grateful for this support.

With this multifaceted support, we developed interactive graphics computer utilities to help students learn special relativity and quantum physics. The relativity programs are complete and are the subject of a short paper and two workshops at this conference. The quantum physics programs are still under development—more on them below.

Those who have studied Project Athena have criticized it for concentrating on advanced technology rather than providing facilities that faculty request to carry out their projects. I myself have been in the weird position of having to struggle to *avoid* using the latest, most powerful computers. I want to program on machines that are generally available, so that colleagues in other schools can use the results. Here is a one-paragraph history of that struggle:

For the science departments at MIT, Athena first provided IBM XT machines, which were just then being adopted nationwide. As soon as the first IBM AT computers were introduced, Athena shifted over to them. These ATs were equipped with the expensive Professional Graphics Adaptor (PGA) and display, which shows beautiful colors. The trouble is that the PGA display did not catch on commercially: few businesses or schools bought them. So the programs my student programmers prepared that use the full PGA colors are useless almost everywhere except at MIT. Happily, the PGA display can also be run in a mode that mimics a less expensive display that many people do have. As soon as the IBM RT became available, Athena moved on up to that computer. By now I had learned my lesson, so I stayed with the AT machines. Sure enough, the RT computers did not sell commercially either and were soon retired from use. Since then MIT has installed work stations—useless to me for developing programs that can be used elsewhere—and is discarding their ATs.

Now turn from machines to people. In the summary of their study of Project Athena, Donald Schon and Sherry Turkle tell us: "[Athena assumed] that MIT faculty would be able...to devise new educational software...as something of 'an exercise for the left hand.'...But in the perception of most faculty members who tried, development of new educational uses of the computer was a demanding activity...It took a great deal of time. They are seen by their colleagues as diverging from the path to success within their departments...In the light of our findings, it is understandable that the developers of Athena projects tended to be either individuals marginal to their departments, or 'short timers' who...had no intention of following up the possibilities created by their work, or long-time champions of computers in education."[4]

I fit into the Turkle-Schon category of "individuals marginal to their departments." I am not a professor; my position of Senior Research Scientist is intended for those funded with soft money. This position has served me well during most of a 25 year career of educational research and development at MIT.

Thus I took advantage of a "double ecological niche" within MIT and Project Athena. As a senior research scientist, I could commit the time needed to pursue a demanding project without losing status as a regular research faculty member. Athena provided some of the soft salary support for this effort and the essential wages for student programmers. My students and I further subdivided the Athena niche by intentionally staying "out of date" with respect to the latest Athena hardware; we wrote programs for computers available to physics teachers in other colleges and universities.

The programs and documentation in special relativity that you can see at this conference are the outcome of our time in the MIT-Athena double-ecological niche. In a final irony, an additional outside grant of a mere $6,000 from Apple Computer for student wages, plus the loan of a Macintosh, funded Macintosh versions of our relativity software that sell much better than the Athena-supported IBM versions.

All that is the good news. The bad news is that recent cutbacks in faculty development funding have closed off the Athena part of the niche, leaving our project in quantum physics only partly completed. Still, we did take advantage of a window briefly open in the dimensions of time, space, funding, and shifting institutional priorities to accomplish something that gives us satisfaction.

Eric Lane: Standing Waves

All this talk of outside grants and institutional support raises questions: Whatever happened to simple creativity? Doesn't anybody develop software individually, as a regular scholarly activity? Why can't you work out ideas on your own, without the heavy machinery of grants and technical support?

The answer is you can, but you need to understand the conditions. Let me introduce Professor Eric T. Lane of the University of Tennessee at Chattanooga. His program *Standing Waves* won the 1987 EDUCOM/NCRIPTAL award in not just one but two categories: best physics software and best simulation software. According to the official description, *Standing Waves* "provides animated simulations of waves and pulses, interference effects, [and] Doppler effects for sound and light."[5] For further details, attend the several sessions that Eric Lane is giving at this conference.

The EDUCOM/NCRIPTAL award culminates nine years of lonely work. When you meet him, you will understand that lonely work is, for Eric Lane, not a punishment, but a joy. If Joe Redish and Patricia Laws show us that there is a place for the public person—the organizer, the collaborator, the public speaker—Eric Lane shows us that there is also a place for the private person—the reflective creator, the meticulous worker, the burner of midnight oil.

Eric Lane began work on *Standing Waves* in 1978—in the dark ages before the Apple Macintosh, even before the IBM PC.[6] The Apple II had just been introduced and few had thought of animation graphics. Rapid animation on the Apple II requires programming in machine language, that arcane code of numbers and letters that directly instructs the processor. (Machine language executes the animation in *Standing Waves*, and a BASIC program controls the menus.) Few people have the combination of physics background and technical knowledge of machine language necessary to do this. Eric Lane says that at the start he was "very technical," bringing to this task 20 years of extensive programming experience on several generations of IBM and Hewlett-Packard machines. The breadth of this expertise has allowed him to reprogram *Standing Waves* more recently for a range of personal computers.

Lane had no collaborators, no technical assistants, no student programmers, no participating fellow faculty members. He estimates that he spent 2,000 hours a year for five years developing *Standing Waves* software and documentation. That averages 40 hours per week—mostly nights and weekends—in addition to a full 15 contact hour teaching load.

Did he get help? Not until much of the work was already done. The Center of Excellence for Computer Applications (CECA) was established in 1984 at the University of Tennessee at Chattanooga. Its main emphasis is in artificial intelligence, assistance for the handicapped, and mainframe computer support, with a budget of $20,000 total for all faculty projects. For six of the past eight semesters Professor Lane has received some time off from his teaching load, partly supported by CECA. In addition, CONDUIT, a software distributor in Iowa City, distributed his material and paid for two one-week stays in Iowa City to complete projects. Lane's department chairman supported his effort and was instrumental in reversing a university policy against personal computers. Fellow faculty members encouraged the project, provided an avenue for seminars on his software, and exchanged classes when he had to be out of town. Some faculty outside his department criticized software development as an inappropriate scholarly activity, the presumption being that it did not replace conventional research.

There was some hardware help from CECA, the British Acorn Computer Corporation, Digital Equipment Corporation, and the Lupton Foundation. All were the result of written proposals. Proposals submitted to IBM, Apple Computer, and others received no response.

That's it. That is all the support Eric Lane received—until he won the EDUCOM/NCRIPTAL award. Now IBM, Apple, and AT&T are interested in supplying equipment. He feels that the award has brought him the possibility of increased resources, "Not enough, but more."

There is a strain of asceticism in Eric Lane's attitude and work. "I don't intend to copy protect any software that I distribute personally," he says. "I write it for people to use, for them to get the benefit of my experience and teaching ability. The more the better, as far as I am concerned...I feel that society provides me with more than enough comforts and necessities. I would prefer to spend my time and effort doing my utmost to improve that society rather than trying to get rich at its expense."[7]

For me Eric Lane's story provides an existence theorem: It is very hard, but you can do it alone if you need to. Inventiveness, determination, and industry can result in a product that benefits teachers and students, and through them society as a whole.

Conclusions

I once asked a well-known historian what traits were shared by the great people of history. He replied, "None. What characteristics could Jesus and Napoleon possibly have in common?" The historian was wrong, at least about Jesus and Napoleon. Both of them were immensely energetic, and both had a vision in which they deeply believed, a vision that galvanized those around them.

None of us would claim to be historic characters, but everyone described in this paper also has a vision and is energetic in pursuing it, each according to his or her own style. In reviewing the vignettes, I discern two distinct innovative styles, styles that I call "public" and "private."

Priscilla Laws, Joe Redish, and Bob Tinker have "public" visions, visions that are more about ways of organizing education than about a product. Fulfilling the visions requires direct participation of at least some of their professional colleagues. The large scale of their visions requires outside funding. Within an academic institution, outside funding serves as much a political as a functional purpose: it legitimizes the project and makes more likely the necessary direct participation of some colleagues and toleration by the rest. The public innovator also takes every opportunity to describe the vision outside the home institution during the time he or she is actually learning how to do it.

In contrast, Steve Koonin, Eric Lane, and I have "private" visions, visions that are more about products—textbooks and software—than about ways of organizing education. Steve Koonin and I work with assistants but not colleagues; Eric Lane works alone. We private innovators use local support from within our own institutions. This support is small in scale and has a practical purpose: to give us time and assistance. Any political advantage internal grants bring us is minor and does not make the difference between success and failure in our local setting. We go public with our products only when they are completed.

If the distinction between public and private innovator is valid, then there are consequences for encouraging these contrasting activities. Are you an administrator working with a private innovator? Then consider providing modest direct support for assistants and released time. This can be done quietly. Are you trying to encourage a public innovator on your faculty or staff? Then think of soliciting an outside grant, giving both the application and the award wide publicity within your organization and its administrative surroundings.

In the melody of discussions with our five academic educational innovators, I hear repeatedly a very old theme: the polarity between education and research as perceived by them and by those around them. I believe it is no accident that all five of the academic innovators except the author have tenure. Tenured professors can afford to take the risk of concentrating on educational innovation, at least for an interval. But even full professors think carefully about the relation between their research and their educational projects. Even full professors look somewhat nervously for encouragement or the lack of it in their annual salary reviews. Of the five academics, only Joe Redish manages an ongoing combination of specialty research and educational innovation. Steve Koonin tolerated it for two years. Eric Lane, Priscilla Laws, and I worry some about the tension, while acting as full-time professional educators. I have little to add to this decades-long dialogue except to testify to its continuing significance for innovators and administrators.

I am struck by how closely the setting fits the style of the innovator in most of the cases we have examined. Most of our innovators are well tuned to their environments. Joe Redish thrives in the stimulating setting of university research while orbiting gently in the curved metric around the AAPT, his local pipeline to the world of physics educators. Bob Tinker has learned the survival skills of the nonacademic educational development organization. It is hard to picture Steve Koonin operating comfortably at MIT or Priscilla Laws striding onto the stage at the University of Maryland. As always, it is the subtle mutual fit between organism and ecology that leads to growth and a rich harvest for the rest of us.

Setting makes a big difference. It is not all-powerful—Eric Lane has shown us that—but setting does make a difference. There must be hundreds of people out there with energy, vision, and good educational ideas who have not been heard from. Which of these could become productive to the benefit of the rest of us through changes in their local surroundings, by seeking out a local ecological niche, or receiving administration nourishment and support? There is much to do in applying computers to physics education, and we desperately need their help.

1. Bitnet mail from George Allan, Academic Dean of Dickinson College.
2. Steve Koonin, *Computational Physics* (Menlo Park, CA: Benjamin/Cummings Publishing, 1986).
3. Sherry Turkle, Donald Schon, Brenda Nielsen, M. Stella Orsini, and Wim Overmeer, "Project Athena at MIT" (Unpublished paper, May 1988).
4. Ibid., section 1. Quoted with permission. Omissions in this excerpt oversimplify a more subtle and comprehensive argument.
5. Robert B. Kozma, Robert L. Bangert-Drowns, Jerome Johnston, *EDUCOM/ NCRIPTAL Higher Education Software Awards*, 1987 (Ann Arbor, MI: Program on Learning, Teaching, and Technology, National Center for Research to Improve Postsecondary Teaching and Learning, 1987).
6. Much of the following information comes from "Developing Standing Waves," by Eric T. Lane (Unpublished paper, 28 March 1988).
7. Ibid. Ellipsis dots in original.

The Impact of the Computer on the Physics Curriculum

Edward F. Redish

Department of Physics and Astronomy, University of Maryland, College Park, MD 20742

The computer has immense power to perform tedious tasks quickly and efficiently. With the wide availability and falling prices of microcomputers, both students and faculty are offered new opportunities for innovative learning, especially in a technical field such as physics. Having a substantial amount of computer power can have a marked impact on what can be taught.

At the University of Maryland, a group of faculty and students has organized the Maryland Project in Physics and Educational Technology (M.U.P.P.E.T.). We have been investigating the implication of microcomputer availability on the content and structure of the curriculum.

I have been primarily concerned with the lecture part of the introductory course for physics majors, so I will concentrate on developments there. Others have worked on the laboratory, on courses for nonscientists, and for upperclass majors and graduate students. That work is reported elsewhere.[1]

Problems with the Traditional Curriculum

In considering what the computer can do for us, we started out by trying to decide what it was we really needed done. We considered the content of the standard curriculum in light of the knowledge and skills required by a contemporary professional physicist, and we decided that the current approach had some substantial defects.

1. The present curriculum does not contain enough contemporary physics.

By contemporary physics we don't mean "modern physics." The latter term means physics developed in the years between about 1887 (the year of the Michelson-Morely experiment) and 1939 (the discovery of fission). The excitement and vitality of the current state of physics is rarely visible in the standard physics major's program. Many introductory students (even top-ranking ones) have no idea what a professional physicist actually does.

2. The present curriculum does not sufficiently take into consideration what students know and how they learn.

In our traditional majors course there are two standard assumptions: (i) Students can be considered a *tabula rasa*. If we present our material logically and coherently, they will understand it. (ii) If students have difficulty with our presentation or approach, then "they are not good enough to be physicists." Current developments in cognitive psychology and the theory of education suggest that both of these assumptions may be wrong.[2] Students come into our classes with understandings and preconceptions about the world, and they often translate and misinterpret our teachings as a result.

Some of the skills we expect our students to pick up quickly, such as facile algebraic manipulation and the ability to attack and structure an approach to complex word problems, may be slow in coming because of previous deficient teaching. Their lack does not imply a "fatal flaw," and we may actually be able to teach these skills instead of simply expecting students to learn them.

3. The present curriculum does not provide appropriate training of relevant professional skills at the undergraduate level.

In the real world of the professional physicist, solving straightforward analytic problems like those that dominate the undergraduate curriculum, plays a small (but important!) role.[3] Many of the most important skills required of the professional physicist are not mentioned in the standard curriculum until the graduate level. This is sufficiently unnatural that I will elaborate on this point.

The Skills of a Physicist

Near the beginning of our project, I considered the question: When a student wants to do thesis research under my supervision, what skills would I like (expect) that student to have? My first brainstorming session on this question produced a very long list of notes—nearly 20 pages. Clearly, one can choose such a list of skills in a variety of ways, and I'm sure every reader will have a different list. The following is the result of much discussion, boiling down, and condensing.

For the nonlaboratory part of the curriculum I propose that every graduating physics major should have an opportunity to develop the following skills:

1. *Number awareness.* This is the *sine qua non* of a physics major. Students must understand that the universe is quantifiable.

2. *Analytic skills.* Students must understand the concept of equations and be able to manipulate them in reasonably complex situations. This includes solving problems with up to a dozen variables, understanding the use of limiting cases, and formulating strategy and tactics for approaching a complex problem.

3. *Understanding of natural scales and estimation skills.* Students should understand what parameters are responsible for governing the natural scales of a problem and should be able to estimate plausible answers and the size of effects to one significant figure.

4. *Approximation skills.* Students should understand when an approximate equation is valid and to what accuracy. They should have some idea of ways to improve approximations by a variety of techniques.

5. *Numerical skills.* Students should know how to solve a variety of problems that are not solvable analytically. Perhaps the two most important aspects of this skill are knowing what one can get out of a numerical calculation, and knowing when to do a numerical calculation and when to do an analytic one.

6. *Intuition and large-problem skills.* This includes a variety of metaskills. By intuition we mean having an understanding of when an answer looks plausible and what to check for. By large-problem–solving skills we mean such things as chunking (breaking the problem into parts), mixing library skills with analytic and numerical ones, etc.

The introductory course usually deals almost exclusively with the first two skills. An introductory physics text may include thousands of problems. An informal survey suggests that of these, 80 to 90 percent are "plug-ins"—problems that deal only with skill 1. Of the remainder, most (often marked as being more difficult) are analytic manipulations—skill 2. The number of problems requiring estimation, approximation, or numerical skills can be counted on the fingers of both hands (if they can be found at all in the immense pile of plug-in problems).

In most versions of the current curriculum, approximation and numerical skills are left for graduate study, giving most students a distorted view of physics as an "exact" science, rather than as a science where we know the range of applicability of our equations. Large-problem–solving skills are rarely encountered in an undergraduate program unless the student does a senior thesis. This is particularly unfortunate, since the approach to a complex, open-ended problem (especially one where the answer is not known beforehand) is the fundamental skill of the professional scientist.

Estimation is the epitome of our view of the ideal physicist (who can solve any problem on the back of a paper napkin—to one-digit accuracy), but we almost never teach it. We seem to expect students to learn it osmotically by exposure to physicists who know how to do it.

It is our view that all of these skills should be introduced in the undergraduate curriculum at every level. Our experience at Maryland is that all of the skills in the list can be introduced to majors at the freshman level in connection with computer use.

The M.U.P.P.E.T. Environment

We have attempted to develop our materials and curriculum modifications so they can be used inexpensively and effectively in a variety of environments. Our own environment at Maryland is currently based on the IBM PC and AT personal computers. A reasonable configuration for using our materials is an 8088 or 80286 IBM compatible with two diskette drives or one diskette and a hard disk. For the 8088-based machine (IBM PC or clone), the 8087 coprocessor is highly recommended, as is 640K of memory. All our packages have been prepared to run with CGA or EGA graphics.

In our experience, one such machine suffices for every eight to ten students. We have two microlabs in the physics department: one with sixteen machines (mostly PCs) networked with an AT master. The second lab is an overflow lab with six machines, also networked to the same master. Our freshman majors' classes (each with ten to twenty students) meet there three to four times each semester for one-hour microlabs with the teacher. The lab is kept open by teaching assistants for eight hours a day on weekdays. Students make extensive use the lab for computer homework assignments and term projects.

The University of Maryland also provides a number of general-purpose micro-labs in libraries throughout the campus, and will begin to provide in-dorm computers next year as part of a major effort to expand computer availability.

We concluded that the appropriate language for freshman majors was Pascal. The reasons for this are discussed elsewhere.[4] In order to minimize the amount of programming we have to teach, we have developed class handouts that distill the basics of Pascal down to a few minimal ideas. We have developed utilities that permit students and faculty to easily write programs with interactive input and multi-window graphics. Simple sample programs can be used both interactively in microlabs to study complex phenomena and as starter programs or templates from which other programs may be built. This makes it possible for the student to learn to write useful programs very quickly.

These materials also help resolve the problem of making the computer usable in class by faculty who are not programmers or who program in a language that is too sophisticated for most freshman to use.

The M.U.P.P.E.T. Course

In the introductory course itself we have made use of the computer in a wide variety of ways. Some of the new features included in the introductory course are:

- reorder the introduction to mechanics by starting with Newton's second law in discrete form;[5]

- teach the power tools of physics—Newton II as used in a realistic physical example;

- train the student's intuition in complex situations by using productivity tools, including *Orbits*, *Thermo*, and *Spacetime Physics*;[6]

- include a unit on nonlinear dynamics—the nonlinear pendulum, fractals, period-doubling, bifurcations, and chaos;

- introduce some quantum concepts—the phase plane and Bohr-Sommerfeld quantization, quantum scales, and wave equations; and

- require independent term projects.

Since there is neither time nor room for a full discussion of all these topics, I have selected two of the new features to present in some detail: the example of treating projectile motion with air resistance, and a brief discussion of our experience with student projects.

An Example: Projectile Motion in Air

Including the effect of air resistance on motion of a projectile is a simple case, just one step beyond what we usually do. But that step is not possible at the introductory level without help from the computer; it requires the use of one of the "power tools" of mechanics—the solution of Newton's second law as an ordinary differential equation using numerical methods. Nevertheless, this example is strikingly rich and provides an environment in which we can introduce the student to a variety of qualitative analytic skills.

To begin our analysis of projectile motion with air resistance, we have to first discuss the form of the force law. We can do this qualitatively by

- identifying the relevant parameters and variables in the problem as ρ (the density of the air), R (a size parameter of the object), and v (the velocity of the object through the air);

- using symmetry principles (unit analysis and vector/scalar properties) to generate a functional form for the force law: $F = -\eta \rho R^2 v v = -b v v$ where η is a dimensionless parameter;

- building a simple analytic model to estimate the dimensionless parameter (air is made up of molecules);[7]

- estimating the relevant time scales by building gravitational and air resistance time scales out of (v_0,g) and (m,b,v_0);

- studying the qualitative and quantitative behavior in the microlab, including the phenomenon of terminal velocity;

- deriving the expression for terminal velocity analytically; and

- doing homework problems for realistic cases.

Here are two sample homework problems that can be done at the end of the unit.

Sample problem 1: A 2-gm sheet of paper is crumpled up into a compact ball with a radius of 4 cm. When dropped, it takes 1.0 sec. to fall a distance of 2 m. Use this to determine the air resistance coefficient b in the force law $F = -b v v$. If a wooden and steel ball are dropped from the same height, how long would they each take to fall? What accuracy would you need in your measurements to see the difference in the rates of fall between the wooden and steel balls?

Sample problem 2: I throw a ball of mass 0.25 kg straight up with a speed of 20 m/sec. It comes down 1 sec. earlier than I expected it to if I ignore air resistance. Find the air resistance coefficient b for this object if the force has the form

$$F_{\text{air resistance}} = -b v v.$$

Find the coefficient γ if the force has the form

$$F_{\text{air resistance}} = -\gamma v.$$

Design a simple experiment (with numbers) using this ball to determine which force gives a better description of the real world.

The example shows that the actual computational work and the programming involved are a fairly small part of the unit. But using the computer permits us to bring in scale analysis and dimensional analysis, and to demonstrate approximation techniques and ways of extracting physics from computer programs. These are all skills that the professional must know, but that we have had little opportunity to teach in undergraduate courses.

Student Projects

The second application I would like to discuss is student projects. The style of the traditional introductory course is exceedingly dry and rigid. The problems are usually narrowly stated and have "right" answers. Although this kind of work is an essential tool in the professional's kit, it is not the only tool. Furthermore, it is not at all clear that training the introductory student exclusively in this class of problems provides the proper filter for selecting those students who want to be professional physicists.

We have been experimenting with requiring independent projects at the introductory level. Because of the availability of our utilities and productivity tools, these projects are not restricted to reviewing and writing a precis of a small number of articles read in the library. An ideal project can include:

• formulating a broadly phrased question of interest;

• doing preliminary library research;

• formulating a sharper problem amenable to a calculation;

• performing the calculation;

• studying the output and extracting the dominant physics; and

• communicating the results in both written and oral form.

The students in selected sessions of our introductory classes for majors (15 to 25 students per class) have carried out projects for the last three years. Although few of the projects satisfy all of the above criteria, many students have performed excellent projects. The range of computational activities is very broad, ranging from none at all to building and debugging Pascal codes of over 1500 lines that take a dozen hours to run on a standard 8088 IBM PC.

Our programs and environment provide a wide variety of options. Students can use *Orbits* as a productivity tool to study cases in the gravitational restricted three-body problem without doing any programming at all. They can use *Thermo* as a simulation to generate the data on interacting particles in a box under a variety of circumstances and write small programs (25 to 50 lines) to analyze those data and study nonequilibrium statistical mechanics. They can modify our sample programs *Pendulum* and *Project* or build their own simple programs around MUP*graph* and MUP*scrn* to study examples in nonlinear dynamics. Finally, they can write their own programs from scratch. We have had successful projects at all these levels.

It is important to have this range of possibilities because our current classes of entering students have a very wide range of computer experience, from none at all to being able to program fluently in a number of different languages. One interesting phenomenon that we have observed is that the best projects are not always done by the students who do the best job on exams or homework. By giving them projects, we not only provide our students with an opportunity to train their research skills, we also give them a wider variety of ways in which to succeed.

Some of the topics studied by students in our introductory course in the past two years include tethered satellites, shepherd moons and particles in Saturn's rings, the collision of galaxies, the interaction of the planet Nemesis and the Oort cloud, Olbers' paradox and the spectral distribution of light in the night sky, starquakes and models of neutron stars, and fractal-basin boundaries in Newton's method of solving algebraic equations.

Conclusion

The computer can have a powerful effect on the way we teach physics, even at the introductory level. It is not sufficient to use the computer to illustrate the existing curriculum; the curriculum must be rethought from the ground up. In particular, the computer can be used to bring in more contemporary physics, give the student opportunity to carry out open-ended investigations, address specific cognitive difficulties that are discovered by detailed observation of student response, and develop more broadly the thinking skills useful in research.

This work has been supported in part by a grant by the Fund for the Improvement of Post-Secondary Education (USDE). I would like to thank the chairman, Dr. C.-S. Liu, and the associate chairman, Dr. Angelo Bardasis, of the Maryland Department of Physics and Astronomy for their support and encouragement. The ideas and work presented here have been developed in close collaboration with my coworkers Charles Misner, Bill MacDonald, and Jack Wilson.

1. Jack M. Wilson, "Combining Computer Modeling with Traditional Laboratory Experiences in the Introductory Mechanics Laboratory for Physics Majors," AAPT Announcer 17, 80 (May 1987); Charles M. Misner, "Spreadsheets Tackle Physics Problems," Comp. in Phys. 2, 37 (May/June 1988); "Spreadsheets for Teaching Non-Specialist Physics," AAPT Announcer 17, 47 (May 1987); "Numerical Experiments in Graduate Mechanics," AAPT Announcer 17, 57 (May 1987).

2. I. A. Halloun and D. Hestenes, "The Initial Knowledge State of College Physics Students," Am. J. of Phys. 53, 1043 (1986); B. Eylon and F. Reif, "Effects of Knowledge Organization of Task Performance," Cognition and Instruction 1, 5 (1984); F. Reif, "Teaching Problem Solving: a Scientific Approach," Phys. Teach., 310 (May 1981).

3. Simple analytic problems provide the research professional with a framework or hierarchical "skeleton" about which more complex research is structured. They are very

important to the way the researcher thinks about the subject. Our emphasis on the deficiencies of the curriculum is not intended to suggest that we replace or denigrate this essential component.

4. W. M. MacDonald, E. F. Redish, and J. M. Wilson, "The M.U.P.P.E.T. Manifesto," Comp. in Phys. 2, 23 (July/August 1988); J. M. Wilson and W. M. MacDonald, "Graphics and Screen Input Tools for Physics Students," contributed paper, this volume.
5. E. F. Redish and E. F. Taylor, "Impulse Mechanics," AAPT Announcer 17, 82 (December 1987).
6. J. B. Harold, K. A. Hennacy, and E. F. Redish, "The Computer and Intuition Building: A Multi-Body Orbit Simulator," AAPT Announcer 17, 58 (May 1987); J. B. Harold, G. Norkus, and E. F. Redish, "An Interactive Statistical Mechanics Program," AAPT Announcer 17, 82 (December 1987); *Spacetime Physics* is available from E. F. Taylor, Department of Physics, MIT.
7. R. Baierlein, *Newtonian Dynamics* (New York: McGraw-Hill, 1983), p. 2.

Workshop Physics: Replacing Lectures with Real Experience

Priscilla W. Laws
Department of Physics and Astronomy, Dickinson College, Carlisle, PA 17013

We are suffering from an uncontrollable burgeoning of knowledge in virtually all disciplines. Nowhere is this more apparent to us than in the grand enterprise we call physics, where whole new areas of theory, application, and investigative technology have flourished in the past decade. As teachers we are challenged to learn and teach fundamentally new theories to describe physical phenomena. Do we understand relativity and quantum mechanics well enough to explain them to introductory physics students? How about chaos, super strings, superconductivity, and the big bang?

Unfortunately, the question of what to teach in introductory physics courses is not the only question confronting us. The issue is not simply what to teach, but how to teach. There is an endless array of new computer-based instructional media, including computer-programming languages, integrated software packages, tutorials, simulations, electronic-mail conferencing, and microcomputer-based sensors. Other new instructional tools include video tapes, interactive videodiscs, and satellite conferencing. We have also developed new understandings about student preconceptions and naive problem-solving strategies. The classroom applications of these new understandings about the learning process constitute part of the growing body of instructional technology, even when they don't involve new hardware.

The desire to cover new ground and use new technology presents us with too many choices—relativity, quantum theory, and chaos versus Newton's laws and classical thermodynamics; new teaching methods based on cognitive theories versus nineteenth-century pedagogy; the digital computer versus the electronic calculator.

The problem of having too many choices has already led us to the creation of the 1,000 page introductory text book. Students complete introductory courses in such a state of cognitive overload that all they retain are a few memorized definitions and algorithms for solving standard textbook problems. The mounting pressure to substitute more contemporary topics for the classical ones runs counter to pleas from physics educators and cognitive psychologists to offer students more concrete experience before dealing with weird, abstract theories about things that don't constitute part of everyday reality.[1]

The syndrome of cognitive overload is rooted in the presentation of far too many topics, the more contemporary of which represent major paradigm shifts away from the basic worldview of classical physics. To adapt a familiar phrase, taking introductory physics is like trying to take a drink from a fire hose.

We are in an exciting period of experimentation with course objectives, content, and pedagogy in introductory physics. We may emerge from this period in the history of physics education with a new intellectual canon that sweeps the nation and eventually the world.[2] On the other hand, the accelerating evolution of the fields of physics and physics teaching may force us to abandon the present uniformity of content and teaching methods for endless experimentation with a host of new approaches.

Workshop Physics: Its Premises

The Workshop Physics project at Dickinson College represents one of a growing number of attempts to forge a new canon or intellectual tradition for introductory physics instruction.[3] Those of us on the Workshop Physics staff[4] want to share our experience with the new approach, but not because it represents the only viable one. Instead, Workshop Physics represents one attempt to address some of the generic problems with introductory physics. Workshop Physics applies recent educational research to the introductory physics curriculum, and it facilitates this application with computer technology. A number of observations and assumptions have guided the development of Workshop Physics.

Reducing Content and Emphasizing the Process of Scientific Inquiry

In developing Workshop Physics we assumed that acquiring transferable skills of scientific inquiry is more important than problem solving or acquiring descriptive knowledge about physics. Arons refers to the acquisition of transferable skills as a way of developing "enough knowledge in an area of science to allow intelligent study and observation to lead to subsequent learning without formal instruction."[5] There were two major reasons for the emphasis on inquiry skills based on real experience. First, most students enrolled in introductory physics at both the high school and college level, do not have sufficient concrete experience with everyday phenomena to comprehend the mathematical representations of them traditionally presented in introductory courses. The processes of observing phenomena, analyzing data, and developing verbal and mathematical models to explain observations afford, students an opportunity to relate concrete experience to scientific explanation. A second equally important reason for emphasizing the development of transferable skills is that a student who is confronted with the task of acquiring an overwhelming body of knowledge must learn some things thoroughly and acquire methods for independent investigation to be implemented as needed. This follows Phil Morrison's adage, "Less is more."

Emphasis on Directly Observable Phenomena

The guiding principle for retaining topics in introductory physics is that they be amenable to direct observation and that the mathematical and reasoning skills needed to analyze observations be applicable to many other areas of inquiry. In choosing topics, we should emphasize the development of operational definitions and empirical relationships before introducing formal definitions and theoretical relationships.[6] We approve of the trend toward motivating students with applications of classical physics to problems of current concern. Jearl Walker's expositions of the physics of everyday phenomena, the physics of human motion, and Newtonian cosmology are splendid examples of contemporary applications of classical physics.[7] We do not consider it advisable to add topics such as relativity, quantum mechanics, and chaos. Such topics require levels of abstract reasoning we believe to be beyond the abilities of beginning students.

Eliminating Formal Lectures

Although lectures and demonstrations are useful alternatives to reading for transmitting information and teaching specific skills, no one has ever proved that they are efficient vehicles for helping students learn how to think, conduct scientific inquiry, or acquire real experience with natural phenomena.[8] In fact, some educators believe that peers are often more helpful than instructors in facilitating original thinking and problem solving.[9] The time now spent by students passively listening to lectures is better spent in direct inquiry and discussion with peers. The role of the instructor is to help create the learning environment, lead discussions, and engage in Socratic dialogue with students.

Using the Microcomputer as a Flexible Tool

When used as flexible tools for collecting, analyzing, and displaying data graphically, computers can accelerate the rate at which students can acquire data, abstract, and generalize from real experience with natural phenomena. The digital computer is an essential tool for any inquiry-based learning experience in physics because it has become the most universal tool of inquiry in scientific research. The computer has had a profound effect on the nature and scope of physics research. However, even computer-aided inquiry takes time, and we believe that students cannot engage in the process of guided inquiry and direct observation, even armed with computers, and still cover the amount of material normally introduced in an introductory physics course sequence.

Workshop Physics: Its Practice

Workshop Physics was first taught at Dickinson College during the 1987–88 academic year to students in both the calculus- and noncalculus-based courses. It is taught in three two-hour sessions each week. There are no formal lectures. The course content in Workshop Physics is about 30 percent less than in our traditional courses. Each section has one instructor, two undergraduate teaching assistants, and up to 24 students. Each pair of students share the use of a Macintosh computer and an extensive collection of scientific apparatus and other gadgets. Among other things, students pitch baseballs, whack bowling balls with rubber hammers,[10] pull objects up inclined planes, attempt pirouettes, build electronic circuits, explore electrical unknowns, ignite paper with compressed gas, and devise engine cycles using rubber bands. The workshop labs are staffed during evening and weekend hours with undergraduate teaching assistants.

The material is broken up into units lasting about one week, and students use an "Activity Guide," which has expositions, questions, and instructions as well as blank spaces for student data, calculations, and reflections. The "Activity Guide" is keyed to a standard textbook. Textbooks that have been used in the calculus-based section include Serway's *Physics for Scientists and Engineers* and Halliday and Resnick's *Fundamentals of Physics*. The noncalculus section uses Faughn and Serway's *College Physics*.[11] In general the curriculum emphasizes the four-part learning sequence described by cognitive psychologist David Kolb.[12] Students often begin a week by examining their own preconceptions; then they make qualitative observations. After some reflection and discussion, the instructor helps them develop definitions and mathematical theories. The week usually ends with quantitative experimentation centered on verification of mathematical theories.

The Role of the Computer in Workshop Physics

The computer is used in almost every capacity except that of computer-assisted instruction. Although the M.U.P.P.E.T. project at the University of Maryland has reported great success at teaching introductory students to program in Turbo Pascal,[13] our attempts to incorporate programming into the introductory lab at Dickinson, led us to feel that we were using physics to teach computing rather than the other way around. We therefore use spreadsheets as the major tool for calculation.

Linearization with Spreadsheets and Graphing

The most popular use of spreadsheets involves entering data directly into the spreadsheet in Microsoft *Works* on the Macintosh SE computer for analysis and eventual graphing with the *Cricket Graph* software package. Linearization is considered to be one of the essential transferable skills associated with Workshop Physics. Using the microcomputer for linearization and least-squares analysis, students discover simple functional relationships empirically or verify mathematical theories. An unusual application of linearization involves a project in which students poke nails through insulation board to create a parallel array of "flux lines." Students then count the number of nails passing through a wire hoop (representing a surface area) to compare the angle of the hoop's normal vector with the direction of the nails. A graph of the number of nails versus $\cos \theta$ allows students to discover that $\Phi = E \cdot A = E A \cos \theta$. A simple linearization and graphing exercise is illustrated in Figure 1.

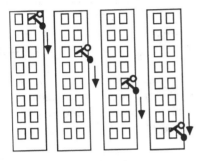

	A	B	C
1	S	t	t^2
2	(m.)	(sec.)	(sec^2)
3	0.82	0.44	0.19
4	6.32	1.15	1.32
5	10.71	1.49	2.22
6	13.55	1.70	2.89
7	17.01	1.91	3.65

(a) An object is dropped from different heights.

(b) The data is entered into a spreadsheet and calculations are performed. This takes a minute or so of a student's time.

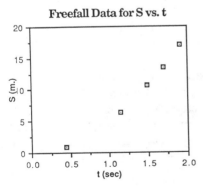

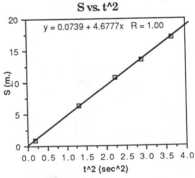

(c) The Data is transferred to Cricket Graph™ and plotted. The parabolic curve indicates that S is proportional to t^2. This takes another minute or so of a student's time.

(d) S is plotted as a function of t^2, and the curve is linearized. The slope of 4.68 m/s (which represents the "best estimate" for g/2) is derived from a built in least squares analysis.

Figure 1. Linearizing data with spreadsheet and graphing software.

Calculations and Modeling with Spreadsheets and Graphing

Spreadsheet calculations are also used instead of integration as a tool to solve numerical problems. In some cases spreadsheet calculations are used for mathematical modeling. For example, spreadsheet relaxation calculations work beautifully for modeling the pattern of electrical potentials surrounding the "electrodes" on electric-field–mapping paper. Mathematical functions representing traveling waves can be plotted in position space at three different times, and the velocity of the wave can be measured on the graph. This helps students explore the real meaning of the expression $y = f(x \pm vt)$.

Microcomputer-Based Laboratory Tools

The so-called MBL tools are used extensively to collect, analyze, and display data. An MBL station consists of a sensor or probe plugged into a microcomputer via an electronic device known as an interface. With appropriate software the computer can perform instantaneous calculations or produce graphs. Sensors that have been linked directly to the computer include the ultrasonic motion detector, photogates, temperature sensors, and geiger tubes. In cases where the user can observe or control changes in a system directly, the microcomputer can be set up to display a real-time graph of system changes. Ron Thornton and others have demonstrated that using MBL to create real-time graphs helps students develop an intuitive feeling for the meaning of graphs and for the characteristics of the phenomena they are observing qualitatively.[14] We have found that a time trace of the position of the student's own body is unparalleled as a tool for learning how the abstraction known as a graph can represent the history of change in a parameter. The real-time frequency distribution that can be produced by using the geiger tube with a radioactive source affords the student the same kind of opportunity to explore and develop intuitive notions about both the meaning of frequency distributions and the nature of counting statistics. A simple counting statistics exercise is illustrated in Figure 2.

The funding for Workshop Physics has allowed us to develop an interface to link MBL sensors including the geiger tube and the photogate to the Macintosh computer. Some of this development has been coordinated with related projects at Tufts University and Technical Education Research Centers.[15] The photogate software is pedagogically oriented and utilizes a raw plotter to allow students to see the times when one or more photogates are switched on or off by real events. The real-time raw plot, which is one of Robert Tinker's many innovative ideas, encourages students to discover for themselves how to use operational definitions in the measurement of velocity and acceleration. A simple photogate timing exercise with raw plotting is illustrated in Figure 3.

Simulations

In select cases where acquiring real data is not feasible or is too time consuming, we have resorted to simulations. David Trowbridge's program simulating posi-

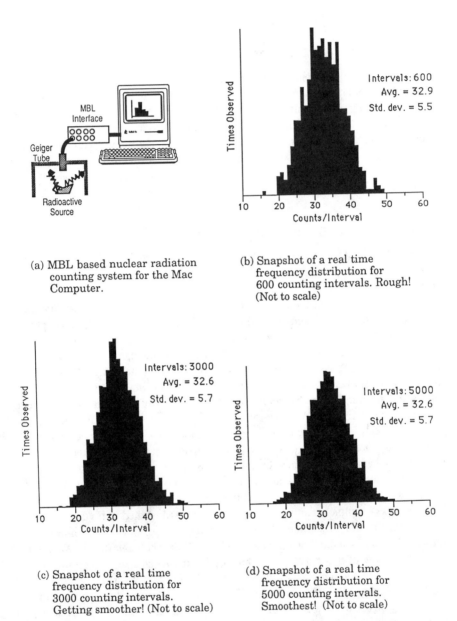

(a) MBL based nuclear radiation counting system for the Mac Computer.

(b) Snapshot of a real time frequency distribution for 600 counting intervals. Rough! (Not to scale)

(c) Snapshot of a real time frequency distribution for 3000 counting intervals. Getting smoother! (Not to scale)

(d) Snapshot of a real time frequency distribution for 5000 counting intervals. Smoothest! (Not to scale)

Figure 2. MBL radiation detection for the study of counting statistics.

tion, velocity, and acceleration graphs for a ball rolling down a set of inclined ramps is one such simulation.[16] Eric Lane's simulation of wave interference is another.[17] A third is the display of electric field lines from a collection of charges developed by Blas Cabrera in a program called *Coulomb*. In *Coulomb* students enjoy creating strange and unique charge configurations on the computer screen and watching the patterns generated by the field lines. This simulation allows stu-

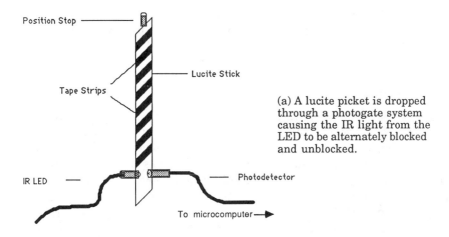

Position Stop

Tape Strips

Lucite Stick

IR LED

Photodetector

To microcomputer →

(a) A lucite picket is dropped through a photogate system causing the IR light from the LED to be alternately blocked and unblocked.

Selected Δt = .252 sec

Raw Plot

Δt_1

$\vdash\Delta T_2\dashv$

Gate 1 Unblocked

Gate 1 Blocked

0   1   2   3   4   5

Time(s)

(b) A raw plot, showing times at which blocking and unblocking of the gate occur, appears in real time on the microcomputer screen. The raw plot concept was developed by Robert Tinker of Technical Education Research Centers. Once an experiment is complete students can move cursors and determine time intervals between various events of interest.

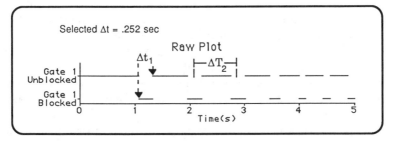

$$\overline{v1} = \text{tape width}/\Delta t_1 \qquad \overline{a1} = (\overline{v2} - \overline{v1})/\Delta T_1$$

$$\overline{v2} = \text{tape width}/\Delta t_2 \qquad \text{etc.}$$

(c) Novice students can use operational definitions of velocity and acceleration to determine the rate of fall of the lucite picket through the photogate.

Figure 3. Measuring acceleration with photogate timing.

dents to discover that in two-dimensional Cabreraland, the flux enclosed by a loop is always proportional to the net charge enclosed by the loop.[18]

Desk-Top Publishing

Since written communication skills are considered important, students are required to hand in several formal lab reports each semester. Students use the computer for word processing and creating apparatus drawings for formal laboratory reports. Students augment the word processing program from the Microsoft *Works* package with diagrams created using *MacDraw*.

Conclusions

Although computers served a vital role in the Workshop Physics program, the central focus of the program is on direct experience-based instruction, not computers. Thus we'd like to think the program could survive without computer. Still, the availability of a computer system for every pair of students is heady wine. We are just beginning to explore the full potential of classroom computing, but every time we design a better approach, another new computer technology beckons.

Our first year of Workshop Physics was at once exciting and exhausting. We eliminated the hour-long lectures, and substituted hands-on experience, but old habits are hard to break. Instructors droned on too long at times during "expositions." We often crammed too many activities into a week. Students discovered that learning by inquiry takes time, and that progress can seem agonizingly slow. At the end of the first semester a significant number of students complained that they had done more work in the course than in all their other courses put together. We learned from our experience. By the end of the second semester, we had calmed down enough to evoke more favorable responses from students. In fact one of the sections got the second-highest numerical rating of any of the hundred-odd laboratory science sections taught at Dickinson in the past three years. Students commented most often that they enjoyed being active and acquiring transferable computer skills. The results of tests on mechanics concepts showed statistically significant gains over those of the pre–Workshop Physics students. A full assessment of the educational gains is underway but will not be completed until the third year of the program.

Our experiment will continue, and we hope many introductory physics teachers will join us in this exciting approach to introductory physics education. But even as we expand and refine our program, we look forward to the outcome of the experiments of our colleagues who seek to use computers in different ways or to incorporate twentieth-century paradigms into the fabric of introductory courses. We are all part of the same exhilarating quest to forge a new intellectual canon that renders science education more meaningful and stimulating for students and teachers alike.

1. There is a growing body of physics-education literature. The focus in most of the literature pertaining to concept development is on topics amenable to concrete experience. An excellent introduction to the problems students encounter in developing fundamental concepts in mechanics is contained in a review article by L. C. McDermott, "Research on Conceptual Understanding in Mechanics," Phys. Today 37, 24 (July 1984).
2. G. Allan, "The Canon in Crisis," Liberal Education 72, 89 (1986). This article provides an insightful analysis of the threats to the intellectual canons in various disciplines. George Allan argues that "contemporary intellectuals have increasingly been driven to abandon their belief in objective truth and to seek refuge in relativism." This argument is consistent with Thomas Kuhn's notions about the cultural relativity of the scientific

enterprise. The applications to physics education are obvious. Quantum theory and special relativity have shaken the primacy of classical mechanics in the realms of the very small and the very fast. More recently the emerging field of chaos has threatened our notion of approximate Laplacian determinism even for the motion of large objects moving at ordinary speeds.

3. The Workshop Physics project began officially in October 1986 with a three-year grant from the U.S. Department of Education Fund for Improvement of Postsecondary Education.

4. The 1987–88 Workshop Physics staff included Robert Boyle, Priscilla Laws, John Luetzelschwab, Guy Vandegrift, and Neil Wolf from the department of physics and astronomy at Dickinson College and Mary A. H. Brown from Troy State University in Dothan, Alabama.

5. A. Arons, "Achieving Wider Scientific Literacy," Dædalus 112, 91 (1983).

6. R. Karplus, "Educational Aspects of the Structure of Physics," Am. J. Phys. 49, 238 (March 1981).

7. J. Walker, *The Flying Circus of Physics with Answers* (New York: Wiley, 1977); P. Brancazio, *Sport Science* (New York: Simon & Schuster, 1984); K. Laws, *The Physics of Dance* (New York: Schirmer, 1984); C. M. Will, "Newtonian Cosmology," Contribution to the Introductory University Physics Project Conference at Harvey Mudd College, March 1988. Professor Will is from the McDonnell Center for the Space Sciences, Department of Physics, Washington University, St. Louis, MO.

8. In *Alternatives to the Traditional* (San Francisco: Jossey-Bass, 1972), O. Milton describes an experiment in which students in an introductory psychology course who are not exposed to lectures do just as well in the course as those who attend lectures. In *What's the Use of Lectures* (Baltimore: Penguin, 1971), D. A. Bligh presents metaresearch on the educational impact of lectures. The author concludes that lectures are as effective as other methods for the transmission of information, but that most lectures are not as effective as active methods for the promotion of thought. "Peer Perspectives on the Teaching of Science," *Change*, 36 (March/April 1986), S. Tobias describes a study in which a group of experienced college teachers outside of physics helped identify the nature of conceptual difficulties in traditional physics lectures and demonstrations.

9. The literature on peer learning is scant. However, Robert Fullilove from the University of California at Berkeley has used peer interaction almost exclusively in helping minority students succeed in mathematics courses by enhancing their study skills and problem-solving abilities. The program has enjoyed incredible success in the past decade. It has been the observation of the Dickinson Workshop Physics staff that students display more thinking capabilities and are far more articulate with each other when they think no instructor is within earshot. Once a student understands a concept he or she often uses a more comprehensible style of explanation and vocabulary.

10. E. Taylor, "Impulse Mechanics" (Unpublished paper for the Maryland University Project in Physics and Educational Technology, University of Maryland, Department of Physics and Astronomy, College Park, MD 20742, 1987).

11 R. Serway, *Physics for Scientists and Engineers*, 2nd ed. (Philadelphia: Saunders, 1986); D. Halliday and R. Resnick, *Fundamentals of Physics*, 3rd ed. (New York: Wiley, 1988); J. Faughn and R. Serway, *College Physics* (Philadelphia: Saunders, 1986).

12. D. Kolb, *Experiential Learning* (Englewood Cliffs, NJ: Prentice Hall, 1984).

13. W. M. MacDonald, E. F. Redish, and J. M. Wilson, "Freshman Physics with the Microcomputer" (Unpublished paper for the Maryland University Project in Physics and Educational Technology, University of Maryland Department of Physics and Astronomy, College Park, MD 20742, 1988).

14. R. Thornton, "Tools for Scientific Thinking—Microcomputer-Based Laboratories for Physics Teaching," Physics Education 22 (1987). Also see R. Thornton's paper, "Tools

for Scientific Thinking: Learning Physics Concepts with Real-Time Laboratory Measurement Tools" in these proceedings.

15. The "Tools for Thinking" project under the direction of Ronald Thornton of the Center for the Teaching of Science and Math at Tufts University is funded by the FIPSE program at the U.S. Department of Education to oversee the use of MBL materials in introductory physics laboratories at a number of other colleges and universities. The "Modeling" project under the direction of Robert Tinker of Technical Education Research Centers was funded by the National Science foundation to develop hardware and software that allow a student to collect real data and develop a mathematical model for the behavior of a system in an interactive fashion.

16. David Trowbridge, of the Center for Educational Computing at Carnegie Mellon University, has developed the software package known as *Graphs and Tracks* for Macintosh, IBM, and Sun computing systems using the cT programming language. Also see D. Trowbridge's paper, "Applying Research Results to the Development of Computer-Assisted Instruction," in these proceedings.

17. Eric Lane's *Standing Waves* software for the Apple II Computer is distributed by Conduit at the University of Iowa. Also see E. Lane's paper, "Animation in Physics Teaching," in these proceedings.

18. B. Cabrera, *Electromagnetism: Physics Simulations II*, available through Kinko's Academic Courseware Exchange in Santa Barbara, CA. Also see B. Cabrera's paper, "Early Experiences with Physics Simulations in the Classroom," in these proceedings.

Computers in Learning Physics:
What Should We Be Doing?

Alfred Bork
Educational Technology Center, Information and Computer Science, University of California, Irvine, California 92717

We have serious problems in the learning of physics that we are not currently facing. The computer could be a major tool in overcoming these problems, but the ways we have used computers so far do not address these problems or do anything substantial to improve the learning of physics. Many current efforts, although well intended, are counterproductive. I put forward a plan in this paper that would lead to substantial improvements and a much better future for all of us. But it will involve substantial changes in policy at the national level.

Problems with Learning Physics

At the beginning of this century most graduates of secondary school in the United States had physics as one of their courses. Currently only a small fraction of secondary school students take physics. The numbers of students taking physics in secondary school has declined all through this century. There was some slight change in this pattern during the development of the new curricula, about 1960, but this was temporary. Many now think that physics should be taken in secondary school only by the few students who absolutely need it for the future. Counselors actively urge students not to take physics unless these students are exceptional or are pursuing careers for which physics would be essential.

The situation has deteriorated to the point that a recent survey done by the National Science Teacher's Association shows that only about half the high schools in the United States currently offer a physics course. Far fewer high schools offer advanced-placement physics courses, and these high schools usually cater to the children of the wealthy. The data also show that about half of the teachers currently teaching physics are not certified to teach physics. As with physics enrollment, these figures have been moving in the same monotonic direction for long periods of time. The situation is a disaster.

We hear less about the problems with beginning university physics courses. But to my mind the situation is quite similar. Research in naive worldviews often shows that beginning physics students, even many of those making high grades, do not understand the fundamental concepts of these courses. They have learned to work set problems with little understanding of what is actually involved.

Our physics courses at the secondary and university level are old in content and in structure. Attempts at producing radically new courses at both levels during the post-Sputnik period have had little lasting effect. The dominance of the Sears-Zemanski Halliday-Resnick text tradition has been almost complete at the university level. New books are extremely similar to the old books. Attempts at new directions, such as the Feynman Lectures, have had little effect on introductory courses.

Although there has been considerable research on the issues of how students learn physics, this research has had practically no effect, statistically, on text and classroom practice. None of the currently available textbooks reflect this research, and the vast majority of courses follow these textbooks closely. We can point to individual examples where there has been considerable influence. But here I am not concerned with examples that affect only a few students, but with the national and international situation.

Computers in Physics Today

Although there is much talk about computers and physics classes, we see little positive net effect. So far the computer as a learning device has made little change in the way courses are being taught. Again there are exceptions, but most of our courses are still essentially lecture and textbook-based courses, with only at best minor additions from the computer. In addition, much of the computer material available in physics is not of high quality; it lacks even the professional standards that we see in the poorer textbooks. Most of the material produced so far must be considered bits and pieces, small individual programs. It makes little difference in the extensive process of teaching physics.

Computer units now available, in all areas including physics, can typically be described as amateurish. What has happened so far should be viewed not as serious production of curriculum material, but rather as an experimental effort; we are trying out different tactics involving the computer to get some idea of the range of what is possible.

Even as an experimental effort, however, production so far has been limited. It has not been done at sufficient volume, or with sufficient controls, to allow experimentation that would be regarded as satisfactory in physics itself. There are some exceptions to this negative picture. But these exceptions are uncommon. We must view the situation realistically if we are to make progress.

Antitechnology

I advocate the use of the computer as a way of solving the problems we currently have in the learning of physics. I also urge caution. Experience shows that many technology approaches are based on technology instead of pedogogy. This is a serious error.

Often the technology is dominant, the developers believe that there is some magical new hardware or software that will lead to solutions of educational problems. I do not believe these strategies are likely to be of any use. They follow the current will-o'-the-wisp. Thus developers are interested in Macintoshes for a while, Suns for a while, then RISC processors, and then some new generation of Crays implanted in human brains at birth. The "solution" always lies, such people conclude, with the next generation of hardware.

On the software side, developers think that they will get magic results from a new authoring system, computer tools, or artificial-intelligence techniques. There is little to support this so far, in spite of the vast amount of money that has gone in this direction.

I insist that our learning problems are pedagogical problems, and that the computer, no matter what is used or how it is used, is not a magical device. There is no ideal technology. Rather, technology should be dictated by pedagogical needs. This suggests that technology should always be in second place, rather than in first place, perhaps a modest interpretation of the title of this section!

In the pursuit of the latest technology, we tend to forget that we seldom use the technology that we have adequately. If one looks at the current generation of inexpensive microcomputers, it becomes clear that many of the things we want to do can be done now.

That is not to say that there will not be uses for new technology. Certainly we will continue to find that the newer technology can be very valuable. But our problems are not fundamentally technological, they are pedogogical. They involve the lack of adequate knowledge and resources for curriculum development. The pursuit of technology in education consumes large amounts of money, but instead of spending money in rational ways, we are always looking for magical new solutions.

What Can the Computer Do?

So far in this paper I have taken a negative tone, pointing out that problems are severe and that the computer has done very little to help with these problems. But my intent in this paper is not to be negative. Rather it is to outline a positive position, to show first how the computer could be used to improve our educational system, including the learning of physics, and to suggest a strategy to achieve this.

What are the advantages of the computer in learning modes? Most of these advantages flow, I claim, from the fact that computer-based learning, in the most general sense of the term, can be highly interactive. Interaction means that there is frequent and meaningful response from both the computer and from students.

Our standards of interaction in our work at the Educational Technology Center consider two factors, the degree of interaction and the quality of interaction. The degree of interaction refers to how frequently interaction occurs. We find, in testing in environments where students are free to leave, that 15 seconds is about the maximum time that should elapse between the time a student has completed replying to one question on the computer (or making some other meaningful input) and the time a student is in a position to consider some different input. There is nothing hard and fast about this 15 seconds, but we find that in testing in environments like public libraries that students tend to leave when interaction becomes much slower than this. This implies that there cannot be much text or visual information presented in one chunk if we are to attain a satisfactory degree of interaction. We do not want to create books or films on the computer.

Quality of interaction is a more difficult issue. The standard for quality should be the standard in human conversation. If a good tutor works individually with two or three students, helping them to learn physics, then we are seeing quality interaction. We can refer to this type of computer use as Socratic, because Socrates taught in such an interactive fashion. He seldom gave information; rather he asked questions, leading students to discover for themselves the fundamental ideas. The process is like the development of science itself.

We can easily point to examples of poor interaction. I have argued in the past, and continue to believe strongly, that multiple choice is a poor form of interaction, not worthy of use in any form with students, and certainly not desirable in computer-based material. While I am sympathetic with the problems of those with very large classes, people who give multiple choice tests should realize that they do so as a matter of expediency; such a practice is not pedagogically desirable. Multiple choice in computer material is ridiculous, a sign of poor material.

Interaction, with the computer interacting frequently and carefully in a meaningful way with users, has at least two important consequences. First, it allows us to do something almost impossible in lecture-based and textbook-based classes, as long as there are more than about ten people in these classes: it allows us to individualize learning to the needs of each student. Very few physics classes are currently individualized to the needs of each student. Some students (about 20 percent) do well because the learning method presented matches their needs, or because they have excellent backgrounds or other advantages. But students who need specialized help, different from that offered in the standard course, do poorly.

There is good evidence from the research on mastery learning that every student can learn material to "A" level. Our experiences in physics with the personalized system of instruction shows the possibilities, although success in that area was very often limited to small classes and to instructors who understood the system. In large classes, there were severe logistic problems with the personalized system of instruction. However, experience indicates that computers can solve these logistic problems.

The second major consequence of interaction is motivation. Our own studies in public libraries show that students will stay for very long periods of time at difficult learning tasks, provided these tasks are interactive. Generally we need to give far more consideration than we currently do to motivational issues in our classes and in our learning materials. The notion (often underlying development in the early parts of the 1960s and still too frequent today) that material intrinsically interesting to the professional will be intrinsically interesting to the student, has been shown many times to be false.

The computer affords us an opportunity to magnify the power of excellent teachers. We have always had a few very good teachers. Our problem has been that the effect of these extremely good teachers has been very limited. They see only a few students, and there are vast numbers of students in the country and in the world. Attempts to spread their tactics to wider audiences have generally failed. Thus we discount the enthusiastic articles in the journals about a marvelous course because we know that course was marvelous because of the individual involved, not because of the particular strategies that individual was using.

The computer gives us a way for excellent teachers to reach not just the few hundred people that they might normally have in their classes in a year, but millions of people. We hope at one point that video material would accomplish this, but video is an inherently noninteractive medium and has seldom produced results better than those of traditional lecture.

Current Courses versus New Courses

Most of the computer use at the present time is in small amounts of material added to courses developed before the computer came into existence. Adding a few computer modules to a Halliday and Resnick course and giving the tests that were given before the computer arrived, is unlikely to lead to any fundamental difference in the course.

Our courses are old, and therefore we need new courses. This is independent of the computer. But the computer gives us a whole range of new possibilities for these courses. Courses using the computer extensively can be very different in structure and in content from our current courses. I have elsewhere described possible new course structures.

Our hope lies in our ability to generate new courses, courses that from the very beginning assume that the computer is available as part of the learning resources. All of the decisions about how the computer is used can be made because of the pedagogical needs of the students, thus leading us to possibilities that could not possibly have been in existing courses, because they were developed before the computer was available.

This is not to say that we will not continue to use the older learning media. Naturally we will use all the learning resources we have at our command. This is not a game we are playing, but rather a serious attempt to assure maximum learning for all students! Hence we would not want to rule out anything that might be helpful to some of the students involved in our learning environments.

The aim of the new courses should be that all students learn. This means that they should learn to the mastery level, typically represented by "A" work in our courses. Experience with mastery learning strategies, such as the personalized system of instruction, shows that this is possible. The new courses should provide individualized aid and assistance to meet the typical student problems that occur within the physics environment. They should be interesting to students. The learning mode should be interactive.

These courses must be developed with a full professional system. We cannot develop the courses we need unless we approach them seriously. We use professional standards in our research efforts in physics, and we should not use amateur standards when it comes to developing learning material. Learning must be considered as serious a problem as research.

We do have, nationally, experience in developing full-scale courses using the computer. Many of these courses were developed many years ago. The Stanford logic and set-theory courses, taught for many years as the standard courses in those areas at Stanford, certainly compare favorably with the traditional noncomputer courses in those areas, as judged from the standpoint of student learning. The introductory physics course developed at the University of California, Irvine, many years ago is another promising example. Unfortunately, funding for developing new courses of this type has been difficult to obtain. Indeed, funding for converting these existing courses to modern machines has not been available.

How Do We Proceed?

If the reader has been following the arguments to this point, the question that comes to mind involves procedure. How do we move on from the current unsatisfactory state to a desirable state? What do we do, for example, with students in the high schools who do not have a physics teacher, or have a physics teacher on emergency certification without the qualifications for teaching physics? What do we do about the fact that even in our best classes, many people do not learn? I suggest the following approach.

The Larger Problem

First, we must realize that we cannot solve this problem for physics alone. The learning problems are too interrelated with other subject areas; many of them occur across the educational spectrum. Furthermore, the efforts needed are such that if we do it for one course we have hardly scratched the surface.

Hence the problem of constructing new physics courses for all levels, from earliest childhood through our advanced university degrees, is not a problem that can be considered by itself. Even if we look only at the most critical needs, the beginning courses in the secondary school and the university, we cannot view this as an isolated problem. Rather we must look at the entire spectrum of education if we are to have success. Physics depends on other subjects such as math. Furthermore, the knowledge we need about how computers can be used effectively must be gathered in the broadest possible environment.

The Experimental Phase

We cannot, I would argue, proceed immediately and wisely to the large problem of full curriculum development. Rather we need to learn more about the effectiveness of the computer as a learning device, more than we have learned with the sporadic, often expensive, efforts that have occurred so far. We need a rational approach.

Before we can start on a full development of the extensive computer-based learning material needed to restructure the educational systems of the world, we need to learn about how to proceed. We have had development of full-scale courses. But this development has been too specialized, with too few courses over too long a period to give us the data we need for the rational planning and implementing of the much more extensive development phase that would follow.

Hence it is essential that we begin as soon as possible a full-scale experiment to develop, evaluate, and use a series of courses that include all that we know about technology and learning in general. These courses should be developed in areas that correspond to our standard courses so that these courses can be used as a basis for testing.

If we are to gather the detailed information required a summary evaluation of each new course should be done with very large numbers, hundreds of thousands of typical students. We wish to discover not which courses are superior, but which components and tactics work best for what situations. Furthermore, we want to study carefully the processes for developing of courses. We should develop different courses with different strategies so that we can gain experience for future development.

How many courses are needed to gain enough information? I believe that about 20 courses should be developed in this stage. The cost for such an experiment, over possibly a six-year period, including extensive evaluation, would be about $200 million. In terms of the educational budgets of the United States and of the world, this is trivial, almost at the noise level.

The Development Phase

The next stage would be to extend the lessons learned in the first stage, the experimental stage, to a much wider basis. We need to work quickly, because our time to accomplish these tasks, given the problems that education faces, is brief. But we will not, I believe, adequately plan and carry out this phase without the results of the experimental effort.

The cost of full-scale development would be sizable, if judged in absolute terms. The cost for the hardware to support it would be similar to the costs for development of the material. As a percentage of the total cost of education the development costs here are still quite small.

I have written elsewhere about this plan, and I will be happy to send additional details. Many of the details would depend on the results of the experimental phase.

The Politics of Education

How do we get this to happen? First, people have to view this plan as desirable. That would be part of the effort of the experimental phase. Second, the money must be available, and the planning must be undertaken. For the experimental phase, existing organizations with existing funds could carry out this activity if they felt it desirable. But for the full development phase we need a sizable commitment of the governments of the world to improving learning for all students. Thus a management structure is essential.

The task of convincing the authorities will not be easy. But it seems essential that we try. What are our alternatives?

Bork, Alfred, Weinstock, Harold, eds., *Designing Computer Based Learning Material* (Berlin: Springer-Verlag, 1986).

Bork, Alfred, *Learning with Personal Computers* (New York: Harper and Row, 1986).

Bork, Alfred, *Personal Computers for Education* (New York: Harper and Row, 1986).

Bork, Alfred, *Learning with Computers* (Bedford: Digital Press, 1981).

Bork, Alfred, *Computer Assisted Learning in Physics Education* (Oxford: Pergamon Press, 1980).

Bork, Alfred, "Computer-Based Instruction in Physics," Phys. Today 34 (September 1981).

Kiester, Sally, "It's Student and Computer—One-on-One," *Change* (January 1978).

Bork, Alfred, Stephen Franklin, and Martin Katz, "Newton, A Mechanical World," Science, NECC81 (June 1981).

Bork, Alfred, "Producing Computer Based Learning Material at the Educational Technology Center," J. Computer-Based Instruction, 11 (Summer, 1984).

Bork, Alfred, "Production Systems for Computer-Based Learning," *Instructional Software Principles & Perspectives for Design and Use*, ed. Walker, D. E. and Hess, R. D. (Belmont: Wadsworth Publishing, Inc., 1984).

Bork, Alfred, "Interaction—Lessons from Computer-Based Learning," *Interactive Media: Working Methods and Practical Applications*, ed. Laurillard, D. (Chicester: Ellis Horwood, Ltd., 1987).

Bork, Alfred, "The Potential for Interactive Technology, BYTE Magazine, 12 (February 1987).

Bork, Alfred, "Let's Test the Power of Interactive Technology," Educational Leadership (March 1986).

Bork, Alfred, "Planning for the Future of Education," Technology and Learning, 1 (September/October 1987).

Bork, Alfred, "New Structures for Technology Based Courses," unpublished.

Contemporary Physics in the Introductory Course

D. F. Holcomb
Department of Physics, Cornell University, Ithaca, NY 14853

Stimulated by a widely perceived need for renovation in the content and style of the introductory calculus-based physics course, a project to seek realistic ways to achieve that renovation is underway. The project is known formally as the Introductory University Physics Project (IOPP), with John S. Rigden as leader. Several study groups have been set up to concern themselves with particular modes of attack on the problem. One such study group, of which I am a member, focuses on seeking natural ways to introduce contemporary physics into the introductory course through a phenomenological approach.

By the term "phenomenological approach" we mean the development of models and connective tissue based upon the organization and analysis of observed and experimental facts rather than upon the development of underlying synthetic models. We envision that the level of theoretical background needed for a phenomenological approach will not differ significantly from the level currently achieved in a typical introductory course.

Those of us working with the IUPP think that the computer will play a strong role in whatever proposed course patterns eventually emerge from our activities. Thus while the phenomenological study group has not specifically examined the way in which the computer will facilitate the presentation of materials, we presume that such facilitation will occur. My purpose here is to make a rather brief progress report of the activities of this study group so that those attending this conference will be aware of one possible future direction for the content and style of the introductory course.

It is important to clarify the way in which the word "contemporary" is used in my remarks. The term "modern physics" has come to denote primarily the content of the great revolutionary theories of the first 30 years of this century: relativity and quantum theory. The last main events of this drama took place approximately 50 years ago with the codification of nuclear physics in the great Bethe and Livingston articles of 1938. I wonder if our students muse upon the fact that to us "modern" means 1900–1938? By "contemporary physics" I mean topics that recently have occupied the attention of physicists and interested bystanders. These topics may involve major theoretical syntheses such as the recent union of particle physics and cosmology, or subjects that are at the moment essentially phenomenological, such as the high T_c superconductors. Topics might indeed be based on quantum theory and relativity, but they could very well rest upon new glories of Newtonian mechanics brought to our consciousness by the power of the modern computer.

Activities of the "Contemporary Physics through Phenomenology" Group

The first substantive meeting of the study group took place at Harvey Mudd College on March 11–13, 1988. Those participating were Judy Franz, West Virginia University; Franz Gross, College of William and Mary; Daniel Kleppner, MIT; Brian Maple, University of California at San Diego; Jonathan Reichert, SUNY Buffalo; Brian Schwartz, Brooklyn College; Clifford Will, Washington University; and myself. Each participant discussed a particular topic in physics of contemporary interest from the standpoint of how it might fit naturally into the fabric of an introductory course. Following these presentations, we found it natural to ask ourselves how the fundamental underlying structure of the introductory course might be changed to accommodate our suggested topics while at the same time keeping the total syllabus to a reasonable length. In fact, another guideline of the IUPP group was that a renovated syllabus for the introductory course must be less encyclopedic than the typical current pattern.

Sample topics in contemporary physics that members of the group feel would be amenable to our approach are listed below. The list is obviously a consequence of the particular array of research backgrounds of study-group members, so it should be viewed as a list of examples rather than a list winnowed through a thorough review process.

Nuclear Physics
Determining nuclear sizes from proton and electron scattering
Stability of matter—nuclei, crystals, neutron stars

Semiconductors—p-n junction in particular
Signal diode
Photodiode
Si solar cell
Light-emitting diodes

Superconductors
Electrical properties
Magnetic properties
Applications

Electrons in Metals

States of Matter near 0 K

Particle Physics
Study of collisions and creation/annihilation phenomena using, for example, bubble-chamber photos
Organizing experimental information through conservation principles

Newtonian Cosmology
Cosmological Principle, Hubble Law
Bound or unbound
Age of the Universe
Evidence for dark matter

General Relativistic Phenomenology (no curved space time)
Weak principle of equivalence
Gravitational redshift
Deflection of light

Short Topics of Special Interest
Single ions trapped in EM fields
Chaos

Material developed for these or other topics must be tested against the following check list:

1. What is the required student background in both physics and mathematics?

2. Where and how might the topic fit into the introductory course as now taught? Appropriate questions would be

   a. How would it depend on previous material?

   b. How might it lead on to other suitable topics?

   c. Does it synthesize earlier ideas?

Thoughts about a New Course Framework

After reviewing the topics, the group realized that it needed to understand how such topics might be integrated into a one-year introductory calculus-based course. Thus we blocked out a framework for the course. This proposed framework is sketched below, with possible locations of the contemporary topics previously listed. This framework is more thoroughly developed in my remarks to the symposium entitled "Physics Student of the 1990s" elsewhere in this volume.

Framework	Contemporary Physics Topics
First Term	
Mechanics, Part I*	
Conservation of p, E	Particle physics, via analysis of bubble, spark, or wire chamber pictures
$F = ma$	Theme: Building conservation laws
$1/r^2$ forces	
Potential energy	
$E^2 = p^2c^2 + m_0^2c^4$	

Waves
  Standard introduction
  Interference phenomena
  Diffraction
  Band-width theorem

Comment: Early introduction of wave physics is key to making contemporary physics available in the first term. Wave physics is actually easier than much of mechanics.

Quantum Ideas
  Wave properties of particles via, e.g., electron diffraction
  $\lambda = h/p$
  $\Delta E = hf$
  Heisenberg
  Standing Q-waves in a box

Nuclear sizes, via scattering
Spacing in crystals
Molecular dimensions
Stability of matter
Nuclei, atoms, crystals

Second Term
Mechanics, Part II*
  Fields
  Gauss's Law
  Gravitation

Newtonian cosmology
Size of universe
Hubble, etc.
Bound or unbound?
Age of universe
Dark matter

Electricity and Magnetism*

General relativistic phenomonology
No curved space time

Statistical Physics*
  To get to the point of introducing the Boltzmann distribution so that one can use the relation: Occupation of nth state $\approx \exp(-E_n/kT)$

Condensed-matter physics
Electrons in metals
Superconductivity
Semiconductors: Exploiting the p-n junction
States of matter near 0 K
Phase transitions

* The appropriate content of these basic sections is the subject of deliberations of other study groups in IUPP.

Physics is a living subject. Its ways of describing the world, its conceptual and theoretical structure, and its subject-matter priorities are eternally provisional. At the same time, the strength of the physicist is confidence in the time-tested models and modes of analysis developed between the time of Newton and the beginning of the twentieth century. The person who plans course content for an introductory course is pulled between two attractive but perhaps mutually exclusive options: to provide a sound grounding in "classical" physics, or to show the liveliness and open-endedness of contemporary physics. An effort to achieve a satisfying synthesis of the two desiderata must involve a rather substantial rearrangement and alteration in emphasis of the paradigm for the introductory course that has prevailed for the past 30 years. Such a rearrangement may permit natural and effective integration of contemporary physics within the fabric of the introductory course.

Changing the Introductory Physics Sequence to Prepare the Physics Student of the 1990s

Jack M. Wilson
AAPT, Department of Physics, University of Maryland, College Park, MD 20742.

Change in physics comes quickly, is readily communicated, and is widely appreciated. Change in physics teaching, however, is long in development, poorly communicated, and generally resisted by various special interests in the physics teaching community.

Academic research physicists resist change because it will place a larger demand on their time. Leaders in research in physics education resist because the academic content may change to what they feel is a more abstract (non-Newtonian) approach. Evaluators look at what's being done and say "You can't prove this works any better, so why should we bother to change?" Some compare our efforts to the "new math" and express great skepticism over any attempts to improve the curriculum. Is it any wonder that such a large disparity exists between teaching physics and doing physics?

The introductory physics course epitomizes this problem.[1] It will also be the most difficult to change because the special-interest groups have well-formulated positions and entrenched bureaucracies from which they will fight proposed innovations. Perhaps there is a positive side to this—all innovations should be tested and criticized—but the effect has been to stifle and inhibit needed improvements in the curriculum—particularly in the introductory course.

Topic selection in introductory physics is often driven by the mathematical complexity of the techniques needed to study the topic. Many topics are "off limits" because the students are not expected to have the requisite mathematical tools. As computing power becomes widely available in physics departments, a new set of tools becomes available for student use. Previously off-limits topics become easily accessible. How will this power be put to use in the teaching of physics? What are some of the topics that could, and should, be added to the introductory sequence? How will that change the courses? How will the rest of the physics sequence change as a result? These questions will be considered in this and several related talks.

The use of the computer has become one of the dominant physics teaching issues of our time. Large questions loom before us. Will computers improve physics teaching or make it worse? What other changes will come after we introduce computers into the curriculum? How can the computer best be used? How can we ensure that less advantaged students will have access to computers? Will the advances of research into physics learning be incorporated into the new curricula? Will the new curricula reflect contemporary physics and the way physics problems are solved today? I like to sum all of this up by stating that there are three forces being exerted on physics today—the three C's: computers in physics, contemporary physics, and cognitive issues.

At the University of Maryland we have formed a group of physicists, most of whom are well-known researchers, who are trying to pull together some of these issues in order to set new directions for undergraduate physics programs. The M.U.P.P.E.T. is directed by Edward F. Redish and me, and includes Charles Misner and William MacDonald as additional principal investigators. Approximately eight other physicists from the department have contributed to some of our projects. Much of what I have to say today will draw upon our experiences with the M.U.P.P.E.T. project while working with a variety of classes ranging from the introductory to the graduate level.

We began by reviewing what students were learning in introductory physics and what we wanted them to learn. Unfortunately, in our opinion, most introductory physics courses were teaching students that physics was hard to understand, mathematically intricate, relentlessly deterministic, and concerned with levers, inclined planes, and projectiles. The body of literature on teaching physics and research in physics education tends to confirm this opinion.

We saw the introductory physics course as the pivotal course in the curriculum. Improvement in this course is a prerequisite for improvements in both advanced and high school courses. Advanced physics courses are built upon the solid foundation of the introductory course, and high school courses try to emulate the university course to prepare students for further study. We felt that the introductory physics course acted as an anchor for the physics community—an inviolable bastion against change. We felt that the outdated and authoritarian curriculum often lost us good physics majors. Students majored in physics not because of the introductory course, but in spite of it. We also felt that the training was inappropriate for today's physics majors.

It is not enough simply to catalog the failings of the present system; we also wished to explore the characteristics such a course should have. What physics skills do we want our students to acquire?

1. Quantification skills. The student should understand the relation of numbers with the real world and with algebraic quantities.

2. Analytic skills. The student should be able to manipulate algebraic equations and extract physical content from them.

3. Scales and estimation. The student should understand what controls scales in a problem and be able to estimate answers to one significant figure.

4. Approximation skills. The student should understand when an equation is being treated approximately and the range of validity of all equations used.

5. Numerical skills. The student should be able to write simple programs in order to solve physics problems that cannot be solved analytically.

6. Intuition and the approach to problems. The student should develop an intuition for when results "look right," the ability to choose the appropriate approach to a problem, and the ability to break down a large problem into component parts.

Gordon Aubrecht, while working with M.U.P.P.E.T. in 1986, made a survey of physics texts to determine how well they would meet our desires. He found that 95 percent of the content of all texts was universal and that 95 percent of the content was pre-1935. The current texts are enormous, with over 1,000 problems; yet those problems emphasize only a very limited class of skills. Over 90 percent relate to analytical skills, only a small number relate to the very important estimation skills, and almost none expect students to use numerical approaches. Indeed, it is likely that a student could study physics for eight years without ever seeing a problem the teacher couldn't solve.

We must engage students in the intellectual process of modern physics much earlier in their training. The microcomputer can help achieve this by permitting students to approach a wider variety of phenomena and problems than is possible with only analytic tools. "Canned" programs can also be used to develop students' physical intuition and ability to estimate. We see great advantages in using these new approaches with physics majors as well as students who will never again take a physics course.

Let me address the issues of the three C's by first introducing what I find to be the mathematical hierarchy of physics: algebra, geometry, trigonometry, calculus, differential equations, linear algebra, probability and statistics, and partial differential equations.

The path through the mathematical hierarchy is generally serial. Linear algebra and probability are taught at various times in the sequence, but the remainder are taught in "lock step." Numerical methods, if taught, generally come late in the sequence.

Selection of topics and indeed the physics sequence itself are largely determined by the expected mathematical level of the student. At the high school level, students are generally expected to know trigonometry. In the introductory university physics sequence, students are expected to start with a little knowledge of calculus and develop more proficiency as they advance through the sequence. Differential equations are rarely introduced until the first advanced courses after the introductory sequence, and partial differential equations are not used until some junior and many senior courses. Topics from probability and statistics are often introduced in an ad hoc fashion when necessary.

Teachers often equate students' mathematical level with their level of conceptual development. Topics requiring a higher level of mathematics are often viewed as being more "abstract." Frequently topics are dismissed as "too abstract" when in reality the problem is not abstraction, but the mathematical level. When one hears an argument that topics are too abstract, one should be suspicious enough to look critically at the argument to determine whether the problem is conceptual or mathematical.

We have made a virtue of necessity by selecting a palette of topics that are mathematically accessible to students and then defining them to be those topics that are necessary to the students' conceptual development. Once again a critical reexamination is called for.

It is fair to ask if the mathematical hierarchy given above (which is roughly the hierarchy as it existed at the turn of the century) remains a valid hierarchy today. I would say "No!" Certainly physics students need to know these topics, but these are not the only, and are probably not the most important, topics in mathematics as used by research physicists today. Research physicists today rely heavily on numerical and computational techniques for solving problems. In the more traditional fields of physics, such as condensed matter, nuclear physics, particle physics, astrophysics, and so on, both theorists and experimentalists have become increasingly dependent on computational methods. Few interesting problems remain that rely solely on the old analytical techniques. This is not to say that these techniques are no longer important. Analytical techniques are often used extensively to put the problems in a form that can be solved computationally, and it surely is an advantage for a physicist to recognize portions of problems that can be readily solved in closed form.

The old mathematics is simply not introducing enough. Even mathematicians recognize that there is a problem. At the Calculus for a New Century conference held at the National Academy of Sciences in 1987, mathematicians called for a reexamination of what was being taught as calculus.[2] Joe Redish wrote the original "think piece" for the physics portion of the study; I chaired the physical sciences discussion group and wrote the final report for that group. There were amazing parallels between the experiences of the mathematicians and our experiences in physics. But mathematicians have been extraordinarily effective in translating their need for curriculum revision into federal dollars to support the effort. We have not been so lucky.

If one begins by assuming that students should learn something about numerical and computational methods as a prerequisite or corequisite for their physics courses and examines which topics contribute most to the logical and conceptual development of physics, one might come up with a very different order for the standard mathematical curriculum. With this new order, new topics become accessible, and old topics can be treated differently. Some abstract topics even become concrete! The language of numerical methods is algebra. Differential equations are replaced by difference equations. Although computational physics has its own set of advanced mathematical techniques that are perhaps even more opaque than some of the analytical techniques, much can be done with the Euler method or simple higher-order Runge-Kutta methods.

Making this assumption should lead one to question conventional wisdom in topic selection, concept development, and approach. Work in each of these areas has been accomplished under a limiting set of rules. Conclusions reached may or may not be valid. Certainly we should at least reexamine them critically.

Consider, for example, the burgeoning cottage industry in student misconceptions in physics.[3] This work has helped us to better understand concept development by students and has led to some tentative efforts to improve science learning. Much of the attention has been directed toward mechanics. In the past, quite a bit of published work dealt with the reconciliation of students' Aristotelian and teachers' Newtonian worldviews. Now, of course, we recognize the naivete of referring to the various levels of concept development as Newtonian or Aristotelian. Some early computer programs dealt with microworlds in which the students might explore these views. One of the most popular was a race track on which students were to race cars by giving each a "kick" or impulse at various times. Through experimentation students were to learn all about inertia and Newton's first law. The only problem was that the Aristotelian interpretation was better than the frictionless Newtonian approach! At the very least, it agreed more with the students' own experiences.

Most of the literature cited above tends to classify students along historical lines, either consciously or unconsciously equating historical development and conceptual development. "Ontogeny recapitulates phylogeny" is alive and well in physics. Classification of concepts into Aristotelian or Newtonian (or later into waves and particles) is one of those nonproductive exercises we should avoid. These are simply two different perspectives on the same world. One is more complete and consistent, but the other is often more useful to nonscientists as a working model!

If students approach physics through analytical techniques, they are restricted from consideration of a consistent model of mechanics. Dissipative forces and nonlinear systems are simply beyond their mathematical ability. The resulting frictionless Newtonian microworld is so different from the highly dissipative world the students inhabit that it is no wonder so many studies have shown students divide things into "physics" and the "real world." We owe them a more consistent picture of physics, and it is not that difficult to deliver.

One body of opinion holds that the historical development of concepts is important to students' own conceptual development. While I'm willing to concede some utility, I think that an excessive emphasis on historical development can create obstacles to better understanding of concepts. I would cite the Bohr model as a perfect example of Gresham's law as applied to physics teaching: "Bad physics drives out good physics!"

As I hinted earlier, wave/particle duality is another example of a useless construction. We needlessly spend one or two years developing students' concepts of waves and particles as distinct entities.

Contemporary physics is not quantum mechanics and relativity. It is especially not particle physics. Many groups are busily investigating which topics from modern physics might be included in an introductory course.[4] Others have become incensed that such things are even considered. Arnold Arons represented this latter group recently when he delivered a stirring polemic condemning those who would presume to include some topics from physics after 1920.[5] Although his talk left the impression that issues he raised had not even been considered by the modern physics group, they had in fact been central to our thinking throughout the process. We still don't know what will work and what will not, but we remain persuaded that we must investigate these questions.

Incorporation of quantum concepts would be quite desirable. Feynman and French have made two notable efforts to introduce quantum physics at an introductory level. More recently Eugen Merzbacher and the group from the Introductory University Physics Project have renewed attempts to do this.[6] The problem is not trivial, and thus far we have found more problems than opportunities!

One aspect of modern physics that is almost totally missing at the introductory level is uncertainty (due both to complexity and quantum limitations). Teaching nonlinear systems helps students to understand why the world is not as relentlessly deterministic as we might expect from Newtonian mechanics. It's certainly not the entire story, but it is an important, untaught portion of the picture. It has also been our experience that these topics are interesting and accessible to students at the introductory level.

It is our view that the curriculum change will be evolutionary rather than revolutionary. It is likely to retain most of the traditional features while adding some items that are on our desired-skills list but are not currently found in our courses. Among these are a stronger emphasis on units, dimensional analysis, and scales; use of numerical techniques for problem solving; use of programming as an integral part of learning physics (perhaps in a way analogous to the use of calculus); and the introduction of problems with more complexity than is found in most illustrative problems in present texts.

This problem of complexity versus simplicity is one of the most important and most controversial issues facing those of us in this field. Most physicists would agree that the ability to find a simple model in a complex situation is one of the greatest strengths of physics. There can be no question that students should be taught to develop a simple model and solve it using traditional analytical techniques. The question is whether it is better to present the simplified model first and then add the complicating factors years later, or to present a problem with all of its real-world complexity, solve it, and then boil it down to the traditional simple model.

Prior to the use of computers in teaching, the first option was the only option. Students could not be expected to have the mathematical skills needed to solve the complex problems. By necessity, students were restricted to the simple models. But using numerical methods to solve complex problems (for example by the simple Euler method) is in a fundamental way easier than the traditional analytical calculus approaches.

Consider the case of the simple pendulum. The usual classroom approach is to write down Newton's second law for a pendulum, make the approximation of small angles, and then solve the resulting differential equation for the simple harmonic oscillator. But what happens at large angles? What happens when a damping force is present? Or what happens if a driving force is added? The teacher is forced to evade such questions because the students do not have the necessary mathematical background to consider such complex problems. Further consideration must be delayed for several years until the students' mathematical skills develop. Actually, even the differential equation for simple harmonic motion taxes most students to the limits of their development.

Consider the alternative of developing the phenomenon through numerical methods. In that case, a simple program (using the Euler method) solves it.

```
For i = 1 to NumData do BEGIN
  alpha = -omega * omega * sin(theta);
  omega = omega + alpha * dt;
  theta = theta + omega * dt;
  t = t + dt;
END;
PlotData(theta, t, NumData)
```

The small-angle approximation can be demonstrated simply by replacing the first line by

```
alpha = -omega * omega * theta
```

and the results compared. The case of driving forces and damping forces is also easily done by adding the appropriate terms to the first line. The introductory physics student is able to consider the phenomenon in all aspects while using mathematical tools at a level well below that of the traditional approach. This "step-by-step in time" approach is also a much more concrete—even transparent—approach to this problem. It is far less abstract than the traditional small-angle differential equation analysis.

This example reopens the important and controversial issue of whether students should be presented with problems with real-world complexity or whether students should consider first, only simplified cases. It is my opinion, based on my experience teaching physics for 22 years and doing computer-related teaching for 12 years, that it is much better to deal first with the realistic problem and then boil the problem down to the simple but powerful model that illustrates the essential features of the phenomenon. This is the way physicists actually solve problems, so why not introduce students to this right away? Why do we have to give the students the impression that all problems in physics come already stripped of interesting detail?

It has also been my experience that the traditional approach has the damaging effect of convincing the student that real-world problems, with all their complexity, are fundamentally different from "physics class" simplified problems. For physics majors this results in a large discontinuity in approach to problems—usually somewhere in the junior year.

Consideration of realistic problems has another advantage that is shown at once by the pendulum problem. Students are able to explore some new phenomena not previously accessible at the introductory level, in this case resonance and chaotic motion. Students quickly discover the phenomenon of resonance as they adjust the frequency of the driving force and observe the response of the system. The approach to chaotic motion is somewhat more subtle. As the amplitude of the driving term is increased, the pendulum may swing "over the top" and enter a new region of oscillation. It is instructive to introduce the students to a phase-plane analysis of the motions. Again this is far more concrete than abstract to the student, although it may seem (at first) much more abstract to professors who are not well prepared on these topics. The actual path of the pendulum is so critically dependent on the initial conditions that it is difficult to predict. This becomes an example of deterministic chaos.

The previous example is but one of many used in the M.U.P.P.E.T. We have used these examples both in classes and in laboratories taught by at least five different professors. One of the largest challenges facing our group was the need to select a "language" for considering such problems. We considered many, including BASIC, Pascal, FORTRAN, C, APL, FORTH, and *Lotus 1-2-3*, among others. We eventually settled on two possibilities, each of which has its advantages and disadvantages, its boosters and detractors. The two were Turbo Pascal and *Lotus 1-2-3*.

We selected *Lotus 1-2-3* because it is so widely used in business. Use of a spreadsheet is therefore a skill that would be valuable in a variety of professions. The spreadsheet also provides a set of very well-tested and easy-to-use tools for data entry, data manipulations, programming, and graphics output. Spreadsheets are particularly well suited for use of the discrete mathematics found in the previous example.

Although sophisticated spreadsheet programs are in wide use, the most popular versions are still expensive and do require students to learn another "language." Recently a number of inexpensive spreadsheets have become widely available. For example, *Quattro* is available for educational uses at about $40, and *Quattro* is a significantly better spreadsheet than *Lotus 1-2-3*. However, because the spreadsheet language does not contribute to the further development of computing skills for physicists, many of our project group feel that spreadsheets are best suited to working with nonmajors.

Turbo Pascal provides an attractive alternative for physics majors. It is widely used, inexpensive, and quite powerful. Programming in Pascal encourages good programming and problem-solving habits. In fact, a Pascal program mimics the solution to simple physics problems. Students are forced to identify the variables and constants in a program and discover which are to be given and which to be found. Students are forced to think about data structures and how to process those structures. We were struck by the similarities between the techniques used in Pascal programming and the hierarchical thinking advocated by Reif and others who have studied formal problem-solving skills.[7]

It is also an advantage that many schools and universities, as well as the Educational Testing Service, have adopted Pascal as the introductory computer language. We expect the student to know Pascal prior to taking the physics class or to learn it concurrently with the class. We have developed a student tutorial for use by the roughly 50 percent of our students who have not been exposed previously. We also present physics applications concurrently with Pascal programming. As the semester goes on, both applications become more sophisticated. Our approach to programming is not unlike the traditional approach to calculus and physics. We are not the only ones to reach these conclusions. Richard Crandell at Reed College in Oregon has published two excellent books based upon the use of Pascal at Reed.[8]

At the beginning of the semester, students are presented with rudimentary programs and asked to run them and make modifications. By the end of the semester, they are developing their own programs. In most classes students work on an individual project for completion by the end of the semester. Some of the students work out particularly elegant projects. These include waves on a hanging rope, the inverted pendulum, colliding galaxies, and sophisticated orbit programs using fourth-order numerical methods. Essentially, 80 percent of our classes perform adequately in the computer area by the end of the semester. For the top 10 to 20 percent of our students, we find that the use of computational techniques opens new vistas into exciting and challenging problems.

Although we have been concentrating on languages as tools for learning physics, we must also recognize that other tools will also become important. One of the most important of these tools is the simple data-acquisition device. John Layman of the University of Maryland, Bob Tinker of TERC, and Ron Thornton of Tufts University have pioneered using the computer as a data-acquisition device, including using an ultrasonic transducer with this interface to measure position, velocity, and acceleration of objects—including students! The group at the University of Munich has developed a simple interface that allows one to use a television camera to locate a single white object against a black background. Thus, more complicated two-dimensional motions may be registered digitally.

Computers can also play a role in preparing students for the laboratory, especially for laboratories that are more advanced and more complex. Just as pilots are expected to put in some time on a simulator before taking the controls of a complex aircraft, physics students should be expected to put in some time becoming proficient in conducting themselves in the laboratory prior to inflicting themselves on laboratory assistants. Prelab lectures and reading assignments are notoriously ineffective, but there is evidence that use of the computer prior to the laboratory experience yields better results.[9]

In order for students to be able to use the computer with a minimum amount of effort devoted to input and graphic output, we have developed a set of Pascal tools that can be used to get data into the computer and to present data graphically. The data-input tools allow the student to design a "form" to be displayed on a screen and to display and edit that screen as a part of a program. This provides a professional "full-screen" entry method that we feel is far superior to the traditional "line-at-a-time" approach. It is very important for students to see all of the input data and questions at once. Otherwise it is like a card game in which one card at a time is revealed. It is difficult to make consistent and reasonable responses under these circumstances. We are also able to provide default responses on these "screen forms." This helps get students started with interesting and realistic situations.

The graphic tools are equally important. I have developed tools that allow students to plot data simply by saying "`PlotData (y, x, NumData);`". This means "plot y versus x for NumData points." Scaling, clipping, and adjustment to screen coordinates is done automatically. Students do not need to know anything about graphics drivers, CGA, EGA, VGA, Hercules, or even pixels. Availability of these tools has reduced the programming and computer-related overhead enough to make M.U.P.P.E.T. possible. On the other hand, the more sophisticated student has control of the graphic environment with commands such as Scale (sets scale manually), ViewPort (selects a window on the screen), Axis (adds an axis system to a viewport), and various color modes, line types, and label positions. Each package also allows the student to include bar or pop-up menus in each application.

The Maryland University group has also developed a number of special-purpose programs including a sophisticated orbit program that can handle five masses, a thermodynamics/kinetic theory program, several nonlinear dynamics programs including a sophisticated "Mapping Machine," fractals, and many others. We have also developed the curriculum materials to be used with these programs. We recognize that these special programs have an important place in intuition building and in exploration of "microworlds."

I have not commented much on the tutorial materials that are becoming more widely available. PLATO was one of the first, and very elaborate, teaching systems, and Alfred Bork at the University of California–Irvine has been a pioneer in this area. Our group does not have a strong interest in using these materials with our students. We are more interested in having students use computers as tools rather than in having computers program students. We have been essentially unimpressed by past efforts to develop tutorial materials, but perhaps the authoring and artificial-intelligence techniques being developed at places like Carnegie Mellon will change our minds.

Using the computer as a tool allows our students to explore physics more deeply and allows us to present current, previously inaccessible applications of physics. Although we have had much success with classes at Maryland, we have not yet tried to integrate this in all introductory classes with a variety of faculty. There are some exciting opportunities, but we still have much to learn.

1. J. Wilson, "What to Teach?" AAPT Announcer 14, 91 (1984); J. Rigden, "The Current Challenge: Introductory Physics," Am J. Phys. 51, 516 (1983); R. Hilborn, "Redesigning College- and University-Level Introductory Physics," Am. J. Phys. 56, 14 (1988); A. P. French, "Some Thoughts on Introductory Physics Courses," Am. J. Phys. 56, 110 (1988); A. A. Bartlett, "Role, Mission, and Change in Our Introductory Collegiate Physics Courses," Am. J. Phys. 56, 204 (1988).
2. *Calculus for a New Century* (Washington, DC: MAA and NAS, 1987).
3. L. C. McDermott, "Research on Conceptual Understanding in Mechanics," Physics Today 37, 24 (1984); I. A. Halloun and D. Hestenes, "Common Sense Concepts about Motion," Am. J. Phys. 53, 1056 (1985); I. A. Halloun and D. Hestenes, "The Initial Knowledge State of Physics Students," Am. J. Phys. 53, 1043 (1985); L. Viennot, "Spontaneous Reasoning in Elementary Mechanics," Eur. J. Sci. Ed. 11, 205 (1979); B. Y. White, "Sources of Difficulty in Understanding Newtonian Dynamics," Cog. Sci. 77, 41 (1983); A. di Sessa, "Unlearning Aristotelian Physics," Cog. Sci. 66, 37 (1982); D. E. Trowbridge and L. C. McDermott, "Investigation of Student Understanding of the Concept of Velocity in One Dimension," Am. J. Phys. 48, 1020 (1980); D. E. Trowbridge and L. C. McDermott, "Investigation of Student Understanding of the

Concept of Acceleration in One Dimension," Am. J. Phys. 49, 242 (1980); A. B. Champagne et al., "Factors Affecting the Learning of Classical Mechanics," Am. J. Phys. 48, 1074 (1980); R. J. Whitaker, "Aristotle Is not Dead," Am. J. Phys. 51, 352 (1983); D. M. Watts and A. Zulbersztain, "A Survey of Some Children's Ideas about Force," Phys. Ed. 16, 360 (1981); M. McClosky et al. "Curvilinear Motion in the Absence of External Forces," Science 210, 1139 (1980); J. Clement, "Students' Preconceptions in Elementary Mechanics," Am. J. Phys. 50, 66 (1982).

4. G. Aubrecht, ed. *Quarks, Quasars, and Quandries* (College Park, Md.: AAPT, 1987).
5. A. Arons, "Uses of the Past: Are We Condemned to Relive It?" AAPT Announcer 18, 82 (1988).
6. R. P. Feynman, R. B. Leighton, and M. Sands, *The Feynman Lectures on Physics*, 3 vols. (Reading, Addison Wesley, 1964); A. P. French and E. Taylor, *An Introduction to Quantum Physics* (New York: W. W. Norton Co., 1978); E. Merzbacher, "Can Quantum Mechanics be Taught in a One-Year Calculus-Based Introductory Physics Course?" AAPT Announcer 17, 74 (1987).
7. F. Reif, "Teaching Problem Solving: A Scientific Approach," Phys. Teach. 19, 310 (1981); J. Larkin et al., "Expert and Novice Performance in Solving Physics Problems," Science 208, 1335 (1980); J. Larkin and F. Reif, "Understanding and Teaching Problem Solving in Physics," Eur. J. of Sci. Ed. 11, 191 (1979).
8. R. Crandell, *Pascal Applications for the Sciences* (New York: John Wiley, 1985); R. Crandell, *Scientific Programming with Macintosh Pascal* (New York: John Wiley, 1988).
9. J. Wilson, "The Impact of Computers on the Physics Laboratory," in *Research on Physics Education* (Paris: Editions du Centre National de la Recherche Scientific, 1984); J. Wilson, "Experimental Simulation in the Modern Physics Laboratory," Am. J. Phys. 48, 701 (1980).

Computers and the Broad Spectrum of Educational Goals

Peter S. Signell
Department of Physics and Astronomy, Michigan State University, East Lansing, MI 48824

Computers can play a vital role in helping undergraduate students gain skills that are general as well as topical. If faculty turn tasks that are repetitious and counter-productive (such as lecturing) over to materials and computer programs, they will be freed to provide an appropriate environment and to work on skills with students. Faculty members might spend part of this freed-up time improving a system of nationally shared teaching materials, but most of their time would be devoted to coaching groups of students in order to ensure learning, improve the academic environment seen by students, and give the students a sense of style in the disciplines they are studying.

Computers might be used: (1) as the medium in which a national community of college faculty continually generates, shares, and updates a coherent system of hypermedia lessons that provide the primary presentation of information; (2) as the medium in which the information is eventually delivered to each student; (3) as the medium that guides the student along planned and unplanned paths through the system of lessons; (4) as the medium that gives each student and faculty advisor continual feedback on the student's progress toward acquiring agreed-upon personal and professional skills.

These ideas came out of a project I have been working on for some 14 years. During the first seven years of that period the project was supported by the Alfred P. Sloan Foundation and the National Science Foundation. During the past seven years it has continued to evolve with university support.

The original idea of the project was to work toward some long-standing educational goals that had been given a new urgency by new demands on this country's industries and government. It was obvious that faculty could not turn their attention to new goals while saddled with current instructional tasks. Fourteen years ago the possibility of success in freeing the faculty from their instructional burden seemed promising because of developments that were even then appearing in the area of personal computers.

We decided to start out on this project even before personal computer technology had matured, because we observed that the development of high-class software traditionally lags far behind hardware, and that effective courseware is particularly difficult to generate. We in the education business seem to need a long "lead time."

In the rest of this talk I will discuss some highlights of the public discussion on goals in higher education; the changes educators must make so we can work toward these goals; an objection and a response; the idea of hypermedia as a critical element in working toward the goals; progress in learning how to use hypermedia lessons; how it feels to be a college teacher whose students are in a hypermedia environment.

Boyer: Short-Changed Goals

For a good summary of goals currently perceived as being slighted by higher education, see *College, The Undergraduate Experience in America*, by Ernest Boyer of the Carnegie Corporation. I am sure that some of you have read this very influential book.

To gather data for the book, the Carnegie staff interviewed faculty, administrators, and students on many campuses. The staff noted the way the processes of college education are currently perceived by the three groups, along with the problems those groups see. Boyer's book pulls all these perceptions together. His conclusions echo those of previous studies: he finds that major educational goals are being grossly slighted in the classroom and curricula, to the detriment of college students and of society.

Boyer identifies some 80 goals that he feels we should start working toward to erase instructional deficiencies. An important benefit of his book is that it collects virtually all known goals in one place. It does not, however, have much to say about what paths might reasonably move us toward these goals. The book tells us where we should go, but it does not tell us much about how to get there.

Active Learning

One of Boyer's goals, very relevant to this conference, is that students should become active learners and not just sit passively in lectures. Students who become habitually active learners will continue to learn when they get out of college and are no longer sitting in lectures.

Encouraging Creativity

Another goal is to foster student creativity. Unfortunately, no one has figured out how to measure creativity, except *ex post facto*. Studies in this area are usually made by interviewing creative professionals, asking them to remember whether they felt challenged to show creativity while taking undergraduate college courses. The response is that college courses emphasize figuring out what the professor is going to put on the next exam, and figuring out which of the presented stuff the profesor really cares about. This is not the kind of mind-set people want when they look for creative people.

A Rational Environment

Boyer also notes that students perform better when they feel that they are operating in a rational system. To create a whole environment of rationality, we need to give students course rules and grading schemes that are obviously based on rationality and obviously not based on a professor's whims. Such course rules and their rational bases should be published. Academics should have no trouble with the concept of publication because of their long-standing tradition of "publish or perish." We have found that only by publication can any other interested person critique something, build on it, challenge it, and help it evolve.

Internalizing of Basic Concepts

Another goal is to get students to internalize the concepts they learn in the classroom, to adopt those concepts as parts of their own ways of thinking. This contrasts to students' associating those new concepts only with the classroom and exams and ignoring them outside the classroom and after finishing the course. In physics, this point is brought home to us in the research recently published in the *American Journal of Physics* by David Hestenes's. Hestenes asks what style of lecturing best gets students in calculus-based and algebra-based sophomore physics courses to internalize the most basic concepts taught in those courses. Is the most effective style the presentation of striking demonstrations? Is it theoretical derivations? Is it going over homework-problem solutions during the lecture? Is it explaining the textbook paragraph by paragraph? Is it a judicious mix of the above, created and delivered by an award-winning lecturer?

Which method causes students to internalize basic concepts? None of the above! Students who have just taken standard lecture courses can work complex problems on exams, but they have not incorporated the concepts into their own ways of thinking. They seemed to treat physics as something that impinges on their lives only in the classroom. Hestenes' study also shows that students who have taken physics in high school or previously in college do no better in internalizing basic concepts than do first timers.

Industrial Research: Graduates Need More Skills

Several years ago the magazine *Industrial Research* polled its readers on this question: "How are the colleges doing in producing the people you need?" Tom Hudson, whom many of you know, ran the poll for the magazine. The results came out under the headline, "Colleges Get Low Grades." The readers of *Industrial Research* felt that we are not producing graduates who have the range of skills that industry needs.

If industry people claim that our graduates are not well prepared, we might wonder whether grades can predict our students' futures. Do grades give students, faculty, and prospective employers a measure of the likely success of a student in a particular professional field? Research in this area typically attempts to determine whether there is a correlation between grades received in college and measures of eventual personal and professional success. Personal success is rated by querying the subjects, who are working professionals. Professional success is rated by querying the subjects' bosses and colleagues. The researchers find that in virtually all fields there is zero correlation between college grades and postcollege success scores. Although grades do seem to predict future grades rather well, they do not appear to predict much else.

The Skills Needed by Industry

If grades are not good predictors of success, is there a set of skills we can measure that would correlate well? We asked that question at two conferences we held with people from industry who were supervisors of college graduates who, when undergraduates, were typical of students in our sophomore physics courses. The supervisors who came to our conference were from both large and small companies.

At our conference we asked the supervisors what kinds of skills we can develop and measure in our students that will correlate well with later professional success. The supervisors agreed that students need to be able to talk about their subject. The supervisors said they couldn't tell if the cat got the graduates' tongues or if the graduates didn't know their subject, but the result was the same.

The supervisors also said that too many graduates appeared to think, with great relief, that learning is over at graduation. The president of a small high-tech company said, "All of my engineers are over the hill; they're all over 30 and I haven't been able to get them to use new software to keep us competitive. We have competition from all over the world. We'll have to keep ahead of the game or we'll go under. We need people who are active learners."

The supervisors said they need people who will keep learning, looking ahead, reading, talking, figuring out where their company should go. They then need to be able to persuade their bosses, by means of a good presentation, when the company needs to switch course in order to survive. They must also make the presentation to higher executives. This calls for sophisticated communication skills.

Professional Learning Skills

If students are to be willing and able to learn from professional materials after they graduate, we must get them to use professional learning skills, and to use them habitually.

By professional learning skills we mean, first, that students should start into material by reading the table of contents and then the overview, making mental hypotheses about the material's contents and conclusions. Then, every few minutes while learning, the students should stop to reflect, review, and critique. This should certainly be done at the end of each paragraph-length piece of material.

Students should continually think carefully about what they've just learned and make connections with their own past experiences. They should check on whether what they have just learned seems reasonable. They should make approximations to equations and check limiting cases, such as zero friction and infinite friction, for reasonableness and for insight. They should try applying what they have just learned to other examples. Only when they are satisfied should they go on to the next paragraph of material. Professors take such professional learning skills for granted, but students need to acquire them systematically.

The Lecture

Wouldn't it be great if a student could practice professional learning skills in live lectures? Suppose a lecture starts, a couple of minutes go by, and the presentation of a concept ("paragraph") finishes. The student presses a button and everything is suspended, right in midair. The student now carries out the appropriate professional learning activity. When satisfied, the student pushes the "go" button and the live lecture resumes right from where it left off. By connecting the material being learned to material learned previously, the student would stand a better chance of internalizing the concepts just learned.

Most lecturers feel compelled to present all details of an explanation so that all students can follow the line of argument. But students benefit by filling in details or completing an argument on their own if they can. They need to retrieve as much as they can from their own brains because such retrieval from memory strengthens the retrieval path in the brain and hence increases the ease of future retrieval. This retrieval capability varies from student to student on any particular topic, and the time needed to perform it also varies, so this very desirable goal is not easily realized in live lectures. Another failing of lectures is that they provide little opportunity to build on individual students' hobbies, sports, and intellectual interests so as to engage their minds and eliminate the "restricted to the classroom" syndrome.

Different students have different preferred learning styles, and these are not accommodated well by lectures. For example, some students seem to do better entering a topic with theoretical deductions, and others seem to do better starting with phenomenology or with a "simple to complex" approach. Some students prefer to go through many topics on a somewhat superficial level, then go back and recycle through on a deeper level; others prefer to go into one topic deeply before going on to the next. Finally, some students like to be studying lessons that are far ahead of the lessons on which they are being tested; others prefer to be tested on each lesson or group of lessons as soon as they finish it.

Then too, physics lecturers seldom take the time to be good orators. All learners do better when a topic is started with a well-constructed overview. Professional writers and orators concentrate hard on section overviews for groups of paragraphs and on topic sentences, which perform the overview function for individual paragraphs. They have learned that this is necessary for success. Physics lecturers, however, just don't take the time. The result is that students do not figure out a good way to organize the material in their heads and consequently cannot retrieve it.

Summary of Goals

Let me summarize. We want students to gain professional learning skills and to use them habitually. We want them to internalize the basic concepts being taught, to relate those concepts to their experiences, to other courses, to their thoughts, and to their conversations. We want them to talk about their disciplines fluently, perhaps even become experts in multimedia presentations. We want them to learn in the manner that is most efficient for them as individual learners. We want them to believe that all of the skills they are acquiring, day by day, are part of a rational scheme leading toward their own life goals, so they no longer view physics courses as just hurdles to get past or cheat in, and so there is a seamless transfer to the new requirements for success in the collegiate "afterlife."

Conclusion: Give Up the Lecture

How can we find time to work on all these new goals? The answer: Give up lecturing and use community-shared presentations produced in hypermedia. Just as in an earlier, time faculty had to get used to the idea of community-shared textbooks, so now we need to go the next step and produce community-shared hypermedia that can take over from lecturing. Sure, each of us gives unique lectures, but that is of interest only to us, not to our students. If we want the faculty to start doing what only people can do, and letting media do what they can do, then there's no choice. We must give up the lecture and spend the saved time on other more important, uniquely human activities with our students.

An Objection and a Response

People say to me, "I think physics instruction is pretty good. I just don't see any big problem to be worrying about—after all, I turned out pretty good. You can't tell me that I got a rotten education. On top of that, if you change things you might lose something precious that you won't know about until it is too late."

One way to examine the introduction of new technology to the classroom is to look at a past introduction, the fast total replacement of the slide rule by the calculator in college science and engineering courses. Letting today's students use calculators does not mean that previous generations who used slide rules received a rotten education; it just means that today's students (and faculty) can use the technological advances available to them. No one worries that today's faculty will be held responsible for some coming disaster owing to the abandonment of slide rules. Similarly, no one need worry about changing the medium of the primary presentation of information to the one that is used throughout the professional world, providing the faculty spend their teaching time with students, keeping them on track in their education.

Producing the Lessons: General Guidelines

Media lessons that replace lectures have to include all of the information normally presented in lectures and in routine "office hours" assistance to students. This material needs to be organized in line with researched results on how people in targeted groups learn best. This greatly increases the efficiency of learning for students. We have produced an author's guide that contains this material.

An Eventual Dynabook

The hypermedia presentations are (eventually) to be made into a Dynabook version of knowledge. The Dynabook will store knowledge in a book-sized computer where the reader can easily construct any desired path through related material. Fourteen years ago, when we started, the Dynabook concept was little more than a dream, but we found that a lot of its features could be implemented on paper. The recent development of Hypercard and related products is a big step toward realizing the ultimate delivery device.

We have been developing our material on word processors and then distributing inexpensive print versions to our students. These print versions retain the discreteness and user-option branchings of the original model. We have learned a lot about what can happen in an instructional system based on such materials.

The breakthrough that allowed us to start replacing lectures with such materials was the idea of hypermedia. This idea developed about 15 years ago and that's why our project started then. However, the idea of replacing lectures with hypermedia presentations is beginning to be considered rational only now. The situation closely parallels early faculty hostility to textbooks, a hostility that gradually subsided.

Hypermedia: A Crucial Idea

The basic idea of hypermedia is nonlinearity. "Nonlinearity" means that the presentation has many branches, many options, all at the discretion of the user. The user can take any branch at any time and can stop the presentation at any time.

To further define the concept of hypermedia, I will list some types of presentation that are inside the concept of hypermedia and others that are outside of it. You might think of a map that has various demarked regions, each corresponding to a different concept. One of the regions is labeled "hypermedia." Some types of presentation are definitely inside the hypermedia region, some are definitely outside it, and some are closer to the edge or closer to the center.

Types of presentation that are outside the concept of "hypermedia," at least in the way they are generally used, are live lectures, videotape and audiotape lectures, the Keller Plan, programmed instruction, computer assisted instruction, and personalized student instruction. But some presentation methods lie inside the concept of hypermedia. These include very short talks at professional meetings, poster sessions, textbooks, and printed modular lesson systems. In each of the media that are inside the hypermedia concept, you can make many choices. The ease of access to the options and the likely consequences of choosing them are better taken care of in some media than in others.

Using Hypermedia

Using a CCH (computer-controlled hypermedia) presentation, you can superimpose a lecturer's head on a physics presentation, with the lecturer's synthesized voice "reading" the lecture while the lips move on the digitized image of the lecturer's head. This talking head need not be in a separate window; it can be directly superimposed on the action just as on television newscasts. The head can be that of the student's favorite professor, or the student can switch it to be the head and voice of, say, Bruce Springsteen, giving out synthesized physics words. The student can switch the head off and just hear the voice, or can switch the voice off too and just watch screen text. The student can even decide to download the text to a printer text.

In a CCH presentation the student sees buttons to press on the screen. Here is what the student might see at the beginning of the lesson:

The student might want to see data on the lesson's use—how much trouble other students have had with this lesson. There might be a prequiz available so the student can test readiness for this particular lesson. The student might want to see how this lesson fits into the whole of physics or of science. The student can use a mouse to move the screen cursor to the appropriate button image and click on it to select any of these, or he can choose none of these options.

When he finishes a lesson, the student might want to select questions to ponder or problems to solve, or take a self-test exit exam. The student might want to review certain topics.

While in the midst of studying the lesson, the student has choice buttons available at various points. Whether those buttons are displayed on the screen at any

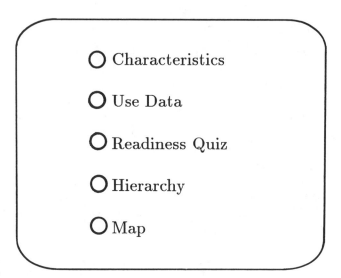

Figure 1. Hypermedia Buttons.

particular moment is also a choice. Here are some option buttons that might appear on the screen at an appropriate point: a filmed or animated demonstration; more examples; illustrative and practice problems; a quiz; references for background; references for further study; or initial help on completing an argument or on working an example. Suppose the student pushes a "help" button but finds the resulting help insufficient. He can push a help button within the help presentation and receive further help. In some of our current lessons we have as many as four layers of help. A student need only go to the depth needed to get past a personal sticking point and then get back to the main presentation. We aim to emulate well-designed office hours or consulting-room help. We give varying amounts of help to various students, and we try not to clutter up the main presentation.

Other buttons, appearing on demand, allow the student to see where this lesson resides in the general scheme of things, or to see buttons next to all words that are in the glossary, or to see the glossary itself with the screen centered on entry for the word at the button. The student can also simply hold down a button that causes a fast backward or forward flipping through the presentation.

Pre-Computer Printed Versions

All of our lessons in physics, covering most of the normal undergraduate curriculum, are currently available to students only in the paper medium, although they are developed on computer-stored word-processor versions. A basic design goal of the word-processor versions has been to hold each paragraph, complete with graphics, to the size of the word processor's computer screen. This should make it easy to convert the lessons, one by one, to computer delivery.

In the printed versions, the CCH "buttons" are represented by alphanumeric codes in brackets. Instead of just pressing a button at some point in the presentation, the student must note the code, turn to the end of the lesson, find the glossary or the right supplement, then find the right numbered window in the supplement or the word in the glossary. Within a help window, a further level of help is provided by a reference (in brackets) to another window in the same supplement. Access to help is obviously easier in the computer version than in the printed version. Note that a print-medium reference to a demonstration on video disk may require the student to go out and find a video-disk player.

Animation and Lap Dissolves

One interesting thing we discovered was that by and large students do not like true animation. They think it is too hard to follow. What the students like better is a series of soft-lap dissolves, where one frame stays on the screen for one second, then dissolves into the next frame over a period of one second. As one frame dissolves into the next, the students can see the change since they're only having to follow a finite number of changes. They feel that they really get a feeling of motion from the dissolves but at the same time they can follow what is happening. We put these frames in the print versions with presentation techniques that assist the eye-brain system in obtaining a modest feeling for the motion.

Lesson Options

We also use a concept called "lesson options." Lesson options let students apply physics to areas of strong personal interest. For example, when studying moments of inertia, a student who is a tennis player can learn how to locate the three sweet spots on a tennis racket and then demonstrate what he has learned to a staff member for credit. The student can also learn the derivation of the tennis-racket theorem and demonstrate the theorem using a tennis racket.

Hypermedia Development: People

To develop hypermedia lessons, we assign one faculty member at an institution to one lesson. He spends several years perfecting that lesson. This is a cottage-industry approach as opposed to an approach where a central project hires media professionals to develop lessons in cooperation with a few physicists.

If we were to hire professionals to produce all of the necessary hypermedia lessons, the cost involved would probably exceed the U. S. gross national product. The level of expertise needed and the minute-by-minute cost are about the same as that of creating good television ads, and for much the same reasons.

Developing hypermedia lessons, with all the optional paths, takes a lot of physics insight, some lesson-design insight, and a lot of dedicated work. We have found that, generally, only regular physics department faculty are able to apply our set of lesson design principles successfully to any particular topic. We have found it far more effective to help physics faculty learn lesson design than to have the lessons developed primarily by professional writers or educators.

The individual lessons in our system have been produced by physics faculty located in various colleges and universities around the country. Our production system has to ensure that the lessons fit together like the pieces of a jigsaw puzzle. One piece of evidence that this happens is that not one of the 10,000 students who have used our lessons has ever complained that successive lessons had different styles or didn't fit together. Another piece of evidence is that not one of the 1,000 or so student reviews of lessons has ever mentioned this as a problem (they have mentioned every other problem we could think of). To the students, at least, the system appears to be seamless.

In our development model, the participating faculty do not get paid directly to develop lessons. This is exactly as in research, where faculty are not paid royalties for published papers, even when these published papers turn out to have enormous industrial spin-offs. However, according to a national poll of department chairpersons, hypermedia development is considered equal to a research paper in decisions on promotion and salary.

Developing the Help Branches

We develop the optional help branches in lessons in two stages. First, the lesson author constructs help branches based on his or her design and insight. Then when students using the lesson come in for office hour or consulting-room help, or make points in their "graffiti reviews," the author receives relevant feedback and adds additional help sequences to the lesson. Over time, as more and more help options are added to the lessons, the need for office hour or consulting-room help drops dramatically.

Proliferation of Courses

The hypermedia system fosters a proliferation of courses. With lectures gone, a faculty member's teaching load depends on the number of students taught rather than on the number of courses. Thus the hypermedia system in our department has become a collecting point for courses whose small enrollments would otherwise cause them to be canceled or eliminated. We are currently using hypermedia to teach an upper-division course required of secondary science–education majors. We also have special courses for special constituencies, such as a calculus-based course taken by hot-shot premeds and all biochem students. This course is somewhat less extensive than the course for physical science and engineering students but it is more rigorous than the noncalculus course. We also have courses that "bridge" students up from lower-level courses taken for another major or at a community college. We have been able to customize courses for individual students with odd deficiencies.

The Lessons' Prerequisite Map

The key to putting lessons together coherently and to exposing the structure of knowledge to view for both students and faculty is the hypermedia system's prerequisite map.

The figure shows only a small section of the physics map. Each numbered oval represents an individual lesson. You can think of the ovals as cities with roads connecting them (indicated by lines on the map). These lines represent paths students can take as they work their way through the system of lessons. Notice that the lines are really arrows, indicating one-way travel. In fact, an arrow pointing from one lesson to a second indicates that the first lesson contains one or more prerequisite skills needed for entry to the second lesson. To determine the prerequisites on the map the lesson's author makes statements of the actual skills that are prerequisite for the lesson. After each such input skill, the author notes in parentheses the numbers of other lessons in the system that adequately teach that skill (in the author's

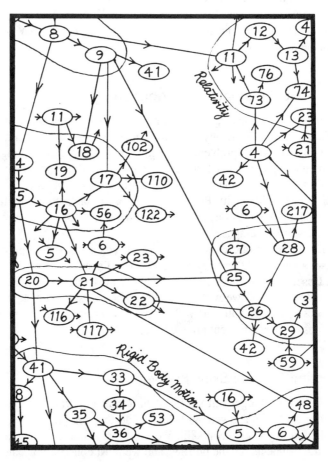

Figure 2. Section of a prerequisite map.

opinion). We use those references to place the lesson on the map and to draw in the prerequisite links.

The figure above shows a portion of the prerequisite map in the area of classical mechanics. Note that the author of lesson 21 found that lessons 16 and 20, together, supply the input skills needed for lesson 21. Notice that lesson 21 has many arrows pointing from it to other lessons, so it must teach an important concept. In fact, lesson 21 is on conservation of energy.

The smaller the lessons, the more structure of knowledge is exposed to view. If one were to make each lesson the size of a whole course, these huge "lessons" would occur in a linear chain and the map would be uninteresting. But each lesson in our hypermedia system currently has about the same amount of material as a lecture; this size of lesson produces a lot of branching in most areas of the resulting map. Once we go to CCH we may wish to make the lessons the size of paragraphs, but we don't know how much extra work this will be for a lesson's author.

One can imagine the prerequisite map extending to all disciplines in the university. Under such a system, a student could take lessons from various departments without recourse to "courses." A student would have no motive to cheat if he or she could see on the prerequisite map exactly where skills learned today will be used down the road.

Printed Version: Method of Distribution

Each of our present lessons has a list of input skills and output skills, and is placed in a delineated region of the physics map. Many lessons are self-standing and are complete with overviews, problem sets, and special sections containing graduated help for completing arguments in the text and for working the problems. The most-used lessons in particular courses are collected into loose-leaf textbooks published by Michigan State University Press. Seven such volumes are currently in print. Other individual printed lessons are purchased by the students from copying stores, like Kinko's, near the campus.

The current version of each printed lesson also exists as a word-processing file, so the lessons are already partway to the hypermedia stage. As we finish transferring each lesson to hypermedia, we will make it available for free use in campus microcomputer labs, and we will make it available for sale on disk in Kinko's or similar places. Thus, we will start with voluntary piecemeal substitution of hypermedia lessons for the equivalent printed ones.

Constructing General-Skills Profiles

After we articulate the general skills we think are important, we must provide instruction in each of those general skills. To do this, we must provide diagnosis and remediation in each general skill. We must test a student's general skills frequently and let students see the profiles, and we must constantly update those profiles.

Here is a general-skills profile for one of our students. This student is one of approximately 700 to 800 students we have each term. These students are spread out among 25 to 35 courses.

In the skills profile the height of each bar represents that student's percentile ranking in his or her class for a single general skill. Thus a bar labeled 68 percent means that the student did better than 68 percent of his peers on that skill.

The first bar, labeled "knowledge," shows the student's ability to acquire and communicate knowledge and to apply rules in physics.

The second bar shows the student's ability to develop problem plans to solve relatively complex homework problems.

The third bar shows the mathematical accuracy with which the student solves problems.

The fourth bar shows the student's percentile rating in being able to do parts of exams and also whole exams "excellently." For example, this bar distinguishes between a student who gets everything half right and one who does half the things perfectly but the others not at all.

The reason to separate "mathematical accuracy" from "problem solving" is that these skills sometimes have separate value in the workplace. For example, a person who is excellent at making plans but is not meticulous in carrying them out might

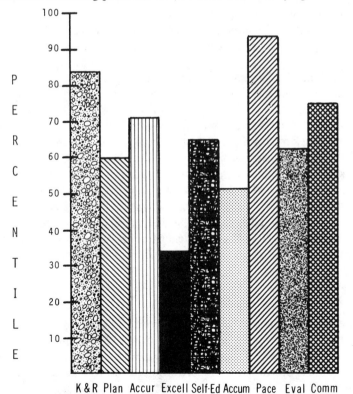

Figure 3. Attribute-profile bar graph.

be a good group planner. Conversely, a person who has trouble constructing plans but is meticulous in executing them might be a good group worker.

The last bar represents communication. To produce a student's communication rating, we score every question on every exam for communication skill (as well as for other appropriate skills). We do not believe in using multiple-choice or fill-in-the-blank questions; the student should formulate the complete response to the exam question and must not include extraneous material in the response. Students' exam responses must be neatly laid out, as though for publication. Students occasionally give lectures, complete with demonstrations, and are rated on them. In fact, most students do become proficient communicators after they become convinced that we are serious about it.

How Staff Are Used

When materials are the primary medium of presentation, the teaching staff is of crucial importance. First, there must be human consulting assistance for students with idiosyncratic problems (these will always be with us). The teaching staff must evaluate complex presentations by students, coach the students, and give them a sense of style in the disciplines they are studying. The staff must also give master classes, where appropriate. Finally, teaching staff must help students acquire group-project skills and admire good output from individual student projects. We are currently doing some of those things and not others.

In our project's current instructional system, the lowest level of staff assistance is provided by hired advanced students who were highly successful when they took courses in this format themselves. Students can come to our consulting room any afternoon or evening. If the student seeks a consultant's help, the consultants' job is to smoke out the student's real problem, whether it be a mistaken concept, a poor background, a lack of pursuit, or poor study habits. If the student's sticking point is a homework problem, the consultant does not to show a student a complete solution to the problem. Rather, he just gets the student past the immediate roadblock so the student can proceed on his or her own. If the sticking point is in an explanation in the materials, then the consultant does not give the student a lecture on the topic, but just gets that individual student past that individual sticking point. If the consultant or the student feels that a higher level of help is needed, the student is immediately passed on to an appropriate higher-level person in the teaching staff.

How Does It Feel?

Before I embarked on this rather large enterprise I was a regular lecturer. Over the years I had taught most of the undergraduate and graduate courses, and I got good ratings from the students. Of course my lectures were works of art. Then I got involved with this system and I stopped lecturing. Well, I suffered. I had lecture-withdrawal pains, and they didn't fade away for over three years.

After five years I decided to try lecturing again, and it was a personal disaster. As I was lecturing I could not help thinking that the students sitting there listening to me should be recalling much of what I was saying for themselves. They should be filling in the arguments. They should be stopping to think things over in light of what they already knew. They should be applying professional learning skills. I felt that I was cheating them out of their educations. I never tried lecturing again.

I also found that it was tough to set up a rational environment. Students constantly asked that I articulate justifications for things we do not normally discuss, such as: "Why do I have to take physics?" "Why do I have to learn these particular things in this particular amount of time?" "How can you justify the exact grades you give?" "How come some physics professors encourage us to use three-by-five inch 'cheat sheets' in their exams but you prosecute us as cheaters if we use them in your exams?" Our *Policies Handbook*, containing the system policies and justifications, runs to 50 typewritten pages and is frequently updated as students challenge or seek further clarification on some point or other.

Having a written book of policies and justifications and having it become relatively debugged after a period of time makes me now feel as though I am operating according to a "rule of law." On points of policy, the staff and, on occasion, I myself now consult the handbook. When I think about it, it seems seems strange to be a professor constrained by rational policies.

One final note: This system seems pretty weird to my graduate TAs when they are first assigned to it. They have spent years being on the receiving end of the educational system and are now eager to show and tell someone else what they know. In this system the students are supposed to tell what they know and the staff are supposed to react. Just before I flew down here to the conference, the telephone rang in my office and it was a new graduate TA in our consulting room. He said indignantly, "This student tells me I have to climb up on this rotating highchair and he'll demonstrate the Coriolis force on me. Do I really have to do this?" I said, "Yes, the student is supposed to do the demonstration on you and explain it to you. Then you give feedback. But remember, the student is supposed to be the active one." There was a sigh from the other end of the line. It was an echo of myself when I first tried doing this 14 years ago.

I would like to thank my long-time collaborators, Jules Kovacs, Tom Burt, Bill Lane and Gene Kales, without whose help the system would be a pale reflection of its present self. Michelle Rademacher has kept the personnel and office humming with her steadiness and infectious enthusiasm. Many of the basic ideas were pushed into being by Gregg Edwards, who was always confident that it could all be done by yesterday.

Problem Solving, Pedagogy, and CDC Physics

Karen L. Johnston
Department of Physics, North Carolina State University, Raleigh, NC 27695-8202

When we describe the calculus-based university physics course to our students, we immediately emphasize problem solving. We might even tell our students rather boldly, "In this course you will learn problem-solving skills." When we are with our colleagues, we nod and agree that developing problem-solving skills in students is the primary goal of the calculus-based physics course. We say these things with such authority that we are likely to convince ourselves that we are indeed teaching problem-solving skills. We teach problem-solving skills by carefully illustrating to students how we work problems. We assign many problems to students for practice because drill and practice are good for them. After all, one of the features of a good textbook is the number, variety, and quality of the end-of-chapter problems. Most calculus-based textbooks have approximately 50 problems per chapter, over1500 per textbook.

We discharge our duty of teaching problem-solving skills by some variation of the following scenario. We plan a well-prepared lecture with sufficient examples applying the principles of physics to give students the opportunity to see how we (professors and obviously expert problem solvers) arrange the solution in our minds. We make the assumption that when students see how logically we've approached the problem, they will do likewise. So, we turn our backs to the class, cock our heads to the side as far left as we can and say, "Suppose you had…" We sketch a diagram and label everything. We ask the class if they know what relationship to use. They mumble something. We turn our back to the board again, write the appropriate relationship, substitute, and solve.

The tension builds as we complete our solution down to the correct units. We turn back to the class and say, "Do you understand?" The students know their part well. They answer, "Yes." And we continue. The delusion is complete.

I contend that the "yes" means nothing more than a receipt of information. When students drop by our office or mention after class how they understand the problems when we work them, but can't get started on their own, we fail to recognize our deficiency of our strategy in achieving the desired outcome for university physics.

Problem solving is complex. Until recently, understanding how people learn to solve problems has been in the realm of unexplainable phenomena. It isn't that way any more. Over the past 15 years, cognitive psychology and the tools of computer modeling have provided insight into how novices and experts go about the business of solving problems.[1] This comparative information has opened our eyes to the some of the strategies that must be utilized in teaching problem-solving skills. The bibliography of interesting and useful research in problem solving is continuing to grow. Likewise, investigations into student misconceptions has produced information that can be integrated with the problem-solving research to improve teaching strategies in courses where analytical problem-solving is important.[2]

The wide-scale availability of microcomputers in universities and schools compels us to change the ways in which we work with students. Computers with artificial-intelligence capabilities offer the widest range of opportunities to alter instructional strategies, but even the one-computer classroom offers new challenges to teachers and new ways to improve our instruction.

The Microcomputer and Problem Solving in Physics

Investigations in the area of problem solving have produced information that can be used well with microcomputer technology. Anecdotal information as well as systematic studies support the notion that the novice problem solver (the student) differs appreciably from the expert problem solver (the teacher). Rief has suggested that students rarely utilize expert problem-solving techniques such as generating a comprehensive initial description of the problem, synthesizing a solution to the problem, and assessing a solution.[3] Without proper guidance, even good students rely on problem-solving methods that follow the path of least resistance.

The microcomputer, however, has the potential to force students to assume an active role in learning. When highly interactive software is employed, the microcomputer can be an effective tutor. Students can participate in a structured dialogue or control a simulation of nature. By intellectually engaging students, the microcomputer in the role of a tutor encourages the student to think.

Students gain useful practice in the art of problem solving when the software is interactive and is centered on carefully constructed dialogues. Research has demonstrated that neither extensive observation of expert problem solvers, nor solving many textbook problems, has noticeable effect on building strong problem-solving skills. However, students can practice problem solving in a way that emulates the strategies of experts by engaging in computer dialogs that capitalize on the unique features of the microcomputer as a learning tool.

Educational software should take advantage of the following features of the microcomputer:

1. The potential for high interaction invites students to become active instead of passive learners.

2. Interactions between computer and student can be planned, not spontaneous. We can build into the interaction the collective knowledge about student misconceptions, and we can pose unambiguous questions that guide students to correct reasoning.

3. The computer has infinite patience for repetitive tasks. A good interaction can be repeated many times with many students.

4. Interactions are private, thus allowing the student to respond incorrectly without psychological penalties. The microcomputer is nonjudgmental.

5. The computer allows for variety in interactions, even those where the student explores nature via simulations.

Obviously, the microcomputer also does other things of value in physics instruction, such as running tool software for data collection and analysis.

When the microcomputer is exploited to its fullest extent as a tutor, we have at our fingertips a way of challenging students intellectually that cannot be achieved in any other way given the constraints that exist in schools and universities for teaching resources. Software that maximizes these features of the microcomputer provides teachers with a way to give students individual attention. On the other hand, software that does not use the interactive features of the microcomputer poses a host of pedagogical problems as ennui overcomes the student who merely "turns the pages of a computer screen." This kind of software will have little, if any, positive impact on education.

CDC Physics 1 and 2 and Problem Solving

Physics 1 and 2 is a comprehensive program in calculus-based physics developed by Control Data Corporation as a part of its lower division engineering curriculum. Initially the program was delivered on 8-inch diskettes using Viking 110 terminals; it is now available on 5 1/4-inch diskettes for IBM PC or compatibles. We use both at North Carolina State University. CDC also offers the option of subscription to its PLATO system via telephone link with online mastery-based testing and record keeping. We no longer use the online features in our integration project.

The courseware is highly interactive and the pedagogical design embedded in the software focuses on the development of problem-solving skills. The content of *Physics 1 and 2* matches well with the standard engineering physics course. A detailed look at one of the modules gives a good indication of the strength of the program in developing problem-solving skills.

Each lesson begins with an introduction describing the content of the lesson, what students should know prior to beginning the lesson, what students will be able to do at the end of the lesson, and what tools students need to do the lesson.

To test the student's prior knowledge on essential concepts, the program presents a review in which students must demonstrate their knowledge. Simulations are frequently used in the review as well as in the main lesson. Skills that students develop in the review may be repeated in the main lesson. Whenever students control a skill-oriented simulation, the program allows students to practice until they are ready to begin the problem posed in the main lesson. The simulations often use gamelike features to keep student interest high and require more actual student input than page turning.

In areas covering problems more like the typical problems in a physics textbook, the design of *Physics 1 and 2* requires that students (1) describe the problem, (2) identify a strategy, (3) implement the strategy, and (4) check the results. In the problem description students identify explicit and implicit numerical information and sketch a diagram of the problem where appropriate. Identification and implementation of the strategy include deciding in advance what the answer should look like, selecting equations that may be useful, isolating unknown variables, and computing an answer. This strategy, which mirrors research findings on problem-solving skills of experts, is repeated throughout *Physics 1 and 2* and results in what I believe is high-quality software. After an extensive study of the pedagogical design of *Physics 1 and 2*, we have modified and embellished the problem-solving strategy into a problem solving template that we use with students in our Physics Tutorial Center. We have been pleased with this as a spinoff of our CDC Project and see the problem solving template as something we learned from good software.

The details of how I have integrated this courseware into engineering physics have been discussed in *The Physics Teacher*,[4] but let me quickly recall some of the information for you. Approximately 40 students each semester from my section of engineering physics select to use computer-assisted instruction as a part of their course. We supplement CDC *Physics 1* with other commercially available thermodynamics software. Students spend three hours per week at a microcomputer station using *Physics 1*. We encourage students to view the three-hour-per-week commitment as structured study time. Table 1 shows a comparison of grading schemes of students using courseware to students following a traditional course with a problem session.

Table 1.
Comparison of Traditional Physics and Physics with Courseware

Physics using Traditional Physics	Credit	Courseware	Credit
Lecture (3 hr./wk.)		Lecture (3 hr./wk.)	
Four tests	56%	Four Tests	50%
Experimental lab (2 hr./wk.)	10%	Experimental lab (2 hr./wk.)	10%
Homework	10%	Homework	10%
Final examination	24%	Final Examination	24%
Problem Session (1 hr./wk.)	0%	Courseware work (3 hr./wk.)	6%

Students keep a notebook of courseware assignments and are graded on this work. Attendance is required, but we also offer students the option of returning to the traditional course and grading scheme at any time.

We monitor student comments and attitudes about the program both formally and informally. We keep the level of interaction high by using the software and by assigning tutors to work with students. Students' comments are positive for the most part. One student summed it up nicely: "It was almost as though I was working with a calm, patient, flesh-and-blood tutor."

The Most Frequently Asked Question

I am frequently asked which students learn more—those using CDC physics or those attending the traditional problem-solving session? First let me tell you what I believe and then tell you what I know. I believe that learning is both difficult to define and difficult to measure. Unless you have repeated measures of achievement on a particular objective over time, you can't be sure if learning has taken place. That is, if someone has "learned" something, there must be evidence that the new information can be applied correctly on a construct-valid measure and replicated on repeated measures. We might find it useful to know that if we apply Topic X to a student via a microcomputer-based lesson, the student will learn more, but this learning is not something we can measure definitively.

So, what do I know about learning problem-solving skills and CDC *Physics 1 and 2*?

1. The authors of *Physics 1 and 2* have taken traditional topics in university physics and developed those topics in such a way that the microcomputer acts as a "good" tutor for developing "good" problem-solving skills.

2. Students who use CDC physics in a structured homework mode for approximately three hours per week are favorably impressed with the problem-solving discipline imposed on them by the microcomputer.

3. In comparing achievement on common examinations between students in CDC physics and those in a traditional problem session, we find indications that achievement may be improved for those using the courseware. The evidence is limited at this point, but this continues to be an interesting problem for investigation.

1. J. Larkin, "Cognition of Learning Physics," Am. J. Phys. 49, 534 (June 1981).
2. L. C. McDermott, "Research on Conceptual Understanding in Mechanics," Physics Today (July 1984).
3. F. Reif, "How Chemists Teach Problem Solving," J. Ch. Ed 60, 948 (November 1983).
4. K. L. Johnston, L. Grable-Wallace, and J. S. Risley, "Integrating CDC Physics Courseware into Engineering Physics," Phys. Teach. 25, 286 (May 1987).

Section 2

Physics Computer Simulations

Early Experiences with Physics Simulations in the Classroom

Blas Cabrera

Physics Department, Stanford University, Stanford, CA 94305

Many concepts in physics involve the dynamic motion of a system. We understand the transmission and reception of radio waves because we understand the motion of electric-field patterns produced by accelerated electric charges. Other examples are planetary orbits, which we understand through Newton's laws, and the kinetic theory of gases, which we understand as the motion of many atoms bouncing in a box. In the traditional instruction of elementary physics courses, instructors introduce these concepts with classroom demonstrations and laboratory sessions. However, some concepts have been difficult to elucidate. Powerful computers have now become sufficiently inexpensive to allow the widespread introduction of simulations as teaching aids for elementary physics instruction at the college level.

Physics Simulations is a series of computer simulations for elementary physics instruction.[1] These programs have grown out of our desire to use the instructional potential of modern, inexpensive microcomputers as a tool for students to explore the structure of physical models. Such programs are intended to supplement introductory-level physics courses. Particularly useful is the graphics capability of the Macintosh, which uses the mouse interface and is capable of high-quality animation. The visualizations provided by these simulations develop physical intuition and allow the use of this software over a wide range of mathematical sophistication, from our Physics for Poets course to intermediate-level undergraduate-major courses.

Philosophy

The Macintosh was the first inexpensive computer to introduce a higher level of animation capability. Using this tool, we have developed simulations that graphically represent simple physical models so that students can understand the exact relation between the screen animation and the analytic descriptions. Our programs, designed to help the student understand physical concepts, should be used together with conventional teaching techniques for elementary physics courses at colleges and universities. In designing the software, we have adhered to the following principles:

1. *No computer literacy should be required of the student.* We do not want to turn the physics class into a computer-programming class. Students use only the mouse and the customized pull-down menus. They can enter precise data with the mouse by using slide bars with numerical displays.

2. *The simulations should maximize the use of graphics.* Graphical representation of data always aids the understanding of physical processes, particularly for physical systems that evolve in time.

3. *The presentation and the format should be as simple as possible.* We are not interested in generating more computer-arcade games, but rather in ensuring that every input by the user will produce an easily understandable result on the screen.

4. *The scope of each program should be maximized.* We want each program to address a large number of problems without making them overly complicated.

5. *No computation should take longer than about five minutes.* Therefore, the total time involved in a student session with each program would be about one to two hours.

Description of Programs

The central programming theme of these simulations is animation. We have used three different techniques, depending on the level of complexity in the screen presentation. If computations and screen updates can be done quickly enough, then each new frame is generated in real time and displayed while the next frame is prepared offscreen. If the computation time is more than about 0.1 sec. per frame, each frame is generated in an initial set-up procedure and stored in memory. About 32 frames are generated and then played back in rapid succession (up to about 15 frames per second) to produce the animation. If the set-up time for frame generation exceeds about five minutes, the frames themselves are stored on the disk and loaded into memory to produce the animation sequence. This last technique is less desirable because it does not allow student interaction.

The following programs are in the *Physics Simulations* software. The program disks include utilities for printing the screen display on an Imagewriter or storing it on the disk for later printing.

Mechanics

Ballistic simulates two-dimensional motion in a constant gravitational field. A drag proportional to velocity can be included with an optional dependence on exponential altitude.

Potential simulates the motion of a particle in a one-dimensional potential well. Students can plot energy, velocity, acceleration, and time-average probability density. Students may choose from predefined potentials (e.g. harmonic) or define their own.

Oscillator simulates simple, damped, and driven harmonic oscillators. Displays include a mass on an elastic band, amplitude versus time, and energy (see Figure 1). The total energy is continuously partitioned into potential and kinetic.

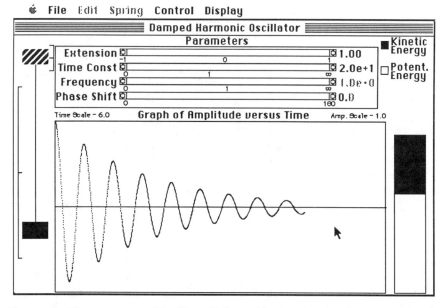

Figure 1. The program "Oscillator" from *Mechanics* simulates the simple, the damped, and the driven harmonic oscillator. The damped oscillator is shown.

Kepler simulates planetary motion. Students study Kepler's laws by using predefined orbits or by setting the parameters of one or two planets around a large central mass.

Einstein demonstrates special relativity. The screen is divided into a stationary and a moving frame that contain clocks and light pulses on a grid. Students can simulate special relativistic effects (including the twin paradox).

Electromagnetism

Coulomb displays the electric-field pattern for up to 15 point charges on a plane. Students set the position and magnitude of each charge.

Laplace calculates and displays the solutions to Laplace's equation on a two-dimensional rectangular lattice. Students can set fixed boundary conditions anywhere on the lattice to simulate a number of physical models.

Radiation simulates the time evolution of the electric field of an accelerated point charge. Linear, circular, and oscillatory motions with user-defined velocities and near-field to far-field magnifications are displayed as animation sequences (see Figure 2).

Ampere displays the magnetic field pattern for up to nine coaxial current rings. Students select the position and magnitude of each ring.

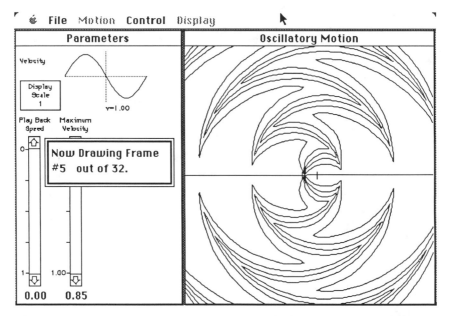

Figure 2. The program "Radiation" from *Electromagnetism,* simulates the time evolution of the electric field produced by an accelerated electric charge. Dipole radiation, such as that generated by a transmitting antenna, is shown.

Monopole simulates the passage of a magnetic monopole through a perfectly conducting ring (superconductor). Students can view the frames individually or as an animation sequence.

Modern Physics

Gas simulates the thermal motion of particles in a box and demonstrates elementary kinetic theory.

Brownian simulates the random thermal motion of a particle in a one-dimensional potential. Students may select from a range of potentials and temperatures.

Wave demonstrates the concepts of group and phase velocity in an animated sequence. The ratio of group to phase velocities is adjustable.

Fourier performs the Fourier transform and inverse transform of any user-defined function. Students can plot results as amplitude versus phase in real and imaginary components.

Hydrogen generates electron density plots for each of the quantum states of the hydrogen atom. Students may select the n, l, and m quantum numbers (see Figure 3).

Diffraction generates density plots for single and double slits and other apertures.

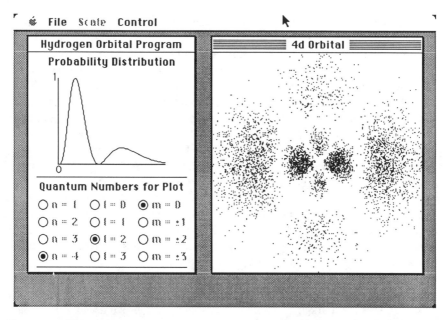

Figure 3. The program "Hydrogen" from *Modern Physics,* simulates the electron density around the proton in a hydrogen atom. Here the 4d orbital is displayed.

Ginzburg introduces an elementary model of a superconductor by displaying the coherent quantum phase on a lattice. Students can set fixed boundary conditions.

Transition is an animated sequence showing the semiclassical transition between the 2p and 1s states of the hydrogen atom.

Course Use and Early Experiences

We developed these simulations specifically for use in the advanced freshman physics course at Stanford University. The one-year course covers mechanics, electricity and magnetism, and modern physics. There is also an accompanying laboratory course. Advanced freshman physics is intended for entering students who have taken more than a year of calculus in high school and who intend to major in physics as undergraduates. The course typically enrolls about 25 students.

To introduce the simulations, we devote some class time to numerical techniques. For example, we cover Euler's approximation in mechanics when we introduce initial-value differential equations in the formulation of Newton's equations, and we cover the lattice-relaxation technique in electricity and magnetism when we discuss solutions to Laplace's equation. In fact, these specific examples aid students in developing an intuitive understanding of formal differential equations.

After we have covered the numerical technique in class, we introduce each program to students with a lecture demonstration using a video projector onto a high-contrast screen. We cover several specific examples. Then students have to answer one or two problems in each weekly problem set (out of a total of about eight) by using the software. Students can complete these problems by checking out the software for two-hour sessions either at the physics department library, which maintains a cluster of four Macintosh computers, or the central undergraduate library, which maintains a 40-Macintosh cluster, or by purchasing the software and using their own Macintoshes. The solution handed in by the student comprises a computer printout and a written discussion of its significance. The questions typically contain a first part that students can solve analytically and check against the simulation, and a second part that is beyond the analytic scope of the class but can be solved numerically. For this second type of question, we typically ask students to explain the qualitative behavior based on inequalities derived from the analytic formulations. We have used many of these programs in the advanced freshman physics classes for nearly two years.

In addition, these programs have proven useful for introducing physical concepts to nontechnical audiences, such as students in the Physics for Poets course, which requires a minimum of mathematical sophistication. We use computer simulations to present the solutions to mathematical equations. We then discuss these solutions in qualitative terms. We believe that this application for the simulations will become more important. We are developing student exercises geared to these less technical uses.

Conclusions

Inexpensive microcomputers are now sufficiently powerful to be very useful teaching aids in elementary physics classes. *Physics Simulations* is a series of such teaching aids that can be used over a wide range of introductory physics courses at the college and high-school levels. For the instructor, these simulations provide easier-to-follow alternatives and supplements to traditional classroom demonstrations. After introducing the program in class as a demonstration, instructors can assign follow-up homework problems. Students can then investigate the structure of these physical models in an interactive environment, even though the mathematical sophistication of the models is often beyond their grasp.

The development of this software, always a time-consuming effort, would not have been possible over these last two years without substantial contributions from the Faculty Author Development program at Stanford University (supported through an Apple Corporation grant). Programming support was provided through FAD by Sha Xin Wei ("Keppler," "Einstein," "Laplace," "Gas" and "Ginzburg"), Jim Terman ("Ballastic," "Potential," "Oscillator," "Radiation," "Monopole," "Brownian," "Wave," "Fourier," "Hydrogen," "Transition" and "Diffraction"), Daniel Schroeder ("Einstein"), Donald Geddis ("Coulomb") and Dian Yang ("Ampere"). These applications were written in Pascal and compiled using the Macintosh Workshop environment. Early contributions from Brian Penprase were also important. Needed support and encouragement has been provided continuously by the FAD staff, including Michael Carter, Ed McGuigan, Barbara Jasinski, Tom Malloy, and Stuart Crawford.

1. *Mechanics*, *Electromagnetism*, and *Modern Physics* were developed by Blas Cabrera and the Faculty Author Development Program at Stanford University for use with the Macintosh computer. The three software packages are available separately from Kinko's Academic Courseware Exchange (a service of Kinko's Copies), 4141 State Street, Santa Barbara, CA 93110 (telephone: (800) 235-6919 or in California, (800) 292-6640).

Cellular Automata

George Marx

Department of Atomic Physics, Eötvös University, Puskin 5, Budapest H-1088, Hungary

Continuous or Discrete?

When Galileo studied the laws of free fall by dropping stones from the Leaning Tower of Pisa, he counted time by his heart beats. As the stones dropped, he noticed that less and less time was needed for the stone to reach the next lower floor. He explained this phenomenon in the following way: something gives small kicks F to the falling stone causing its speed v to increase:

$$(t+1) = v(t) + F. \tag{1}$$

Half a century later, Newton made a differential equation from Galileo's idea:

$$d^2x/dt^2 = F. \tag{2}$$

Newton described free fall (without or with air resistance) as $dv/dt = g$ or $dv/dt = g - kv^2$. The analytical solutions of these equations could be obtained by evaluating integral:

$$v = gt \quad \text{or} \quad v = \sqrt{g/k} \, \tanh(\sqrt{gk}\, t).$$

In more realistic problems, however, such an elementary analytic solution does not exist. Let us consider a pendulum with air resistance $d^2\varphi/dt^2 = -(g/l)\sin\varphi - kl\,(d\varphi/dt)^2$. In order to force an elementary analytical solution (for blackboard use) we truncate this differential equation to a linear one by brute force: $d^2\varphi/dt^2 + kl\,d\varphi/dt + (g/l)\varphi = 0$, which leads us to the usual analytical formula

$$\varphi = A \exp(-kl\,t/2) \sin(\sqrt{(g/l) - (kl/4)}\, t + B).$$

Education has gotten used to offering a sterile linear world picture, but we lose a lot by such linearization and oversimplification.

The advent of microcomputers has eliminated the need for such simplification. Microcomputer clocks work in discrete ticks (in microseconds). If the coordinates of a mass point at the times $t-1$, t, $t+1$ are $x(t-1)$, $x(t)$, $x(t+1)$, then the average values of speed in the time intervals $t-1$, t and t, $t+1$ are $v(t-1/2) = x(t) - x(t-1)$,

$v(t+1/2) = x(t+1) - x(t)$. Therefore we can estimate the acceleration at time t as $a(t) = v(t+1/2) - v(t-1/2) = x(t+1) - 2x(t) + x(t-1)$. Let us assume that we know the explicit force function $a = F[x]$. In this case we can write $x(t+1) - 2x(t) + x(t-1) = F[x(t)]$. Thus we can calculate the future location of the body from its present and past:

$$x(t+1) = 2x(t) + F[x(t)] - x(t-1). \tag{3}$$

By choosing a small enough unit of time (if the frequency of ticks is high enough) we can integrate the Newtonian differential equation (2) as accurately as we wish (or the speed and capacity of our computer allow us to do) for any force function $F[x]$.

When the raindrop starts falling, it certainly does not look at the integral

$$\int \frac{dv}{1 - \dfrac{v^2}{g}}$$

and say: "Aha, I remember! Its primitive solution is the area of the hyperbolic tangent!" in order to find out how to increase its speed while falling. As physics sees it, nature works more in the way of equation (1) and equation (3).

If one takes quantum mechanics into account as well, one may say that a stone in Earth's gravitational field has discrete quantum states (discrete energy levels) and the classically continuous trajectory of a "mass point" is an (acceptable) approximation. John von Neumann has said that "the continuity that we observe everywhere in the macroscopic world is the misleading outcome of averaging in a world which is discontinuous according to its inherent nature."

We used to describe the manifold of discrete atoms with a continuous variable, as in the case of radioactive decay: $dN/dt = -kN$, giving $N(t) = N(0) \exp(-kt)$. We still describe population curves with smooth lines (in spite of the fact that population can change only by discrete numbers). We forget that our heritage is inscribed in DNA molecules by an integer of bits, that our body is made of an integer of cells, that our heart beats once a second, that our thinking happens in discrete space time (by firings of neurons). We approximate integers by continuous variables in order to be able to describe physical laws with analytical formulas such as (2). Nowadays, we solve these (mostly nonlinear) differential equations by approximative methods such as (3) in the discrete space time of the computer. Is the "deviation"

discrete reality → continuous model → discrete computer

a necessity? Or is the differential calculus (and all those analytical functions of continuous variables) just the product of Newton's efforts before the arrival of the computer era to compute how things move.

This seems to be a good question, but it is not so deep as it looks. Continuum models—like space time and differential equations of motion—are useful models of reality, just as discrete models—atoms, cells, species—are. Which model is more appropriate depends upon what we are aiming at. But if we decide to begin with a discrete model, let us put it directly into our discrete computer.

Deterministic Cellular Automata

The modern world has discrete space time. Our digital watches, our computers, our movie projectors and our television sets count time by ticks. The television screen, the photo film, and the moving picture depict space by discrete spectra. This digital technology enables us to handle information free of dissipation losses.

Two extraordinary mathematical talents of the twentieth century, Stanislav Ulam and John von Neumann, have introduced the concept of cellular automata. I am convinced that in the era of television, video games, and computers, cellular automata may offer educational advantages worth exploring. They are less abstract, more concrete and visual—like the thinking of teenagers. You are now invited to explore this discrete universe.

Ulam and von Neumann had their roots in central Eastern Europe but they worked for the Manhattan Project in the United States during World War II. Both of them had access to the first electronic computers. Stanislav Ulam enjoyed generating beautiful patterns by simple recursive rules (e.g. "Symmetry" on the *Games Nature Plays*[1] diskette for the Apple II), thus imitating crystal growth. Von Neumann dared to raise more provocative questions. He became increasingly worried by the fact that in the computer—built from thousands of vacuum electron tubes—almost one tube a minute burned out. The search for the spoiled tube and its replacement consumed more time than the useful running time of the computer. He began to wonder why a living organism—even the human brain—is able to work well for decades, in spite of the fact that some of its cells die every minute. He started using computer-oriented cellular models to investigate some basic properties of life.

The space of a cellular automaton is like a huge checkerboard, a two-dimensional array of squares called cells—like the monitor screen. A state of the cellular automaton is characterized by the states of its cells. A cell may have a finite number of (quantum) states like the set of colors on the screen. In the simplest cases a cell may be dead (off, black) or alive (on, bright). Time passes in finite units (ticks, beats, generations).

The state of a cell in the next generation depends only upon the state of its immediate environment in the present generation, according to a simple deterministic law. That is, it may be influenced by the present states of its four immediate neighbors:

$$C_{i,j}(t+1) = F[C_{i-1,j}(t), C_{i,j-1}(t), C_{i+1,j}(t), C_{i,j+1}(t)]. \qquad (4)$$

(Q=quart-type environment.) The law may be as simple as this:

$$C_{i,j} = 0 \text{ or } 1 \text{ for each cell,}$$

$$C_{j,j}(t+1) = 0 \text{ if } C_{i-1,j}(t)+C_{i,j-1}(t)+C_{i+1,j}(t)+C_{i,j+1}(t) = \text{even,}$$

$$C_{i,j}(t+1) = 1 \text{ if } C_{i-1,j}(t)+C_{i,j-1}(t)+C_{i+1,j}(t)+C_{i,j+1}(t) = \text{odd.}$$

(See "Reproduction" on the *Games Nature Plays* diskette or "Q13" on the *Cellular Automata*[2] diskette.) Simply said, c cell will live if there is now an odd number of live cells in its environment. The cell will die if there is an even number of live cells around it (this law was introduced by Ed Fredkin at MIT). Even this simple law has different visualizations. We may consider it to be a simple model of elementary wave propagation:

We can add of a sort of interference: two waves may annihilate each other by interference:

This law offers a manifestation of the superposition principle. The law can be reformulated in this way: Imagine an array of Petri dishes. We put an amoeba in some of the dishes. In the next generation each amoeba produces four offspring in the four neighboring dishes and itself dies. Each dish contains indicator dye that becomes colored if there is an odd number of amoebas in the dish but becomes invisible if the number is even (different amoeba families coexist, they do not influence each other).

Von Neumann in 1948 created a more sophisticated system to show that the behavior of atoms may be described by simple laws, but appropriate ensembles of these atoms can realize an organism that is able to make its own replica. In his classical paper he introduces a cellular automaton in which the atoms have twenty 29 states 29 different colors on the screen). The state of an atom in the next generation is influenced by the state of the atom and the states of its four immediate neighbors in the present generation. ("V" environment, "V" for Von Neumann and "V" for s/o in Latin.) These laws are rather sophisticated. Von Neumann has proved in this special case that self-replication is possible if one takes enough atoms.

Von Neumann's organism has been constructed of two parts. There is a tape (a linear sequence of colored atoms) that carries a finite amount of information. (we would now call it software. The tape contains a machine (a sophisticated arrangement of atoms), that uses an arm to read the message written in the tape and another arm to place an atom to the appropriate square of the screen. We would now call this hardware. If the tape carries the instruction to produce a replica of the machine, and to copy the tape, this arrangement will be able to produce a perfect copy of itself. Later on the copy can make a new copy, and so on.

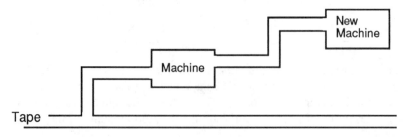

Watson and Crick discovered the molecular base of life in 1955, seven years after Von Neumann's self-reproducing automaton—the genetic information of a living organism. It is carried in the DNA molecular chain (the software). This message is copied by a system of enzyme molecules (the hardware). The main part of the genetic information is a blueprint for the enzyme molecules.

Von Neumann gave a mathematical proof for the possibility of such a self-reproducing machine, but he did not implement it in a specific program on a computer. It is too complex even for the computers of today. But Fredkin's automaton offers a simplified version of self-replication. If you start with any pattern, the code "Reproduction" will produce four or 16 or 64 replicas of this pattern on the screen after a sufficient number of generations.

Fredkin's model is oversimplified; in some sense it is linear approximation. Two patterns may produce transient interference, but they do not influence each other in the long run. The cellular models that are interesting enough are essentially nonlinear.

Conway's Game of Life

John Horton Conway of Cambridge University had the intention to create a cellular automaton with the following properties:

1. the laws of the game are simple;
2. most junk configurations disappear soon;
3. some structures survive;
4. some structures perform unexpected evolution.

In Conway's *Game of Life* each cell is either dead (empty) or live (full). Its fate is influenced by its state and by the states of its four neighbors:

Law I.   Birth: a cell will be born if the empty place has three neighbors.

Law II.   Survival: a live cell will survive to the next generation if it has two or three neighbors.

Law III. Death: a live cell will die (of isolation) if it has fewer than two neighbors or it will die (of suffocation) if it has more than three neighbors.

You may try the fate of some three-cell configurations with three buttons on a chessboard:

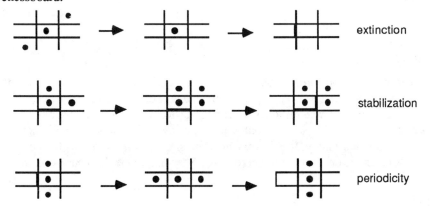

extinction

stabilization

periodicity

An important simple structure is the glider, made of five cells. It performs loco-motion diagonally across the screen with 1/4 of the "speed of light" (which is ≠ equal to unity, the maximum speed on screen).

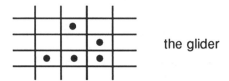

the glider

This cellular automaton has become an all-time favorite. It is played on thou-sands of computers (costing millions of dollars, according to *Time* magazine). Young hackers compete in making faster and more extended versions of it. Thousands of people explore the strange world of "Life." Books describe the "physics" and "biology" and "technology" of this nonlinear universe. (The excel-lent book *The Recursive Universe*[3] is fully devoted to the exploration of the "Life" environment. Some simple but interesting creatures are described in the book *Games Nature Plays*.[4]) On the Apple II diskette *Games Nature Plays* you find the code for "Life," e.g. how a tissue (type 4) can heal a wound (type 28, 20) or how the tissue may die like a tumor due to a misplaced cell (type 29, 20). It is worth experimenting with an initial state of randomly distributed cells (type 2). Some junk dies away, but some structures stabilize, propagate, and interact. The interest-ing feature of "Life" is that favorable initial patterns lead to unlimited spreading of cells, e.g. 13 colliding gliders make a "mother" (type 3), which produces new glid-ers that fly away in infinite sequence. (The *Cellular Automata* diskettes offers fast versions of "Life.")

With the help of gliders one can send signals across the screen. The trajectory of these signals can be interrupted by blocks; the simultaneous input of two gliders can be processed by appropriate gates. Conway has shown that one can build NO, AND, OR, NAND, and NOR gates in the world of "Life." (See Conway's book *The Winning Ways*.[5]) In this two-dimensional world even high technology can be developed: universal computer can be constructed.

The universe of "Life" is far from being completely explored. With its simple nonlinear laws and with its unexpected behavior, it is an amazing model of our real world.

Phase Transition

We can explore populating problems by taking the "Life" screen and scattering cells randomly onto it to cover a given percentage P of the screen. (For the "Life" on the *Cellular Automata* diskettes we can fabricate appropriate initial states by the "percent" code.) If P is small initially, junk dies away (by isolation) and later the screen gets rather empty. If P is very high, most cells will die again (by suffoca-tion). What is the optimum initial percentage P leading to the most survivors in the long run? If we experiment with this problem we begin to understand the present problems of our globe a bit better.

Such "population problems" can be treated with a wide class of cellular automata introduced in the computer science lab of MIT under the name "Majority." (See "Majority" on the Apple II diskette *Games Nature Plays*, "Additive Law" on the *Cellular Automata* diskettes.) The law of these games is of the type

$$\text{Cell } (t+1) = F[\text{Environment } (t)].$$

There are different options for the environment:

Q (quatre=4 in French): 4 neighbors;
V (Von Neumann): 5 (4 neighbors + central cell);
H (huit=8 in French): 8 neighbors;
M (Moore environment): 9 (8 neighbors + central cell).

The fate of any square in the next generation depends only on the number of present live cells in the environment of this square. E.g. "Q13" means Fredkin's Law: we shall have a live cell on the square in the next generation if and only if the number of live cells is one or three in the four (Q) neighboring squares.

Let us consider the M56789 automaton (we type M56789 when we select the "Additive" law of the game on the *Cellular Automata* diskette). We fill P percentage of the squares randomly using the "percent" option for initial state. If P is small, isolated cells perish. Connected formations fill up the concavities and get rid of sharp corners. (A cell will survive for the next generation if and only if it has at least four neighbors, offering a model for surface tension.)

M456789 is interesting as well: Convex droplets stabilize, and their growth stops (phase 0). If, however, P is larger than $P_0 = 0.25$, capillary action makes independent droplets fuse by coalescence, filling the whole field with live cells (phase 1). This automaton imitates the condensation of droplets from supersaturated vapor. The fate of a small high-density region in an extended low-density region (making average $P < P_0$) is worth our specific investigation. P_0 is called the spinodal point in the theory of phase transitions.

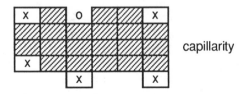

capillarity

In the case of the Q234 automaton $P_0 = 0.133$, for the V2345 $P_0 = 0.0822$ and for the H45678 $P_0 = 0.333$. These critical values have been obtained by computer experimentation; no one was able to derive them by theoretical (analytical) methods.

Percolation

By compressing a mixture of metallic and insulator grains we can make a solid block. Then we can connect the two opposite faces to a voltage. If metallic grains touch each other in a continuous chain, the block behaves like a conductor. If they are interrupted, the block is an insulator. Thus we can try to find the critical metal percentage P_0. (See the "Forest Fire/Woods" programs on the *Games Natures Plays* diskettes.) This phenomenon is called percolation, coined from the name of the coffee machine (which has been stuffed with coffee so the hot water can come through). Percolation is a subject of active research in materials science.

Several cellular automata show the phenomenon of percolation after a transient "recrystallization" period: V345 with $P_0 = 0.500$, M546789 with $P_0 = 0.500$.

Reversible Cellular Automata

The laws of games mentioned above are irreversible in time.

$$C_{ij}(t=1) = F[C_{environment}(t)] \tag{5}$$

corresponds to the diffusion equation

$$\partial C/\partial t = D\nabla^2 C, \tag{6}$$

which leads to irreversible phenomena. For example, in the game of "Life" small fragments disappear. From the obtained emptiness one cannot reconstruct the scattered live cells. One can say that in "Life" and in other irreversible cellular automata there is an entropy sink that helps in the formation of sensible structures.

The laws of microphysics are, however, reversible in time. Let us see, for example, the "Reverse" model on the Apple II diskette *Games Nature Plays*. Here a special law describes the fate of 2*2 blocks:

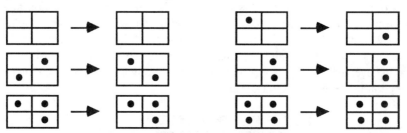

We can take also the rotated patterns into account in the same way. The trick is that the 2*2 block network is different in even and in odd generations, respectively:

if t is even   if t is odd

In this way we can model inertial locomotion and collision of billiard balls. This law is reversible. If one omits an odd generation, i.e. if one takes an even generation after an even generation, the ball begins to roll backwards!

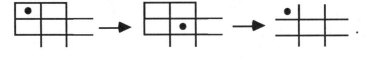

t is even   t is even again

With this "Reverse" model we can watch how the second law of thermodynamics works in the world. We begin with a simple pattern (like the character E, type P). If we let its eight atoms roll on the screen for ten to 20 generations, the pattern decays. If we stop the pattern and reverse the direction of time (type T!), we experience a movie played backward: the pattern E will be restored.

Since the second law of thermodynamics is of statistical origin, it is better to study this law on the *Cellular Automata* diskettes where we may get better statistics. An irreversible law can be made reversible by the simple trick of adding a new term.

$$C_{ij}(t+1) = F[C_{environment}(t)] - C_{ij}(t-1). \tag{7}$$

This can easily be reversed: $C_{ij}(t-1) = F[C_{environment}(t)] - C_{ij}(t+1)$. The new term makes a second-order equation (7) out of the first order equation (5). We can see that the Newtonian equation (3) is of the same type as (7): $x(t+1) = \{2\,x\,(t) + F[x(t)]\} - x(t-1)$. Equation (5) is a discrete model and equation (6) is a continuous model of an irreversible world; to calculate $C(t+1)$ the requested initial condition is $C(t)$ in case of equations (5-7) is a discrete model, equation (2) is a continuous model of a reversible world; to calculate $C(t+1)$ the requested initial conditions are $C(t)$ and $C(t-1)$.

The computer is able to demonstrate how irreversibility operates in the macroscopic world in spite of the fact that the laws in the microscopic world are reversible. If we load the initial state "Pic.Einstein" or "Pic.lion" from the *Cellular Automata* diskette and choose the law "Reversible reproduction(+)" on the Apple II or program R13 on the IBM PC diskette (this is Fredkin's Q13, made reversible by the extra term of [7] we can run ten generations. When nothing changes anymore (ten generations have passed), if we press "escape" on the Apple II ("R" on IBM), the computer will ask: "Change of time direction?" If we answer "Y(es). CONTINUE," even/odd generation pictures will be exchanged, which means that the movie will run backward. Einstein's face (or the lion's body) will build up again.

In the AND gate there are two inputs and one output (A and B→C). Common logical gates lose information, and therefore their use increases the entropy in the computer. To get rid of it, computers emit heat, and therefore they consume electric-free energy. This energy consumption can be decreased by using a reversible logical gates. Norman Margolus has proven that in the "Reverse" automaton the rolling balls can transfer information. We can construct RAND and ROR (reversible AND and reversible OR) gates in this model (by using much larger screen). In the two-dimensional universe of reversible cellular automata we can construct dissipation-free technology. No wonder: digital data processing knows only yes/no states; slight dissipative changes are not tolerated. Loss of information can be avoided. Therefore digital electronics is more powerful in computation and audiorecording than analog electronics.

Stochastic Cellular Automata

Fundamental particles and atoms obey laws of motion that are reversible to a very high degree of accuracy. But due to the (almost) continuous spectrum of states, dissipative phenomena occur on a large scale. Larger "cells" (like atoms in a crystal, coupled to invisible photon or phonon fields; dust grains performing a random walk because of collisions of invisible air molecules; neurons of the brain subjected to thermal fluctuations) may behave according to "digital" patterns, their laws of motion may contain dissipative tendencies. This is why the deterministic but irreversible rules of the popular cellular automata (like "Life") can be used to reflect reality. And this is why randomness is tolerated in the more relaxed rules of some scientific simulations. (See "Random Walk," "Vacancy," "Domain," "Struggle," "Shark," "Survival" on the *Games Nature Plays* diskette.) The computational possibilities of deterministic and stochastic cellular automata are far from being exhausted.

Parallel Processing

In nature, events happen simultaneously in each space point: all the atoms interact and accelerate. In the Von Neumann computer the central processor decides about each change in a deterministic time sequence. It has to scan the dots of the whole screen, one after the other; this makes the screen simulation of varying fields very slow. Nowadays there is a tendency to build computers that work with several processors. These enable the parallel evaluation of the fate of each "cell," (at least each group of cells). That means an enormous gain in the speed of computation. These computers often are called "non-Von-Neumann-computers," even though the idea for them appeared first in Von Neumann's papers.

Cellular automata are mathematical models for parallel processing: time ticks simultaneously at each square. Therefore they offer a good analogy for the parallel processors of modern computers. Actual microcomputer codes of cellular automata are not yet parallel processing, but if they are speedy enough, they give the impression of parallel evaluation. Thus these codes educate our mental vision to understand how nature works.

The first high speed card for IBM PC was constructed by T. Toffoli and others at MIT. "CAM—A High-Performance Automaton," Physics **10D**, 195 (1984). It uses the pipeline architecture to economize the work of the central processor. Pipeline firmware has been adapted to the software of common microcomputers by Zsolt Frei and Andrew Juhasz and is used on the *Cellular Automata* diskettes. The author is indebted to Gerard Vichniac for discussions.

1. George Marx, *Games Nature Plays* (1983–87), diskette for the Apple II Plus, IBM PC, BBC Micro (Department of Atomic Physics, Eötvös University).
2. Zsolt Frei and Andrew Juhasz, *Cellular Automata* (1986), diskette for the Apple II Plus, IBM PC (Department of Atomic Physics, Eötvös University).
3. William Poundstone, *The Recursive Universe* (New York: Oxford University Press, 1987); Gerard Vichniac, "Simulating Physics, Taking Computers Seriously," in the Microscience proceedings, Hungary (1985).
4. George Marx, *Games Nature Plays* (Veszprem, Hungary: National Center for Educational Technology, 1984); Nora Menyhard, "A Phase Transition in Life," in the Chaos in Education Proceedings, Hungary (1987).
5. E. R. Berlekamp, J. H. Conway, R. K. Guy, *The Winning Way* (London: Academic Press, 1982).

Animation in Physics Teaching

Eric T. Lane

Department of Physics and Astronomy, University of Tennessee at Chattanooga, Chattanooga, TN 37403-2598

In the laboratory, animated simulations on the microcomputer ease the burden on students to grasp difficult concepts in experiments. They allow students to simulate experiments that are too expensive, too delicate, or too dangerous to do in real life. Microcomputers should complement the usual laboratory offerings rather than replace them.

Like the simulators that airlines used to train airline pilots, the microcomputer allows simulations of difficult experiments to save us time, effort, and money in doing more and better experiments. As a result, the students' laboratory experience will improve significantly at a fraction of the cost that would otherwise be necessary.

High-quality, interactive simulations have to be written in machine language today. The fast pace of computer development means that anyone will soon be able to write these simulations in high-level languages. This paper presents some of the techniques used to generate elementary simulations of waves and particle motion. It also discusses some possible simulations for physics and science teaching.

Why Simulation?

If a picture is worth a thousand words, then a dynamic animation may well be worth a thousand pictures. Dynamic, interactive presentations make difficult concepts easy for people to see, to use, and to understand.

One can tell students all about wavelength, amplitude, and phase of a wave. And these concepts can be illustrated with diagrams. But the concepts of wave speed, frequency, period, and many other dynamic concepts requires direct, moving experience for most people to understand easily.[1] Ordinarily experiments in the laboratory or experiences in everyday life provide the necessary background for understanding these and other dynamic concepts. But too often, students miss the necessary observations because they don't know what to look for.

We can diagram the idea that a standing wave can be thought of as the linear sum of two traveling waves going in opposite directions, but most students fail to see the connection from the still picture. Yet an animated presentation makes the traveling waves easy to see and the two-wave addition is clear almost immediately.

The exponential decrease of the number of particles with height in a gravitational field shows the basic idea of the Boltzman energy distribution. Diagrams and still pictures encourage students to assume that once the particles are in position, they must stay there to maintain the exponential distribution. Only a dynamic presentation can show the active change in position of each of the particles as they move independently of all the others while retaining the exponential distribution.

The old Bohr model of the atom is used to illustrate the wave nature of the electron around the nucleus in an atom. When I first started teaching science and modern physics, I wanted to show the dynamic properties of this wave. This desire was one of the driving forces in my efforts to animate science concepts. The Apple II microcomputer was the first to allow sufficient graphics resolution with reasonable speed at a low enough price to make such animations possible.

Prospects

Recently it has been stated that there is no place for the "teacher developer," that we should let the professionals handle it all. Yet I submit that the professionals must follow the market. Rarely do they provide really new, innovative software. It remains for the individual developer with a unique idea and little to lose to come up with true innovations. We as teachers can see the needs of our disciplines. It is up to us either to develop the programs or to encourage others to do it for us. We will soon have machines fast enough to do this.

Fairly extensive animated games have appeared in the last few years. With a little help from hardware, simulated airplanes and space ships move smoothly and with fair realism. Unfortunately the educational market does not provide the incentive for companies to develop such hardware for educational use, much less the more difficult software to run on today's popular microcomputers.

Yet these games do show what can be done with today's technology. Fortunately, with the steady decrease of chip and system prices, we can look for equivalent capability in just a few years. Recent reports suggest that today's microcomputer has roughly the same power as a mainframe computer of 20 years ago at one-thousandth the price. With the development of dedicated VLSI graphics chips, the next generation of microcomputers will provide the level of speed and memory capacity required for really good animated simulations.

Wave Algorithms

The next generation of machines will have enough power to do simple, interactive animations written in high-level languages such as Pascal, BASIC, etc. (Currently, animation on microcomputers must be done in machine language to get the speed and interactivity required.) In the following, I present some of the algorithms and tricks to do fast animations.

The basic approach to animating waves on a computer is to generate a table for one full period of the wave. This avoids having to calculate sines and cosines for each point during the animation, which takes a relatively long time in most computers.

First, we erase the old point and plot the new point. Stepping through the table with a variable increment allows us to control the apparent wavelength of the animated wave. To make the wave appear to move, we simply increment the starting point in the table. Moving the starting point to the left in the table makes the wave appear to move to the right, etc.

To produce a standing wave, we use two pointers in the table. We move one to the left and one to the right, corresponding to the two traveling waves. Adding the values at the pointers generates the standing wave. Thus animations like those in *Standing Waves* are quite easy to generate.

To see group velocity or relativistic electromagnetic wave effects, we use two separate tables with different amplitude sine waves. Adding values from two tables allows us to simulate a wide variety of physical phenomena.

The tables, once calculated, can be used to produce any wave shape imaginable. We can demonstrate pulses, triangular waves, square waves, or any other wave shape we desire.

To generate a circular wave such as is used in the *Electron Wave* animation, we also need tables of x and y conversion factors for each angle. We take the appropriate factors from the table rather than calculate them "on the fly." The rest is just wave animation all over again.

Particle Animations

Particles are extremely easy to animate. We set up a calculation loop that successively operates on each simulated particle. In the loop, we erase the old particle, calculate, and plot the new particle.

In the calculation, the acceleration of each particle is added to the velocity and the velocity is added to the position for both the x and y dimensions (and the z dimension, if you're doing 3-D). Ordinarily, one multiplies the acceleration by the time increment before adding to the velocity, and multiplies the velocity by the time increment before adding to the position. To avoid four (or six) time-consuming multiplications per point, we scale the acceleration and the velocity so the time increment is unity. This increases the conceptual difficulty of programming but makes animation much quicker.

Choosing the velocities from a true Maxwell distribution may seem difficult, but in fact is quite easy. If we look at the Maxwell velocity distribution, we see that each component of the velocity, x, y, and z, is chosen from a normal or Gaussian distribution. A simple way to obtain Gaussian random numbers is alternately to add and subtract twelve uniform random numbers. This gives a Gaussian random number with an expected average of 0 and a standard deviation of 1. Then we simply multiply the Gaussian random number by the square root of the simulated temperature to obtain the appropriate Maxwell distribution.

All the rest is keeping track of where the particles are, making sure that they stay inside the walls of the simulated box, etc.

Work in Progress

This section summarizes some of my work in progress and ideas for the future. They are arranged roughly in order of the amount of work that I have already put into them.

Current Development

Kinetic theory. Animation of moving particles under varying conditions of temperature, pressure, volume, gravity, etc. Shows Maxwell's Demon in action. Can display Bose-Einstein, Fermi-Dirac, and other distributions. Calculates entropy, energy, etc.

Diffusion. One-dimensional concentration plot shows exponential diffusion from various initial conditions: delta function, edge, dipole layer, sines of decreasing wavelength, etc. Diffusion coefficients can be varied. Program is expandable to include other interesting effects.

Chromatography. Real-time display of concentration along a gas or liquid chromatography column. Shows partitioning of various components between mobile and stationary phases as well as separation of components at end of column. Can also simulate diffusion, nonequilibrium, instrumental, and other effects.

Lissajous Figures. Animation of crossed sine waves or other wave forms of arbitrary frequency and phase, with or without amplitude variation. Has possible artistic application. Can animate thousands of points.

Holograms. Making holograms, also called kinoforms. Can be output to a matrix printer for photographic reduction to view with a laser. Would be good for high school science projects, and for showing the phase relationships of waves, Fresnel diffraction, etc.

Elliptic waves. Animation of particle motion in waves.

Earthquake waves. Rayleigh surface waves, Love waves, compressional waves, transverse waves, etc.

Water waves. Capillary waves, deep-water waves, variation of amplitude with depth. Will be of interest to geologists, engineers, physicists, etc.

Fermat's principle. Real-time display of Fermat's least-time principle.

Reflection and refraction. Could be presented as a game.

Multislit interference. Draws either Fraunhofer or Fresnel diffraction patterns in real time for an arbitrary number of slits with variable widths. Shows quantum probable buildup of photons as they are diffracted by the slits.

Relativistic shapes. Animation of simple shapes as they move past an observer moving at speeds near that of light.

Orthogonal functions. Real-time display of Hermite, Legendre, Laguerre, and other orthogonal functions. Would be useful in quantum mechanics to show hydrogenic atom probability clouds, in engineering to show antenna radiation patterns, and in chemistry to show atomic orbitals.

Algorithms Developed

Rotating three-dimensional display. Animated 3-D perspective display of any object with 256 points rotating about a vertical axis. More points slow the animation. Considers 3-D chemical molecular models, crystals (hexagonal, face-centered, etc.), Platonic and Keplerian solids and others. Can have different objects rotating at different rates at various places on screen.

Curve drawing. Curves with second-order differential equations can be drawn in real time. Can go to third and higher orders with little effort. (Real-time calculation should be done in less than a second.)

Fourier transform. Real-time display of periodic wave forms and their Fourier transforms including amplitude, phase, real or imaginary components, power spectra, etc., in rectangular or polar plots.

General relativity. Animates the three basic effects: frequency and amplitude increase of light waves falling into a black hole: light bending in a gravitational field, and precession of the perihelion of mercury.

Wave group. Animates particlelike wave groups to illustrate Heisenberg uncertainty, Feynmann diagrams for elementary particles, etc.

Compton scattering. Simulates photon collision with wavelike electron, shows change in wavelength of each, or tries for simulated interference effects.

Dashed lines marquee effect. Animation of dashes moving along arbitrary curves would be useful to illustrate charge motion in an electric or electronic circuit. Considers animation of charges moving in a vacuum tube or in a semiconductor, cars moving on a highway, traffic flow in a hospital, etc.

Ideas under Development

Four-dimensional rotations. Real-time display of 4-D objects in perspective under interactive control. Could do rotating hypercube, spheres, molecules, etc. Would be useful in mathematical discussions, relativity, crystallography, group theory, etc.

Doppler effect. Animation of circular wave fronts of light or sound, moving outward from a source as the source and observer move across the screen at various relative or absolute velocities. Could also show shock-wave effects of objects moving faster than sound.

Vibrating beam. Animation of the vibration of a solid beam under various end conditions: clamped, fixed, free, etc.

Reactor simulation. Animation of the neutrons in a nuclear reactor as they collide with uranium atoms, which causes them to fission and generates more neutrons. Would also show slowing of neutrons in moderator, absorption by damper rods, poisoning, delay effects, etc.

Wave reflections. Animation of waves being reflected from hard, soft, or intermediate boundaries with different wave velocities. Would show both transmitted and reflected waves including components.

Cardiac simulation. Animation of an electrocardiogram showing abnormalities of the heart under interactive control. Could also show vector plot, etc.

Quantum mechanics. Real-time solution of Schrödinger's equation with interactive parameter changes. Would solve square well, simple harmonic oscillator, hydrogen atom, and other potentials.

Crystal defects. Animation of the motion and behavior of defects produced by inclusions, vacancies, slip planes, etc. in crystalline and amorphous solids.

Relaxation techniques. Real-time solution of Laplace's equation for heat flow, current flow, temperature or voltage distribution, etc.

Fluid flow. Real-time display of the vector velocity field around various objects in a moving compressible or incompressible fluid.

Sound generation. Would produce the sounds of a given Fourier series or generate tones with fixed ratios. These would be of interest to musicians (especially those who work with synthesizers), engineers, etc.

Particle scattering. Animation of particles scattering off a nucleus or other object. Would also generate the scattering pattern in real time. Could be set up as a game with various levels of difficulty. Particle-beam collisions, center of mass conversion, and cross-section concepts.

Quantum scattering. Real-time generation of the scattering pattern produced by particle waves using S-matrix theory, etc.

Accelerators. Animation of the charges in a cyclotron, synchrotron, linear accelerator, or others. Would show bunching, phase effects, quadruple focusing, etc.

Supercollider. Geographical dimensions, goals of supercollider, particle beams, experimental areas, detectors, etc.

Possibilities

Hanging chain oscillation. Bessel functions, solution of finite partial differential equation.

Advanced waves. Block waves in a crystal, quantum electronic devices, Huygen wavelets, hologram formation.

Green's function simulation. Would show point effects, illustrate summation, delta function, impulse, etc.

Electron motion. Electron beam in television tube, oscilloscope, drift-tube amplifier, magnetron, vacuum tube, or in a semiconductor diode, junction transistor, or field-effect transistor.

Maxwell's equations. Would show differential forms, integral form; would animate qualitative interpretation; would show plane current increasing, radiating dipole; would have animated, interactive displays of accelerated charges.

Bell's theorem. Would review theory and quantum evidence; would give interactive examples from classical and quantum cases.

Superconductor. Wave properties, energy gap, flux quantization, interference.
Superfluid. Zero viscosity, fountain effect, second sound, lambda point.
Nuclear reactions. Conservation rules, reaction lab, energetics, decay schemes.
Elementary particles. Group theory, conservation laws, reactions, quark theories, other theories.

Conclusion

My present goal is to develop animation modules for the Macintosh, the IBM PC, the DEC Rainbow, the BBC Acorn, and other microcomputers. The modules should be accessible from BASIC, Pascal, and any other language that allows parameter passing to machine-language subroutines. This will allow teachers to create their own demonstration, tutorial, and CAI materials using these animations. It will also allow students to use the animation modules in their own projects.

Experience suggests that I concentrate on animation graphics design. Given the limitations of time and support that I face, I must concentrate on those efforts that will yield results in projects that no one else is attempting. This optimizes the gain of teachers and students alike by giving them a number and variety of animations that they would not otherwise have.

I welcome ideas and suggestions from the physics community. I especially need input on what teachers would like to see in the way of simulations. What is important? What is useful? What mainframe programs have you seen that you would like your students to use on microcomputers?

The possibility of interactive, animated simulations holds great potential for providing high-quality education to the public as well as to our students in classroom and laboratory.

I wish to thank the physics department and the Center for Excellence for Computer Applications at the University of Tennessee at Chattanooga for their continuing support. *Animated Waves and Particles* may be obtained through the Physics Courseware Laboratory, Department of Physics, North Carolina State University, Raleigh, NC 27695-8202, or telephone (919) 737-2524. *Animation Demonstration* is published by CONDUIT, which also publishes the related programs *Group Velocity* and *Standing Waves*. To order these packages, or a free catalog, contact CONDUIT, Room 4557, Oakdale Hall, University of Iowa, Iowa City, IA 52242, or telephone (319) 335-4100.

1. Since a written discussion cannot illustrate the dynamic nature of animations, the reader may wish to obtain a copy of the author's programs from the sources listed to see the effects described in the paper.

Physics Simulations for High-Resolution Color Microcomputers

Richard A. Arndt and L. David Roper
Department of Physics, Virginia Polytechnic Institute and State University, Blacksburg, VA 24061

Since the mid-1970s we have had many physics simulations running on DEC-VAX11/780 (now VAX-8800/2) and IBM (currently IBM-3084 and IBM-3090 with vector facility) mainframe computers. We have used monochrome graphics terminals that emulate a Tektronix terminal. Students have been able to use our simulations because we have made strong efforts to register all entering freshmen and first-year graduate students as VAX users and because monochrome graphics terminals are available at various locations on campus. We also have taken a terminal into classrooms for lecture demonstrations, connecting it to a VAX by a telephone line and modem in the early years and by a Sytek local-area network in recent years.

We have a restricted log-in user ID on the VAX and freely advertize it to all interested students. Anyone who logs into the VPI&SU VAX cluster on this PHYSICS user ID can do nothing but run our set of physics simulations. (Since VAX allows multiple log-ins to a user ID, many students can simultaneously log into the PHYSICS user ID.)

About 1,500 freshmen engineering students at VPI&SU are currently required to buy an IBM PS/2 when they enter the university; there were over 10,000 personal computers at VPI&SU in January 1988, and about 2,000 are being added each year. The free Kermit terminal emulator software for IBM PC/AT/(PS/2) (version 2.31) emulates a monochrome Tektronix graphics terminal. Thus all engineering students at VPI&SU are able to use our physics simulations on a VAX computer, either from the data connection in their dormitory rooms or through modems in their apartments.

In 1983 we decided that it was time to buy a color-graphics microcomputer for our research and teaching. Our criteria were that resolution had to be about what we had had on our monochrome-graphics terminals, with the addition of at least eight colors; the characters had to be easy to read (no discernible jagged edges or spaces between pixels); the graphics had to be fast; software for doing scientific word processing had to be available; software for Tektronix color-terminal emulation had to be available; color-graphics software callable from one or more standard MS-DOS programming languages had to be available; and, of course, the cost had to fit our budget. After a thorough study, we found only one microcomputer that met our criteria in 1983: the NEC-APC, two of which we purchased. After 1985 the choice was greater; then our choice became the cheaper and faster NEC-APCIII; our department, faculty, and students have purchased more than 25 of these machines. At that time it cost at least $2,000 more to buy an IBM PC with

EGA graphics than it cost to buy the APCIII. Now, of course, EGA and VGA machines are very inexpensive and we are buying such machines.

With the advent of the NEC-APCIII in 1985, all of our simulations were instantly available on it because the ESC140 VT100/Tektronix color terminal emulator was available for both the APC and the APCIII. (Several Tektronix color terminal emulators are available for the IBM PC/AT/PS/2, the least expensive of which is VTEK.)

We have converted many of our VAX physics simulations to run on the NEC-APC, NEC-APCIII and IBM PC/AT/PS/2, both as color terminals with a VAX mainframe and as microcomputers.

The physics department has an NEC-APCIII on a wheeled table for easy transport to a classroom. We use the APCIII as a standard classroom tool, similar to slide and film projectors, lecture demonstrations, etc. The use of color greatly enhances viewing a computer screen in a classroom. In large classrooms we use a Sony large-screen projector, which automatically adjusts to the resolution of various microcomputers used with it. When the APCIII is not being used for lecture demonstrations, it resides in the department's SAIL (Student Audiovisual Interactive Laboratory) room for student use as a microcomputer and a color graphics terminal.

Physics Simulations at VPI&SU

Many of our physics simulations have been converted to run on the NEC-APC, NEC-APCIII, and IBM-PC/AT/PS/2 as stand-alone microcomputers, using MS-FORTRAN and the QCAL graphics package. QCAL emulates CALCOMP graphics, has graphics cursor capability, and allows several special features of high-resolution (not CGA) microcomputers (e.g., eight colors, four drawing modes, and quick-graphics save and restore). CALCOMP graphics run on Tektronix terminals and their emulators; we have been using them for two decades on VAX and IBM mainframes. Thus we can download FORTRAN graphics programs from a mainframe to a microcomputer and quickly get them running on a microcomputer. (A site license for educational institutions is available for QCAL.)

There are at least two types of physics simulations: those that draw a "picture" of a physical situation and then show how it changes with changed parameters (e.g. time); and those that draw mathematical functions that show how some physical function varies with changed parameters (e.g. time). Both types are used in some of our simulations. Students also appear to enjoy game aspects in simulations, so several of our simulations are designed so that they can be used as games.

Some features of our physics simulations are as follows:

1. Seven colors are used for text as follows: magenta (purple) for questions to user, red for warnings and errors, yellow for standard text, green for help information, cyan (greenish blue) for user input, blue for incidental information (e.g., copyright notice), and white for titles and general information. These color guidelines follow recommendations based on the physics, physiology, psychology, and sociology of color vision.[1] We sometimes deviate from these color guidelines when colors are needed to differentiate several parts of a screen; e.g., parts of a menu.

2. Seven colors are used in graphics, usually in rainbow order (magenta, red, yellow, green, cyan, blue, and white), for curves with increasing values of time or some other parameter, for example, rainbow-ordered colors are used for third-dimensional values instead of hidden-line removal, thereby showing more information and saving calculational time.

3. Brief help information is usually present on a screen, but extensive help is quickly available by pressing the "?" key; all graphics and text on the screen are quickly saved and then quickly restored after the help screen is read.

4. Many of our simulations use a graphics cursor for input; for example, the graphics cursor is often used to enter position and/or velocity for a particle.

5. Run-time branching allows the user to skip around in a simulation by using two-letter codes. This includes restart (RS) or quit (QT) at most points of user input.

6. Reasonable default parameter values are available so users can repeatedly press return to run an example before choosing their own parameter values. Some of our simulations allow much leeway for the user to enter variable parameters, so we use an easy, fast, and error-checking input procedure described in table 1.

7. Typical interactive sessions can be demonstrated using a "data-stream" feature whereby user input can be saved in a "replay" file. The replay file is an ordinary

---

Table 1.

Data-Entry Rules

There are two kinds of data entry: alphanumeric and graphics cursor.

Alphanumeric Data Input
- Separate numeric entries by spaces, end of line, + or -.
- A real entry does not need a decimal point (see examples below).
- An entry can be repeated: .2R6 or 34R13 or 5.3 R 7
- Entries can be incremented: 315R13 or 2.5I.6R5 or 11.3 I 3 R 7
- The remainder of an array can be skipped: (array X(19)) → 5.3 2 -4.5 S
- The remainder of an array can be zeroed: (array X(19)) → 5.3 2 -4.5 Z
- An element for an array can be entered at any next element by using *: (array X(19)) → *5 3.2 5 *10 5 S → Sets elements 5, 6, and 10 only.
- The remainder of an entire entry call can be skipped: (call for X,Y,Z,N) → -3.2 4 Q
- Transfer codes may be available for some applications.

Graphics Cursor Input
- When a full-screen graphics cursor is displayed on the screen, move it around with the cursor keys and then stroke a single key for entry as directed. The INS key toggles the speed of cursor movement.

Table 2

Replay of Physics Simulations

Recording into a File

(Default file name: INPUT.INP or run program with PROGRAM FILENAME)
DSSR: Switch on/off recording into replay file. (DS = 'data stream').
Enter Program Data and Commands
DSPS: Insert a pause at non-graphics-cursor input.
DSPS n or # n: When pause is disabled, insert a pause or a delay of *n* seconds at
    non-graphics-cursor input.
#: Insert a pause or a delay at any input (at graphics-cursor input you will be
prompted for a delay time or a pause).
DSRF: Rewind replay file.
DSEF: Go to end of replay file.
DSRS: Set random-number generator seed from system clock.
DS?: Display this data-stream information.
DE?: Display data-entry rules information.
(Only # is entered at graphics-cursor input.)

Starting Replay from a File

(Run PROGRAM FILENAME).
DSRF: Rewind replay file.
DSSB: Switch off/on input-attention beeps.
DSSI: Switch input from keyboard to replay file.
DSSI *n*: Switch input to replay file and set a common delay of *n* seconds for all
    pause locations.

At Pauses when Replaying from a File

DSSB: Switch off/on input-attention beeps.file
DSNP: Disable pauses.
DSNP *n*: Disable pauses and set a common delay of *n* seconds for all pause loca-
    tions (*n* = –1 to return to specific pauses).
DSSI: Switch input from replay file to keyboard.
DSRF: Rewind replay file.
DS?: Display this data-stream information.
DE?: Display data-entry rules information.

While Replay is Running

<ESC>: Quit program while running replay.
Q: Stop replay and return to keyboard entry.
<KEY>: Actuate any pauses while running replay without pauses
    (<KEY> = any key other than <ESC>, q, or Q).

ASCII file, so it can be edited by any editor. One can insert pauses or time delays at any input point in the replay file, or one can cause the replay to ignore all such pauses and time delays or insert a common time delay at all input points. The replay can continuously recycle for demonstrations. See table 2 for recording/replay details.Brief descriptions of our physics simulations are given in Table 3.

Some representative screens available in our simulations are shown in Figure 1. Of course, these static black-and-white pictures cannot adequately show the information available in an interactively changing color display. They are unretouched graphics screen dumps from a NEC-APCIII microcomputer to a Hewlett-Packard-500-Plus laser printer. Since the APCIII screens sometimes contain nongraphics text as well as graphics text (an excellent feature of the NEC-APC and NEC-APCIII, not available on IBM-EGA or VGA), the graphics screen dumps do not show everything that a user sees on the screen.

QCAL programs can run on the IBM PC/AT/PS/2 with either the color-graphics adaptor/monitor (CGA) or with the enhanced graphics adaptor/monitor (EGA). Of course we prefer the latter because of its higher resolution (640x350 pixels) and greater number of colors (8 + high intensity = 16). Our simulations are linked such that they will run on a microcomputer with or without the 8087 mathematics coprocessor installed, but we strongly recommend installing the 8087 (or 80287/80387 in 80286[AT]/80386 machines). Floating-point calculations and graphics are extremely slow without it (by a factor of over 10 for PCs).

Some papers have been published about our simulations.[2] Our physics simulations for the VAX computer can be obtained from the authors on a VAX back-up

Table 3

Physics Simulations at VPI&SU

Name	Description
ART	Trajectory motion (artillery game)
LEM	Planetary landing (lunar lander)
TRAJ	Orbits and scattering in a spherical potential
SCAT	Two-dimensional scattering game
TURT	Relativistic view of moving object
CLOSC	Classical oscillator
QNT	One-dimensional quantum mechanics
WAVEPAC	Traveling wave packet
QCLOCK	Quantum mechanical clock
ECB	Motion of electron in crossed E and B fields
EFIELD	Fields, equipotentials, and motion in E and B fields
LRC	LRC circuit
GRATE	Diffraction grating patterns
BEATS	Two frequencies
GEOPTIC	Ray tracing for mirrors, surfaces, and lenses
FOURIER	Fourier representation of wave forms
MATHF	Orthogonal functions

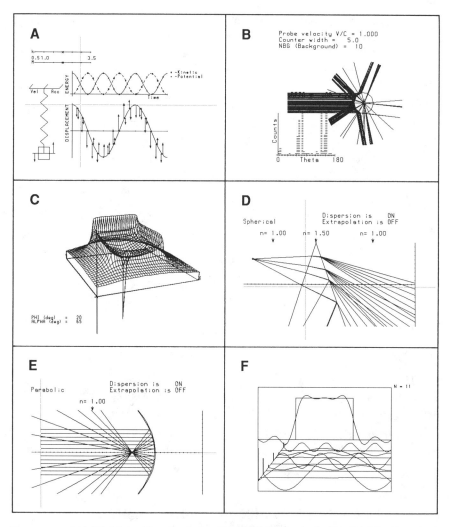

Figure 1. (A) CLOSC: Classical Oscillator Simulation, (B) SCAT: Classical
Scattering Simulation, (C) EFIELD: Simulation of Charge
Distributions, (D) and (E) GEOPTIC: Geometric Optics Simulation,
(F) FOURIER: Fourier Representation of Functions.

tape. The microcomputer physics simulations for the NEC-APCIII and the IBM-PC/AT/(PS/2) are available commercially at low prices from GlobalView, Inc., Rt. 1, Box 282, Blacksburg, VA 24060.

Future Physics Simulations at VPI&SU

We plan to continue improving our current physics simulations and continue converting more of the VAX simulations to run on high-resolution color microcomputers and to create more simulations in all areas of physics.

We plan to transfer our programs to the new IBM VGA graphics and the new Macintosh II in the future (see table 4). We have chosen the Macintosh II as our future research teaching machine; however, it is much harder to program than

Table 4

Comparison of High Resolution Microcomputers

	NEC-APC	*NEC-APCIII*	*IBM-PC/AT*	
Pixels	640x480	640x400	320x200 (4 colors)	CGA
			640x200 (mono)	
CGA				
			640x350	
EGA				
Colors	8	8	4	CGA
			2 (mono)	CGA
			16(*+high int.)	EGA
CPU	8086 (5 MHz)	8086 (8 MHz)	8088 (5 MHz)	PC
	(1.3×IBM PC)	(2.5×IBM PC)	80286 (6/8 MHz)	AT
FORTRAN			16 s PC	
Whetsone	12 s	8 s	8 s 6/8 MHz AT	
(with 8087)			5 s 12 MHz AT	
FORTRAN			208 s PC	
Whetsone	141 s	99 s	69/49 s 6/8 MHz AT	
(w/o 8087)			36 s 12 MHz AT	
	IBM PS/2/MCGA	*IBM PS/2/VGA*		
Pixels	640x480	320x200	640x480	720x400
			(graphics)	(text)
Colors	2 (mono)	256	16	16
CPU	8086 (8MHz)	8086 (8 MHz)	80286 (10 MHz)	
	(2.5×IBM PC)	(2.5×IBM PC)	80386 (16-25 MHz)	
	(Model 30)	(Model 30)	(Models 50, 60, 70, 80)	
	Macintosh II	*Hercules Incolor*	*VaxStation 2000*	
Pixels	640x480	720x348	1024x864	
Colors	256	16	16	
CPU	68020 (16 MHz)	—	VAX	

8086- family machines, if one allows all Macintosh features to be constantly available to the user of a simulation. The advent of the much faster Macintosh II and machines using the Intel 80386 cpu, will enable us to expand our simulation set to include simulations that require more complicated numerical calculations.

We wish to thank many students who have helped us develop and refine our physics simulations over the past decade. Some who were especially helpful were Brent Landers, James Finegan, Randall Price, Charles (Chet) Perkins, Kuryan (Oby) Thomas, Mark Mathews, and Craig Burkhead. Special gratitude is due to Roger Link in the physics electronics shop at VPI&SU for much help with hardware and software.

1. "The Effective Use of Color," *Eurographics 86 Tutorial* (Gerald Murch, Tektronix, Inc., P.O. Box 500, Beaverton, OR 97077).
2. R. A. Arndt and L. D. Roper, AJP **51**, 418 (1983); R. A. Arndt and L. D. Roper, AJP **54**, 614 (1986); R. A. Arndt and L. D. Roper, Computers in Phys. **2**, 61 (1988); R. A. Arndt and L. D. Roper, "Scattering in a Spherical Potential: Motion of Complex Plane Poles and Zeros," VPI&SU preprint (to be published in Computers in Physics).

Apple Simulations

R. H. Good
California State University, Hayward, CA 94542

What should we do with computers? We can do almost anything:

1. *Word processing and number crunching.*

2. *Data acquisition and analysis.* For example, with an Apple II and A/D converter we can pick up data at the rate of 10,000 bytes per second and carry out real-time frequency analysis.

3. *Text programs.* In such programs the screen displays text material in much the same way it might appear in a textbook. Such programs are still being written, even for the new microcomputers, but they represent an abuse of the medium, because for this purpose a book is more legible and more flexible.

4. *Tutorial programs.* Text programs often are mislabeled as tutorials, but real tutorials change their presentation depending upon the user's response. Surely such programs are destined to play the major role in CAI (computer-assisted instruction). However, the biggest problem they currently face is how to get students to use them.

5. *Simulation programs.* Tutorial programs are intended to replace human teachers (quite properly), whereas simulations enhance the teacher's resources. Quantitative simulation of physical phenomena that cannot otherwise be demonstrated easily (or at all) constitutes a revolutionary new resource for the physics teacher. For example, in no other way can one conjure up quantum-mechanical probability amplitudes with real and imaginary parts, and at the

same time display probability densities quantitatively, and moving across the screen in real time, all subject to interactive adjustment by teacher or student. Such programs are of value in concept formation from high school through college and into graduate school. They can be used briefly, for a few minutes, as a lecture demonstration, or they can support an hour or two of individual study.

This paper discusses simulation programs and the microcomputers that run them.

In presenting the Doppler effect and the related sonic boom, many a teacher has filled the blackboard with circles and then had to say, "Now imagine it moving." The Apple II provides a kind of magic blackboard on which the circles really do move. And in this case the circles are practically perfect (user adjustable) and quantitative work can be done with measurements on the screen images.

Radiation from a dipole antenna is of interest over a broad range of class levels, and corresponding still pictures abound in textbooks. The 48K Apple II can store a sequence of 64 full-screen high-resolution pictures and present them at eight frames per second, showing how field lines "break off" and so forth.

At the most fundamental level, electromagnetic radiation begins with acceleration of charge. A highly interactive Apple II program enables the teacher or student to move a single charge arbitrarily in the vertical direction and observe in real time the effects on the electric field, and in particular the formation and growth of the transverse components that represent electromagnetic radiation.

You may never have seen a Maxwell's demon, but an Apple II simulation allows you to be a Maxwell's Demon. And it's no bed of roses: you have to be quick and clever, and kids do it better than their teachers. You can open and close a trapdoor to select large atoms, or fast ones, thus lowering the entropy, in flagrant violation of the second law of thermodynamics. It's like a video game, and your score is the (negative) entropy you achieve.

Nothing like these programs is available for microcomputers other than the Apple II. This is of course a tribute to the impressive simplicity, flexibility, and performance of the Apple II with Applesoft BASIC. But it also constitutes an indictment of the new, more "sophisticated" microcomputers.

For example, the new Tandy 5000 personal computer advertises 256,000 different colors, 16M RAM, an 84M hard disk, and 20 MHz. But we simply don't need all of this performance, and in fact the wonderful variety of options not only complicates the programming but also tends to make us lose sight of what we're really doing.

Eric Lane informs me that the Macintosh is about ten times as hard to program as the Apple II, and the IBM PC has problems of its own. This is why we do not yet have any of his excellent programs available for these machines, even though they have been around for years. And Bruce Sherwood paints a poignant picture of a top-flight programmer in front of his microcomputer, with ten manuals open in front of him, struggling to put together a simple system that others will find usable for the "sophisticated" micros. I believe that the cT approach (formerly Andrew and CMU-Tutor) is sensible for making the new machines accessible to educators generally. But I question whether the new machines themselves are really appropriate for the field of education.

We need to be able to program the machines ourselves because Apple corporation is not going to be writing our physics programs for us. And especially if we want to bring students into the operation, simplicity of programming is of utmost importance. That is what the supposedly more "sophisticated" machines do not have. Sophistication without simplicity is barren. The promise is there but the programs are not.

It is not true that increased performance requires increased complexity. An Apple II at 4 MHz would possess formidable performance and yet be precisely as easy to program as it now is. And such performance is now easily attainable; but unfortunately the Apple corporation is not currently disposed to make it available to us. The Apple IIGS is not an upgraded Apple II; it is a downgraded machine of greater intrinsic complexity, and at its higher speed it is relatively incompatible with the old Apple II and is harder to program. I know of none who is programming in physics for the GS, and that in itself suggests that the machine's premise is foolish.

For educational purposes, the ideal machine would be something like an Apple IIc with DOS 3.3, operating at 4 MHz with 128K RAM, and with one channel each of A/D and D/A. This is simply an implementation of current technology. It would cost no more than now. It would have plenty of RAM; indeed, it could hold half a dozen of my programs simultaneously, or several hundred high-resolution pictures. It would also be a powerful and inexpensive laboratory instrument capable of 12 frequency analyses per second. Most important, it would be easily programmable. With its size, power, flexibility, and economy, it could represent the general-purpose computer of choice for educators well into the twenty first century.

That is the direction we should try to go.

Macintosh Demonstrations for Introductory Physics

Irvin A. Miller and Richard D. Haracz
Physics Department, Drexel University, Philadelphia, PA 19104

Each Drexel University student owns or has access to a Macintosh computer. The engineering and science freshmen are introduced to Pascal and some of the graphics capability of the Macintosh in their first quarter. The two-quarter introductory calculus-based physics course, which starts the following quarter, covers mechanics and wave motion. We have written short Pascal programs to illustrate physical principles or their applications.

Each program illustrates a single concept or topic, such as the difference between static and kinetic friction, or provides a graphical solution to a problem, such as the comparison of two linear motions. The programs are used as short demonstrations during course lectures. They immediately follow the presentation of a concept or the analysis of a problem. Many use graphic animation to illustrate sequences of events or motions.

Concentrating on a single concept or idea reduces the complexity of the physical interactions and allows variation of the input data to show how changes in one or more variables affect the result. Thus the student gets a visual representation of the numerical solution and a better understanding of the relationships between the relevant variables. Since the data set is small and a single execution of a program usually takes less than two minutes, programs can be run in the middle of a lecture without disrupting the flow of the overall presentation.

The programs are written in MacPascal and have been converted to a stand-alone application using the *PSHELL* software in MacPascal. To execute the applications on the Macintosh, the other software needed is Macintosh system software and *PSHELL*. When a program is opened, either the drawing commences (as for the wave reflections) or a prompt is displayed in the "Text" window on the screen for the input of some data. After all data are given, the program executes and presents the graphical solution in the "Drawing" window of MacPascal, usually with the input data displayed.

The software is also available to students for individual use. Because the programs run quickly, they do not intimidate the student. The short time commitment encourages reruns with varied input data. An additional benefit is that some students reinforce their Pascal experience by delving into the program code.

In our presentation, we demonstrate some examples as stand-alone applications. Two programs illustrate the reflection of a wave pulse at a fixed and free end, one program shows the difference between static and kinetic friction, one shows the locations of two vehicles moving with constant accelerations, and one shows the trajectories involved in the classic projectile problem of a hunter shooting at a monkey. We are currently trying to simplify user interaction so that students will be able to concentrate on the relationships without being distracted or inhibited by data entry.

The software programs *Fixed and Free Reflection, Friction, Car & Truck*, and *Monkey & Hunter* are part of the collection *Computers in Physics Instruction: Software*, which can be ordered by using the form at the end of this book.

Teaching High School Students to Write Physics Simulations

Joel Goodman
Cedar Ridge High School, Old Bridge, NJ 08857

For four years now, the science department of Cedar Ridge High School has offered a course that uses computers to help teach science. Prospective students are required to have a working knowledge of BASIC and the approval of their current science teachers.

The written materials for this course consist mainly of lecture notes, though we will be developing a text over the next year. So that students may complete work on their home computers, we have acquired Vic-20, C64, Apple II, and IBM-compatible computers. This course is quite different from the computer science courses offered by the math or business departments. By including introductory-level topics such as statistics, CAD and three-dimensional wire-frame drawings, and digital electronics, the course exposes students to a broad range of computer fields related to science that they may wish to pursue after high school.

As part of the curriculum, students write physics simulations. Writing simulations has much educational value. To write successful simulations, students must learn the fundamental physics, use a certain amount of reasoning to write and debug programs, and study the results for accuracy. In the process, students develop an appreciation of how simple simulations can be used to obtain numerical answers to problems beyond their physics or mathematical ability.

The unit on simulations begins with a review of random-number generation and simple graphics. The first program is a simulation of Brownian motion. This is followed by a simple particle-in-a-box model for a gas. Students are asked to change parameters in these simulations, observe the results, and discuss deviations from the real systems.

Next, students develop a projectile-motion simulation, using a set of formulas to find the height and range as a function of launch velocity, launch angle, and time. Students must then improve the simulation so that the projectile is shown moving at a scale set by the user. The simulation must run in real time (within hardware limitations) and must include an option to draw scale targets (walls, inclines, etc). Since about half of the students taking the course have not yet had physics, they cannot solve problems like determining the angle to launch a projectile at a given velocity to achieve a maximum range. Still more difficult, is answering the same question applied to landing the projectile on a plane inclined as specified by the user. Using simulations, students can easily find numerical approximations to both of these problems.

Students who have already had physics experience often want to attempt something more difficult. Sometimes they try to simulate the action of a rigid pendulum. This involves torque and motion in angular terms. This simulation can then be used to determine how long it takes for a rigid pendulum to fall through a given angle. Although this problem is difficult to solve with only an introductory background in calculus, students can apply numerical integration techniques learned earlier in the course, to obtain results that closely agree with the simulation.

Depending on the available time, students can go on to develop simulations on topics such as electron orbitals, sound waves, and light waves.

The Evolution of Computer Simulations in Physics Teaching

Harvey Weir

Department of Physics, Memorial University, St. John's, Newfoundland, Canada

There are many physical phenomena that are too dangerous, too expensive, or simply impossible to demonstrate with real equipment in the physics classroom. The potential of the computer to simulate such phenomena has been recognized since its invention. This paper traces the evolution and benefits of computer simulations in physics teaching both in Great Britain and North America.

Some fear that simulations will be used in lieu of real experiments and demonstrations, and that physics education will suffer. I argue that the marriage of computer and interactive videodisc technologies can enhance many traditional simulations, create opportunities for whole new classes of simulations, and help students bridge the gaps between computer simulations and real phenomena.

Simulation Laboratory for General Education in Natural Science

Martin H. Krieger

School of Planning, University of Southern California, Los Angeles CA 90089-0042

General education courses in the humanities require students to read the great works of literature and politics, and do not mediate this initial reading with secondary sources. It is possible for general education in courses in natural science to use the same approach. Such courses might set up situations in which students confront reasonably complex and untamed phenomena, and then try to figure out not only what is going on, but what might be of interest at all.

On these pedogogical principles, we have developed a simulation laboratory for teaching the general education science course for honors students at the University of Southern California.[1] The laboratory covers two-dimensional percolation, an Ising model, random walk, differential equation in the plane, cellular automata, and a Polya ball-and-urn process. The software is reasonably friendly and allows students to conduct a variety of parametric changes so that they can experiment in each of these worlds and look for what might be called "interesting phenomena."

Our simulation laboratory gives students an Ising model or percolation and tells them little about the expected phenomena. If the students play with the model—and that willingness to play is what needs the most encouragement—they will begin to get a feeling for how hard it is to find reproducible effects, and for the relationship of the simple to the complex and of randomness to order—crucial themes in modern physical science.

The experiences students encounter are archetypal of modern physical science. The linear differential equation is the archetype of Newtonian science. Brownian motion is the archetype of random processes. The Polya ball-and-urn (selecting balls with replacement determined by the color of the ball picked) models nonergodic processes, that is, processes that are sensitive to initial conditions and that settle into a final state. The cellular automaton shows how complex phenomena may be a product of simple rules iteratively applied. Percolation shows one way a seemingly smooth process can have discontinuous consequences, and how small changes in a parameter (say, infectiousness) can change the behavior of a system radically (from no spread of disease to epidemic, for example). Pruning a percolation lattice or tree is a good way of introducing algorithms and programs. And an Ising model is a story of phase transitions and of the coherence available in a random coupled thermodynamic system.

The models typically are about $N = 20$ on a side, if they are grids (Ising, percolation), or hundreds of steps if they are sequential random processes (random walk, Polya process). In the long run, I hope to model several of the processes in three dimensions, so that dimensionality will become a parameter of interest.

The simulations are actual event-by-event, not analytically smoothed, approximations. Each step is separately generated and accumulated; their evolution is displayed in a real-time scale. What students see on the screen is the whole story. The simulation on the screen is all there can be.

The computer environment has other benefits. Students who feel alien to apparatus and equipment seem to be more comfortable in the clean world of the PC. We do give up the rite of passage represented by "wet" or mechanical laboratories, but this sacrifice allows us to give students the difficult experience of finding something interesting in this tame but varied laboratory. Students also learn that a computer is not just a number cruncher or a big file manager. As Zabusky suggests, there are synergistic effects from experimenting on the machine—surprising effects you would not see in the analytic mode (including fluctuations and idiosyncrasies).[2]

The laboratory works reasonably well. Students eventually understand that there is no right answer; their problem is to find something interesting. I should note that at my university some of the best undergraduates are in cinema and journalism, and their universe is rapidly being populated by graphics-intense computer environments.

The software *Simulation Lab* sets up the variety of experimental worlds in which students can vary parameters, and for random processes, do many nonidentical runs. Though the laboratory is designed for freshman undergraduates, it might well be useful for seniors majoring in physics. In any case, the laboratory is quite friendly, and the graphics are straightforward and helpful.

The software is completely self-contained with the six models described above and includes built-in help, including explanations of the models. At the prompt, one types `simlab`, a menu appears, and the rest is self-explanatory. A PC XT with 512K and either a text monitor or a color graphics monitor is all that is needed.

Acknowledgments: Richard Martin did all the programming and is responsible for the friendliness and look of the laboratory. The research was supported by Exxon Education Foundation and by Project Socrates, a grant from IBM to the University of Southern California, as well as the Thematic Option Honors Program of the University. The course syllabus and a more detailed paper are available from the author.

The software program *Simulation Lab* is part of the collection *Computers in Physics Instruction: Software*, which can be ordered by using the form at the end of this book.

1. W. B. Arthur, "On Historical Small Events and Increasing Returns Scale," mimeo, Stanford University Food Research Institute, 1986; D. Bell, *The Reforming of General Education* (New York: Columbia University Press, 1966); M. H. Krieger, "The Physicist's Toolkit," Am. J. Phys. **54**, 1033 (1987); M. H. Krieger, "The Elementary Structures of Particles," Soc. Stud. Sci. **17**, 749 (1987).
2. N. Zabusky, "Computational Synergetics and Mathematical Innovation," J. Comp. Phys. **43**, 195 (1981).

Simulation of a Rigid Pendulum

Joel Goodman
Cedar Ridge High School, Old Bridge, NJ 08857

To help high school and college physics students understand the variables affecting the motion of a rigid pendulum, I have developed a microcomputer program that simulates the action of the pendulum graphically. In this graphic simulation, the student has control over a number of parameters including gravitational acceleration, the length of the pendulum, and the distribution of mass. The student can measure parameters that are difficult to measure in an actual experiment, such as the interval of time between any two angles, or the tangential velocity. The student can also change factors that cannot be controlled in the laboratory, such as the gravitational acceleration. The student can opt to continually display digital values of time, angle, and other data, and can step the motion of the pendulum manually at single intervals of time or make it move in real time. Operating instructions for the simulation are available at all times through a help key.

To start the simulation, the student specifies various parameters involved in a rigid pendulum in a data-entry screen. These result in a graphic depiction of a pendulum. The computer allows students to make changes quickly, to use values that have a much greater range than in real life. Accurate results are easy to obtain and are not influenced by unwanted extraneous factors or poor lab technique. No equipment (other than the computer) is required.

This graphic simulation offers numerous advantages over a laboratory setup, and it can be used in a number of educational situations. The way it is used is entirely up to the teacher or the student. When the teacher uses the simulation to demonstrate the behavior of a rigid pendulum to a large class, the program allows for quick changes to be made in response to student suggestions. When the simulation is used for smaller lab groups, the groups can be asked to explore what factors affect the period of the pendulum. This could be a qualitative or quantitative experiment. Individual students can use the computer to check results obtained from a lab, performed with a real pendulum. Students can also compare the action of the rigid pendulum to that of a pendulum in simple harmonic motion. The simulation can also be used to provide numerical solutions to various problems, which can range from finding the kinetic energy at a given height to determining the time required for a long pole to fall to a horizontal position.

Like many simulations, the rigid-pendulum simulation is not always an ideal pedagogical choice. Since no equipment is required, it diminishes the importance of laboratory technique, a skill that students frequently need to develop. The simulation also tends to diminish the opportunity for discovery or inventiveness, because it provides all the simulated equipment, along with a list of parameters.

Future additions to the program will include options for damping the motion, applying a periodic torque, using different mass distributions, and graphing various results versus time or angle.

The software program *Rigid Pendulum Simulation* is part of the collection *Computers in Physics Instruction: Software*, which can be ordered by using the form at the end of this book.

Graphs and Tracks

David Trowbridge
Center for Design of Educational Computing, Carnegie Mellon University, Pittsburgh, PA 15213

Researchers in physics and mathematics education have documented several common difficulties that students have with drawing graphs of motion.[1] These difficulties include not representing continuous motion by a continuous line, not distinguishing the shape of a graph from the path of the motion, and not representing the continuity of velocity (i.e., kinks in a position-vs.-time plot).

Graphs and Tracks is a two-part educational software package dealing with graphs of motion. In the first part, students are shown graphs of position, velocity, and acceleration vs. time. They then try to create arrangements of sloping tracks on which a rolling ball executes the motions represented in the given graphs. In the second part, students watch demonstrations of animated motions and then sketch graphs on the screen that correspond to those motions.

Part 1, "From Graphs to Motion," displays graphs of position, velocity, and acceleration for a number of examples of rectilinear motion. The student is presented with an adjustable set of sloping tracks, and with scales of initial position and initial velocity. The student is asked to set up an arrangement with suitable initial conditions so that when a ball rolls on those tracks, the graphs it generates match the given graphs. Students use a mouse to adjust the incline of the ramps and to set the initial position and initial velocity of the ball. They may use either the collection of built-in examples, or can define their own examples. A "help" facility provides specific suggestions for correcting errors, and works equally well for any problem that can be entered using the arrangement of the tracks.

Part 2, "From Motion to Graphs," poses the complementary task: given a motion and a verbal description of that motion (generated automatically by the computer), sketch the corresponding graphs of position, velocity, and acceleration vs. time. The graphs need not be precise to be judged as correct, but they must represent the most salient features of the motion correctly. For instance, an object traveling in the positive x-direction and speeding up should be represented by an upward curving x-vs.-t graph. The program looks for common kinds of errors and gives appropriate feedback. The facility for giving feedback is completely general; specific help can be given that refers to particular segments of the student's graphs, and this help can be automatically generated for any problem that can be created using the *Graphs and Tracks* simulation.

Students may ask for feedback on their graphs at any time. The program is able to evaluate both the overall qualitative correctness of the graphs, and the detailed attributes such as differentiability of x-vs.-t, continuity, and correspondence of segments among different graph types. The evaluation facility of "From Motion to Graphs" was built using the specific research results on student difficulties with graphs. For example, the program anticipates and gives feedback on student graphs that do not represent continuous motion by a continuous line, that do not distinguish the shape of a graph from the path of the motion, or that do not represent the continuity of velocity. These errors have been documented as common errors of students drawing graphs of motion.

Working in pairs, participants in my workshop will examine the main features of the *Graphs and Tracks,* including interactive documentation, context-specific help, and feedback keyed to predictable student difficulties, and will work through a few problem examples.

Graphs and Tracks is an example of research-based software. The program was produced by incorporating results of research conducted by the physics education group at the University of Washington. It was implemented using the cT programming environment developed by the Center for Design of Educational Computing at Carnegie Mellon University. *Graphs and Tracks* illustrates a way in which specific student difficulties, uncovered by research, can be targeted for remediation using a highly interactive, graphically oriented computer program.

1. L. C. McDermott, M. Rosenquist, and E. H. van Zee, "Student Difficulties in Connecting Graphs and Physics: Examples from Kinematics," Am. J. Phys. **55**, 6 (1987).

Using Computer Simulations of Orbital Motion

Charles E. Peirce
Physics and Chemistry Department, Carson City–Crystal High School, Carson City, MI 48811

Computer simulation of orbital motion can allow interactive observation of orbiting objects. Using a heliocentric setting and a period short enough to maintain interest, the simulation facilitates the display of force vectors at any time during the orbit. After the observation, students can test the simulation of the orbiting objects on hard copy to reinforce problem-solving techniques without being restricted to expressing their knowledge of orbits to formulas or a set of numerical data.

I have developed an Applesoft BASIC program, *Orbital Motion*, that allows the user to interact with the program and test the program's simulations by making actual measurements off the hard copy and substituting those measurements into proportions made from relevant variables.[1]

The student uses Kepler's third law to predict the correct Vy (vertical velocity) required to produce an "orbit" when a new "radius" is selected. Kepler's third law demonstrates that the force of attraction (F) between two objects is inversely proportional to the square of the distance (R) between those objects. It also demonstrates that the cube of the average radius of orbit is proportional to the satellite's period squared, and that the satellite's orbital velocity is inversely proportional to the square of its average radius.

Graphic printouts enable the student to perform the analytical evaluation of the above objectives or predict new values of those same variables for a new experiment.

The software program *Orbital Motion* is part of the collection *Computers in Physics Instruction: Software*, which can be ordered by using the form at the end of this book.

1. Harold J. Bailey and J. Edward Kerlin, *Apple Graphics Activities Handbook* (Bowie, MD: Robert J. Brady Co., 1984); David Halliday and Robert Resnick, *Physics*, parts 1 and 2 (New York: Wiley, 1966); Franklin Miller, Jr., Thomas J. Dillon, and Malcolm K. Smith, *Concepts in Physics* (New York: Harcourt Brace, 1980); Lon Poole, Martin McNiff, and Steven Cook, *Apple II User's Guide for Apple II Plus and Apple IIe* (Berkeley: Osborne/McGraw-Hill, 1985).

Simulation Programs for Elementary Physics Courses

Harold Salwen
Physics Department, Stevens Institute of Technology, Hoboken, NJ 07030

Since September 1983 every entering student at the Stevens Institute of Technology has owned a microcomputer. This was once a Digital PRO/350, but is now an AT&T 6310 (IBM AT–compatible).

I distribute supplementary material for my introductory physics courses through the Stevens computer network. Programs are placed on a file server and students bring their diskettes to the computer center to copy the files, using one of many networked microcomputers. We are now extending the network to the dormitory rooms so that students will be able to access the server from their own machines.

My programs give students a chance to see graphic representations of physical phenomena that they would not be able to see without a computer. In general, the student enters a number of parameters for the phenomena and the computer carries out calculations and presents results for the given parameters. To overcome the time constraints of most plotting routines, which take a long time to draw a large number of curves, I have written several of my programs with "fast graphics" (direct access to the video screen) to speed up the drawing. At the present time, only PRO/350 versions of these fast-graphics programs are available, but most or all of them will have been translated to IBM AT–compatible programs by the time of the conference.

I will demonstrate five programs.

Orbit is a simple BASIC program that integrates the orbit equation for a general attractive power-law potential, $V = Cr^v$, and plots the orbit. The user may input the power, v, and the (scaled) maximum or minimum radius as well as the step size for the integration (an issue of waiting time vs. accuracy) and the number of loops. This program gives students a chance to see that most orbits are not closed, despite the fact that Kepler and harmonic-oscillator orbits are.

Efield draws lines of force in a plane for a user-specified set of point charges in that plane.

Wave uses fast graphics to produce moving curves representing the superposition of two sinusoidal waves moving along the same line. The apparent movement is produced by rapidly erasing the curves and redrawing them in their changed form and position. (On the PRO/350, the curves are erased and redrawn about six times a second.) The user has control of the wavelengths, amplitudes, and speeds of the two component waves. Five examples with already-determined parameters are also available.

Fourier uses fast graphics to illustrate the convergence of Fourier series and show the forms of the sums. A large number of terms may be added and all intermediate sums are plotted and then replaced. In addition to preset examples demonstrating some well-known Fourier series, input may be taken from a user-prepared file listing the parameters for the terms of the series. I also find this useful to demonstrate results in some of my more advanced courses.

Damp is a BASIC program that plots the solution for damped harmonic motion for given frequency, damping constant, and initial conditions. The user may also see phase-space plots (*x*-vs.-*p*).

I have found using these programs to be an enjoyable and intuition-building experience for me. I hope it will be for you as well.

Animated Waves and Particles

Eric T. Lane
Physics Department, University of Tennessee at Chattanooga, Chattanooga, TN 37403

The dynamic concepts of waves and pulses, the kinetic theory of particles, and the behavior of electron waves cannot easily be demonstrated in classroom demonstrations or laboratory experiments. The guided simulation *Animated Waves and Particles* is a convenient way for students to learn such concepts. The software provides an interactive environment with immediate feedback, which allows students to learn more effectively. It may be used at any educational level—from grade school through introductory university courses. It is effective in several different educational modes: as a dynamic demonstration in the classroom, as a complement to laboratory experiments, or as an individual tutorial.

Animated Waves and Particles is divided into three parts. The first part, "Waves and Pulses," covers standing waves, pulses, group velocity, Doppler effects, and electromagnetic waves. The treatment of standing waves demonstrates the concepts of sine and cosine waves, wavelength, amplitude, traveling waves, and a standing wave composed of the superposition of two equal amplitude waves traveling in opposite directions. The treatment of pulses illustrates the addition and cancellation of pulses, including the effects of reflection of pulses at fixed and free ends of a string or spring. Group velocity is shown by the motion of waves within a group, the motion of the group alone, and two real-world applications: deep-water waves and capillary waves. The treatment of Doppler effects for sound shows the apparent shortening of wavelength as a source of sound moves toward us, compared to the apparent increase in frequency as we move toward a source. Also shown is the effect of high-speed motion. The treatment of electromagnetic waves shows the same relativistic Doppler effects of shorter wavelength, increased frequency, and increased amplitude regardless of how the source and observed move toward each other.

"Waves and Pulses" provides an attractive, interactive, random display of waves and pulses. This display motivates the student to experiment with waves of different wavelength, speed, and amplitude. The student controls the number of waves on the screen, their velocity, their amplitude, and their appearance on the screen. The student can explore the possibilities of any combination of two moving waves.

The second part of *Animated Waves and Particles*, "Particle Motion," covers kinetic theory, Brownian motion, gravitational effects, Maxwell's Demon, electric conduction, and magnetic effects. The treatment of kinetic theory shows the effects of particle speed on temperature and pressure, including sound effects proportional to the pressure of a simulated ideal gas. The treatment of Brownian motion demonstrates the effect of density on the mean free path of diffusing particles and also the effect of diffusion of particles in a closed room. The treatment of gravitational effects on a collection of free particles shows the exponential density distribution as well as the escape of high-speed particles from low gravity. Maxwell's Demon is shown as the equilibrium of particles diffusing through a hole in a wall. There is also an elementary demonstration of the entropy concept. The treatment of electric conduction, demonstrates the drift velocity of charged particles in an electric field when collisions occur. The treatment of magnetic effects shows the charges moving in circles in a magnetic field. Also shown is the Hall effect, when an electric field is imposed and collisions are allowed.

The third part of *Animated Waves and Particles*, "Electron Waves," tells an illustrated story of the development of the concept of the wave nature of the electron, its constructive interference in a stable orbit, and its destructive interference as it makes transitions from one energy level to another when light is absorbed or emitted. The treatment of constructive interference shows how the wavelike nature of the electron requires an integral number of wavelengths around the atom. The treatment of destructive interference demonstrates what happens when we try to use a nonintegral number of wavelengths for the electron around the atom. Also shown is the transition as the number of waves changes as the electron goes from one energy level to another, absorbing or emitting light of the proper energy.

Animated Waves and Particles gives the instructor a selection of animated demonstrations to use in class, in the laboratory, or for tutorials. The dynamic nature of the demonstrations provides motivation for students to view the animations and experiment with the interactive wave displays and the electron-wave interference displays. The wide variety of the animations offers the student a rich experience in areas that are otherwise difficult to understand. No computer skills are required, except that the user turn on the microcomputer, insert the disk properly, and use a keyboard to prompt a menu of choices.

The software program *Animated Waves and Particles* is part of the collection *Computers in Physics Instruction: Software*, which can be ordered by using the form at the end of this book.

Animation on the IBM PC

Eric T. Lane and Daniel B. Lyke
Physics Department, University of Tennessee at Chattanooga, Chattanooga, TN 37403

The IBM PC program *Animated Waves and Particles* allows students to see the dynamic simulation of waves and pulses, particle motion in kinetic theory and electric conduction, and electron waves around an atom. The interactive animations allow the student to try out the behavior of the waves, particles, and electron waves. The animations are coded in the C language using an optimizing compiler for speed. The menu and text displays are all written in a version of BASIC to allow teachers and others to write their own versions of the programs using the animations.

A Potpourri of Color Graphics Enhancements of Classroom Presentations

Clifton Bob Clark
Department of Physics and Astronomy, University of North Carolina at Greensboro, Greensboro, NC 27412-5001

To show students how to draw principal rays for thin lenses and spherical mirrors in the absence of spherical aberration, I have written two software programs in BASIC for a Heath/Zenith Z-100 microcomputer.[1] Each program is available in a version for students to use in a tutorial outside of class.

For each of the programs, the student gives focal length, object distance, and height, and the program calculates the image position and height. For a lens, the scale drawing shows the object, lens (line), principal axis, and focal points. Each of the three principal incident and transmitted rays is drawn and erased four times, and then drawn a fifth time. The instructor controls the interval between drawing the rays, so he may describe as much or as little as he wishes. Each of the rays is in a different color (red, green, blue). Any extensions or virtual paths are shown in yellow. The program then draws the image.

For a mirror, the object, principal axis, focal point, center of curvature, and the bending line are shown.[2] There are four rays in red, green, blue, and magenta, with extensions in yellow.

The programs accommodate any single lens or mirror with real or virtual, and upright or inverted object. The student works out multiple-lens or mirror problems one at a time.

Another program for the Heath/Zenith Z-100 is a sequence of drawings using the commercial software package *Palette*.[3] Successive drawings show the production of a Lissajous figure for equal amplitudes and frequencies of the two rectangular component oscillators, with phase difference described as "x leads y by 45°."

I have also written a number of programs that will run on any IBM-compatible microcomputer. Teachers may use existing commercial spreadsheet software (*Lotus 1-2-3* and *SuperCalc4*) for classroom presentations concerning oscillators, beginning with Feynman's simultaneous introduction of simple harmonic motion,[4] and numerical integration by the "half-step" method. The use of the spreadsheet eliminates the tedium of the calculations, makes the technique clear, and allows rapid and relatively painless production of graphs that can be altered to show effects of changing parameters. My software presents examples of damped oscillators (overdamped, critically damped, and underdamped), and the driven-damped oscillator (resonance phenomena and phase-angle relations).

The notation is consistent with

$$a + 2\gamma v + \omega_0^2 x = f_0 \cos (\omega t),$$

for the equation of motion, with A used for displacement amplitude, subscript m denoting maximum, and P for the phase difference angle.

The undriven ($f_0 = 0$) overdamped oscillator ($\omega_0^2 < \gamma^2$) and the critically damped oscillator ($\omega_0^2 = \gamma^2$) spreadsheets allow the user to see the graphs for initial conditions, where the oscillating mass is released from rest and a positive displacement and asymptotically approaches the origin. In textbooks, these are usually the only graphs shown. The spreadsheets also show graphs for other initial conditions in which the mass crosses at most once through the origin.

The underdamped oscillator ($\omega_0^2 > \gamma^2$) spreadsheets allow the usual plots of displacement vs. time, with damping envelopes $-A \exp(-\gamma t)$ and $A \exp(-\gamma t)$. Here one can easily point out the difference in the points of tangency to the envelopes and the relative minimum and maximum points. The spreadsheets for the driven-damped oscillator present relations between the ratio of the steady-state displacement amplitude to the maximum amplitude and the driving frequency. Specifically, the relation

$$(A / A_m)^2 = \{[(\omega_0 / \gamma) - (\omega / \gamma)]^2 + 1\}^{-1}$$

yields the familiar bell-shaped resonance curve for discussions of half-width, Q of an oscillator, and the resonance, damped, and undamped frequencies. The abscissa is (ω / γ), and the parameter (ω_0 / γ) is easily changed.

Also available is a spreadsheet for the relation between the phase difference between the driver and the displacement, as given by

$$P = \arctan \{2(\omega / \omega_0) / [(\omega_0 / \gamma)\{1 - (\omega / \omega_0)^2\}]\}$$

where the abscissa is (ω / ω_0) and the parameter (ω_0 / γ) is easily varied.

Spreadsheets by C. B. Clark for *Lotus 1-2-3* are part of the collection *Computers in Physics Instruction: Software*, which can be ordered by using the form at the end of this book.

1. Clifton Bob Clark, "A Color Computer Program for Thin Lens Ray Diagrams," AAPT Announcer **15**, 51 (May 1985); "A Color Computer Program for Spherical Mirror Ray Diagrams," AAPT Announcer **16**, 62 (May 1986).
2. Clifton Bob Clark, "Ray Diagrams for Spherical Mirrors in Introductory Texts," AAPT Announcer **14**, 94 (May 1984).
3. *Palette* (St. Charles, MO: Software Wizardry, Inc.).
4. Richard Feynman, Robert B. Leighton, and Mathew Sands, *The Feynman Lectures* (Reading, MA: Addison-Wesley, 1963).

Software Packages for High School Optics

David Singer and Uri Ganiel
Department of Science, Weizmann Institute of Science, Rehovot 76100, Israel

We will demonstrate three software packages designed to enhance the study of geometrical optics in high school. The packages include all necessary instructions, so they can be used by individual students without any additional guidance. The packages are available for Apple II (+, e, c) and for IBM PC microcomputers. The Apple versions can run with either monochrome or color monitors. The IBM version is for a color monitor (CGA card necessary).

Hide-and-Seek in Mirrors

One of the basic principles in geometrical optics is the law of reflection. The law itself is simple, and by using it, the student can understand the formation of images in plane mirrors.

In a plane mirror, the image is located behind the mirror at a location that is symmetric with the source with regard to the plane of the mirror. Many studies show that students have difficulty understanding the concept of an image's location behind the mirror; they think that the image is located on the plane of the mirror. Our package enables the student to broaden and consolidate his/her understanding.

The program includes two parts: a tutorial and a game. In the tutorial, a cross-section of a plane mirror appears on the screen. The location and size of the mirror change randomly. Two random points appear in front of the mirror. The student has to decide whether a person located in one of the points can see the image of the other point in the mirror. The student develops a strategy for making this decision by using the graphic representations that appear on the screen.

In the game, the computer "hides" at a random location in front of a plane mirror. The student has to find this location in an optimally small number of steps. In each step the student locates himself in one of 220 possible points in front of the mirror. The computer tells the student whether it "sees" the student's image in the mirror.

After each step the student receives some feedback. Using this feedback, the student can use the law of reflection and the various symmetry properties that follow to develop an intelligent strategy to proceed successfully. Important advantages of the computer are the instant feedback and the possibility of playing the game repeatedly, where each time a new situation—and therefore a new challenge—is presented. The game is designed to improve students' motivation, interest, and understanding. We are presently evaluating its success.

Fermat's Principle

The laws of reflection and refraction can be derived from Fermat's principle, that the time light takes to travel between two points has a stationary value, which usually is simply a minimum. Proving this principle for the case of reflection is simple; but for the case of refraction, proving the principle is too mathematically complex for high school students and is usually avoided. The principle is so important, however, that it ought to be introduced.

In this module the student is engaged in a simulated experiment that leads him/her to Fermat's principle. The program starts with an everyday problem (reach a drowning person as quickly as possible). To solve the problem the student is guided through a simulated sequence during which he/she tries to "enter the water" in different locations on a beachline. Through trial and error he/she finds the answer and then the law that describes the requested path.

In the second part of the simulation the student learns about reversibility: the path that takes the shortest time from point A to point B is also the shortest path from point B to point A. Having established the rule for determining the paths of minimum travel time, the law of refraction (Snell's law) can be introduced in a more meaningful way.

This package illustrates some important advantages of the use of microcomputers in science teaching. It demonstrates phenomena that cannot be shown in an actual laboratory experiment, and it solves problems that are conceptually simple but need a level of mathematical ability many students do not have.

Thin Lenses

This tutorial emphasizes elements that are sometimes neglected in lecture-textbook presentations. The program enables the student to examine any type of thin lens, choose its parameters, and study any variety of image-formation situations. Thus, the student controls the size of the lens, its focal length, and the distance between it and any object. Performing such a variety of experiments in a real laboratory would be practically impossible.

Another feature of the package is that it deals with beams of light rather than rays. Although light sources actually emit beams of light, for reasons of convenience most textbooks deal with rays, and usually only a few special rays. This package allows the student to engage in situations that are closer to reality, and consequently to avoid serious misconceptions. There are six parts to the package: (1) classification of thin lenses, in which the student learns how to distinguish between converging and diverging lenses; (2) focal lengths of different lenses, in which the student learns about the focal length and its dependence on the shape of the lens; (3) the image of a point source in a converging lens, in which the student observes the behavior of different beams of light emerging from point sources located on the axis of a converging lens and passing through the lens; (4) the image of a point source in a diverging lens, which is similar to the previous part, but deals with diverging lenses; (5) the image of an extended object in a converging lens, in which the student observes beams of light emerging from different points on an object that is perpendicular to the axis of a converging lens, and by choosing different points on the object observes how the image (real or virtual) of the object is constructed; (6) the image of an extended object in a diverging lens, which is similar to the previous part, but deals with diverging lenses.

Although this package deals with standard material that can be found in any textbook on geometrical optics, the approach is different. The package specifically treats subjects that are traditionally difficult for students (the ray-beam confusion, the effect of lens size, the use of "special" rays to construct images, etc.). For activities like simulating the motion of the source approaching a lens and observing how the image moves as a consequence, the package makes effective use of the dynamics of computer simulations. Such simulations help students internalize the topic much more clearly than do mathematical representations by formulae.

Interactive Optics Software for the General Student

Michael J. Ruiz
Department of Physics, University of North Carolina–Asheville, Asheville, NC 28804

I have written programs for the IBM PC to illustrate principles of optics for the general student enrolled in Physics 101: Light and Visual Phenomena, a conceptual optics course taught at the University of North Carolina at Asheville. The course is offered each spring and has an approximate enrollment of 50 students. The software programs were developed to help students master material in their excellent text, *Seeing the Light: Optics in Nature, Photography, and Holography.*[1]

Students use the software in a departmental computer facility, partially funded by National Science Foundation College Scientific Instrumentation Program Grant

No. CSI-8750 195. The departmental computers are used by majors for data analysis and by general students for programs like those described in this abstract. Charles A. Bennett is project director of the grant ($17,400, 50 percent UNCA matching), through which four XTs, one AT, and two laser printers were acquired.

The above textbook gives the background physics. The pedagogical approach of the program is similar to that of the textbook: diagrams and illustrations teach basic principles of light.

In many cases the software is an interactive tutorial where students can see how optics problems can be solved with graphical schemes. The student can often pick the initial arrangement. This enables them to study and review material discussed in their text and in class. They are then given homework assignments in which they can apply the learned principles on their own, e.g., ray tracing to locate images formed by mirrors and lenses.

The individual software programs are accessed through a main menu. Programs include *Rainbow, Kaleidoscope, Ray Tracing with Spherical Mirrors, Ray Tracing with Lenses, Camera Lenses, Optical Illusions,* and *Practice Exam.*

Rainbow constructs a large raindrop and shows what happens when a white beam of light enters. Ray diagrams are given in color for primary and secondary rainbows.

Kaleidoscope illustrates the formation of virtual images by two plane mirrors with one common edge. The student can choose from a variety of vertex angles.

Ray Tracing with Spherical Mirrors allows the student to investigate ray tracing for concave and convex mirrors. The student chooses object distances and the program gives three key rays leaving the tip of the object and locating the tip of the image. Image characteristics (real, virtual, smaller, larger, upright, inverted) are noted. One mode sweeps the object in from far away toward the mirror and illustrates the shifting image position as the object nears the mirror.

Ray Tracing with Lenses offers the student an opportunity to visualize ray tracing for converging and diverging lenses. The student chooses an object distance and uses the function keys to see three key rays leave the tip of the object and locate the tip of the image. Another function key produces a window with image properties: type (real or virtual), orientation, and size. The student watches as an object gradually approaches a lens; the image location and size changes as a function of the position of the object.

Camera Lenses allows the student to choose the focal length of a camera, and then shows the resulting different angles of view for a given two-dimensional scene. It also illustrates the use of compound lenses in cameras by covering the principles behind the close-up lens and teleconverter.

Optical Illusions illustrates several traditional optical illusions associated with ambiguous depth clues. Also included are computer graphics to illustrate effects of lateral inhibition, eye movements, and color afterimages.

Practice Exam is a multiple-choice exam on the material for the entire course.

Each program has its own instructions and/or help screens to assist the user. The programs were written for students with no prior background in computers and are therefore very user friendly.

The software was developed with Borland's *Turbo BASIC* on a compatible IBM XT running with MS-DOS 3.1. Although the first version of the programs requires EGA and a math coprocessor, a future version is planned that will run on EGA, CGA, or HGA, with or without math coprocessors. However, the loss of color with HGA will limit the use of some of the programs.

1. D. S. Falk, D. R. Brill, and D. G. Stork, *Seeing the Light* (New York: Harper and Row, 1986).

The Computer as a Teaching Aid in Physics: A Case Study in Optics and Wave Theory

Frank Bason
Silkeborg Amtsgymnasium, Oslovej 10, DK–8600, Silkeborg, Denmark

During the past decade we have been privileged to witness the evolution of remarkably versatile computers. The challenge to us as physics teachers has been to select from among the large number of available options in order to develop useful learning tools for our students.[1] Incompatibility of equipment and operating systems previously hindered efficient development and exchange of programs, but several clear standards have emerged within the past five years, making the choice of a relevant computer system much easier. One of these standards is the IBM personal computer and the MS-DOS operating system. The programs presented in this paper operate in this environment.

This contribution to the conference is a combined poster and display demonstration. Using the theme of optics and wave theory to illustrate each point, I will address the following questions: (1) What material within the discipline of interest is pedagogically appropriate for presentation on a microcomputer? (2) What is required of the programmer and the program development environment to develop a program package quickly and successfully? (3) What factors affect student response to the program package? (4) Do students really benefit from the use of the microcomputer as a teaching aid?

It is, of course, essential to consider the first of these questions very carefully before embarking upon a major programming effort. After the main features of a project have been established, it can be difficult to make major changes.

The subject of optics and wave theory is particularly suited to illumination by means of a modern microcomputer because it makes good use of the extensive graphics facilities currently available. Computer graphics make possible rapid dynamic interactions between visual information and quantitative algorithms that could scarcely be contemplated just a generation ago. With this in mind I have developed the following program modules:[2]

Huygens illustrates Huygen's principle, showing the propagation of a plane wave, reflection from a specular surface, refraction at an interface, and interference between waves emitted from two sources.

Waves illustrates several phenomena associated with the interference of waves. Standing waves and beat notes can be generated, allowing the student to experiment with various combinations of parameters and to see the consequences on the screen.

Doppler illustrates the Doppler effect by showing the emission of waves from a point source. It is possible to move the source or the observer at a given speed and to observe the consequences. The program can also provide insight into the development of a shock wave when the velocity of a point source exceeds the propagation velocity of waves in a medium.

Fermat illustrates Fermat's principle of least time and displays results to demonstrate the equivalence of Snell's law of refraction.

Prism simulates the optics of a prism on the graphics screen and provides tables of results for direct comparison with the same experiment performed with a prism and a laser in a laboratory. The passage of a ray of light through a prism illustrates several interesting aspects of optics.

Lenses simulates the passage of light through a selection of lenses to supplement the experience gained by direct experimentation in the physics laboratory.

The student can select these demonstration programs from a convenient menu program. They will run on any IBM PC XT or compatible, provided that it has a color-graphics or enhanced-graphics adapter. We developed this demonstration diskette using the UniComal COMAL development system.[3]

This demonstration concludes with a description of teaching situations in which these programs have been used. Student comments and critiques are summarized and presented in tabular form with a view to improving future program packages for use in physics teaching.

1. F. Bason and Leo Højsholt-Poulsen, "Computer Use in Undergraduate Physics Teaching," 3rd World Conference on Computers in Education, WCCE 81, Lausanne, Switzerland, 1981.
2. Several programs have been developed just for this conference. Others have been adapted from earlier work and from descriptions appearing in the literature. Complete references will be available in the final paper.
3. For further information, contact Unicomal A/S, Trarmarksveg 19, DK-2860 Søborg, Denmark (telephone: Int. + 45-167-3511). Ask for their free demonstration disk.

Making Waves: Software for Studying Wave Phenomena

Joseph Snir

School of Education, University of Haifa, Mount Carmel Haifa 31999, Israel; and Educational Technology Center, Harvard School of Education, Cambridge, MA 02138

The phenomenon of waves is central to classical as well as modern physics. Wave theory is the basis for our understanding of sound and light, the duality of waves and particles, and quantum mechanics. But teaching this major topic at all levels of education is a very difficult task. We have developed a computer program, *Making Waves,* for learning and teaching about waves. In developing this software tool, we have paid special attention to students' commonly held conceptions about waves. In our presentation we will mainly discuss waves in elastic media, but we have designed *Making Waves* so that it is also a useful modeling tool for describing electromagnetic waves.

Making Waves is composed of three parts, which can be entered from the main menu: pulses, waves traveling in a one-dimensional array, and waves traveling in a two-dimensional array. In each of the three parts the student can learn about the waves that he creates on the screen by making waves. The student can choose to explore one or two pulses and one or two waves and their sums. He can change any of the parameters that define the waves, collect data, perform measurements of time and length, and change the number of points by which a wave is represented on the screen. Student worksheets that suggest specific activities accompany the software and can be used to structure student interaction with the package.

The program provides three modes of experimentation. In the first, the student observes qualitative features of waves. In the second, the student solves a problem by making use of the software to create a particular situation on the screen that meets the requirements of the problem. In the third, the student performs experiments with real-life materials and uses the computer to reflect and model phenomena that he observes in the laboratory.

The student can select various options of the program from menus without having to leave the created wave phenomena. Thus, the student can manipulate parameters or change options in any of the four windows while observing the effect of these changes in the other windows. The movement between the menus is simple and options can be freely selected in any desired order. This modularity between options and windows allows the student to create and study almost every wave phenomenon. The synchronized, multiple, and modular representation of the waves makes this program unique among other wave programs available today.[1]

From a pilot study in which high school students were interviewed, we have identified several obstacles to understanding waves. One such obstacle is the need to differentiate between the local behavior of individual particles and the integration of the movement of these particles in a global wave pattern. Our program can easily correct this basic misconception.

Making Waves has been specifically designed to help the student see the trees and the forest—to be able to integrate the global and the local in wave phenomena. Among the options that are offered to the student for manipulating the waves on the screen there is one that allows a change in the number of points by which the wave is described. When the number of points is maximum, the wave is seen as a continued front. As the number of points is gradually reduced the movement of each individual point in the now-discrete chain of points moves up and down around a fixed point. When the shape of the wave is made up of only a few points, it starts to fade, but can still be traced. Most students are startled to discover the movement of the individual points that occurs when the number of points is reduced, and are led to a new understanding of the phenomenon. The flexibility of this dynamic simulation of the phenomenon makes feasible the teaching of this aspect of waves in a unique way.

The modularity of the program allows the student to alter the number of points in many other activities. For example, in studying the phenomenon of waves interference, the student can create two waves and watch how their sum changes from constructive to destructive interference when the phase difference between them changes. At the stage when the phase difference reaches 180°, the number of points in the waves can be reduced gradually, thus allowing the student to observe the asymmetric movement of individual points in the two waves. This observation clarifies the destructive nature of the combination.

Making Waves allows the student to understand the concepts and process that underlie waves without using mathematical formalism. Comprehensive research to find optimal ways to use this software is now under way.

1. E. T. Lane, *Group Velocity Demonstration* (Iowa City: Conduit, 1980); E. T. Lane, *Standing Waves* (Iowa City: Conduit, 1984); J. Tesh, J. Kimmitt, *Transverse Waves* (London: Heinemann Computers in Education, 1982); D. Vernier, *Wave Addition* (Portland: Vernier Software, 1984).

Educational Software: Oil Drop, Schrodinger Equation, and Size of a Molecule

John Elberfeld
EduTech, Rochester, NY 14609

I have developed a series of highly interactive problem-solving activities to be run on an Apple II or compatible computer. A single disk drive is required. A printer is optional, but helpful for keeping records. In addition to the software, the packages include a comprehensive manual containing teacher notes, student instructions, and a backup disk.

Oil Drop

This simulation of the Millikan oil-drop experiment provides students with the feel and excitement of performing an important experiment that is often considered too expensive or complicated to implement in the lab. In a computer animation, an oil drop is shown dropping between two plates. The student varies the voltage between plates by turning the dials on game paddles or pressing arrow keys on a keyboard. When the student obtains a voltage that suspends the drop motionless between the two plates, he/she leaves the computer to calculate the charge on the drop. (The mass or radius of the drop is given.) This strategy makes the computer available to other students, which reduces the number of computers needed per lab.

The student must correctly enter the charge of the stopped drop before repeating the experiment. Individual files are kept of each student's calculations and a master file is kept of the entire class, which permits a histogram of the pooled data.

Schrodinger Equation

Schrodinger Equation permits students to calculate and plot numerical solutions to the one-dimensional Schrodinger equation for a variety of standard potentials (harmonic oscillator, square well, and Coulomb) and for any potential that can be drawn with a game paddle.

The student specifies a trial value of the energy and the initial conditions of the wave function. The display of the resulting solution shows graphically whether the solution is convergent. *Schrodinger Equation* also allows investigation of reflection from and transmission through a barrier.

The program permits solutions to be stored on the disk and printed on a printer equipped with a graphics card. The disk includes demonstration solutions. Game paddles are required to draw your own potential.

Size of a Molecule

Developed for use in a benzene-free laboratory, *Size of a Molecule* is a simulation of an experiment to determine the length of an oleic (or stearic) acid molecule from measurements of a monomolecular layer.

Using animated high-resolution graphics, the program takes the student through the experiment in a self-paced, step-by-step fashion, generating reasonable data as needed. From these data, the student performs a step-by-step analysis resulting in a solution of the value of the length of the molecule. The student must work with relationships between volume and density.

This program can also be used to analyze data obtained from an actual laboratory experiment. Thus it is valuable as a prelab, make-up lab, or simulation when the actual experiment cannot be performed. Data can be saved on disk and printed.

Electric Field Lines
Simulation Lab

Johnny B. Holmes
Physics Department, Christian Brothers College, Memphis, TN 38104

Using the graphics of an Apple II–family microcomputer, *Electric Field Lines Simulation Lab* allows students to investigate the electric field lines resulting from various charge distributions. It introduces students to the concept of an electric field line and demonstrates the behavior of the electric fields produced by point charges.

At Christian Brothers College, we use *Electric Field Lines Simulation Lab* as our first lab experiment in the second course of the introductory physics sequence that covers electricity and magnetism. The students in this course are normally first-semester sophomore engineering and science majors.

Although we do supply a handout that specifies five charge distributions for the student to investigate, the pedagogical approach is that of experimentation. It is up to the student to discover the shape of the field lines. We require the student to specify at least two additional charge distributions and investigate their electric fields. (The disk will hold up to ten plots.)

The student enters the charge and the position of up to six point charges on a grid of 280 by 160. When all the charges are specified, the computer asks for a beginning field point (which must be on the grid). After the student specifies the field point, the computer calculates the electric field (magnitude and direction) at that point, and draws a five-unit line segment in the direction of the field. The computer then recalculates the field at the new point (i.e., at the end of the line segment) and draws another five-unit line segment in the direction of the field at the new point. The process continues until it runs off the screen, comes within five units of a negative charge, or is aborted by the student.

After completing one field line, the computer asks for another starting field point or allows the student to end the plot. When the student ends the plot, he can save the plot on disk for recall later and start another plot with another set of charges. When all the plots are reviewed near the end of the lab session, a grade is assigned.

The program is self-booting and menu driven. There are four choices on the menu: run the plotting program, review the plots stored on the disk, cancel all plots on the disk, and end the program. The program returns the user to the menu after each application.

The software runs on any of the Apple II–family microcomputers and uses Applesoft BASIC and the Apple DOS 3.3. It is completely self-contained and needs no software or hardware other than the Apple, a 5.25-inch floppy disk drive, and a monitor. The plots are stored on the disk in a bit-mapped form usable by graphics printing programs (e.g. Beagle Brothers *Triple Dump*) so hard copies of the plots can be printed. The names of the plots (for use with the hard-copy programs) are: DR #1-n where n is the number of the plot (from 0 to 9).

The software program *Electric Fields Lab Diskette* is part of the collection *Computers in Physics Instruction: Software*, which can be ordered by using the form at the end of this book.

United States Energy Simulator

Richard W. Tarara
Department of Chemistry and Physics, Saint Mary's College, Notre Dame, IN 46556

The complexity of our modern energy system was well demonstrated by the Energy-Environment Simulator, an analog computer developed in the early 1970s at Montana State University for the Department of Energy. Although working versions of this computer still exist, there is now a microcomputer software package designed to cover the same subject matter in an updated format that is more widely accessible.

The *United States Energy Simulator* is in the form of a game in which the class (in demonstration mode) or an individual tries to pilot the country into its future by maintaining a balance between the demands and supplies of energy. The pedagogical objective is for students to grasp the complexity and interdependence of the task, the finite nature of the fossil fuels, the effect of various actions on the environment and standard of living, and the importance of early research on new forms of energy.

In demonstration mode, the decisions on how to supply needed energy or what demands to increase or decrease is done by consensus of the class; in the individual mode, all choices fall on the operator. Tutorials provide the student with background information that puts the game into sharper focus.

The software uses a number of screens to present the rate of use of the various resources, remaining reserves, the condition of the environment, and the status of supply and demand in food and both chemical and electrical energy. The user can view the numerical values of all important parameters or change the supplies and/or demands. The goal of the game is to make it as far into the future as possible while maintaining the environment and standard of living. There are various starting conditions based on different assumptions about the available reserves. Unlike the analog computer, the program limits the changes permitted to ± 25 percent of any parameter in any given year. The program also evaluates the effort once the player quits.

The exercise can be valuable at both the secondary and undergraduate levels. All energy parameters are given in kWh and shown in computer-exponential notation (xE09). This might require some explanation when used in precollege classes, but a full understanding of the magnitude and units of the energy quantities is not essential to play the game or to grasp the main concepts.

During play, the user must not let the supply fall below 70 percent of demand. If this happens, the player cannot progress until the imbalance is corrected—at least to the point where supply is 90 percent of demand. If the imbalance occurs because one of the resources is exhausted, it may be impossible to correct the problem. It is therefore important to plan ahead.

Long-term supplies are determined by new or future energy technology, but the timing of the appearance of this technology is tied to early investment in research and development. Since such new technology will most likely provide only electrical energy, the user must plan ahead so that the energy demands of the future will be electrical. Heavy use of fossil fuels adversely affects air quality, and heavy use of nuclear energy presents a waste problem. Overall heavy energy use involves waste-heat problems. A crude accounting of these factors is indicated on the main screen. The condition of the environment, the number of shortage years (supply less than 90 percent of demand), the number of critical shortage years (supply less than 70 percent of demand), and the standard of living as measured by the per-capita energy use are the factors that determine the player's final rating.

Versions of the simulator are available for the Commodore 64/128 (disk drive and color monitor), Apple II (DOS 3.3, monochrome monitor), and MS-DOS (BASIC and monochrome monitor). The C-64 version has a more extensive tutorial section than the others and includes short animations to elaborate certain points. Each version includes instructions for running the program and the play itself is guided by on-screen instructions.

Use of one of the surviving analog computers has been a highlight of our thermal segments in all physics courses for a number of years. Students find the analog-computer exercise both exciting and educational (as indicated by essays on the topic). The microcomputer version now provides a way of experiencing this learning exercise in almost any educational setting.

Simulation Software for Advanced Undergraduates: Electrodynamics and Quantum Mechanics

P. B. Visscher

Department of Physics, University of Alabama, Tuscaloosa, AL 35487-1921

Most existing software for partial differential equation (PDE) courses such as electrodynamics and quantum mechanics relates to time-independent problems: computation of electric field lines for point charge distributions,[1] relaxation-method computation of potentials,[2] or computation of time-independent one-dimensional wave functions in potential wells. The connection with physics and intuition is closer if the fundamental time-dependent PDEs are used directly. I will describe several programs for direct time-dependent simulation using discretizations of the fundamental time-evolution equations (Maxwell's equations or the Schrödinger equation). I argue that such programs make it possible to reverse the usual order of introduction of discrete and continuum methods in mathematical physics.

Most physicists think of discrete equations as approximations to continuum partial differential equations, but I argue that this is a matter of conditioning—fundamentally, the derivative is defined as a limit of discrete differences, so that vector operators (divergence, curl, etc.) are ultimately defined as limits of discrete operators. It has been customary in teaching these to go immediately to the abstract continuum limit, because it is hard to do problems that are nontrivial enough to describe important physical phenomena by hand with discrete equations.

Just as Newton had to invent the integral calculus to determine the phenomenological consequences of his theory of universal gravitation, all physics students have had to use calculus to get an intuitive understanding of the phenomenology implicit in PDE systems like Maxwell's equations. It has always been clear that the abstractness of PDEs introduces a formidable barrier to such understanding, but this has been accepted as the price a student has to pay to see how such diverse phenomena as wave propagation, electromagnetic induction, electrostatic fields, and magnetostatic fields arise from a simple set of underlying Maxwell equations. But the advent of the computer has changed all this. It is now possible to study all phenomena using discrete Maxwell equations with high school algebra as the only mathematical prerequisite. The continuum equations can be returned to their natural roles as the small-cell limits of the discrete equations and studied after the phenomenology is clear.

This approach may seem inelegant to someone who is used to thinking of discrete equations as lacking the beautiful symmetry and conservation properties of the continuum Maxwell equations. I will demonstrate, however, that there is a maximally simple discretization of the Maxwell equations that possesses all of these properties. In particular, energy is exactly conserved, Gauss's law (expressed in terms of a discrete area integral) is exactly true,[3] Ampere's law and Poynting's theorem are exactly true. Although the discrete equations in a cubic lattice do not, of course, have the rotational symmetry required to derive Coulomb's law from Gauss's law, the discrete equations can be written for any coordinate system, and in a spherical coordinate system Coulomb's law is easily derived. Thus it is possible for a student to learn all of these laws and gain an intuitive understanding of electrodynamics by experience with simulations, which are not dependent on calculus. Calculus may be a crystal-clear window to a mathematically sophisticated instructor, but to the average student it is more than a little rain-streaked.

Specific simulations that can easily be done with a two-dimensional Maxwell equations solver include antenna radiation and plane-wave propagation. Static fields may be calculated by evolving the system until it stops changing—one can calculate the Coulomb electrostatic field of a point charge brought from infinity, the magnetostatic field of a solenoid of arbitrary cross-section (whose evolution shows the induced electric field along the way), and the solution of an electrostatic conducting-boundary problem (showing the migration of induced charge).

The software program *Sample Simulations from Fields and Electromagnetism* is part of the collection *Computers in Physics Instruction: Software*, which can be ordered by using the form at the end of this book.

1. Harvey Gould and Jay Tobochnik, *Computer Simulation Methods* (Reading, MA: Addison-Wesley, 1987).
2. Steven E. Koonin, *Computational Physics* (Menlo Park, CA: Benjamin/Cummings, 1986).
3. P. B. Visscher, *Fields and Electrodynamics* (New York: Wiley, 1988).

Microcomputer Simulations in Statistical Physics

Jeffrey R. Fox
Thermophysics Division, National Bureau of Standards, Boulder, CO 80303

Because visualization is a key part of the learning process, one of the most important tasks in science instruction is creating effective visual metaphors for natural phenomena or model behavior not otherwise readily observable. The problem of visualization is particularly acute in statistical physics, where the cooperative dynamical behavior of systems of very many microscopic degrees of freedom is the subject, and the abstractions of a higher mathematics is the present instructional strategy.

In this report I will show a microcomputer-based simulation of one of the most important model systems in statistical mechanics, the Ising, or lattice-gas, model. This model may be interpreted as a system of molecules interacting through very simple pairwise forces that are repulsive at short range and attractive at longer range. It is also widely interpreted as a model of a magnetic solid with a structure composed of microscopic magnetic dipoles arrayed on a fixed lattice and interacting with their nearby neighbors through highly simplified force. Computer enthusiasts will perhaps prefer to think of it as a statistical version of Conway's *Game of Life*, a popular microcomputer simulation of population growth.

The model has a number of interesting behavior modes, but the most important is the splitting into two phases at low temperatures (temperature being one of the control parameters). The phases are interpreted as liquid and vapor in the language of the lattice gas, and as two distinct magnetic domains in the magnetic language.

Whether used as a classroom demonstration or laboratory exercise, this simulation effectively shows the connection between the simple energetics of the model particles and the macroscopic thermodynamics of phase transitions, spinodal decomposition, and critical phenomena. Because it is a Monte Carlo simulation, it is based on a "statistical dynamics," rather than Newtonian mechanics. But this means that the central theoretical construct in statistical mechanics, the ergodic hypothesis, is directly exercised by the algorithm. That such an algorithm, which may be clearly and simply stated, can produce a wide variety of behavior, including the discontinuous behavior of the phase transition, is the principle instructional point.

Such simulation programs are neither commercial nor directly pedagogical; they are feasibility studies designed to show the scale of research practicable using microcomputer resources. This type of simulation is, to my knowledge, the only way that phase transition phenomena can be effectively demonstrated with microcomputers. Indeed, the same phenomenon, when simulated with, for example, Lennard-Jones particles, requires extensive supercomputer resources. The reduction in computational complexity represented by the lattice-gas model is due to a digitization or "coarse graining" of the special coordinates of the gas molecules. This digitization is such that the entire simulation algorithm may be parsed into primitive bit-manipulation operations and thus encoded directly into the machine language of any microprocessor.

The principle design goal of the present program is speed of equilibration. The program must be coupled with a system size sufficiently large to allow macrosopic behavior to be readily inferred by visual observation. The simulation is of a two-dimensional system for the sake of unambiguous and high-speed display, but the behavior is qualitatively like that of three-dimensional systems in almost every respect. The program accepts input for two control parameters, which in the magnetic language used in the prompts are the temperature and an applied external magnetic field. The physical system to which the simulation may most easily be compared is thus a magnetic film, like the recording surface of magnetic tape or computer disks, under the action of the recording magnet and temperature.

Because I hope to stimulate the further development of statistical simulation in physics instruction, I have placed the programs, in both source and executable forms, in the public domain. The source contains some assembly language code specific to the target microprocessor, and makes use of the hardware features of the target machines, but the algorithms can be easily ported to any microcomputer.

The software program *Two-Dimensional Ising Model Simulation* is part of the collection *Computers in Physics Instruction: Software*, which can be ordered by using the form at the end of this book.

Spectrum Calculator: Results from Student Use

R. S. Peterson and Eric T. Lane
Physics Department, University of Tennessee at Chattanooga, Chattanooga, TN 37403

The Apple II program *Spectrum Calculator* helps students identify spectral series and devise energy-level diagrams. After experimentally measuring the emission wavelengths of a particular atomic specimen, students enter these wavelengths into the program and interactively search for two or three consecutive lines of a series. The program computes additional wavelengths for the series so that students can compare and evaluate the agreement with the experimental lines. The program provides the energy levels, quantum numbers, and the computed Rydberg with each calculation. From these, the students can design an energy-level diagram.

Spectrum Calculator thus transforms a simple comparison of wavelengths with accepted values into an interactive laboratory in which students discover the atomic structure for themselves.

Computer Simulation of the Scattering of a Wave Packet from a Potential

J. E. Lewis

Department of Physics, State University of New York, College at Plattsburgh, Plattsburgh, NY 12901

This software package illustrates an important aspect of quantum mechanics that has been neglected or treated only briefly in standard texts: the development in time of a wave packet in the presence of a potential. There are several reasons that this problem has been neglected. First, the problem does not readily adapt to textbook presentation. It can be fully understood only by seeing the wave packet develop in time, for example, in a movie, or on a computer monitor. Second, the general solution to the problem cannot be expressed analytically and requires fairly extensive numerical computation.

In the past, teachers have relied either on expensive film loops, which offer a limited and fixed number of solutions (and a rather class-disrupting method of viewing), or on static diagrams culled from various textbooks, which illustrate only the salient points of the progress of the wave packet. Today a new approach is possible. The new class of high-powered microcomputers, which are equipped with excellent graphic and animation capabilities and relatively fast yet accurate floating-point arithmetic, allows in-house production of high-quality animation sequences of the wave packet's progress. We have developed software that produces a library of data files, one for each scattering problem. These data files can be modified with an editing program to fine trim the resulting animation sequence. Since a typical animation sequence lasts less than two minutes, many kinds of behavior can be viewed in a class period, totally under the control of the instructor. Because they are easy to use, the files are also suitable for student instruction in self-paced tutorials.

The software illustrates the basic wave behavior of quantum particles in the presence of a potential. The student learns fundamental facts of quantum mechanics by seeing the development in time of various solutions of the time-dependent

Schrodinger equation. The program uses easily understood animation sequences to illustrate the spatial broadening of a wave packet, oscillations in the packet envelope induced by abrupt changes of potential, tunneling phenomena, transmitted and reflected packets, and metastable resonance particles. Because the software can be easily reprogrammed to compute data files for any user-defined wave packet or potential profile, it can be tailored to the exact interests of a particular instructor, student, or course.

The intended audience is intermediate and advanced undergraduate students in physics or any other discipline where quantum mechanics or modern physics or their equivalents is a requirement.

The software assumes that the viewer has already been introduced to the topic of wave packets, the steady-state solution of the Schrodinger equation, probability density, etc., and it covers only the abstract and poorly understood aspects of quantum mechanical scattering. Besides the opening title and a legend displayed during the animation, the program offers no explanation of events; these must be supplied by the instructor, either as comments during a formal demonstration, or as notes in a self-paced study project.

The package consists of three programs and a library of data files. The library of data files covers the common scattering problems encountered in quantum mechanics. The first program, *Scatter*, displays on a monitor the data file associated with a particular wave packet and potential. The easiest use, therefore, is to run *Scatter* with appropriate data files.

The program *Makewave* is used to display other scattering situations. Here the user must modify the source code *Makewave.c*, supply code to specify the new potential and/or wave packet, change three constants that specify the output, and recompile the program. The sections to modify are clearly labeled in the source code, and should pose no difficulty to a person knowledgeable in the C language. As an aid, six different potentials are included in the source code, and the user can choose the one required. They are a square, triangular, and Gaussian barrier and well, and can be used as templates for other potential profiles, if needed.

The output of *Makewave* can then be input to *Modwave*, an editing program. *Modwave* allows the position and scale of the animation to be changed, as needed, and the number of skips between iterations to be adjusted. This dramatically reduces the size of the data files and speeds up the animation, but with a loss of smoothness. The title and subtitle to the animation can also be edited in *Modwave*.

As supplied, the library of data files can be run either from the command line interface (CLI) or from Workbench (the window environment) with a mouse. A typical use from the CLI would be `scatter df0:data / wave1.rec lo .3 .6`, and the data file <wave1.rec> on drive <df0:> in directory <data> would be run in 600 by 200 pixel resolution between the limits $x = 0.3$ and 0.6. For 600 by 400 resolution, the argument `lo` is dropped. The limits on x are 0 to 1. The minimum input is `scatter data file`, resulting in a high-resolution display between 0 and 1. Using the mouse to click on the appropriate icon will invoke the program with the selected data file. The data file selected can be varied by modifying a script file with the screen editor. Full details on the software are included as comments in the sourcecode, and as a "README" file.

To run the software, an AMIGA 500, 1000, or 2000 is needed, with at least 512k memory, using AMIGADOS 1.2. To speed things up a little and cut down on disk access, *Scatter* and the data files can be read into RAM and called by *Scatter* from there, if memory permits. The software takes over the display completely while running, returning the initial display upon termination. It releases all used memory, multitasks satisfactorily, and has no known bugs.

The software program *Scattering of a Wave Packet from a Potential* is part of the collection *Computers in Physics Instruction: Software*, which can be ordered by using the form at the end of this book.

Spacetime Software: Computer Graphics Utilities in Special Relativity

Edwin F. Taylor
Department of Physics, Massachusetts Institute of Technology, Cambridge, MA 02139

Although no one can experience directly the world of the very fast described by special relativity, *Spacetime Software* provides interactive graphics displays that help students visualize this world.

The *Spacetime* program combines four displays, all of which share the same data. The first is the position-vs.-velocity view of a multilane highway (see Figure 1) on which move clocks, rods, light flashes, and a shuttle that can change lanes.

The second display is the position-vs.-time view of a conventional spacetime diagram, showing events and the world lines of objects that move along the highway. The operator places objects and events on the highway and places events, light cones, and invariant hyperbolas on the spacetime diagram. The user can step time forward and backward, ride on any object on the highway (except a light flash), and transform the spacetime diagram from one reference frame to another. A third display splits the screen to show both the highway and the spacetime diagram as time is changed. The final display is a table of numerical data on events and objects, whose entries can be annotated by the operator.

Collision is a program that helps students analyze the relativistic collisions, creations, transformations, decays, and annihilations of particles that move in one or two spatial dimensions. The program shows three interrelated displays. The first is a table on which the operator enters what he/she knows of the mass, energy, and momentum of each incoming and outgoing particle. On command, the program attempts to complete the table, giving messages about what law or equation is being used at each step. The completed collision can then be played as a movie, which can be run continuously or stepped forward or backward frame by frame. Perspective three-dimensional plots show energy vs. x- and y-momentum of each particle and the total energy and momentum of the system before and after the interaction. All three displays can be transformed to the rest frame of any particle (except a light flash), to the zero-total-momentum frame, or to a frame moving with an arbitrary less-than-light velocity in the two-dimensional spatial plane of the collision.

An additional demonstration program, *Visual Appearance,* is available on some hardware configurations of IBM personal computers. It shows a one-eyed view of a rudimentary landscape seen through the windshield of a rocket ship moving with any speed up to that of light. At high relative speeds, objects appear to distort and rotate. One version also shows Doppler-shifted colors; no version attempts to present predicted changes of intensity.

I have used sequential versions of this software for three years in classes at the Massachusetts Institute of Technology and for two years at Harvard University Extension School. Professor Gregory Adkins used the programs in a fall 1987 class at Franklin and Marshall College. Students carry out about one-third of the homework exercises with these programs. At MIT and Harvard they also complete two take-home projects that make considerable use of the computer utilities.

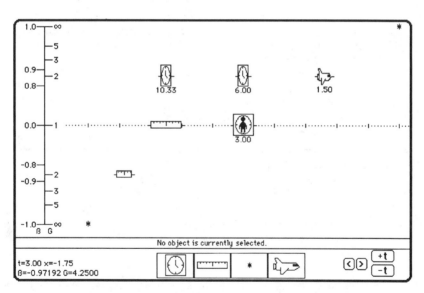

Figure 1   Highway display in the Spacetime program, showing clocks, rods, light flashes, and the shuttle (the only object that can change lanes). In this version the operator drags objects onto the highway from "object wells" at the bottom. Two scales at the left calibrate the velocity beta = v/c of the different lanes and the corresponding time-stretch factor gamma (printed as G). Mouse buttons at the lower right step time forward and backward and move the screen right and left. Printed at the lower left are frame time and highway location of the cursor (not shown).

Details of student responses to the programs are presented elsewhere.[2] In brief, with few exceptions students responded enthusiastically to the *Spacetime* program, treated *Collision* with the respect due to a powerful tool, and found the demonstration program *Visual Appearance* "interesting, but not particularly useful." Overall, students said they found the programs helpful to their understanding of the subject, easy to learn, and user friendly. Many mentioned the usefulness of the displays in visualizing the results of special relativity. There were few significant differences among the responses of students in different institutions or over the years in which the classes were taught.

Our overall conclusion from these trials is a modest one: computer graphics utilities provide a useful additional tool that helps students engage and master the subject. Graphics utilities are important because they require student manipulation of the models, an involvement that enhances learning. They are particularly valuable in visualizing relativity, where no one has direct experience of the primary phenomena under study.

The software programs *Spacetime* and *Collision* are part of the collection *Computers in Physics Instruction: Software*, which can be ordered by using the form at the end of this book.

1. For information on availability of IBM PC/XT/AT and IBM PS/2 and Macintosh versions of these programs, write to me at room 26-147 at the address above. Bitnet: ETAYLOR @ MITVMA.
2. Edwin F. Taylor, paper submitted to the *American Journal of Physics*. Preprint available on request.

PPP: A Physics Program Pool for Undergraduate Level

J. Becker, H. J. Korsch, and H. J. Jodl
Fachbereich Physik, Universität Kaiserslautern, D-6750 Kaiserslautern, West Germany

Some students take more interest in solving special physical problems than in programming a computer. The Physics Program Pool (PPP) is a collection of physics courseware developed for IBM PCs. PPP courseware is addressed to students who want to use comfortable software to investigate special physical problems without writing their own programs. During a two-year project (1987–1989), we are developing PPP programs for theoretical physics as well as for experimental physics in student labs. These programs are at the disposal for other physics departments on request. The pool presently contains about 25 programs.

Membrane. Discrete mass points coupled by springs form a model for an oscillating membrane. The student can vary the form of the membrane, the position, and the frequency of excitation. The program calculates the propagation of the excitation and shows the oscillating membrane perspectively in the manner of a film.

Billiard. The student can study the motion of a mass on a billiard table limited by an arbitrary convex curve in configuration space as well as in phase space. Depending on the initial conditions and on the form of the border periodic, quasiperiodic, and chaotic motion can be investigated.

Linear Optics. The student can set different optical elements (such as thin and thick lenses, plane and spherical mirrors, diaphragms, and screens) on an optical bench. The program presents the path of several light beams.

Wavepacket. The program calculates the dynamics of a one-dimensional wave packet in an arbitrary potential by iteration. Several import potentials are predefined, but any other analytical potential can also be used. One typical problem to be investigated by this program is the quantum mechanical tunnel effect for which transmission and reflection coefficients can be calculated.

Hydrogen Atom. The hydrogen wave functions in the xy-plane can be presented for arbitrary sets of quantum numbers ($n < 50$) as a multicolor "map" or as a perspective "mountain."

Milne. The program allows the student to determine energy levels and wave functions in arbitrary boundary potentials; in addition, he can calculate expectation values and Franck-Cordon factors.

Tabula. This is a spreadsheet program designed especially to support analysis of experimental data. For example, it allows the student to calculate one column as a function of others, to calculate mean values and standard deviations and to plot columns against other columns.

Swing. The real motion of a torsion pendulum is registered by the computer. By the program, the student can investigate harmonic and anharmonic oscillations of the pendulum in configuration space as well as in phase space.

Semiconductors. The student can record and display complete characteristics of transistors or diodes.

Radioactivity. The computer registers pulses of a Geiger-Muller counter. The program allows the student to measure the characteristics of the counter tube, to analyze counting statistics, and to determine half-life of radioactive sources.

Demonstration of Physics Programs Written in cT

David E. Trowbridge, Ruth Chabay, and Bruce Arne Sherwood
Center for Design of Educational Computing, Carnegie Mellon University, Pittsburgh, PA 15213

A number of interesting physics programs[1] have been written in cT (formerly called CMU Tutor), a programming language especially suitable for the rapid creation of interactive programs for modern graphics- and mouse-oriented environments.[2] This poster/demonstration session will provide an opportunity to see such programs operate in an integrated cT programming environment. The physics materials to be shown represent several different types of programs: lecture demonstration tools, tutorials, simulations, exploratory activities, and combinations of these forms.

The Center for Design of Educational Computing (CDEC) and the department of physics at Carnegie Mellon University are developing educational physics programs that use the cT language developed by CDEC. Programs already developed include the following:

Graphs and Tracks, by David Trowbridge. This program comprises two activities dealing with graphs of motion. Students are shown graphs of position, velocity, and acceleration vs. time, and they create arrangements of sloping tracks on which a rolling ball executes the motion represented. Students watch demonstrations of motion and sketch the corresponding graphs. This program draws on the work of the Physics Education Group at the University of Washington.

Waves, by David Trowbridge. This tutorial describes wave motion. The student engages in a dialogue on concepts of waves, including ideas of amplitude, frequency, and wavelength. The program uses a variety of graphical techniques to illustrate its points. The student is asked to enter simple mathematical expressions that describe oscillations. The aim is for the student to develop skills in writing expressions for sinusoidal traveling waves. At the end of the dialogue, a brief quiz is given. Problems are stated in words, with numerical parameters generated at random.

Efield, by David Trowbridge. This is a graphical simulation of electric fields, developed for a course on electricity and magnetism. It allows students to assemble a collection of point charges of varying magnitudes. The student can explore the resulting electric field at any point by clicking the mouse. There are three alternative views of the field: vectors, directional arrows (vectors of fixed length), and field lines. An optional section challenges the student to use these field-measuring tools to locate hidden charges and determine their magnitudes.

Pluck and Bow, by Ned Vander Ven. This program animates the motion of plucked and bowed strings for a course on the physics of musical sound. The program demonstrates the detailed motion of vibrating strings such as those on stringed instruments and offer Fourier spectra for the motion.

Quantum Well. This program allows the student to use a mouse to build an arbitrary square-cornered well and choose a trial energy. The program then integrates the Schrödinger wave function for that energy, starting from minus infinity, and plots it (typically without satisfying the boundary condition at plus infinity). The original program was written by Bruce Sherwood. Brad Keister and Harry Stumph added additional options that offer a selection of analytical well shapes and the ability to have the computer scan for wave solutions.

Robert Schumacher has developed a suite of programs used in an advanced undergraduate physics laboratory to perform various kinds of analysis.

For those interested in how such programs are written using cT, we will demonstrate the cT programming environment, which features incremental compiling, an integrated graphics editor that produces cT code, direct use of multifont text, easy control of keyset, mouse, and menu inputs, and unusual portability across diverse machines.

1. D. Trowbridge, "Quick Generation of Lecture Demonstrations and Student Exercises," in *Proceedings of the IBM ACIS University Conference, Discipline Symposia: Physics*, Boston, June 1987 (Boston: IBM Academic Information Systems, 1987), pp. 2–7.
2. B. A. Sherwood and J. N. Sherwood, "CMU Tutor: An Integrated Programming Environment for Advanced-Function Work Stations," in *Proceedings of the IBM Academic Information Systems University AEP Conference*, San Diego, April 1986 (San Diego: IBM Academic Information Systems, 1986), IV:29–37; M. Resmer "New Strategies for the Development of Educational Software," Academic Computing, 22 (December–January 1988).

Microcomputer Tools for Chaos

Alan Wolf
Cooper Union School of Engineering, New York, NY 10003

Scientists look for fundamental physical laws that can predict natural phenomena. A long-standing problem is that such laws, whose mathematical expression is often concise and therefore "simple," have been believed to have simple solutions. This is in dramatic contrast to the diverse and often erratic behavior observed in nature.

In the past decade, "chaos theory" has had great success in clarifying the relationships between physical laws and the properties of their mathematical solutions. We now know that a deterministic nonlinear system with only a few active variables may exhibit a wide range of behavior ranging from highly ordered to highly disordered. Chaos theory is now well developed, and applications have been found in such areas as fluid flow, chemical kinetics, population dynamics, structural analysis, cardiac and brain studies.[1]

My research has concerned the detection and quantification of chaos in experimental data. Often one has a system in which a time series of a single dynamical variable is available: the concentration of a chemical species, a voltage trace from an EEG, or perhaps the time-dependent location of a node in a structure. We may know very little about the system or the nature of the observable (which could be a nonlinear function of one or more "underlying" system variables), and there is likely to be experimentally introduced "noise" in addition to intrinsic dynamical disorder. I have made some progress in this real-world data-analysis problem.

I will present software that serves the pedagogic function of illustrating chaotic systems and provides the researcher with tools for analyzing chaotic data. Such data could come from numerical simulations, or from an actual experiment. Most of these programs are written in *QuickBASIC*, a few are written in FORTRAN; and all run on IBM PCs and compatibles.

Programs that illustrate chaotic systems solve for the dynamics of systems of varying complexity: "geometric chaos" on a stadium-shaped billiard table, the *n-*

body gravitational problem, coupled pendula, one- and two-dimensional mappings, and sets of ordinary differential equations. These programs are all graphically oriented and display such phenomena as bifurcation sequences, fractal structure, and sensitive dependence to initial conditions. Some of these programs generate data files that may be used with research tools software.

The research tools include software for data acquisition (with LAB-40 hardware by Computer Continuum), power-spectral analysis and filtering, rotating delay-reconstructed data sets, creating Poincare sections, and quantifying chaos. Two methods are provided for the latter: calculation of the fractal dimension of a time series, and calculation of the largest Lyapunov exponent of a time series.

A system may be defined as chaotic if it exhibits "sensitive dependence to initial conditions." This means that almost all identically prepared systems show, on the average, an exponentially fast divergence in their future behavior. Lyapunov exponents quantify this exponential divergence and set the time scale on which such systems are "predictable" in the usual sense. I have developed an algorithm for calculating the dominant Lyapunov exponent. This has been used successfully to quantify chaos in a variety of systems. Its graphical demonstration of "sensitive dependence" is of great pedagogical value.[2]

The potential audience of the software ranges from high school students (some of whom have used and/or developed it in research competitions), to undergraduate and graduate courses in mechanics, to researchers.

Hardware requirements vary somewhat between programs, but most use CGA or EGA graphics and benefit from (but do not require) an 80 by 87 math coprocessor. QuickBASIC source codes and executable codes will be provided. A few of the research tools are written in Microsoft FORTRAN 4.01. Programs are interactive and somewhat self-documenting, but additional documentation will be provided.

1. Predrag Cvitanovic, *Universality in Chaos* (Bristol: Adam Hilger, Ltd, 1984).
2. Alan Wolf et. al., "Determining Lyapunov Exponents from a Time Series," Physica **16D**, 285 (1985).

Chaotic Dynamics:
An Instructional Model

Bruce N. Miller and Halim Lehtihet
Department of Physics, Texas Christian University, Fort Worth, TX 76129

During the first half of this century classical and quantum physics was sufficiently powerful to explain in detail most linear phenomena. Nonlinear problems, however, such as the stability of the solar system, the evolution of turbulence in the atmosphere, and changes in the global population, could not be intelligently addressed. The work of the Russian mathematicians Kolmogorov, Arnold, and Moser, which actually started in the middle 1950s but was not widely known for some time outside of the USSR, and of Mitchell Feigenbaum at Los Alamos, which started in the late 1970s, altered our understanding of nonlinear phenomena for all time. Their work on conservative and dissipative systems showed that chaos and uncertainty frequently characterize deterministic dynamical laws.

During the past decade we have witnessed a revolution in our understanding of dynamical phenomena. Although the focus of the new understanding has been on theoretical and applied physics, the concepts have found important applications to biology and medicine, chemistry, and astronomy.

As usual, pedagogy is behind research, and there are few texts available in the field of chaotic dynamics. Because it is the nature of nonlinear analysis that many results cannot be expressed by simple equations, carefully designed interactive computer simulations are essential tools of pedagogy in this field. The central objective of this software is to provide graduate and undergraduate students of physics with an effective tool for mastering the basic concepts of nonlinear dynamics.

A by-product of research in gravitating systems at Texas Christian University[1] was the development of an apparently simple dynamical model that a freshman could easily visualize, but that nonetheless possesses all of the features of conservative chaotic dynamics understood to date, as well as others that are beyond our present grasp. The model consists of an accelerating body that interacts elastically (bounces) from a wedge-shaped boundary. We published a research paper on this system before we realized its potential for instruction.[2] When we demonstrated some preliminary software at the International Conference of Physics Education in Balotin, Hungary, the audiences were enthusiastic and we received numerous requests for copies of the software.[3]

We have incorporated this model into an interactive software package entitled *Wedge* dealing expressly with conservative systems. The software can essentially stand alone for typical physics undergraduates: little, if any, additional resource material is required. The package consists of an introduction describing the main physical characteristics of conservative dynamical systems, a thorough description of the model, instructions regarding the use of the program, a demonstration run illustrating its essential features, and the interactive program itself in which the student can completely control the initial conditions and all system parameters.

The software is a useful addition to the regular undergraduate and graduate courses in analytical mechanics required in all physics curriculum, as well as to special topics courses in chaos and nonlinear phenomena.

Instructions are contained on the diskette. MS-DOS is required for the PC.

The software program *Billiard in a Gravitational Field* is part of the collection *Computers in Physics Instruction: Software*, which can be ordered by using the form at the end of this book.

1. For a review, see Charles J. Reidl and Bruce N. Miller, "Gravity in One Dimension: Selective Relaxation?" in Astrophys. J. **318**, 248 (1987).
2. H. Lehtihet and B. N. Miller, "Numerical Study of a Billiard in a Gravitational Field," Physica D **21**, 93 (1986).

3. "Falling Body in a Wedge: What Happens When the Floor Tilts," in *Proceedings of the International Workshop on Teaching Nonlinear Phenomena*, edited by George Marx (Veszprem, Hungary: National Center for Educational Technology, 1987), pp. 318–24.

Exploring Nonlinear Dynamics and Chaos Using an Interactive Program with Graphics Animation

R. W. Rollins
Department of Physics and Astronomy, Ohio University, Athens, OH 45701

We will describe and demonstrate a set of utilities (*The Ohio University Chaotic Dynamics Workbench*), consisting of over 5,400 lines of code written in *Turbo Pascal*. This set of utilities perform interactive numerical experiments on nonlinear systems modeled by ordinary differential equations. Using techniques recently developed in the field of chaotic dynamics it analyzes the behavior of the systems. The program emphasizes direct interaction with the user and provides an animated graphical display of the solution in phase space as it is being calculated. A user with minimal programming experience and *Turbo Pascal 3.0* can modify one of the demonstration program source files so that all the capabilities of the *OU CD Workbench* can be applied to the user's own system of ordinary differential equations.

The objectives of *OU CD Workbench* are: (1) to provide a tool that can be used to demonstrate the physical meaning of some of the central themes of the new and developing field of chaotic dynamics including fixed points, limit cycles, strange attractors, sensitive dependence on initial conditions (Lyapunov exponent spectra), fractal dimensions, period doubling route to chaos, quasi-periodicity, the effect of damping on the dimension of the attractor (mode reduction), etc.; (2) to provide a tool that is powerful and flexible enough to do exploratory research quickly and reliably on any system that may be modeled by a set of ordinary differential equations; and (3) to provide high-quality graphics in an interactive environment that will give qualitative insights into, and stimulate conjectures about, problems that appear to be analytically intractable.

To show the capabilities of the *OU CD Workbench* utilities, we will demonstrate the program Duffing. Duffing numerically integrates the set of differential equations that describes a unit mass moving in one dimension under the influence of a double-well potential, viscous damping, and a sinusoidal driving force (motion described by a Duffing equation). This simple nonlinear system is very rich in its dynamic behavior and can be used to demonstrate many concepts of chaotic dynamics.

OU CD Workbench's demonstration programs, such as Duffing, are intended to be used in conjunction with an introductory monograph on chaotic dynamics.[1] The programs will be of interest to those who are interested in learning or teaching about the behavior of nonlinear systems and the new concepts used in the field of chaos. Because *OU CD Workbench* can be used to explore parameter space rapidly, those with a professional research interest in chaos will also find the program useful.

OU CD Workbench has many attractive features, including the following, which are important to the demonstration program Duffing.

The system uses either a fourth-order, Runge-Kutta method with fixed step size, or a Hamming predictor-corrector method with variable step size to integrate the differential equations.

The system allows the user to change model parameters (such as drive amplitude or frequency, damping, etc.) "on the run." In fact, many features occur or are initiated while the program is integrating away. This provides a very interactive environment in which to play (or work).

In the "Orbit" mode, the user may switch instantly back and forth between a "live" graphic display of a projection of the trajectory in phase space and a Poincare section of the trajectory. The user may select the particular projection in phase space and type of Poincare section (subject to some limitations). Graphics display areas may be panned and zoomed.

The "Flip" mode shows, with animation, the consequences of "sensitive dependence on initial conditions." The user may switch instantly to the "Flip" mode at any point in the calculation. The "Flip" mode calculates and displays 20 Poincare sections at 20 equally spaced phase intervals as the driving force completes each period. Each of these Poincare sections is a "snapshot" of the cross-section of the trajectory (or attractor) at a particular phase of the driving force. The animated flip-card effect of "Flip" demonstrates, in a dramatic fashion, the folding and kneading 20 disk and displayed at a later time using the *DSPFLP.COM* program, which is included with the *OU CD Workbench*, or called back into the main program for further calculation and display.

At any time while in the "Orbit" mode, the user can calculate Lyapunov exponents, which give a quantitative measure of the sensitive dependence on initial conditions. The system uses the method described by Wolf, et al.[2] The results of the Lyapunov exponent spectra are also used to estimate the fractal dimension of the attractor by calculating the Lyapunov dimension as proposed by Kaplan and Yorke.[3]

The user can save ten different setup conditions in situation keys and recall them instantly to illustrate any point desired. The user can also save input data, situation-key setup, and results. Thus, if the user starts again later, he can pick up the calculation or demonstration where it left off.

The program requires an IBM PC/XT/AT or compatible with at least 512K of memory, two floppy disk drives or one floppy and a hard disk drive (preferred). It also needs an 8087 math coprocessor, IBM color graphics adapter, graphics monitor (color or B/W) (EGA with CGA emulation is supported but Hercules graphics is not), DOS 2.0 or higher, and an EPSON FX or IBM graphics printer (optional).

If you want to recompile using your own set of differential equations, you will need *Turbo Pascal 3.xx* (8087 version). We are developing an updated version of the *OU CD Workbench* which will use *Turbo Pascal 4.0*.

The two *OU CD Workbench* demonstration diskettes contain instructions on "README" files. The first demo diskette contains the Duffing program and the second contains a set of 20 previously calculated Poincare sections that can be displayed using "DSPFLP."

The program *The Ohio University Chaotic Dynamics Workbench* is part of the collection *Computers in Physics Instruction: Software,* which can be ordered by using the form at the end of this book.

1. Francis C. Moon, *Chaotic Vibrations* (New York: Wiley-Interscience, 1987); Pierre Berge, Yves Pomeau, and Christian Vidal, *Order and Chaos* (New York: Wiley-Interscience, 1986); J. M. T. Thompson and H. B. Stewart, *Nonlinear Dynamics and Chaos* (New York: Wiley, 1986); Heinz Georg Schuster, *Deterministic Chaos* (D-6940 Weinheim, Federal Republic of Germany: Physik-Verlag GmbH, 1984).
2. A. Wolf, J. B. Swift, H. L. Swinney, and J. A. Vasano, "Determining Lyapunov Exponents from a Time Series," Physica **16D**, 285 (1985).
3. J. Doyne Farmer, Edward Ott, and James A. Yorke, "The Dimension of Chaotic Attractors," Physica **7D**, 153 (1983).

Bouncing Ball Simulation System

Nicholas B. Tufillaro and Al Albano
Department of Physics, Bryn Mawr College, Bryn Mawr, PA. 19010, and Tyler A. Abbott, Walnut Creek, CA 94598

Bouncing Ball is a program written for the Apple Macintosh computer that accurately simulates the dynamics of a ball bouncing repeatedly on a vibrating table. The *Bouncing Ball* program, along with a recently developed undergradute lab in nonlinear dynamics and chaos[1] illustrates many of the ideas and methods used in describing nonlinear dynamical systems.

Bouncing Ball simulates the dynamics of a ball bouncing on a sinusoidally oscillating table. The user can vary two control parameters (the tables forcing frequency and amplitude) along with the initial conditions to explore different periodic and chaotic motions. He can display the resulting trajectories in configuration or phase space, and display Poincare maps in the window-based system. The *Bouncing Ball* program can be used in undergraduate courses and is a valuable aid for lectures and demonstrations in nonlinear dynamics.

The software program *Bouncing Ball* is part of the collection *Computers in Physics Instruction: Software,* which can be ordered by using the form at the end of this book.

Development of this program was supported in part by a grant-in-aid of research from the National Academy of Sciences, through Sigma Xi, the Scientific Research Society.

1. N. B. Tufillaro and A. M. Albano, "Chaotic Dynamics of a Bouncing Ball," Am. J. Phys. **54**, 939 (1986); T. M. Mello and N. B. Tufillaro, "Strange Attractors of a Bouncing Ball," Am. J. Phys. **55**, 316 (1987); N. B. Tufillaro, T. M. Mello, Y. M. Choi, and A. M. Albano, "Period Doubling Boundaries of a Bouncing Ball," J. Physique **47**, 1477 (1986); Kurt Wiesenfeld and Nicholas B. Tufillaro, "Suppression of Period Doubling in the Dynamics of a Bouncing Ball," Physica **26D**, 321 (1987).

Automata on the Mac

Nicholas B. Tufillaro
Department of Physics, Bryn Mawr College, Bryn Mawr, PA 19010

Jeremiah P. Reilly
Philadelphia, PA 19103

Richard E. Crandall
Physics Department, Reed College, Portland, OR 97202

Current research on cellular automata suggests that they may provide insight into the behavior of numerous physical phenomena.[1] These computer-simulated universes, governed by simple rules, demonstrate complex and unpredictable behavior. Further, there is a captivating beauty inherent in many cellular automata.

The means to explore—to play with—cellular automata is finally within reach of most computer owners.[2] We have developed a popular cellular automata program, *Automata*, that runs on the Apple Macintosh computer and features a simple user interface allowing one to explore rules and initial conditions by pointing and clicking.

Automata has many applications in education. We have, for instance, used it to teach binary addition to children (six year olds). Somewhat older children learn the art of experimentation by exploring the cellular-automata universe. *Automata* can also be used by college students to explore the elements that go into the making of "lattice gas models" of fluid flows.

We will begin our discussion of cellular automata assuming no previous knowledge of the subject and end by presenting the current theories about cellular automata and what the future may hold for these totally discrete models of natural phenomena.

1. Stephen Wolfram, *Theory and Application of Cellular Automata* (Singapore: World Scientific, 1986).
2. N. B. Tufillaro, "Cellular Automata Program," in *Who Got Einstein's Office?* by Ed Regis (New York: Addison-Wesley, 1987).

Demonstrations of Diffusion-Limited Aggregation and Eden Growth on a Microcomputer

G. W. Parker

Department of Physics, North Carolina State University, Raleigh, NC 27695-8202

The diffusion-limited aggregation (DLA) model provides a simple method of simulating real processes in nature such as the aggregation of small neutralized metal particles in a liquid, driven by their van der Waals attraction to form an open fractal-like structure.

The DLA model releases particles at random locations some distance from a central seed particle and lets these particles execute random walks. They execute these random walks until they make "contact" with the seed, in which case they stick to it; or until they move some distance farther from the seed than where they were started, in which case they are "lost" to the system. The injection of many such particles forms a structure that exhibits branching, treelike growth.

Counting the number of particles $N(R)$ inside a sphere of radius R centered on an arbitrarily chosen particle of the cluster, it is found that $N(R) = \text{const} * R^D$ where D is the fractal dimension. In two dimensions $D = 1.7$, which satisfies the theoretical bounds $d - 1 < D < d$, where d is the dimension of the space ($d = 2$ in two dimensions). In three dimensions, $D = 2.5$.

I have written a small program, *Grow/DLA*, to carry out the DLA process in two dimensions on a Macintosh microcomputer using Microsoft BASIC. The interpreted language is slow, but this is an advantage; it allows the uses to follow the diffusion process on screen as it unfolds. After a number of hours a reasonable structure is obtained having roughly 1,500 particles. The structure can then be saved to disk for subsequent replay and analysis using a second program, *Replay*. Both programs carry out the analysis of counting particles about three randomly chosen centers at ten different radii. Averaging these numbers for several runs and making a log N-vs.-log R plot, one finds the fractal dimension to be about 1.7, as expected.[1]

My third program, *GROW/EDEN*, contrasts growth on the DLA model with another model, the Eden model. The Eden model also uses a seed particle to initiate the growth. However, no particles move to join it. Instead, the particles materialize at random positions next to the seed. Once materialized, they bond to begin the formation of a cluster. In contrast to the DLA structure, these clusters form with a uniform density, i.e. they have $D = d$. There is no branching.

The branching in the DLA process is understood in terms of the amplification of local inhomogeneities.[2] The random walking of the incoming particles produces local inhomogeneities on an initially smooth seed. These protrusions in turn tend to trap incoming particles more readily than other nearby areas, thus leading to a branching type of growth.

In my presentation I will show and analyze several DLA structures.

1. T. A. Witten and L. M. Sander, "Diffusion-Limited Aggregation," Phys. Rev. **27**, 5686 (1983).
2. T. A. Witten and M. E. Cates, "Tenuous Structures from Disorderly Growth Processes," Science **232**, 1607 (1986); L. M. Sander, "Fractal Growth," Sci. Am. **256**, 94 (January 1987).

Star Cluster Dynamics Simulation with Dark Matter

John J. Dykla
Department of Physics, Loyola University, Chicago, IL 60626

During approximately the last decade, evidence from several areas of observational astronomy has supported the concept that a large proportion of the mass of the universe is associated with so-called dark matter. Dark matter, which may be dominant by an order of magnitude or more, contributes to gravitational dynamics but not to the electromagnetic radiation we have detected so far. Rubin has recently completed a survey of the literature on dark matter, which is accessible to nonspecialists, including upper-division undergraduate physics majors.[1] Among the most important conclusions of this survey are the ideas that the dark matter is clustered in association with visible matter and that it is probably significant on scales both somewhat smaller and much larger than the size of a typical spiral galaxy.

I have developed a microcomputer program, *Solar Neighborhood Cluster Dynamics Simulator,* that simulates the behavior expected theoretically of the orbits of a small system of the 18 nearest stars as they would be if the system were independent of the Milky Way galaxy. Mass and initial position and velocity data were adapted from those presented by Dewdney by correcting for a systematic error in his length scale.[2]

The star cluster evolves according to Newtonian gravitation and dynamical laws applied in three-dimensional space. The plot represents a cubical volume initially enclosing the stars projected onto a square region of the viewing screen. Although displacement and velocity components normal to this plane are not obvious in the visual display, they are used in the calculations. The gravitational interactions of stars that leave the viewing volume are also included.

The user begins the simulation by choosing the time interval between plots of positions for the stars in the evolving cluster. Next, the user chooses one of two rest frames: either a frame in which the sun is at rest, or an automatically calculated frame in which the center of mass of the cluster is at rest. Finally, the user chooses the factor by which to multiply the mass of each star in the cluster. This factor represents the gravitational influence of dark matter. One option is for the program to multiply the masses by the automatically calculated factor required to satisfy the virial theorem relation, as presented, for example, by Marion, between kinetic and potential energies in a relaxed, bounded system.[3]

Unless the mass of dark matter is within a very narrow range of that required by the virial theorem, the cluster will disperse outside the initial volume or collapse to a very small fraction of the initial volume in a relatively short time.

The software program *Solar Neighborhood Cluster Dynamics Simulator* is part of the collection *Computers in Physics Instruction: Software*, which can be ordered by using the form at the end of this book.

1. Vera C. Rubin, "What's the Matter in Spiral Galaxies?" in *Highlights of Modern Astrophysics: Concepts and Controversies*, edited by Stuart L. Shapiro and Saul A. Teukolsky (New York: Wiley, 1986), pp. 269–97.
2. A. K. Dewdney, "How Close Encounters with Star Clusters Are Achieved with a Computer Telescope," Sci. Am. **254**, 24 (January 1986).
3. Jerry B. Marion, *Classical Dynamics of Particles and Systems*, 2nd edition (New York: Academic Press, 1970), pp. 233–35.

Educational Software for Undergraduate Astronomy

J. C. Evans
Physics Department, George Mason University, Fairfax, VA 22030

As authors of a widely used textbook in astronomy[1] and designers of sky-simulation software,[2] we hold the view that educational software should not duplicate the function performed by textbooks. Textbooks are the most economical means of delivering large volumes of words. Software that simply displays text on the screen is nothing more than an electronic page turner; software that only displays static illustrations is but an electronic viewgraph. Educational software should avoid booklike serial presentations; it should take advantage of the possibilities of the medium to engage users in an interactive fashion that books cannot duplicate. To accomplish this, the design of the software must take advantage of these pedagogical strengths of microcomputers: (1) They provide multiple and dynamic pathways from one point to another; (2) They simulate the behavior of physical phenomena through animation; (3) They present and demonstrate physical concepts graphically; (4) They perform extensive calculations as needed to illustrate; and (5) They require the user to think and then to perform a series of actions based on his understanding of the concepts and phenomenology.

If software does not provide variety in the way in which a user progresses through it with each use, it becomes boring and will be used very sparingly. Properly designed educational software will supply continuing interest to users by appealing to their sense of exploration.

We have designed a set of learning modules for an undergraduate course in astronomy. Astronomy is a highly visual subject and lends itself to being taught through graphic simulation and animation. Such techniques can help illustrate astronomical phenomena, concepts, and processes. Reviews and user feedback on our existing sky-simulation software convince us that our basic premise is sound, and we believe that our pedagogical philosophy is adaptable to all the biological and physical sciences.

In pedagogical design, our learning modules are suitable for use in lecture demonstrations, in conjunction with laboratory exercises, or in independent study. They cover topics that are normally part of an introductory course. They engage users to both "think and do" by providing multiple paths to the same end point. They avoid testing comprehension by verbal questions on the screen, since companion workbooks can do this much more efficiently. They are not to be keyed to any specific book, but rather supplement any college-level text.

In technical design, our modules make extensive use of data from the Data Center of the NASA Goddard Space Flight Center. They include rigorous calculations of such aspects as position, motion, astrophysical processes, and recurring phenomena; and compile an MS *QuickBASIC* program. In order to support laboratory courses, three of the modules are data files with simple display interfaces.

The first data module, based on the work of Bretagnon and Simon, provides both heliocentric and geocentric positions of the sun and planets for any time between 1 January 1950 and 31 December 2050.[3] The second data module is a modified version of the well-known Yale Bright Star Catalogue.[4] The final data module displays data on nonstellar objects, such as galaxies.

In this meeting we will report on and demonstrate the first module in the series, a sky-simulation module. This module is a revision and adaptation for the IBM PS/2 series of our earlier HP design. It allows users to explore the skies without outdoor observation, which can be hampered by bad weather. The module simulates the appearance of the sky as seen from any position on the earth for any time between 1950 and 2050. It simulates the relative brightness of stars down to fifth magnitude for approximately 5,000 stars and 100 nonstellar objects. It simulates the motions of the sun, moon, and eight planets, and displays their orbits relative to the background stars. It permits the showing of on-off overlays of constellation asterisms, galactic and celestial equator, and ecliptic.

Users may explore the skies at will or ask for locations and rise-set times for any of the following: the sun, the moon, the planets, any of 88 constellations, any of the 100 brightest stars, any of 100 nonstellar objects.

This work is supported in part by a contract from the IBM Corporation, which is acknowledged with gratitude.

1. L. Berman and J.C. Evans, *Exploring the Cosmos*, 5th ed. (Boston, MA: Little, Brown, 1986).
2. J. C. Evans and L. A. Dreiling, *NIGHTSKY/150, Sky Simulation Software for the Hewlett-Packard 150* (Fairfax, VA: Cosmographics International, 1985).
3. J. Bretagnon and J-L. Simon, *Planetary Programs and Tables from -4000 to +2800* (Richmond, VA: Willmann-Bell, 1986); J. C. Evans and L. A. Dreiling, *Software for Planetary Positions from 1600 to 2800 A.D.* (Richmond, VA: Willmann-Bell, 1987).
4. E. D. Hoffleit and W. H. Warren, *Yale Bright Star Catalogue*, 5th ed. (Greenbelt, MD: NASA Goddard Space Flight Center, 1988).

Computers in the Physics Laboratory

Computer-Based Tools: Rhyme and Reason

R. F. Tinker

Technical Education Research Center (TERC), 1696 Massachusetts Ave., Cambridge, MA 02138

The Rhyme: Experimenting and Theorizing

The words "rhyme and reason" are most often used with the negative connotation of the phrase "without rhyme or reason." And too often that is what happens in technically oriented enclaves like this conference: we pursue technology without rhyme or reason. In this talk I want to give rhyme and reason to our pursuit of a set of powerful tools.

General-Purpose Hardware

Microcomputer-based labs (MBL) seem finally to have arrived, judging by the number of contributed and invited talks, by the number of different products currently available, and by the increasing interest in large-scale implementation of MBL. After 12 years, we seem to have finally broken the chicken-and-egg problem: there are enough products on the market to interest educators, and there are enough educators ready to purchase MBL products that there is significant commercial interest. As a result, a broad spectrum of sensors and interfaces running on all the popular microcomputers are now available. You can measure temperature, light, force, pressure, magnetic field, distance, acceleration, timing, pH, dissolved oxygen, heart rate, blood pressure, muscle tone, skin conductance, sound, and much more. You can turn your computer into a oscilloscope or a counter/timer; you have the flexibility to make it into virtually any instrument.

In spite of all this progress, there is still much work to be done. The MBL world is chaotic, with special-purpose interfaces and idiosyncratic software. For reasons of simplicity for educators and intellectual coherence for students, it would be far better if there were only a few very general-purpose MBL software packages that worked the same way in all the different computers and interfaced to each of a small number of general-purpose interfaces. These interfaces would all accept the widest possible range of probes and actuators. In some of our more recent work we are beginning to move in this direction. I will begin by showing some general-purpose hardware, move on to some general-purpose software, and then discuss some of the strengths of the MBL approach.

To design a laboratory interface that can plug into any computer, you must use the standard RS-232C serial port. Though its limited data-transfer rate makes the serial interface far from ideal for this work, its universality makes it difficult to ignore. Almost every computer—Macintosh, IBM, clones, lap tops, even the inexpensive Radio Shack 100—has a built-in serial interface. Of popular computers, only the Apple IIe does not come with one built in (the IIc and IIGS have it). This means that a lab interface that communicates with the host computer over a serial line can be made to work with any computer; invest in a lab interface today and it will still be useful next year when you buy a better computer. Furthermore, some of the trickiest code having to do with real-time interfacing has to be in ROM in the interface. This means that application software that uses a serial lab interface is simpler than software for corresponding special-purpose interfaces.

I designed the first inexpensive serial interface back in 1982 and convinced CDL to market it as the universal analog lab (UAL). This three-chip circuit (shown in Figure 1) is still an extremely attractive unit and a very nice student project. From the perspective of the host computer, it could not be simpler. The host sends it a number n from zero to seven. The interface responds with a number from zero to 255 that represents the digital equivalent of the applied voltage on input number n. The numbers are sent in 8-bit binary format, which means that you cannot use

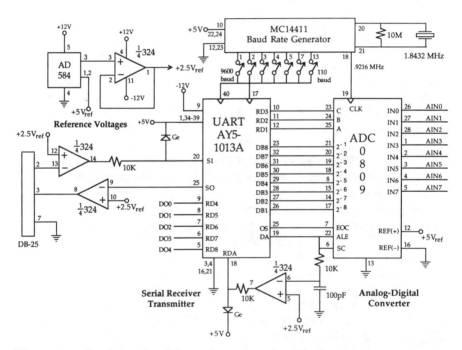

Figure 1.   A schematic of the universal analog lab. The 0809 is an 8-channel, 8-bit analog-to-digital converter. When the universal serial asynchronous receiver/transmitter (USART) receives a complete byte from the computer, it starts the analog-to-digital converter, using one of the eight analog inputs. When the conversion is complete, the USART is signaled to transmit the resulting 8-bit value back to the computer.

the usual ASCII-oriented interface drivers. But aside from that minor complication, it is easy to use and has found wide application. Unfortunately, the UAL was a commercial flop. After CDL invested heavily in bringing it out, they tried to recoup much of that cost in the first few units sold. As a result the price was too high, and not many sold. In addition, CDL had not yet solved the chicken-and-egg problem; users did not know they wanted the product and so they did not buy it.

The next serial lab interface developed in our laboratories was the universal serial interface (USI), conceived as an Apple game-paddle port on the end of a serial line. Humble as it is, the Apple game-paddle port has proven to be a marvelously versatile lab interface with four resistance-analog inputs, four digital-output lines and three digital-input lines. A very wide range of laboratory inputs and outputs have been developed for the game-paddle interface. This game-port interface played an important role in solving the chicken-and-egg problem; it made it possible to create MBL products with dirt-cheap transducers. The only probe electronics required for a temperature or light interface is a one-dollar thermistor or a fifty-cent phototransistor. I hope that the USI will play a similar role in breaking the chicken-and-egg cycle for serial interfaces.

In keeping with the game-paddle analogy, the USI has been designed for low-cost applications; we have tried as much as possible to keep the price down, primarily by resisting "creeping featurism." Creeping featurism is a highly contagious technological disease that involves giving in to the impulse to add features, no one of which makes much difference in the price, but the sum of which defeats the original design goal of low cost.

The USI was designed as part of our modeling project to provide a laboratory input for Macintoshes. Hearing of the USI, Priscilla Laws joined forces with us and applied the USI to a Macintosh-based project at Dickinson College. As a consequence, most of the present laboratory software for the USI is Macintosh based, but that is really a historical accident; the USI has been used with the IBM and can be used with any other computer that has a serial port.

The USI was specifically designed for backward compatibility. Many of the resistive probes designed for the Apple game port plug directly into it. We included a voltage-input feature so we could put on two six-conductor telephone jacks that would accept any of the probes developed under our MBL project and marketed by HRM Software. This means that the ultrasonic motion detector, temperature, and sound inputs, as well as a heater controller currently on the market, will plug directly into the USI. Another dozen probes and actuators currently under development will be compatible.

The commercial status of the USI is still very much in doubt. We distributed about 35 surplus interfaces to colleagues at the end of our modeling project. We hope to generate enough demand that we can find a commercial distributor who will take over the burden of manufacturing and supporting them.

IBM's Phil Smith recently caught the MBL bug, and he has created far and away the best-designed MBL interface, next to which any other MBL interface appears amateurish. As is often the case, IBM profits from what others have learned. From an engineering view, their interface, the personal science lab (PSL), is unparalleled among low-cost interfaces.

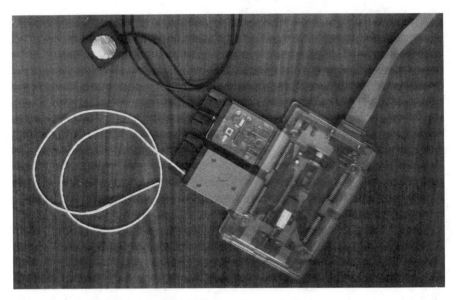

Figure 2. The personal science lab (PSL) with temperature and motion
detectors plugged into two of its four inputs.

The basic PSL unit (shown in Figure 2) accepts four cartridgelike interfaces
that support a wide range of probes. A general-purpose analog cartridge has plugs
for temperature, light and pH sensors. Another cartridge accepts microphone and
high-speed voltage inputs. A third cartridge contains RAM that can be used to
buffer high-speed signals that are funneled into the host computer through the
slower serial line. A fourth cartridge is for prototyping, which allows users to create special interfaces for the PSL. An ambitious range of additional interfaces has
been prototyped or planned.

Perhaps the most exciting feature of the PSL is its expandability. You can stack
up to 64 interfaces and interface a different computer to each level of the stack.
Suppose you instrumented an expensive apparatus, such as an automobile engine,
with ten stacked units; each of ten computers could have access to all the probes on
that engine. Better yet, a battery pack would be available for the bottom of the
stack and a keyboard and readout unit available for the top. This would allow you
to use the PSL away from the host computer. Biologists should love the battery
pack for field measurements, and anyone who wants to free up the host computer
while gathering data over a long time period will find it a convenience.

The commercial status of the PSL is also uncertain at this time. Although IBM
has developed it and is showing it at meetings like this one, the PSL is not yet an
announced IBM product. It seems that IBM is not yet certain whether the chicken-
and-egg problem has been solved. I trust they will find it has been. The PSL is an
excellent product and will increase IBM's stature in the educational community.

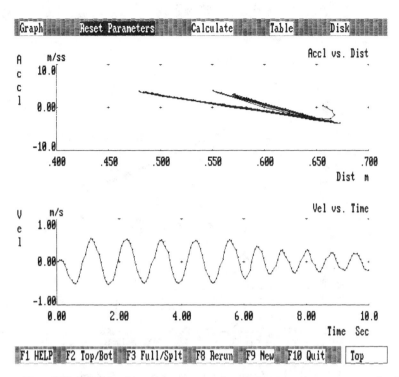

Figure 3.   *PSL Toolkit,* showing two graphs of the same event, one displaying acceleration against position and the other velocity against
time.

Integrated Software

To parallel the development of universal hardware solutions to MBL, we also
need to develop universal software solutions. Universal software should be able to
take data from any probe or combination of probes, display the data in a natural
way in real time while it is being collected, and then perform a wide range of general-purpose analysis functions on these data. The software should be able to identify the probes plugged into the interface and give the user choices appropriate to
these probes. In this environment, there can be so many options that it is essential
to have some menu-driven macro facility so that beginning students can perform
complicated setups and analysis functions by selecting named setups or functions.

The *PSL Toolkit* has some of the generality needed in MS-DOS computers for
the PSL. One of the nice features of the *Toolkit* software is that computations can
be done in real time as data are being gathered in order to generate multiple variables, any one of which can be graphed against any other in either of two graphing
windows. As an example, Figure 3 shows acceleration being graphed against position in the upper graph, and velocity against time in the lower graph. These data
were recorded and displayed as a spring-mass system was oscillating. Partway
through the data run, the spring was grasped approximately at center, causing the

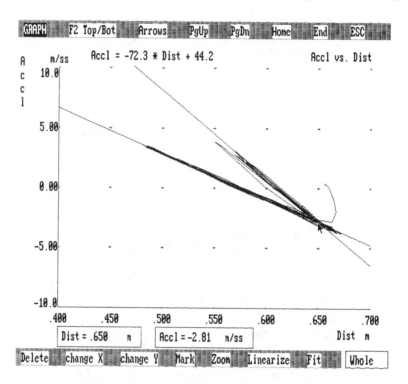

Figure 4. An analysis of the upper graph of figure 3, showing two least-square fits to different parts of the run. The absolute value of the slope should be proportional to the spring constant, which was changed during the run.

frequency of oscillation to increase. Assuming an ideal spring, the slope of the acceleration-vs.-distance phase plot should be $-k/m$. You can use the software to make any points on the graph and fit those points to a straight line. Points generated by the two conditions of the spring are identified and marked, and straight-line segments fit to each of those regions in the display shown in Figure 4.

Figure 5 shows the same software package measuring the cooling of the sample to room temperature. When the equilibrium temperature is subtracted from each of the values, points near the very end of the run are eliminated, and the logarithm of the resulting values is taken and fit to a straight line, the graph in Figure 6 results.

We are interested in extending the range and generality of this software package, particularly in how students might use this software and whether they can easily master it.

Theory-Building Tools

As part of our modeling project, we have been exploring ways to complement the data-gathering capacity of MBL tools with powerful tools for theory building.

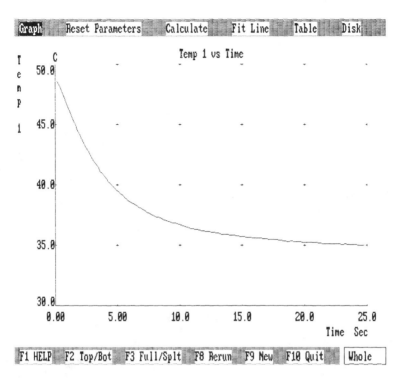

Figure 5.   A cooling curve taken with the *PSL Toolkit*.

An environment that combines data gathering with theory building allows a student to move easily between theory and experiment, greatly enhancing learning.

The model-building environment we are investigating is based on systems dynamics and educational strategies developed by the systems dynamics group at MIT. Systems dynamics is designed to solve problems involving coupled, possibly nonlinear, differential equations where time is the only independent variable. Certainly not all problems one would like to model fall into this category, but a very large number do—and in all the scientific disciplines. MBL is particularly appropriate to systems dynamics because it is often used to collect time-dependent data, which is just what systems dynamics models.

The MIT systems dynamics group learned to teach business students who had no understanding of calculus to set up and solve complicated dynamics models. The key to the MIT approach is a hydraulic metaphor that gives an intuitive representation of integration and differentiation. This metaphor is based on a valve and tank: fluid flows through the valve at a rate determined by the valve and accumulates in the tank. The amount of the fluid in the tank is the integral of the rate through the valve. Conversely, the rate of flow through the valve is the derivative of the level in the tank. Users can control the valves by using functions that depend on constants and levels in the tanks. Originally the valve-and-tank metaphor was simply a learning aid used with paper and pencil. More recently, the software pack-

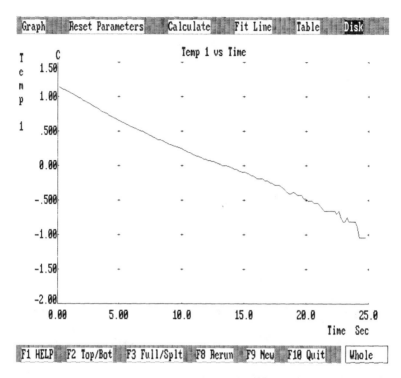

Figure 6.   An analysis of the cooling curve data from figure 5, showing a
graph of $\log(t\text{-}t_0)$.

age *Stella* provides a flexible, graphical environment on the Macintosh that allows
the user to specify a system using this metaphor completely. The system then
quickly generates graphical solutions for the resulting systems.

Our modeling software combines MBL with a simplified modeling environment using the valve-and-tank metaphor, but modified by what we have learned
through extensive use of *Stella* with students. Our system allows us to display both
real-time MBL data and output from models on the same graph. The MBL data
comes from the universal serial interface that was developed for this project. The
model data is generated from valve-and-tank templates. Instead of developing the
model from scratch, we found it much more instructive to have students select the
appropriate model from one of several templates provided and then adjust constants within the model to match the data.

A Reason: Student as Scientist

The rich set of MBL and modeling tools could have a major impact on physics and
teaching. It may make less important our present distinction between calculus-
based and noncalculus-based physics. We can use these tools to make physics far
less formidable and confusing for students whose mathematical abilities are not

strong. We can use these tools to place more emphasis on intuition, and at the same time give students the ability to solve interesting, complex problems.

Perhaps the most important and revolutionary option created by this technology is that it allows students to undertake their own original investigations. Much of what is wrong with science education is that students only learn about science, they do not participate in a meaningful way. Students at every grade level should have an opportunity to undertake physics experiments, to participate as fully as they can in learning something new about the natural world. Hands-on participation provides a motivation and focus that is hard to equal, and more important, it is the only way to give students an accurate understanding of what science is about, whether their careers lead them into science or their responsibility as voters forces them to deal with technological issues.

Our experience with the NGS Kids Network project shows that meaningful participation in science is both a possible and extremely valuable form of science education for students at the elementary level. Kids Network is important because it suggests what might be accomplished at the high school and college levels.

The Kids Network project is targeted at average fourth- through sixth-grade students. The project offers curriculum units that require four to eight weeks of classroom time. Each unit is based on one or more measurements that kids perform and share nationwide among themselves and with a participating scientist who has a research interest in the measurements. The first unit developed focuses on acid rain. In this unit students measure the pH of rain that they collect. In such an exciting environment, student learning is phenomenal. Not only do they learn how to measure pH and what some of the effects of acids and bases are, but they also frequently demonstrate deep understanding of the scientific method. After one group of students observed a high variability in their data (they were estimating the standard deviation), they decided on their own to repeat the measurement (thereby reducing the deviation by increasing the number of observations). John Miller, a deputy director of NOAA's acid deposition study, is extremely interested in the project; when fully implemented, the Kids Network project will have many times the 200 stations in this present network.

We expected the Kids Network to be interesting, but we were surprised by the enthusiasm and seriousness and deep learning that the project generated. The anecdotes that come from our extensive field testing are heartwarming. Asked whether they know any scientists, most students said "no" beforehand, but afterward many said, somewhat hesitatingly, "Well, we are all scientists." Teachers write us to say they are going to postpone their early retirements in order to continue teaching in the more exploratory vein that we encourage. Many teachers use the Kids Network not only as a way of teaching science but also as a focus to an integrated classroom experience because of its value in writing, reading, social studies, mathematics, and current events. Many of the participating teachers—remember, this is elementary school—reported that they had never before felt comfortable with science or technology. One principal devoted almost his entire commencement address to the project, using it as an example of how progressive his school was. And, of course, the potential for worldwide communication (eight sites in Canada, and one each in

Hong Kong, Argentina, and Israel) helps us all to get to know each other better and may hasten the day when war is unthinkable.

If we can provide a structured but meaningful scientific experience for elementary school students, even more exciting forms of collaborative projects are possible at the high school and college level. I dream of a time when students worldwide work together to collaborate on important and original projects, monitoring the environment, observing the stars, studying their cultures, and teaming with scientists for data collection and analysis. If every major high school had a team of students measuring radioactive fallout, then when there is another Chernobyl or Three Mile Island incident, not only will there be interesting data gathered, but there will also be believable, local experts who can talk reasonably about the level of radiation, background radiation, and biological hazards. What an opportunity to educate the public and stamp out some of the incredible misinformation that surfaces!

We will soon launch a new project, LabNet, that is designed to create such a network. Combining telecommunications with MBL and modeling technologies, LabNet will focus on giving teachers the background and tools necessary to foster and support student experimentation. We do not expect students to become scientists overnight on their own, but we can provide laboratories and scaffolding to encourage them to develop their own projects. One such offering is the "divergent laboratory." The divergent laboratory is a set of exercises that introduce a technique or kind of measurement in a way that is initially very structured but then opens out, or diverges, into many different possible measurements. We will also create a series of monitoring networks for measuring radioactive fallout, seismic activity, and weather. We will provide instructions for constructing simple stations in such networks. Simply setting up and calibrating such a station would be an important project, but we will also encourage students to improve on the designs and share those improvements, or to branch out on their own and create their own networks and experiments.

One of the critical aspects of this project will be a link of these students to professional scientists. In many other countries, professional scientists are much more involved with education than in the United States. To the extent that professional scientists' neglect of education stems from logistical problems and time constraints, telecommunications offers hope for the future. From the comfort of home, the participating scientists can make a major contribution by logging on to LabNet and assisting students any time they have a few available minutes. I appeal to all of you to help us in this effort and to recruit others.

Some of the materials incorporated in this work were developed with the financial support of TERC through the National Science Foundation grants DPE-8319155, MDR-8550373, and MDR-8652120 and with equipment donated by Apple Computer Corporation.

The Macintosh Oscilloscope

Elisha R. Huggins

Department of Physics, Dartmouth College, Hanover, NH 03755

We have been using the computer in our introductory physics courses since the late 1960s, when we used it mainly to calculate satellite orbits. (Students calculated Apollo orbits as the spacecraft were going to the moon.) The computer allows students to work realistic problems, which encourages students to work in areas not traditionally associated with introductory or even undergraduate physics.

A popular area for student projects has been the harmonic analysis of the sounds of musical instruments. The student compares the harmonic structure of notes from similar instruments of different quality in order to understand why one instrument sounds better than another. One pair of students compared high C on a spinet piano, an upright piano, and a grand piano. The harmonic structure beyond the fundamental was disorganized for the spinet, somewhat organized for the upright, and had a very definite structure for the grand piano. The students could see the progression to a larger, better-defined instrument. Another student analyzed the whale sounds from the Judy Collins record *Sounds of the Humpback Whale*. Although these sounds seemed nothing more than squeaks, harmonic analysis showed that they were produced by an instrument even more structured, more grand, than the grand piano. (Whales use their blowholes as organ pipes.)

The process of doing a harmonic analysis was horrendous in the late 1960s. Students took a Polaroid photograph of an oscilloscope, enlarged the photograph with an opaque projector, traced the outline on a large sheet of graph paper, located 100 or so data points, typed them into the computer on a Model 33 Teletype, and then wrote a program in BASIC to do the analysis. Students completing this procedure deserved considerable credit.

To streamline and simplify this process, we introduced graphics terminals, minicomputers, microcomputers, and finally a Macintosh. Our next-to-last system, a $12,000 minicomputer system, grabbed the data and automatically sent it to the time-sharing system for analysis. This system had several flaws that prevented its widespread adoption. First, it cost too much. Second, data could get lost in the transmission to time sharing. The addition of a microcomputer could have overcome either of these flaws. But a microcomputer could not overcome the most important problem: in the lab, it is more convenient to turn knobs and read dials than type commands. In short, a computer that is controlled by typing commands on a keyboard is not a good laboratory instrument. Because a computer in the laboratory feels like a computer, not a laboratory instrument, many colleagues would not use the system.

The Macintosh computer is different. The scroll bars, combined with accurate scales printed on the screen, provide outstanding instrumental control. And there are also various kinds of buttons for discrete controls and the menu for major changes in the mode of operation.

Four years ago we decided to turn the Macintosh into a first-class oscilloscope that had the ability to grab, store, and analyze data. Our main objective was to make the Macintosh feel like a laboratory oscilloscope so that our colleagues would use it. We closely modeled the Hewlett Packard oscilloscope that had been used successfully in our introductory labs for the past 20 years. At this point we can go up to a frequency only of around 10 Hz (50,000 points per second), but we have two differential inputs, similar time and trigger controls and amplification ranges, just as the HP scope does. We also have dual-beam capability with A and B, A versus B and A minus B, triggered either on A and B independently or on A alone. (In a chopped mode where we grab points alternately from the two curves, we have to run nearly ten times slower because the input amplifier and offsets have to be reset for each point.)

Our system has many capabilities that go beyond the HP scope. Our sweep time can be adjusted anywhere in the range from one millisecond to one month. When data are taken at a rate slower than 1/40 of a second per point, all 60-Hz noise is eliminated. When any curve is grabbed, in most cases we automatically get 15,000 data points (equivalent to 18 feet of Macintosh pixels). One of our main efforts has been to develop convenient ways of scrolling through and adjusting these data for convenient viewing.

The principle reason for turning the Macintosh into an oscilloscope is to have the experimental data in the computer so that they can be analyzed. We use the Fourier analysis package as our major analysis package in the program, but we have also built in a complete editor so that data files can be saved in text format for analysis by other programs such as *Excel* or *Cricket Graph*.

The original data file, along with all the conditions of the oscilloscope, can be saved as a "MacScope" file. When a "MacScope" file is opened, the Macintosh oscilloscope returns to precisely the same settings that were selected when the file was saved. We have used this feature to increase the reality of lecture demonstrations. During the demonstration we grab the data as a "MacScope" file and put that on our public network. There is also a copy of the "MacScope" program on the network. At night the students download both "MacScope" and the data file, run "MacScope" with the lecture-demonstration data, and carry out homework assignments on the actual data of the demonstration they saw in class.

The accompanying screen dumps in figures 1 through 12 illustrate some of the capabilities of the Macintosh oscilloscope.

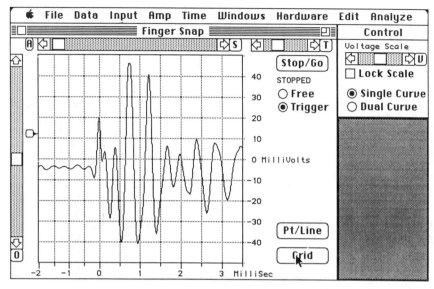

Figure 1.   Oscilloscope trace of the sounds of a finger snap. Notice the pre-
trigger data at times less than zero. The "S" scroll bar changes
starting times, the "T" scroll bar expands or contracts the time
scale, and the "O" scroll bar shifts the voltage offset. The trigger
level is set by the box with a pointer on the left side of the screen.

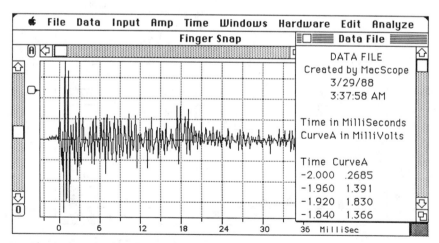

Figure 2.   Here we have contracted the time scale for the finger-snap data
of figure 1 and stretched the oscilloscope window to look at more
data. We have selected and saved data in the region from –2 ms
to 36 ms as the text file seen in the "Data File" window. This text
file can be used by programs such as *Excel* or *Cricket Graph* for
further analysis.

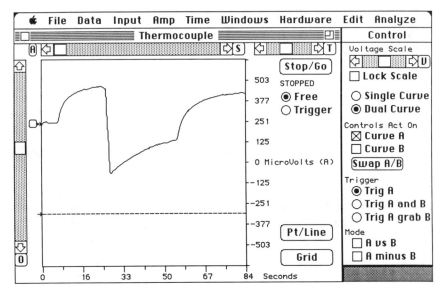

Figure 3.   With amplification up to a factor of 64,000 times, we do not need
an external amplifier to make direct readings of a thermocouple,
seen here as curve A. We have displayed the dual beam controls,
but here there is no input for curve B (dotted line).

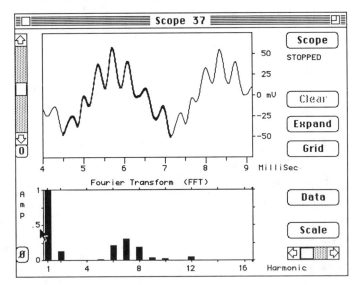

Figure 4.   Fourier analysis of the vowel sound "eee." We tell students that
the bottom window is a "mathematical prism" that separates the
sound waves into their different frequency or wavelength components. The mathematical prism, however, assumes that the wave
is periodic. That is why we have selected one cycle of the wave in
the upper window.

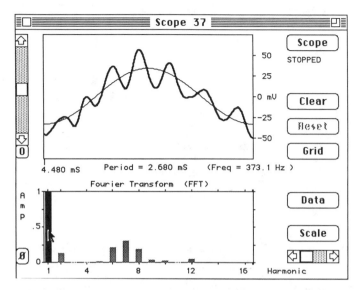

Figure 5.   The selected section of the sound wave of figure 4 is expanded to fill the upper display window, and we see that the wave has a frequency of 373.1 Hz. Clicking on the fundamental harmonic in the lower FFT graph shows how closely our selected section of sound wave can be represented by a single sine wave.

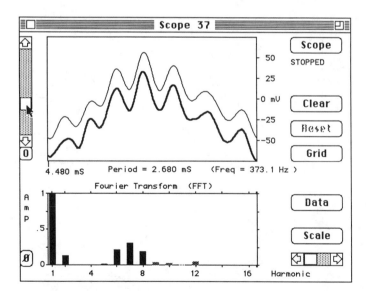

Figure 6.   Here we have selected all the harmonics up through the eighth, and we obtain a fairly accurate representation of the sound wave. Because we can select any combination of harmonics, we can study the effects of low-pass, high-pass, or notch filters.

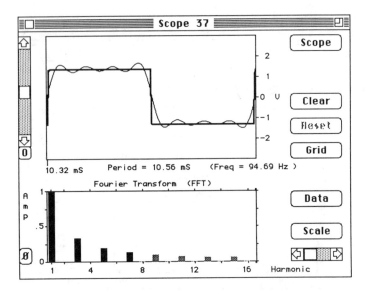

Figure 7. Analysis of an experimental square wave. We grabbed the square
wave from an HP oscillator, selected one period of the wave, and
then reconstructed the wave from the first four nonzero harmonics. Notice that the even harmonics are missing and the odd ones
go as 1/N. When you click on the button labeled Ø, the phases
and then the intensities of the Fourier components are displayed.

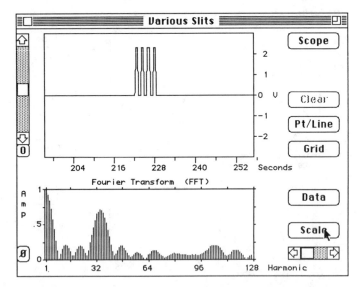

Figure 8. When a laser beam passes through a slit structure, it makes the
resulting diffraction pattern in the Fourier transform of the slit
pattern. Here we made a four-slit structure using a two-volt battery and a switch. The Fourier transform accurately resembles the
four-slit diffraction pattern seen and recorded in our diffraction
laboratory experiment.

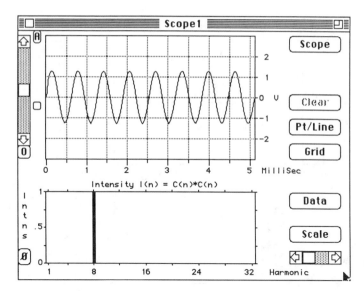

Figure 9. The uncertainty principle ΔE ΔT ≥ h. If we have a pure, infinitely long sine wave (ΔT = infinity), then we have a singleharmonic and the frequency (or energy) of a particle is known precisely (ΔE = 0).

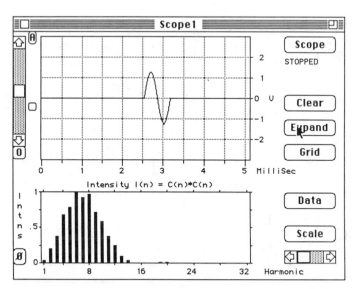

Figure 10. The uncertainty principle continued. If we look at a wave for only a short time ΔT (one cycle in this case), there is a fair spread of harmonics and the uncertainty in frequency or energy of the wave is large.

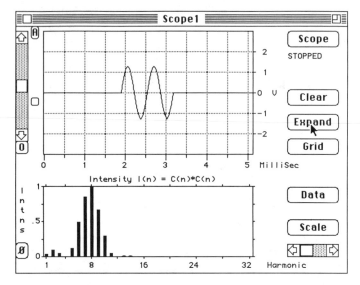

Figure 11. If we look at the wave for twice as long a time as in figure 10 (double ΔT), the spread in harmonics, and therefore ΔE, is cut in half.

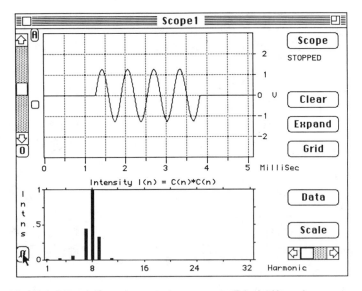

Figure 12. Doubling ΔT again cuts the spread ΔE in half again.

Tools for Scientific Thinking: Learning Physical Concepts with Real-Time Laboratory Measurement Tools

Ronald K. Thornton
Center for Science and Mathematics Teaching, Tufts University, Medford, MA 02155

Learner-controlled explorations in the physics laboratory with easy-to-use real-time measurement tools give students immediate feedback by presenting data graphically in a manner that they can understand. Using microcomputer-based laboratory (MBL) sensors and software, students can simultaneously measure and graph such physical quantities as position, velocity, acceleration, force, temperature, light intensity, sound pressure, current and potential difference. These MBL tools provide a mechanism for more easily altering physics pedagogy to include methods found to be effective by educational research. The ease of data collection and presentation encourage even badly prepared students to become active participants in a scientific process that often leads them to ask and answer their own questions. The general nature of the tools enables exploration to begin with students' direct experience of the familiar physical world rather than with specialized laboratory equipment. The real-time graphical display of actual physical measurements of dynamic systems directly couples symbolic representation with physical phenomena. Such MBL tools and carefully designed curricula based on educational research have been used to teach physics concepts to a wide range of students in universities and high schools. Data show that students learn basic physical concepts not often learned in lectures when they use MBL tools with carefully designed curricular materials.

Some Problems with Physics Education

There is strong evidence that introductory physics students in the usual college and university lecture courses are not learning concepts necessary to their understanding of the physical world. This paper presents some additional evidence. Traditional science instruction in the United States, refined by decades of work, has been shown to be ineffective in altering student misconceptions and simplistic understandings. Even at the university level, students (even science majors) leave physics courses with fundamental misunderstandings of the world about them essentially intact: their learning of facts about science remains in the classroom and has no effect on their thinking about the larger physical world. The ineffectiveness of these courses is independent of the apparent skill of the teacher, and student performance does not seem to depend on whether students have taken physics courses in secondary school.[1]

One of the reasons that traditional science courses have failed in this respect is that they do not make strong connections between the everyday experiences of the students and the concepts these students learn in the classroom. At the same time these courses do not address the incomplete "understandings" that serve students well within limited domains but do not lead to the general principles underlying deeper scientific understanding. Unless a course is planned carefully to examine simple understandings and to introduce general principles while addressing misconceptions, students' ideas will not change.

Developing Physical Intuition in the Physics Laboratory

Even successful physics students who can solve all of the problems at the end of the chapter generally lack physical intuition—a reliable, accurate response to the physical world based on an understanding of its underlying principles. In fact, physics honors students have been shown to have fundamental conceptual difficulties.[2] In contrast, students who develop physical intuition have a conceptual, qualitative understanding that can be applied outside of the classroom in their everyday interactions with the physical world. Thus it is likely to be fruitful to alter the way we teach science students: we should begin with what students can learn by arranging their interaction with the physical world around them, and then connect that learning to the underlying principles that constitute scientific knowledge.

A well-designed science laboratory can provide the sorts of experiences necessary to correct misconceptions and to develop useful physical intuition. The laboratory is one of the few places where students can truly participate in the processes of science by gaining firsthand knowledge of physical phenomena, constructing the theories necessary to understand the physical world, and formulating their own questions. Altering misconceptions generated by interaction with the physical world requires additional interaction. Recent developments in cognitive science and education substantiate the importance of empirical, heavily phenomenological experiences in learning science skills and concepts.[3] One of the ways Arnold Arons suggests to increase student learning is to provide the means for students to form concepts from concrete experience.[4]

In fact, laboratories are often omitted from or deemphasized in physics courses because the teaching laboratory is thought to be a place not where students learn physics, but a place where they develop laboratory skills of limited academic usefulness. The reality is that many science laboratories do not encourage exploration. They offer "cookbook" instructions that often seem unrelated to physics concepts, and they require time-consuming calculations. Many laboratory instruments are hard to use and unreliable. In addition, the results that emerge after an enormous amount of effort are what one would expect anyway. Because of these things, the traditional physics teaching laboratory is often ignored by faculty and disliked by students. Such laboratories are better omitted from courses because they discourage students. They provide no new information about nature and give an incorrect view of the process of science. Yet science laboratories are not a frill that can be discarded without consequence. The absence of the direct experience afforded by the laboratory can make science less interesting and less accessible.

Scientists rarely "preach what they practice." Science is exciting to scientists because they are engaged in discovery and in creatively building and testing models to explain the world around them (the practice). In most courses, students "do" no science; they hear lectures about already validated theories (the preaching). Not only do they not have an opportunity to form their own ideas, they rarely get a chance to apply the ideas of others to the world around them.

It is especially important for students in introductory courses to have direct experience with physical phenomena, but inexperienced students are particularly vulnerable to the problems with teaching labs described above. Such students have not developed sophisticated laboratory techniques, honed investigative skills, or become familiar with analytical skills, so it is difficult to construct laboratory experiences where they can successfully ask and answer questions that interest them. The effort to ensure that students with minimal laboratory skills get the "right" answer has led to exercises with overly explicit instructions that direct students through routine steps to confirm known answers to uninteresting questions.

Tools for Scientific Thinking

The Tools for Scientific Thinking project of the Center for Science and Mathematics Teaching at Tufts University is addressing some of the problems outlined above by introducing microcomputer-based laboratory (MBL) tools and curricula for colleges and high schools. Students need powerful, easy-to-use scientific tools to collect and display physical data it in a manner that can be remembered, manipulated, and thought about. Such tools can go a long way toward making teaching laboratories engaging and effective for developing useful scientific intuition. This sort of MBL tool was first developed at the Technical Education Research Centers (TERC),[5] and is now readily available.[6] MBL tools can eliminate the drudgery associated with data collection and display and allow students to concentrate on scientific ideas. MBL tools can be structured to encourage inquiry and thereby avoid "cookbook" laboratories. Because of their ease of use and pedagogical effectiveness, well-designed MBL instruments are especially well suited to the revitalization of science laboratories. Such tools make an understanding of physical phenomena more accessible to naive science learners and expand the investigations that more advanced students can undertake.

MBL instruments give science learners unprecedented power to explore, measure, and learn from the physical world. They do not simulate physical phenomena but change inexpensive computers into instruments for student-directed exploration of the physical world. MBL instruments of the type developed by TERC and Tufts make use of inexpensive microcomputer-connected probes to measure such physical quantities as temperature, position, velocity, acceleration, sound pressure, light, and force. These tools can also measure physiological indicators such as heart rate. Measurements taken by the probes are displayed in digital and graphical form on the computer monitor as the measurement is taken. Data can also be transformed and analyzed, printed, or saved onto disks for later analysis. Carefully developed software makes these laboratory tools easy to use the first time. MBL tools dictate neither what is to be investigated nor the steps of an investigation. Consequently, students feel in control of their own learning. Moreover, these general tools can be used with many different curricula by both physics majors and nonmajors.

As part of the Tools for Scientific Thinking project, we have tested laboratory curricula and MBL tools at varied institutions including California Polytechnic State University, Dickinson College, Massachusetts Institute of Technology, Muskingum College, University of Oregon, Tufts University, and Xavier University. The project is making available a number of microcomputer-based laboratory modules and curricula that emphasize the role of the physics teaching laboratory. The materials are designed to give students the means to build physical intuition (a conceptual, qualitative, usable understanding of the physical world). Such laboratories are accessible to the naive science learner and provide a foundation for the restructuring of science courses for nonmajors as well as majors. The project is funded by the Fund for the Improvement of Postsecondary Education (FIPSE) of the U.S. Department of Education. This project is linked to another FIPSE-funded project, Workshop Physics, which is developing an entire introductory physics course sequence that will replace the traditional physics course with a workshop format designed to enhance student interaction with the physical world.[7]

Learning Kinematics Concepts with MBL Tools and Curricula

Using a motion detector designed by TERC and curricula developed by the Tools for Scientific Thinking project at Tufts, university and high school students have successfully learned kinematics concepts that they did not learn in lectures. The motion detector (hardware and software) is able to measure, display, and record the distance, velocity, and acceleration of any object. The hardware was developed from a sonic transducer used in Polaroid cameras. The motion probe is essentially a SONAR unit that sends out short pulses of high-frequency sound (50 kHz), and then detects and amplifies the echo. A microcomputer is then programmed to measure the time between the transmitted and received pulse and to calculate the position, velocity, acceleration of the object causing the reflection (much as a bat is able to do). Any one of these quantities may be displayed on the computer screen as the data are taken, and all are available after the measurements are completed. The motion detector can accurately detect objects between 0.5 and 6 meters. It detects the closest object in roughly a 15° cone. The motion detectors are connected to Apple II computers.

The kinematics laboratory curriculum was designed using guidelines and beliefs common to curriculum in all of the various subject areas on which the project has been working. A fundamental belief is that physical concepts are best learned in a laboratory setting. The curriculum is heavily based on research and uses a guided- discovery approach that makes use of student predictions, pays attention to student alternative understandings, supports peer learning, and provides opportunities for students to construct knowledge for themselves. Students use their own motion to learn kinematic concepts.[10]

Student Understanding of Simple Kinematics

Early on we tested students' knowledge of simple concepts in kinematics and classical mechanics so that we could address difficulties in the laboratory. Work by other researchers and our own teaching, had made us aware of the standard alternative understandings and student difficulties.[9] In spite of this knowledge, we were not prepared for the large percentage of university students who had, with even the most basic qualitative concepts. To confirm our findings, we decided to collect results from a larger sample of physics students. Because it is difficult to convince physics professors to give up any course time and because we wanted to make evaluation less subjective, we decided to use short-answer questions. From earlier work with students we evolved a set of short-answer questions that give a reasonable indication of students' basic knowledge about kinematics and its graphical representation. The questions were such that most university physics professors were sure that no more than 10 percent of their students would miss them. Most professors also agreed that students who could not answer questions such as these did not have a basic understanding of kinematics. In fact there was considerable concern that giving such questions was a waste of time and an insult to the intelligence of students.

For this study the sources of the data were college and university physics-teaching laboratories of members of the Tools for Scientific Thinking project. The student population was primarily students enrolled in introductory algebra or calculus-based physics courses. In general students were given two kinematics labs lasting between two and three hours. These labs replaced standard university laboratories. In one case, students were given one laboratory lasting only one hour and 15 minutes. The motion detector connected to an Apple II series computer was used with curricula designed to guide students to explore fundamental motion concepts. Additional characteristics of the curriculum are described above. Students worked in small groups (two to four students). Homework related to the lab was assigned. The labs do not depend explicitly on knowledge gained in lectures or textbooks. The labs covered the position and velocity of moving objects and introduced acceleration. The relationship between force and acceleration was not covered.

The following figures give some of the results of this study. All data presented here, pre- and post-MBL, were taken after students had had the usual kinematics lectures and (in most cases) had done the usual problems. (Other data from smaller student samples seem to show that lectures have no effect on how well students do in the labs. If the labs are done first, class discussions in smaller classes seem to indicate more interest and understanding.) We almost always give the average error rates on a question. Error rates greater than 15 percent on simple conceptual questions such as these are probably cause for altering the curriculum.

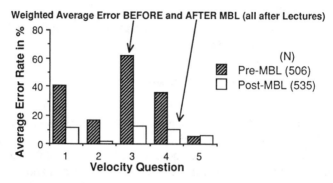

Figure 1.   Average error rate on velocity questions after lectures and before and after MBL.

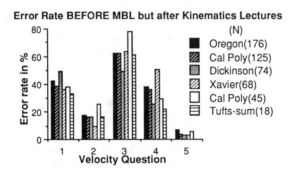

Figure 2.   Error rate on velocity questions before MBL but after lecture.

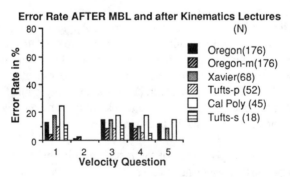

Figure 3.   Error rate on velocity questions after MBL and after lecture.

The dark bars in Figure 1 show the weighted average error rate for 506 physics students from five different universities who answered the simple questions shown in Figure 4. The large error rates are especially surprising since these students had recently heard the standard kinematics lectures and worked the problems associated with them. Figure 2 shows the error rates for the questions above for each of the six student populations making up the weighted average. The pattern is remarkably stable across different student populations.

The large error rates associated with questions one and three are not the result of the wrong choice of "sign," which only accounts for a few percent of the answers. The most common error is the choice of the "distance analog" graphs A and B. But the students do not consistently make such a choice, as the different error rates for questions one and three show.

The error rates for the same five questions after one or two MBL motion laboratories are shown in Figure 1 as the white bars. Note that the pre- and post-MBL populations are not identical. Figure 3 shows the six different populations making up the post-MBL results. The pre-MBL sample is 45 percent physics-with-calculus students, while the post-MBL sample is only 28 percent physics-with-calculus students. Calculus students do better than noncalculus students on the pretest but the effect is only about ±10 percent from the average error rate. Post-test results are much closer. In addition, the 52 Tufts students included in the post-MBL sample were in a "Physics for Humanists" science course that did not generally emphasize kinematics. Their MBL lab exposure was only one short lab (about seventy minutes). No pretest was given because most students had not had physics and many could not read graphs.[8] The post-MBL data also include two different post-tests on the Oregon students. One of these tests is delayed (the midterm). Either one can be removed without greatly changing the weighted averages shown in Figure 1. These Oregon data will be examined more closely in what follows.

Smaller sample studies have shown that post-tests given as part of the homework, produce essentially the same results as post-tests given in class for the kind of question above. In general, students do slightly better when the post-test is given in class, which is the reverse of what one might predict. Smaller sample studies also show that students given the same questions for a pretest and post-test do not do better than students given only the post-test.

The Oregon Kinematics Results

Thanks to Professor David Sokoloff, we were able to do more intensive studies with students at the University of Oregon. Figure 5 shows the results on these questions for 172 Oregon students (36 percent were enrolled in a physics-with-calculus course and 64 percent in an algebra-based course. All students had finished kinematics when they took the pre-MBL test. Students were given two labs using the motion detector. Students also received correlated homework questions. The labs primarily concentrated on the distance and velocity of moving objects (includ-

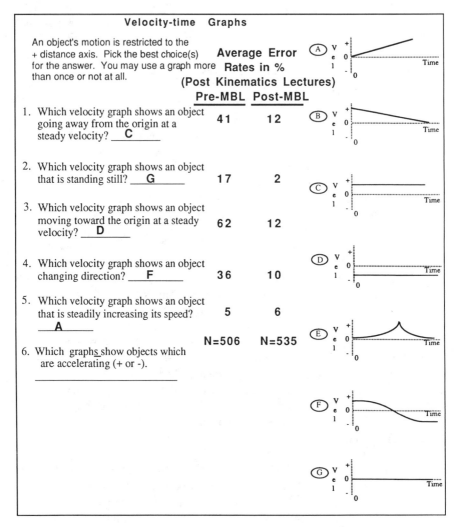

Figure 4. Questions on velocity-versus-time graphs. Questions one through five are the questions referred to in figures 1 through 3. The order of the questions and the graphs was sometimes changed.

ing the students' own bodies). Acceleration was introduced, but not systematically. The post-MBL questions were given as part of the homework, which was graded and returned (answers, however, were not posted).

About three weeks later, the same questions, rearranged, were given as part of the midterm. The results of this delayed-post-test are also shown in Figure 5. The results indicate that the homework helped them to understand kinematics. Is it possible that students have "memorized" the questions after seeing them twice before? Evidence exists that giving a pretest does not help students get better scores on a post-test. Besides, the rearrangement of the questions would prevent strict memo-

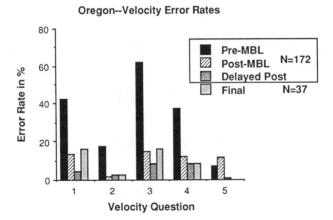

Figure 5.   Error rate on velocity question for Oregon class.

rization and the students also did as well on other, different, questions. Therefore, it seems likely that they now understand these simple concepts.

We were able to give the first four questions (again rearranged) to a small sample of 37 noncalculus students as part of their final exam at the end of the term. The results, shown also in Figure 5, seem to indicate that their understanding is retained over a substantial period of time.

We also asked questions on acceleration even though acceleration had only been introduced and not systematically covered in the labs. Some results are shown in Figure 6. The questions are shown in Figure 7.

Figure 8 shows the student error rates for 11 questions on the final exam for 90 students who attended the same lecture session of the algebra-based physics course. The 53 students who did not do the two MBL kinematics laboratories had substantially higher error rates. The lectures were the same for all students. The table below summarizes the average error rates for each category of question.

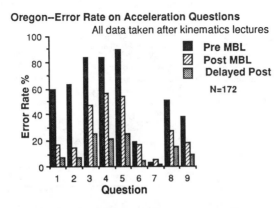

Figure 6.   Error rate on acceleration questions.

Table 1.

Average Error Rates for Students in the Algebra-Based Physics Course

Topic	Questions	with MBL	without MBL
Velocity	1,3,4	13%	39%
Acceleration	5-9	36%	56%
Force	10, 11	55%	55%

Acceleration-Time Graphs

The graph below is an acceleration-time graph for a car, the motion of which is restricted to the + distance axis. Choose the letter of the section of the graph which <u>could</u> correspond to each of the following motions. Choose the <u>one</u> best answer. If you think that none correspond, write **N**.

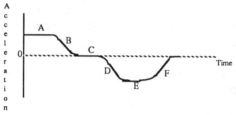

	Error Rates in %		
Correct Answer	Pre- MBL	Post- MBL	Delayed Post
__A__ 1. The car could be speeding up at a steady rate, moving away from the origin.	59.	16.	7.
__C__ 2. The car could be moving at a constant speed away from the origin.	63.	14.	7.
__E__ 3. The car could be slowing down at a steady rate, moving away from the origin.	84.	47.	25.
__C__ 4. The car could be moving at a constant speed toward the origin.	84.	56.	21.
__E__ 5. The car could be speeding up at a steady rate, moving toward the origin.	90.	54.	25.

Acceleration

Consider the acceleration in each of the following situations. For each of the following descriptions of the motion of an object, write a "+", "-" or "0" in the space to indicate that the acceleration is positive, negative or zero.

__-__ 6. A car is moving in the positive direction and comes to rest.	19.	17.	4.
__+__ 7. A car starts from rest and begins to move in the positive direction.	3.	5.	2.
__+__ 8. A car is moving in the negative direction and comes to rest.	51.	27.	15.
__-__ 9. A car starts from rest and begins to move in the negative direction.	38.	18.	10.

Figure 7.   Comparison of students who <u>had</u> MBL laboratories to those who did not.

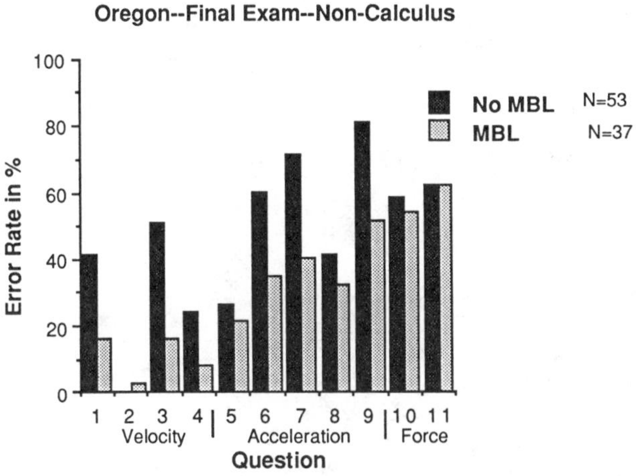

Oregon--Final Exam--Non-Calculus

Figure 8.   Error rate on each question for Oregon group.

The "with MBL" results are quite consistent with the coverage of the MBL lab-oratories. The MBL labs covered distance and velocity well. They introduced but did not emphasize acceleration, and they did not cover the relationship between force and acceleration. The velocity questions were selected from the ones described earlier. The error rates of the MBL students increased a few percent compared to the delayed post-test, but retention over the two-month period is very high. The error rates for the no-MBL students are unfortunately consistent with the pre-MBL error rates taken earlier in the course, except that students apparently did learn through the standard treatment to recognize the velocity graph of an object standing still (question 2). Questions 5 through 11 are shown in Figure 9.

Is MBL Software Pedagogically Successful?

There has been considerable discussion about whether computer-based learning offers substantial advantages over other methods, but there has been very little evi-dence published. Can the MBL software discussed in this paper be considered ped-agogically successful? This paper shows evidence of substantial persistent learning of very basic physical concepts by students using a particular set of curricular materials. These same simple physical concepts were not learned by large numbers of students when they listened to good traditional physics lectures, read respected textbooks, and did the traditional algorithmic problems at the end of the chapters. The laboratory curriculum used by these students was made possible (or at least practical) by the use of a microcomputer-based motion tool that used the power of the computer to allow students to see actual measurements of physical phenomena displayed in real time as graphs. The immediate coupling of the graphs to the phys-ical phenomena seems to lead students both to understand graphing as a useful sci-entific symbol system and also to understand physical concepts from an examination of appropriate phenomena.

Questions 5-11 Given to All Students on the Final Exam

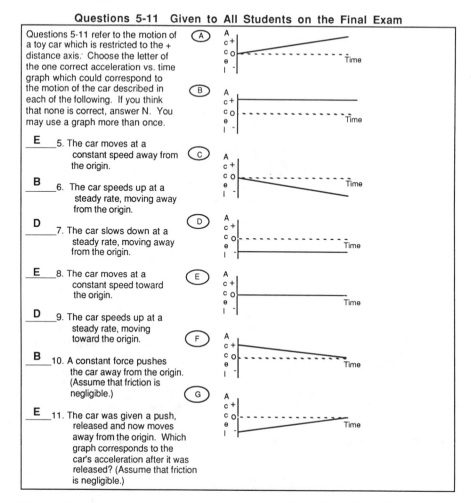

Questions 5-11 refer to the motion of a toy car which is restricted to the + distance axis. Choose the letter of the one correct acceleration vs. time graph which could correspond to the motion of the car described in each of the following. If you think that none is correct, answer N. You may use a graph more than once.

__E__ 5. The car moves at a constant speed away from the origin.

__B__ 6. The car speeds up at a steady rate, moving away from the origin.

__D__ 7. The car slows down at a steady rate, moving away from the origin.

__E__ 8. The car moves at a constant speed toward the origin.

__D__ 9. The car speeds up at a steady rate, moving toward the origin.

__B__ 10. A constant force pushes the car away from the origin. (Assume that friction is negligible.)

__E__ 11. The car was given a push, released and now moves away from the origin. Which graph corresponds to the car's acceleration after it was released? (Assume that friction is negligible.)

Figure 9. Questions on acceleration versus time. Questions five through eleven were given to all students on the final exam.

What happens when beginning students use MBL tools to do more or less traditional physics experiments? The students are pleased. They may do additional exploration, and they do a better job in general. Preliminary evidence shows that they do not, however, learn fundamental physical concepts unless the means of teaching them are built into the experiments. We can certainly say that the MBL software and hardware tools are pedagogically successful for teaching such physical concepts when used in combination with research-based curriculum materials.

This work was partially funded by the Fund for the Improvement of Postsecondary Education (FIPSE, U.S. Department of Education) under the Tools for Scientific Thinking Project at Tufts University.

1. I. A. Halloun and D. Hestenes, "Commonsense Concepts about Motion," Am. J. Phys. **53**, 1056 (1985); I. A. Halloun and D. Hestenes, "The Initial Knowledge State of College Physics Students," Am. J. Phys. **53**, 1043 (1985).
2. P. C. Peters, "Even Honors Students have Conceptual Difficulties with Physics," Am. J. Phys. **50**, 501 (1985).
3. L. C. McDermott, M. L. Rosenquist, and E. H. van Zee, "Instructional Strategies to Improve the Performance of Minority Students in the Sciences," New Directions for Teaching and Learning **16** (1983).
4. A. Arons, "Achieving Wider Scientific Literacy," Daedalus **112**, 91 (1983).
5. Technical Education Research Centers (TERC), 1696 Massachusetts Ave., Cambridge, MA 02138.
6. *HRM Software* (Queue, 562 Boston Avenue, Room S, Bridgeport, CN 06610).
7. Personal communication, Priscilla Laws, Director, Dickinson College, Carlisle, PA 17013.
8. R. K. Thornton, "Access to College Science: Microcomputer-Based Laboratories for the Naive Science Learner," Collegiate Microcomputer, **5** (February 1987).
9. L. C. McDermott, "Research on Conceptual Understanding in Mechanics." Phys. Today **37** 24 (1984).
10. R. K. Thornton, "Tools for Scientific Thinking: Microcomputer-Based Laboratories for Physics Teaching," Phys. Ed. **22**, 230 (1987).

A Collection of Laboratory Interfacing Ideas

David L. Vernier
Vernier Software, 2920 S.W. 89th St., Portland, OR 97225

Computers are now widely used in many physics labs and classrooms for graphing, data analysis, timing, and temperature measurement. This paper presents some laboratory applications that are less widely used. Most of these applications require hardware to allow the computer to do voltage measurement. There are several types of devices that can accomplish this task, including analog-to-digital converter boards, voltage-to-frequency converters, and serial output analog-to-digital converter ICs that can connect to the game port. For some of the applications described, the voltage signal changes quickly and the voltage-input device must be capable of collecting data at a rapid rate. Also note that some of the circuits described require +12 V and -12 V power supply lines. Others can operate off of only a +5 V power supply lead. The AC control circuit is specific to the Apple II computer and requires no additional hardware.

Measuring Magnetic Fields

A voltage-to-frequency converter or an analog-to-digital converter can be used with a linear Hall effect sensor to measure magnetic field intensities. The circuits

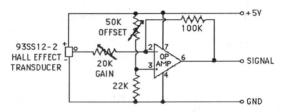

Figure 1. Variable-gain Hall Effect amplifier.

below allow you to measure the field near a small magnet or even the earth's magnetic field.[1] They use Micro Switch 93SS12-2 linear Hall Effect sensors.[2] An operational amplifier steps up the voltage to the range where it can be monitored by the computer. The output voltage is proportional to magnetic induction.

The first circuit (Figure 1) uses one sensor and is designed for use around rather strong fields (laboratory magnets, etc.). The voltage gain of the amplifier circuit can be changed by adjusting the 20-kΩ potentiometer. The other potentiometer is used to set the offset voltage. Note that this circuit requires only a +5 V power supply lead.

The second circuit (Figure 2) uses two back-to-back sensors and provides more amplification. This circuit can detect weaker fields, including the magnetic field of the earth. This circuit uses an amplifier with a gain of 100. The 100-kΩ potentiometer allows you to set the offset voltage. It can be connected to either the +12 V positive or -12 V negative lead, depending on which way you need to offset the voltage. This circuit produces about 0.8 V when aligned with the earth's magnetic field.

Accelerometer

Sensym has recently introduced a series of small, general-purpose accelerometers.[3] These devices are made of micromachined silicon wafers. They contain a "seismic mass" mounted on a beam equipped with piezoresistive elements. Acceleration of

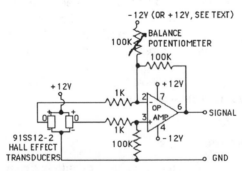

Figure 2. High-gain Hall Effect amplifier.

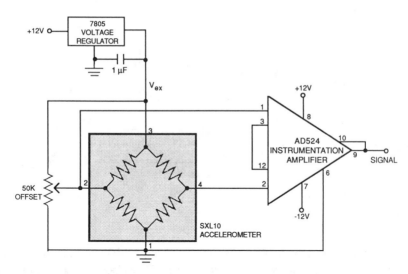

Figure 3.   Accelerometer circuit.

the sensor causes the beam to bend and the resistance of the piezoresistive elements to change. The resistive elements are arranged in a Wheatstone bridge circuit within the accelerometer. The output-potential difference is proportional to the acceleration of the sensor. These accelerometers are referred to as the Sensym SXL series. They are manufactured with maximum acceleration of 10 g, 20 g, or 200 g. The price of one sensor is $50 to $75.

We have interfaced an SXL10 accelerometer to an analog-to-digital converter using the circuit shown in Figure 3.

Four thin, flexible wires are used to connect the accelerometer and amplifier, which are mounted on a small circuit board, to the A/D converter. The excitation voltage for the Wheatstone bridge is provided by a regulated 5-V supply. The data sheet for this particular SXL10 accelerometer lists its output as 2.361 mV per g with a 5-V excitation voltage. The instrumentation amplifier shown in the diagram above is set up for a gain of 200. The output voltage should therefore be 0.472 V/g. When the accelerometer is dropped, the voltage produced by the circuit changes approximately the expected 0.47 V. You can also use the accelerometer to measure and demonstrate centripetal acceleration by spinning it around in a circle by its connecting wires. It can also demonstrate the acceleration of an object in simple harmonic motion if it is allowed to bounce at the end of a spring.

Studying sound waves

If a microphone and amplifier circuit are connected to an analog-to-digital converter with a sufficiently high sampling rate, the computer can take the place of an oscilloscope in studying sound waves. The circuit diagram in Figure 4 uses an inexpensive electret microphone to produce a voltage signal that can be displayed.[4]

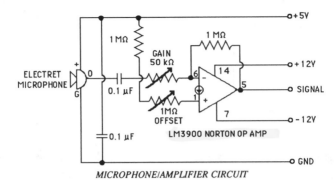

MICROPHONE/AMPLIFIER CIRCUIT

Figure 4.   Microphone/amplifier circuit.

The amplification is done by using one fourth of an LM3900 Norton op amp. The potentiometers are used for sensitivity control (the 50-kΩ pot) and setting the voltage level with no sound (the 1-MΩ pot).

Using a computer to take the place of an oscilloscope for sound-wave studies allows you to store the data in the computer and use it for further numerical analysis. Students can learn a lot about frequencies, periods, amplitudes, beats, overtones, and harmonics. The new electronic keyboards, which provide a variety of simulated sounds and allow synthesis of sounds, make excellent demonstration tools.[5] The graphs in Figure 5 show computer displays made while an experimenter played a fife at about 580 Hz (left graph) and whistled at one octave higher (right graph).

A Quantitative Momentum-Impulse Experiment

Textbook problems involving impulse nearly always assume a constant force. Unfortunately, real-world forces are rarely constant. In this experiment (Figure 6) you study a real collision and compare the impulse with the change in momentum.

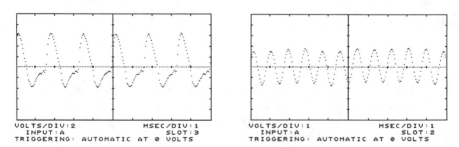

Figure 5.   Sample oscilloscope simulation screens.

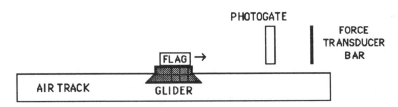

Figure 6. Collision experiment setup.

A glider on a level air track collides with the metal bar of a force transducer. The bar flexes and the glider bounces off of it. A photogate is mounted so that the glider will pass through it just before it hits the force transducer bar. The purpose of the photogate is to provide a triggering signal and to measure the speed of the glider before and after the collision. The force exerted on the bar of the force transducer is measured during the collision. Software can plot a force-versus-time graph. It also does the integration during the time of the collision to calculate the impulse. The times measured by the program also allow you to calculate speeds and momentum change.

The equipment required for this experiment is as follows:

1. A force measurement device (either a homemade strain-gauge system or a PASCO dynamic force transducer[6]).

2. A photogate connected to the Apple game port.

3. An air track and one glider. The experiment can also be done using a dynamics cart, with slightly larger errors introduced by frictional effects.

4. An A/D converter capable of sampling at a minimum of 100 Hz.

5. Software to collect force data during the collision and to measure the speed of the glider before and after the collision. The Vernier Software program *Voltage Plotter III* includes features that allow it to be used for this experiment.[7]

Before the experiment is performed, the force transducer should be calibrated so that its voltage output can be converted to force. During an actual experimental run, the program does the following:

1. It waits until the photogate is first blocked and then begins timing.

2. It continues timing until the photogate is unblocked. The time it takes the flag on the air-track glider to pass through the photogate is therefore measured.

3. When the photogate is unblocked, the program begins taking force measurements. A graph of force versus time is displayed on the screen.

4. The program then waits until the photogate is blocked again (by the rebounding glider). The time it takes the flag of the glider to pass through the gate is again measured.

The two times are displayed and used to calculate the change in momentum. The graph of force versus time can be used to calculate the impulse by integrating over the time of the collision. The integration is another task that can be done efficiently by the computer.

A sample graph, along with the data and calculations, is shown in Figure 7.

Integrating under the force-versus-time graph (Figure 7) from the start of the collision to the point where the curve goes negative yields

$$\text{Area} = 185 \text{ N-ms or } 0.185 \text{ N-s.}$$

The change in momentum is calculated as follows:

The velocities of the air track glider measured by the program before and after the collision were:

$$v_{before} = 0.400 \text{ m/s and } v_{after} = -0.269 \text{ m/s.}$$

The change in velocity is the vector difference between these two is

$$\Delta v = v_{after} - v_{before} = -0.269 - 0.400 = -0.669 \text{ m/s.}$$

The mass of the glider was 0.274 kg. So the change in momentum is

$$\Delta \text{ momentum} = \text{mass } \Delta v = (0.274 \text{ kg}) (-0.669 \text{ m/s}) = -0.183 \text{ kg-m/s.}$$

Notice that this agrees very nicely with the change in impulse.

Pressure

Several manufacturers produce pressure sensors that can be interfaced to a computer.[8] The Sensym SCX series of pressure transducers is "temperature compensated" to minimize the drift caused by temperature change. This series of pressure transducers includes sensors for measuring either absolute or differential pressure with ranges from 0 to 1 psi up to 0 to 100 psi. These sensors cost about $25.

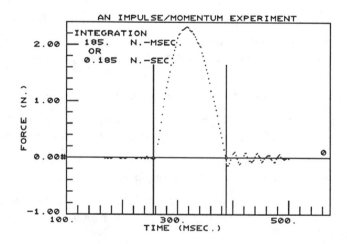

Figure 7.   Impulse measurement.

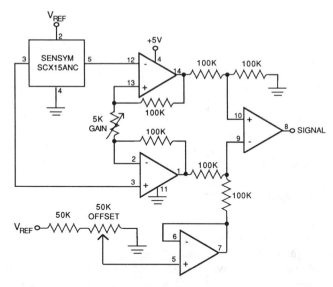

Figure 8.   A variable-gain pressure sensor circuit.

The circuit in Figure 8 uses an SCX15ANC sensor. An excitation voltage (V_{REF}) is applied to the pressure sensor. Internally, it uses a Wheatstone bridge circuit to produce a potential difference between its output terminals that is proportional to the pressure. For the SCX15ANC, the output is 2 mV per psi when a 4-V excitation voltage is used.

The amplifier circuit above uses only a +5 V power supply lead. This allows it to be used with a variety of voltage measurement devices (+12 V and -12 V power supply leads are not required). The four op amps in the circuit above are all contained in one TLC274 quad op amp. The pin numbers on the TLC274 are indicated in the diagram. The 5-k potentiometer controls the gain of the amplifier. The overall gain is $1+2\times100k/R$, where R is the resistance of the gain potentiometer. The bottom part of the circuit provides an offset adjustment.

To minimize the error caused by changes in the excitation voltage, a voltage reference circuit is recommended to produce a very stable V_{REF} from the +5 V power supply voltage. One possible circuit is shown in Figure 9.

We have begun testing these circuits with the goal of producing an inexpensive barometer. Measuring atmospheric pressure changes is difficult to do accurately because the pressure changes are relatively small. The circuits described here seem to be very stable and reliable. Additional offset circuitry and more amplification will probably be needed to produce a computer-interfaced barometer.

AC control

Controlling a 120 VAC circuit with a computer is a fairly simple task, thanks to solid-state relays. These electronic components are available from many mail-order

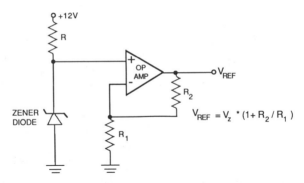

Figure 9. Voltage reference circuit.

suppliers.[9] Prices vary with the current rating of the relay. They start at about $10.00 for new (nonsurplus) solid-state relays.

Most solid-state relays have four terminals. The control (input) terminals are usually designed to operate on a 3-V (minimum) DC signal. They require only a few milliamps of current to turn on the solid-state relay. The output terminals are normally open, but conduct AC current when the input terminals have a voltage applied across them.

Because the input terminals require only a small DC voltage and draw a small current, a computer can easily be used to control the solid-state relay. In the circuit below, which is designed for use with the Apple II game port annunciator lines, a 7404 inverter IC is used as a buffer to make sure that the circuit does not sink too much current into the annunciator line.

On an Apple II, the circuit shown in Figure 10 (using the AN0 line) can be controlled with the following BASIC statements.

To turn the AC circuit on, POKE 49241, 0. To turn the AC circuit off, POKE 49240, 0.

Note that there are four annunciator output lines on the Apple II internal 16-pin game port, so four different circuits could be controlled. The annunciator output lines are not available on the Apple II 9-pin external game port connector.

Computer-controlled AC lines have many uses in the physics lab. Combined with other circuits and software, they could be used to turn on an alarm bell or

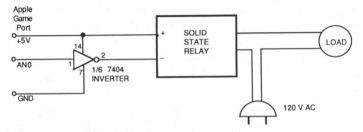

Figure 10. Apple II AC control circuit.

light, operate lab equipment automatically, or maintain a constant temperature in a region by controlling a heater.

1. These circuits are based on an article by P. A. Bender, Am. J. Phys. **54**, 89 (1986).
2. Micro Switch manufactures several Hall effect transducers. For information contact Micro Switch, 11 West Springs Street, Freeport, IL 61032. The Hall effect transducers cost about $7.00 each. Texas Instruments and other manufactures also produce linear Hall effect transducers.
3. Sensym1255 Reamwood Avenue, Sunnyvale, CA 94089, tel. (408) 744-1500.
4. The electret microphone is Radio Shack part number 270-092.
5. The Casio SK1, for example, has five preset tones and digital sampling. It also allows you to synthesize a sound by adding harmonics.
6. PASCO Scientific, 1876 Sabre Street, Hayward, CA 94545 tel. (800) 772-8700.
7. *Voltage Plotter III* (Vernier Software, 2920 S.W. 89th St., Portland, OR 97225, tel. (503) 297-5317).
8. Two sources are Sensym, 1255 Reamwood Avenue, Sunnyvale, CA 94089, (408) 744-1500, and Nova Sensor, 1055 Mission Court, Fremont, CA 94539, tel. (415) 490-9100.
9. Two mail-order electronics suppliers who usually carry solid-state relays are All Electronics, tel. (800) 826-5432 and Mouser Electronics, tel. (800) 346-6873.

Five Years of Using Microcomputers in Basic Physics Laboratories at the U.S. Naval Academy

David A. Nordling
Physics Department, U. S. Naval Academy, Annapolis, MD 21402

Early in 1983 the physics department of the U. S. Naval Academy began using the microcomputer in all teaching laboratories. All students at the Naval Academy must take at least one year (two semesters) of introductory physics. This means that each academic year approximately 1,100 sophomore students take introductory physics. Approximately 30 to 40 of these students will have chosen physics as their major and therefore will take additional physics courses.

The physics department at the Naval Academy has always used the teaching laboratory to supplement the traditional classroom presentation of physics. By 1981 many of us felt our laboratory program had fallen short of its intended objectives, particularly in the large introductory physics course. Instructor enthusiasm for these laboratories was waning, and this lack of enthusiasm carried over into students' participation. One major cause for the lack of enthusiasm was the poor state of much of our lab apparatus and measuring equipment.

It became clear in 1982 that if we were to teach viable physics courses at the Naval Academy, we would have to repair or replace most of our existing laboratory equipment, including measuring devices (stop watches, volt meters, ammeters, etc.).

In January 1983 we began to investigate the application of the microcomputer as a laboratory tool. By the end of that spring semester the concept of a laboratory microcomputer work station began to emerge.

Laboratory Microcomputer Work Station

We chose the following equipment for our work station:

1. Apple IIe microcomputer with 128K or 192K of memory.

2. Monochrome CRT display monitor.

3. One 5 1/4 diskette drive.

4. Dot-matrix printer with parallel interface and graphic dump capability.

5. Analog-to-digital converter with eight double-ended (or 16 single-ended) input channels, and three independent counter/timers, each driven independently from the system clock or other source.

6. Interface box for easy connection to the A/D and counter/timers. The interface box also contains ±5 and ±12 volts DC power supplies for experimental hookup.

7. Two photogates interfaced through the Apple IIe game-port interface, used for timing.

8. Portable rolling cart to hold all components associated with the microcomputer work station.

The approximate cost per work station was $3,500.

In the fall of 1983 we were able to get 15 microcomputers and with about half, were outfitted with A/D converters. We began immediately using the microcomputer as a laboratory measuring device in the lab course devoted to teaching basic physics (heat, sound, and light) to the 30 or 40 students selecting physics as a major. One year later (fall 1984) we had approximately 75 work stations in the basic and advanced physics lab spaces. Because of scheduling and space limitations, three- or four-student lab teams are assigned to one microcomputer work station during a laboratory period.

Experiments

Students use the microcomputer work station in the following experiments:

1. Free fall. A study of accelerated linear motion.

2. Motion on an inclined air track. A study of accelerated linear motion and conservation of energy.

3. Motion on a rough inclined plane. A study of friction and conservation of energy.

4. Collisions on an air track. A study of conservation of linear momentum, elastic and inelastic collisions.

5. Impulse and change in momentum. A study of the relation between the impulse and change in momentum as an object collides with a force transducer.

6. Rotation. A study of rotational kinematics and dynamics.

7. Simple pendulum. A study of a simple pendulum with both large and small amplitudes. Also an investigation of damped periodic motion.

8. Viscous drag. A study of viscous drag on objects as they move through air.

9. Viscosity of water. A measure of the viscosity of water using Poiseulle's law at room temperature and near the ice point. Also a study of exponential processes.

10. Specific heat. A measure of the specific heat of a solid.

11. Latent heat. A measure of the latent heat of liquid nitrogen.

12. Standing waves on a string. A study of standing waves on a string.

13. Standing waves in an air column. A study of standing waves in a column of air.

14. Newton's law of cooling. A study of the cooling of heated objects near room temperature.

15. Electric fields. A study of the electric field between various electrode configurations in a tray of water.

16. Simple electrical circuits. A study of Ohm's law and Kirchoff's rule in relation to simple electrical circuits.

17. RC circuits. A study of electric current, charge, and potential in a simple series RC circuit.

18. LRC circuits. A study of electric current, charge, and potential in a simple series LRC circuit.

19. Faraday's law. An observation of the emf generated as magnets fall through coils of wire.

20. Simple lenses. A study of thin lenses and lens combinations.

21. Diffraction grating. An investigation of the helium spectra using a diffraction grating.

22. Prism spectrometer. Measure of the index of refraction of a glass prism for various wavelengths in the helium spectrum. Shows the dispersion curve for flint glass.

Software

In 1983, when I became involved with our efforts to use the microcomputer in the lab program, it seemed clear to me that carefully written software was essential. I have tried to incorporate the following key features in the software written for our laboratory work stations:

1. The software is written so the work station becomes a laboratory tool capable of reading, recording, and displaying voltages from user selected A/D channels; recording timing data based on when voltages are read or the interruption of a light path in a photogate; saving, mathematically manipulating, analyzing, and graphing data that has been gathered; and outputting user-selected frequency signals.

2. With only three or four obvious exceptions, the software is not tied to any specific experiment.

3. The software does not control the flow of the experiment.

4. The student must decide if (and how) the data should be manipulated or graphically represented.

5. The software on a given floppy disk is integrated into a unified package that is completely menu driven.

6. The software is organized on each disk so that only one disk is needed for any specific lab experiment.

7. User input and control are maintained to a minimum of keyboard inputs. Generally two types of inputs are required: (a) arrow keys and return key for selecting menu items; (b) keyboard input of file names for data storage and recovery from disk or ram.

All of the software for the laboratory work station has been placed on eight floppy disks. A given laboratory experiment requires only one of the eight disks. The software is organized so that during the initial boot operation all (or most) of the software is loaded into RAM. A student's own disk may be used for data storage (if desired). Space is also made available in RAM for some data storage. Data accessed to and from RAM is, of course, faster.

These are some of the software programs that have been written:

1. *Apple Graph*. A general-purpose graphing software. Used in every experiment listed earlier except collision on an air track, LRC circuits, and Faraday's law (available on all eight floppy disks).

2. *Data-Input/Editor*. A four-column general-purpose spreadsheet data editor. Used in every experiment except Faraday's law (available on all eight floppy disks).

3. *Calculator*. A "pocket-type" calculator. Used if students do not have their own personal hand calculators (available on each floppy disk).

4. *Polynomial Fit.* Provides up to sixth-order least-square polynomial fit of data. Generally used in four experiments: free fall, simple pendulum, viscous drag, and prism spectrometer (available on all eight floppy disks).

5. *Timer.* Multipurpose timing software used in conjunction with photogates. Used in six experiments: free fall, motion on an inclined plane, collisions on an air track, rotation, simple pendulum, and viscous drag.

6. *Impulse Experiment.* Software written specifically for the impulse experiment.

7. *Electric Fields.* Software written specifically for the electric fields experiment.

8. *Multi–Channel Voltammeter.* Eight different storage volt meters that use eight double-ended inputs from the A/D converter. Any channel may be used as an ammeter by installing a shunt resistance. Used in four experiments: impulse and change in momentum, viscosity of water, Newton's law of cooling, and simple electric circuits.

9. *Multi-Channel Read/Plot/Store.* Up to 256 voltages from one to four A/D channels are read at sampling rates ranging from 16 Hz to 3000 Hz. Times at which readings are recorded along with the A/D values are collected as data for each channel selected. Used in the experiment RC circuits.

10. *Fast Graph.* Voltage readings from one A/D channel are read at sampling rates ranging from 700 Hz to 45000 Hz. The program records, displays graphically, and stores 1024 data values. Used in two experiments: LRC circuits and Faraday's law.

11. *Specific Heat and Latent Heat.* Written specifically for an experiment involving the measurement of the specific heat of a solid and the latent heat of liquid nitrogen.

12. *Sine Wave Generator.* Written to accommodate experiments involving standing waves on a string and in an air column.

13. *Utilities.* Several general-purpose utilities associated with graphic display, operating system, sorting, etc.

Experimental Procedures and Tasks

The software was written, as much as possible, so that the experimental procedure and specific tasks are under the control of the students (or by direction from the instructor). A printed hand-out and/or oral instructions are usually handed out by the instructor at the beginning of the laboratory period.

Advantages

The microcomputer work station has a number of advantages over the conventional laboratory setting:

1. The microcomputer work station can replace many conventional measuring instruments, i.e., timers, voltmeters, ammeters, signal generators, oscilloscopes, thermometers, etc.

2. The students can handle large amounts of data.

3. In many instances data can be gathered faster.

4. There is less error in data gathering.

5. The work station provides easier and faster data analysis and graphing.

6. Usually a given experiment can easily be repeated several times during a given lab period (with or without modifications).

Disadvantages

1. There is a limit to how many students can make observations on a microcomputer CRT screen. With more than two students per work station, the excess students will tend to stand idly by while the other two collect and analyze data using the microcomputer keyboard input and CRT (or printer) output. The ideal number of students per work station is two. No more than three should ever be assigned to a work station.

2. Students tend to lose sight of what is involved and the significance in certain data analysis (i.e., least-square fit, logarithmic plotting, etc.). This disadvantage may also be present when students use the hand calculator.

3. Extra time and effort must be spent by faculty and staff to write, organize, coordinate, and document software in a particular lab experiment.

Application softwares

1. All application software used in a laboratory experiment should be integrated and linked by a main driving program that requires a minimum of keyboard input for operation. Software execution should not require complicated keystroke combinations or extensive user-manual references.

2. If the application software is provided on several different disks, the procedure for operating the software should be the same from disk to disk. After a one- or two-hour session using the software on a disk, the student should be able to use other laboratory-related software with minimal difficulty.

3. The application software should not (as much as possible) be tied to any specific laboratory experiment. This allows a given piece of software to be used in several different experimental situations. This also allows changing the experimental procedure without rewriting (or revising) the software.

4. The software should not perform the experiment and display the experimental results. The microcomputer should function as a tool to gather and analyze data under the direction and control of the student. Finally, the output of experimental results are the responsibility of the student.

Future Plans

We plan to continue with the present microcomputer (Apple IIe) and associated hardware at each work station for at least the next three to four years because of the large amount of effort put into generating software for the present hardware configuration.

We plan, however, to add some hardware enhancements to the existing basic work stations at an approximate cost of $1,000 per work station. We will:

1. Extend the memory of the microcomputers to one megabyte. This will allow all software to be loaded into RAM. All subsequent calls for software can then be made to RAM rather than floppy disk. This will allow more memory space for fast data storage and recall.

2. Add one high-capacity 3 1/2 inch disk drive to each work station. This will allow all physics lab software to reside on one 3 1/2 inch floppy disk instead of eight different 5 1/4 inch floppy disks. Since all software will be loaded into RAM, both the 3.5- and 5.25-inch disk drives will be available to store experimental data and results on the students' disk.

3. Add a speed-up card to the basic computer. This will speed-up most operations by a factor of approximately 3.5.

Upper-Level Experimental Physics

Lee E. Larson
Department of Physics, Denison University, Granville, OH 43023

Computers are used in the upper-level laboratory course for two primary reasons. Most important, they can enhance the quality of experimental physics. Such techniques as signal averaging, measurements of transient phenomena, automating and controlling experiments, very slow measurements over a long period, and precision timing can improve the quality of data and extend the range of possible experiments. Second, students must learn to work in the type of environment they will encounter in graduate school or in industrial or governmental laboratories.

The experiments described below are interfaced to computers that are equipped with commercially available plug-in boards to provide a number of channels of analog-to-digital (A/D) and digital-to-analog (D/A) conversion, digital input-output (I/O), GPIB (IEEE-488) capability, and timer-counters.[1] We make short sample codes illustrating the basic operation of these boards available to our students, and also a series of machine-language programs needed for special applications (fast acquisition, acquisition with a known time interval between data points, etc.). These machine-language routines are either purchased from board manufacturers[2] or adapted from published routines.[3]

The experimental physics course is normally taken by students in their junior year. The background of these students includes a year of modern physics and semester courses in electronics, classical mechanics, and electricity and magnetism. During the one-semester course in experimental physics each student completes about four open-ended experiments chosen from a list of about 40 possibilities. About 15 possible experiments involve the use of a computer for interfacing and acquiring data. Students build each experiment from the ground up, beginning with a library search, the basic experiment-specific equipment, and cabinets of general apparatus. When the experiment is complete, the student writes a report in a form suitable for submission as a journal paper.

For each experiment that involves the computer, students are given some modest direction in how to use the computer for data acquisition (suitable measuring techniques, appropriate interface capabilities) and, the sample acquisition codes mentioned above), but are expected to do signal conditioning, connection to computer-interface boards, and development of both acquisition and data analysis programs. In practice, this procedure seems to work reasonably well; the experiments take longer than they would if detailed information was provided, but students really seem to end up with the ability to do this on their own. At present the student-written analysis programs are very simple or nonexistent. Students much prefer to write their raw data to a disk file and then to load them directly to a spreadsheet program for graphing and analysis.

Experiments

This section describes four experiments that were developed and performed by recent students. The data shown are taken directly from student lab reports.

Fast Fourier Transform Optical Spectroscopy

A standard homemade Michelson interferometer is equipped with a stepper motor and worm reduction gear assembly to drive the movable mirror under computer control. A reversed biased silicon photodiode is positioned to record the intensity of the fringe pattern. Since the analysis of a symmetric interferogram is easier for first experiments, the movable mirror is carefully positioned to center the interferogram on the white-light fringe pattern. The mirror is then scanned over equal distances each side of the central fringe, and data of light intensity versus mirror position are gathered by the computer. A Fourier transform of these data by the computer gives the spectrum of the source, i.e., the intensity as a function of wavelength. For information on designing the experiment, an excellent reference is Bell.[4] A primary limitation is building a mechanical system for moving the mirror that will provide steps of equal size. If the step size varies, the transformed spectrum will appear noisy.

The spectrum of a He-Ne laser illustrated in Figure 1 clearly shows the spectral line but also gives an indication of the noise present. In general, low-resolution sources are less demanding in terms of mirror positioning and noise. An easier experiment is done with an incandescent source. Figures 2 and 3 respectively show an interferogram and resulting spectrum. The spectrum does not bear much similarity to the continuous spectrum expected, especially with the sharp cutoff at higher wavelengths. However, recall that the response of the silicon photodiode decreases rapidly above 1.1 microns, and we should expect that the spectrum determined will actually be a product of the incandescent source and the detector response curves. Combining the response of a typical photodiode with a 2000 K black-body spectrum gives the theoretical curve shown in Figure 4. Note that this curve is at least qualitatively similar to the transformed spectrum (Figure 3).

Other measurements possible with this system include finding the profile of an optical filter and attempting to resolve closely spaced spectral lines (the sodium doublet, for example).

Short Half-Life

A Geiger counter is connected to a counter-timer interface board through a suitable interface (a carefully chosen comparator) that "squares up" the Geiger pulses.

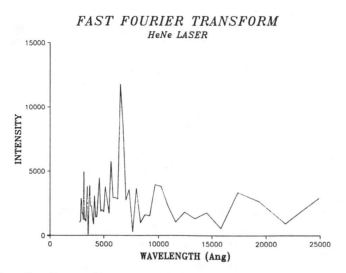

Figure 1.   Fast Fourier transform of a HeNe laser.

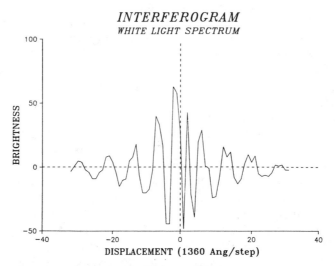

Figure 2.   Interferogram of white-light spectrum.

The circuit is shown in Figure 5. The computer is programmed to record count rate versus time, providing the same function as a multichannel analyzer run in multi-scale mode. Once the data are acquired, a least-squares fit is made to determine the half-life. The computer control of the data acquisition makes it possible to measure and analyze sources with quite short half-lives.

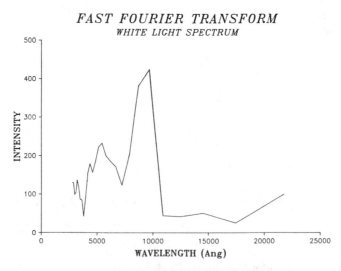

Figure 3.   Fast Fourier transform of white-light spectrum.

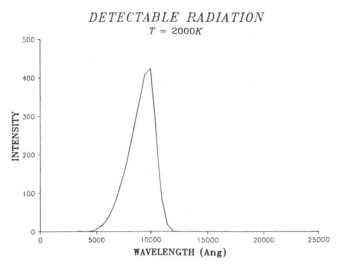

DETECTABLE RADIATION
$T = 2000K$

WAVELENGTH (Ang)

INTENSITY

Figure 4.   Response function of a photodiode to an incandescent source.

A minigenerator[5] can be used to make a barium 137 source with a half-life of 2.55 min. Typical data taken for such a source are shown in Figure 6. Notice that the half-life computed from the raw data is longer than the accepted value. When proper corrections for detector dead time were made, the half-life based on five sets of data was 2.54 ± 0.04 min, in excellent agreement with the accepted value.

For another source, silver can be activated by a neutron howitzer to make sources with half-lives of 24.6 seconds and 2.37 minutes. The analysis is interesting because it involves effectively separating the counts from the two sources. Results of a typical experiment are shown in Figure 7. The count rates are quite low and therefore noisy; however, the least-squares fits, shown in the Figure as straight lines, give results for the half-lives that are very close to the accepted values.

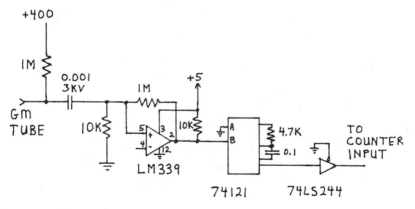

Figure 5.   Circuit diagram for a Geiger-counter interface.

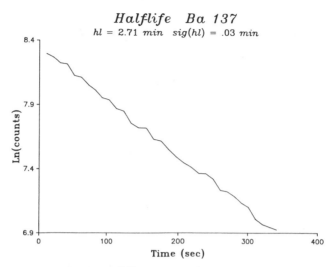

Figure 6.   Decay curve for Ba^{137}.

This work was adapted from an experiment designed for an AAPT workshop by Barney Taylor from Wright State University in Dayton, Ohio.

High-Resolution Spectroscopy

We recently acquired, with the assistance of the NSF CSIP (now ILI) program,[6] an Instruments SA model HR-320 grating spectrometer[7] and a 1024 element photodiode array-detector system[8] that was mounted on the spectrometer as shown in

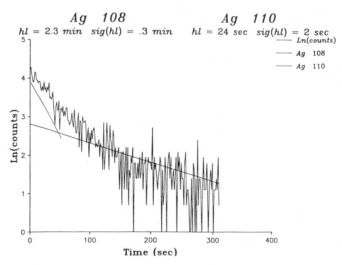

Figure 7.   Decay curve for Ag^{108} and Ag^{110}.

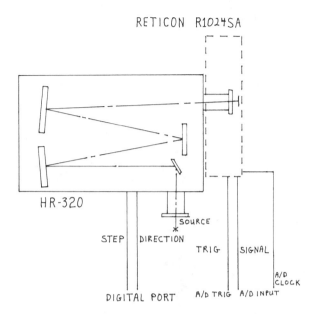

RETICON R1024SA

HR-320

STEP | DIRECTION

SOURCE

TRIG | SIGNAL

A/D CLOCK

DIGITAL PORT   A/D TRIG   A/D INPUT

Figure 8.   Spectrometer with reticon detector.

Figure 8. The initial experiment was to get this system operating under computer control and to make an attempt to measure its resolution.

The spectrometer is easily controlled by the computer. The wavelength is adjusted by rotating the grating. The grating is turned by a stepper-motor system. Two digital signals are needed, one to set the direction of rotation, the other to step the motor.

The output of the detector system is designed to be viewed easily with an oscilloscope. A signal on a trigger output is followed by signals on an analog output from each detector element in rapid succession. This method provides a good check on system operation. However, for quantitative analysis it is much handier to have information of intensity as a function of wavelength in a computer array.

To interface the detector array to the A/D computer board we used three signals. The trigger signal mentioned above was used to initiate a data-gathering cycle. By probing around on the detector circuit board, we found a digital signal that pulsed before the signal from each detector element was presented. This signal was used as an external clock input for our A/D system. The detector analog output was connected directly to the A/D input. The minimum readout rate of the detector system was 33 kHz, which slightly exceeded the maximum acquisition rate of our A/D board. We padded a capacitor on the detector system clock to slow the readout rate slightly.

Figure 9 illustrates data taken of the sodium doublet at 588.9 and 589.5 nm. An expansion of the region of interest is shown in Figure 10. We see that when the detector is placed at the exit slit location, each pixel covers about 0.03 nm. The

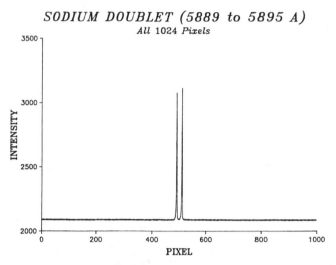

Figure 9.   Visible spectrum for Na showing the doublet lines.

theoretical resolution of the spectrograph with the grating supplied, is about 100,000. This translates to being able to separate spectral lines that are about 0.006 nm apart in the vicinity of the sodium doublet. This is greater than the detection system allows by a factor of five. To make full use of the resolution of the spectrograph requires a more sophisticated optical setup, or a more traditional exit slit and detector, and scanning the line across the slit. For many experiments, however, the detector system as used here provides more than adequate resolution.

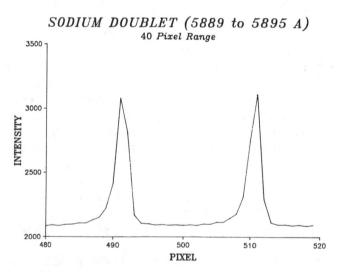

Figure 10.   Visible spectrum for Na showing the doublet lines, with an expanded scale.

Faraday's Law

Small coils are wound on coil forms that pass over a long glass tube. A cylindrical bar magnet is dropped down the glass tube, and the induced emf in the small coils is measured as a function of time. Parameters that can be varied include the velocity of the magnet, the number of turns of wire on the coil, and the strength of the bar magnet (by using different magnets). This experiment is fully described by Nicklin.[9] The apparatus is illustrated in Figure 11. The upper circuitry consists of an optical source and sensor package (TIL149) wired to an invertor that creates a TTL pulse. When the magnet comes by, light from the source is reflected to the sensor, and a trigger pulse is generated for the A/D board. The vertical position of this source-sensor pair is adjusted vertically so that the A/D converter starts just before the magnet arrives at the coil.

The lower circuitry is a simple amplifier to boost the signal from the coil to make better use of the resolution of the A/D system.

We now describe a sample of the extensive results acquired during the course of this experiment. Figure 12 shows data taken with coils of differing numbers of turns placed at the same vertical location. The signal in the second (positive) peak is slightly higher than the first (negative) peak because the magnet is traveling faster when the second pole passes the coil. The data, if carried to one more significant figure than that shown in the diagram, confirm that the signal depends linearly on the number of turns in the coil. Figure 13 shows data taken with a single coil placed at various vertical positions. The velocity of the magnet was computed, and maximum signal as a function of magnet velocity was plotted. Once again, the data show a linear relationship, indicating that the rate of change of flux and therefore the induced emf varies linearly with the speed of the magnet.

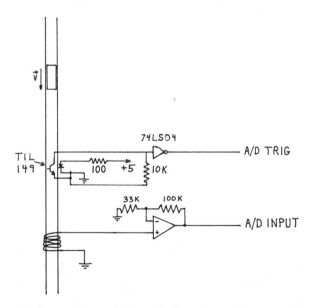

Figure 11. Apparatus and circuit diagram to measure Faraday's law.

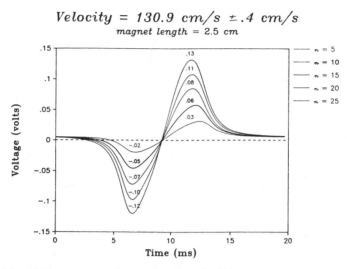

$$Velocity\ =\ 130.9\ cm/s\ \pm.4\ cm/s$$
$$magnet\ length\ =\ 2.5\ cm$$

Figure 12. Voltage versus time using Faraday's law apparatus.

Partial support for equipment purchase to carry out this program was obtained from the NSF CSIP Program, Grant CSI-8552289, and from the Hewlett-Packard Foundation. The interferometer was constructed by W. Alan Brewer, and experiments were performed by Paul R. Loring and Michael Bait.

1. Boards used for the Apple II were the Mountain Computer, A/D-D/A (Mountain Computer Inc., Scotts Valley, CA) and John Bell Parallel Interface Card (John Bell Engineering, San Carlos, CA). Boards for IBM-PC compatibles were the Data Translation DT2801A and DT2806 (Data Translation, Marlboro, MA). Other boards

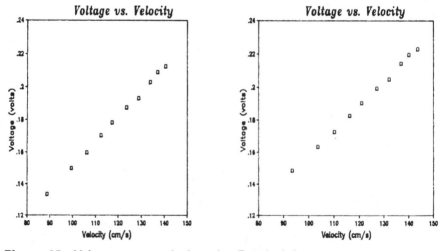

Figure 13. Voltage versus velocity using Faraday's law apparatus.

with similar capabilities are available. For a list and review of those available for IBM PC compatibles contact the author.

2. For example, PC Lab, Data Translation, Marlboro, MA; Soft500, Keithley Instruments, Cleveland, OH, Metrabyte, Taunton, MA offers software provided with DASH boards.

3. M. L. DeJong, *Apple II Applications* (Indianapolis, IN: Sams, 1983).

4. R. J. Bell, *Introductory Fourier Transform Spectroscopy* (New York: Academic Press, 1972).

5. Cs137-Ba137 source was used here. This source and others are available from Redco Science, Inc., 11 Robinson Ln., Oxford, CT 06483, tel. (203) 888-1266.

6. The NSF Instrumentation and Laboratory Improvement Program (ILI) provides matching funds for undergraduate science equipment. National Science Foundation, Office of Undergraduate Science, Engineering and Mathematics Education, Washington, DC 20550, tel. (202) 357-7611.

7. Model HR-320 0.32 meter Czerny-Turner spectrograph available from Instruments SA, Inc., 173 Essex Ave., Metuchen, NJ 08840, tel. (201) 494-8660.

8. Model RC-1024SA detector array available from Reticon Corp., 910 Benicia Ave., Sunnyvale, CA 94086, tel. (408) 738-4266.

9. R. C. Nicklin, "Faraday's Law—Quantitative Experiments," Am. J. Phys. **54**, 422 (1986).

Online Data Analysis: Smart PC Work Stations

Jordan A. Goodman
Department of Physics and Astronomy, University of Maryland, College Park, MD 20742-4111

The idea of using a personal computer as a terminal to a mainframe is at least as old as the first PC. Countless programs are available that allow the PC to emulate all sorts of traditional "dumb" terminals. Most of these programs use very little of the power of the PC to compute and generate graphics. This paper discusses new use of the PC as a very smart terminal. Data from online computers are passed in ASCII to a PC, allowing the user to reconstruct, manipulate, and study data independent of the data-taking computer.

Since interfacing data-acquisition equipment with computers is a difficult and time-consuming task, many special-purpose analysis systems have been developed over time. These systems may be small, self-contained systems such as multichannel analyzers or fullscale systems such as Fermilab's *Multi* or Los Alamos' Q system. These systems provide fundamental histogramming and data-storage services. Programs like *Multi* and *Q* run on machines such as PDP 11s or VAXs. These machines offer limited graphics capability and are not generally as user friendly as a PC. Using a PC as an interface to these online systems can add specialized features and capabilities to existing data-acquisition systems.

Reconstructing and displaying complicated events can be farmed out to PCs, which are better suited to the task than the mainframes that acquire the data. PC work stations can be set up to interact with both the online data-acquisition system and the experimenter. This type of system has been employed at both Fermilab and Los Alamos.

The Cygnus Experiment at Los Alamos

The Cygnus experiment at Los Alamos is designed to look for ultra-high energy (UHE) extensive air showers from astrophysical point sources. When it enters the earth's atmosphere, a UHE particle creates a cascade of particles, mostly electrons, positrons, and photons, that travel downward toward the ground at the speed of light. A typical shower initiated by a 2×10^{14} eV primary particle will contain 10^5 particles at mountain altitude. The Cygnus experiment has over 100 plastic scintillation counters that sample the density of shower electrons and record the relative arrival time of the shower front with nanosecond accuracy. The direction of the shower propagation is determined from the arrival timing. The shower size and core location are found from the density profile. Data are read continuously at a rate of about 4 Hz.

The data-acquisition system is based on CAMAC hardware and a Los Alamos built interface called multibranch driver (MDB). The software system is built around the Los Alamos Q system. The experiment originally used a PDP 11/34 for data taking. This machine had a limited address space and required that all major software be heavily overlaid. Data acquisition later was shifted to a Microvax II.

One of the objectives of the online analysis was to provide for event reconstruction and display. Color was important, as was the ability to rotate the display and inquire about individual counters. It was also important that these tasks could occur at sites other than in the experimental trailer. The online package also utilized the same code and subroutines used for offline analysis.

Analysis software for reconstructing event directions and shower size was written in FORTRAN and run on Vaxes. This software formed the basis for the PC-based online system. Reconstruction code from the Vax was compiled onto the PC using Microsoft FORTRAN 77. Device-independent graphics software was developed using the IBM Graphics Development Toolkit. This utilized a virtual device interface that interacted with device drivers loaded at run time. The precise nature of the graphics hardware (number of colors, resolution, etc.) was not referenced directly. This allowed the same executable image to run on machines with different levels of graphics support.

The Cygnus online software displays a three-dimensional view of the detector locations with either the electron density profile in each counter or the relative arrival time of the shower at each counter shown as the vertical axis. The density plot, shown in Figure 1, represents the electron number recorded in each counter as a scaled vertical line. Fits to the shower density can be displayed superimposed on the actual detected values. The location of the fitted shower core (C) is also shown. The user interacts with the software with either keyboard or mouse. The signal in

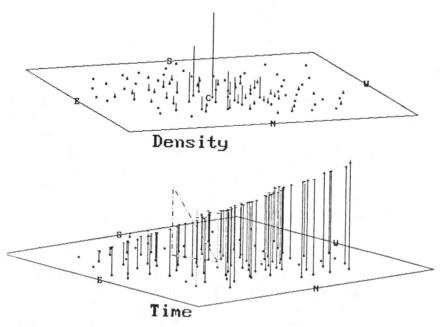

Figure 1. Cygnus experiment online displays of electron density and relative arrival time.

an individual counter and its deviation from the fit may be found by positioning the mouse close to its location and pressing a button. The sensitivity to fitted parameters can be studied by moving the core to a new location and refitting. This too is done with the mouse.

The time display, also shown in Figure 1, shows the relative arrival time of each counter as a vertical line. (Time is converted to the distance scale by multiplying by the speed of light.) The tick marks near the top of each line show the fitted value of the arrival plane. The software can rotate the viewing angle of the display so that the arrival plane is clearly visible. Counters removed from the fit are displayed in a different color. The history of individual counters can be studied by using the mouse. Problem or bad counters can be found quickly and corrective action can be taken. A third display is available to show the location in the sky from which each event has arrived.

The PC software can log events it receives to its local disk for later replay or further study. This allows the same event to be analyzed again with a variety of algorithms and still have the full online display capability. This feature has been valuable in studying reconstruction techniques.

Experiment E710 at Fermilab

E710 at Fermilab is an experiment to measure the total cross-section of protons on antiprotons at a center of mass energy of 2 TeV (10 eV). This experiment, at the Tevatron collider, uses 48 ring counters located around the beam pipe on either side of intersection region E0 to measure inelastic interactions. In addition, four sets of drift chambers in Roman pots that move into the beam pipe are located downstream from the intersection region to measure elastic scattering. The online acquisition system is Fermilab's RT-Multi running on a PDP 11/45. There is no room on the existing PDP system for interactive color display.

A PC-based system similar to the one used in the Cygnus experiment has been employed to provide online interactive graphics. Multi provides for several terminals to interact with the data acquisition system. A string sent from the PC requests that data from an event of a certain type be sent to the terminal line that is connected to the PC. The PC then fits and displays the event. The user selects the event type from a menu using a mouse. The user also has the option of zooming in on the central region or expanding out to study the pot signals. Figure 2 shows an event display for a typical inelastic trigger. The numbers and horizontal lines represent the time deviation from the fitted time in nanoseconds.

Description of the Software System

An important feature of this type of online system is that the PC is linked only serially to the actual data-acquisition machine, so the online program can be run remotely. The data on each counter are sent as an integer in ASCII exactly as if being displayed on a terminal (i.e. they are just printed to a terminal line). The PC can be connected via a modem or over a DECNET link through another machine.

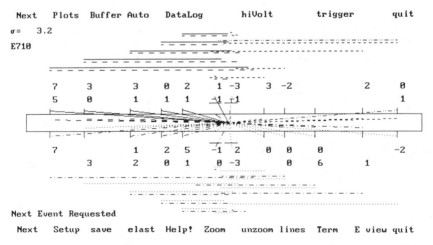

Figure 2.   Online display of an inelastic event taken at Tevatron collision region EO.

The data associated with a single event in the Cygnus experiment is about 800 bytes, which at 2400 baud takes only about four seconds to transmit. Since all calculations and display control as well as all user interactive features are handled locally by the PC, the rate of communication has little effect on system performance. Additionally, no mainframe cpu is required for screen control and fitting because they, too, are done locally.

Perhaps the most advantageous feature of this type of online system is independent software development from the mainframe acquisition system. Once the acquisition software is working, no changes are required to modify or update the PC display and analysis package. Software updates to the PC software do not affect the logging of data or require experiment down time. Source code for updates can be sent to various machines over networks such as BITNET, then downloaded and locally compiled and linked.

The key to the entire system is a flexible interface between the PC and the mainframe. This is provided by an asynchronous communications subroutine callable from FORTRAN. This routine provides a simple protocol for the exchange of information over an RS232 connection. This routine, written in assembly language, consists of two parts. The first part handles the interrupts from the communications port and receives and stores data in a circular buffer. It also sends appropriate Xon and Xoff commands as the buffer fills. The second part of the routine handles the interaction with the FORTRAN program and communications from the PC to the mainframe. The call to the routine has two parameters: a mode flag, which can take on many values, and a data word (or array). The routine has several functions: first, whenever an integer (i.e. any number between ±32767) is sent from the mainframe to the PC, its value is returned to the calling program in the data word; second, a string from the PC may be put into the data array and be sent to the mainframe. In addition, while the program is waiting for an integer to be returned, all keyboard input is sent to the mainframe. The mode flag selects the operation to be performed as well as whether or not the text from the mainframe is displayed on the screen. Other modes allow the user to check the buffer contents or reset the communications port.

The simplest use of this routine is as a terminal program. The FORTRAN code for this program is only four operational lines. (The routine is called *BEG* for historical reasons).

```
c Program Terminal
c call routine "beg" with mode =1 this will
c put all communications from the mainframe on the
screen.

1 CALL BEG(1,idata)

c Beg returns when an integer is found.
c Test to see if integer is 9999 which is returned
c when user typed ctrl-A in which case stop.
```

```
IF (idata.eq.9999) STOP
```

`c Return to terminal session`

`GO TO 1    END`

This program will work as a standard VT100-type terminal passing all typed commands and special characters (arrows) and displaying all the standard symbols. Each time a number is sent from the mainframe it is returned to the calling program in the variable "idata." Whenever Control-A is typed, the routine returns the value 9999 to the main program. This allows the user to exit. If instead of using mode 1, we choose mode 0, no output appears on the screen. In this manner the program can allow the user to establish communications with the mainframe and then switch to a mode where the communications session is not visible. The format in which numbers are passed from the mainframe is completely free-form since an integer is formed whenever a number surrounded by non-numeric fields is sent.

Under software control the PC can use the communications routine to send strings to the mainframe requesting data or terminating data requests. Since the interrupt software buffers data, many entire events can be sent while the user looks at the first one. This reduces the wait time between events to the time required for processing the event and allows for no apparent delay even at low communications rates.

Conclusion

The PC as a smart work station has been demonstrated successfully in experiments at Fermilab and Los Alamos. The interactive graphics available on the PC can be used to simplify and enhance the user interface to existing data- acquisition systems.

Design, Implementation, and Performance of a Microcomputer- Based Laboratory for General Physics

Richard Bobst, Edwin A. Karlow, and Ivan E. Rouse
Physics Department, Loma Linda University, Riverside, CA 92515

The use of computers in the laboratory is popular and growing very rapidly. Many schools want to implement computer-based laboratories, but if the technology is to enhance the educational process and not just confuse or frustrate students, several serious issues must be addressed. We have carefully designed a microcomputer-based laboratory (MBL) from the ground up and feel that our experiences over the past three years will be helpful to others planning or actually in the process of implementing MBLs.

Our top priority in our MBL program is to help students do more physics and consequently learn better. Since most of our students are pursuing careers in the sciences, a benefit of the program is that students are exposed to experiences by which they can realize some of the exciting potential uses of computers in their own disciplines. To accomplish these aims we equipped each lab station with a computer to be used as an instrument to collect data and analyze it in meaningful ways.

Decisions about hardware are critical and costly, so we considered ours carefully. We chose to use inexpensive home computers with good graphics and interfacing potential (Commodore 64) and in fact were able to get a compact portable version. The portability enables us to put them away when they are not needed or to move them to a different lab room when they are needed elsewhere. It is too expensive and takes up too much space for each student to have a plotter and printer, so we chose to have a central printing and plotting facility for our general physics labs. We decided that we needed at most two analog inputs, but that one of them should be able to read voltages from a few millivolts to a few hundred volts. After some research we decided that we could not buy what we needed for a price we could afford, so we decided to build our own.

In order to run an MBL program one must have software, which in our case needed to be flexible enough to collect and analyze data. We decided to design our own software instead of buying commercial software and associated hardware. We could have asked students to write their own programs, or we could have designed general-purpose computer programs that would be used throughout the year. Instead, we designed lab-specific software to be used with specific experiments. We emphasized graphical displays and least-squares fits to enhance the students' understanding of functional relationships. In addition, we had the computer help out with some very tedious calculations. We wanted the students to be able to get plots on a plotter and data-summary printouts, so we opted for a system with hardcopy output capabilities. We chose to include some information on the computer aspects of the experiments in our lab manuals, but not a lot.

Implementing such a major change in our lab program was a significant challenge. The first year we computerized only a certain number of the experiments we were already doing, but we added more each succeeding year. Two of our faculty members took on the task of designing, producing, and refining the lab software. They were aided by numerous able student programmers. One of our faculty designed the computer interface and students assembled it. To meet the expenses of this computerization, we raised funds from a Commodore grant, alumni, and departmental equipment and supply budgets.

We are now using the computers in some way in nearly every experiment that we do. In some experiments the computers only analyze data, but in others the computers are used as instruments to collect and then analyze the data. Using computers in the laboratory has several positive aspects: functional relationships are seen immediately, measurements and lab results are more reliable, bad data are easily identified, lab time is reduced, tedious experiments are more enjoyable, and there is less dishonesty. The computers have certainly enhanced the physics laboratory experience for our students in many ways.

However, having more electronic equipment in the lab also means that many more things can go wrong. This increases the load on the lab instructors. Some students come in without computer experience and are a bit apprehensive, but that soon disappears. The faculty and students have reacted positively to this venture into MBL. All have learned a great deal by being involved in this challenging experience. It would have never been successful without the involvement of nearly everyone in the department.

In the future we would like to be doing more with modern and nuclear physics in our labs. We are putting together plans for a second phase, during which we will implement a number of experiments in these areas. Since this will involve considerable expense for detectors and electronics as well as more software, we have written a proposal to seek major funding outside the institution.

Microcomputer-Based Laboratories in Introductory Physics Courses

Vallabhaneni S. Rao
Department of Physics, Memorial University of Newfoundland, St. John's, NF, Canada A1B 3X7

We have designed several introductory-level physics laboratories using MBL probes developed by Technical Education Research Centers (TERC) and other commercially available software packages. These microcomputer-based laboratories replace conventional laboratory experiments.

In the area of mechanics, our MBL experiments use the motion probe of TERC to allow students to study the graphical nature of motion, Newton's second law, and simple harmonic motion.

In one of our laboratory experiments, students use TERC's motion detector to create graphs (displacement vs. time, velocity vs. time, and acceleration vs. time) of an object in uniform motion and uniformly accelerated motion. This experiment really helps the students to understand the graphical representation of motion, which plays a key role in understanding kinematics.

Our experiment on Newton's second law uses a dynamic cart and TERC's motion detector. The computer is used to give the corresponding acceleration for each load on the cart. We have extended this laboratory to study the acceleration of the cart when the surface is smooth and when the surface is rough. On a specially created ramp in the laboratory the students explore the motion of the cart and establish the relationship between the acceleration of the cart and the angle of inclination. This experiment saves a lot of time in establishing the relationship between the force and acceleration.

The motion probe is used to study simple harmonic motion in two different ways. First, from the distance-vs.-time graphs of the object undergoing SHM, the students determine the frequency and time period for various masses at the spring or for the same mass on different springs with different force constants. This graphical analysis is then used to demonstrate the relation of velocity and acceleration to the displacement of the object. In the conventional experiment on SHM, it is very difficult for the student to understand these relationships graphically. This experiment also helps students understand the effect of damping on the motion of the object.

As part of a general experiment on transducers in one of our introductory physics courses for life science majors, we have developed a lab in which students used HRM's *Body Electric* to explore bioelectricity, to get the print out of an ECG, and to identify the individual waves of the heart's electrical wave. In addition, they observe the effects of inhalation and exhalation on heart rate by making appropriate measurements of their heart rates using the time "ruler." Students note individual variations in the shape of the heartbeat.

Another simple but very interesting laboratory was designed using the TERC's sound probe. In this experiment students produce a nice pattern of beats using tuning forks to establish the relationship between the beat frequency and the two frequencies of the source. They first use the individual display of the wave form of each tuning fork to determine its frequency, and then they measure the beat frequency from the pattern of the beat wave form. This experiment allows students to hear and see the beats displayed simultaneously on the monitor. In our view, this provides the solid experimental background for the concept.

In this paper, I will give the details of these laboratories and will review student feedback. The initial response from the students is quite positive. They regard the laboratory part of their introductory physics course with an enthusiasm that was not there in the conventional labs.

SAKDAC-Based Physics in Introductory Labs at Miami University

Don. C. Kelly
Department of Physics, Miami University, Oxford, OH 45056

Stewart A. Kelly
Department of Electrical Engineering, Stanford University, Stanford, CA 94305

Introductory physics laboratories at Miami University use Apple IIe computers equipped with a locally developed data-acquisition card (SAKDAC), photogates, and other home-brew apparatus. In our presentation, we will describe problems and opportunities associated with computer-based labs. We will show a video of students performing a number of experiments using the SAKDAC-based system. The following experiments are shown:

Acceleration of Gravity. Students measure the gravitational acceleration of a ball bearing using a converted Behr apparatus and the SAKDAC timer. The bearing serves as a switch that starts the timer. The timer is stopped when the bearing interrupts a photogate beam. Students identify the consistent minimum time of fall, and determine g to three significant figures.

Conservation of Linear Momentum. An air track is inclined very slightly to allow an air car to move at constant velocity. The air car has a platform that "catches" a small mass released by the student. The SAKDAC timer is used to measure the velocity of the car before and after the mass catch. Momentum conservation is typically confirmed to within 2 percent.

Conservation of Angular Momentum. A plastic puck rotates in an air pond. The student drops two identical masses in succession onto the puck. The period of rotation is measured continuously using the SAKDAC timer and a photogate. Measurements of the period of rotation before and after each of the two masses strike the rotating puck, allow confirmation of angular momentum conservation without reference to moments of inertia.

Precession. The student measures the precessional period of a bicycle wheel directly and compares the results with the equation where T_{spin} is the spin period of wheel, as measured with SAKDAC timer and hand-held photogate, T_{pend} is the period of wheel as a physical pendulum, about an axis parallel to the spin axis, h is the distance between spin axis and physical pendulum axis, and r is the distance from support rope to center of mass.

Learning Physics Concepts with Microcomputer-Based Laboratories

Ronald K. Thornton
Center for Science and Math Teaching, Tufts University, Medford, MA 02155

High school and college physics courses have little carryover to students' understanding of the larger physical world. They do little to connect coursework to students' everyday experience, to connect simple understandings to underlying principles, or to address directly, student misunderstandings (which are often grounded in everyday experience). Instead of connecting coursework to everyday experience, physics courses often mire the student in the details of data collection and presentation. Microcomputer-based laboratory (MBL) tools ease the tasks of data collection and presentation, thus encouraging students to become active participants in scientific process. Led to ask and answer their own questions, they learn physics concepts more easily in MBLs than in lectures.

Learning is enhanced when students are able construct knowledge for themselves, but MBL tools are not enough. Experience in universities and high schools around the country has shown that carefully constructed curricula are also necessary. The curricula developed under the Tools for Scientific Thinking Project at Tufts, and used in collaborating institutions around the country, makes use of the guided discovery approach. Guided discovery curricula pay attention to student alternative understandings, support peer learning, and do not explicitly depend on physics lectures. Using MBL tools and guided discovery curricula, students more easily acquire an understanding of the scientific principles that underlie their experience, a sound physical intuition, and competence in the use and interpretation of graphs.

Data gathered about student learning by members of the Tools for Scientific Thinking in collaboration and in high schools around the country show substantial improvement in learning concepts related to motion (kinematics), and heat and temperature by students who use MBL laboratories as compared to those in ordinary laboratories. Considerable evidence suggests that many of these simple concepts are not learned by listening to lectures—no matter how good the lecturer is. The concepts learned in MBLs also seem to be retained (at least as long as the final exam). When using MBL tools, students of all sorts (badly prepared, well prepared, even the science anxious) are very enthusiastic.

Learner-controlled laboratory explorations with easy-to-use computer measurement tools give students immediate feedback by presenting data graphically in a manner that can be immediately understood. Such MBL tools provide a means of altering physics pedagogy by enabling teaching to begin with the students' direct experience of the familiar physics world. Using MBL sensors and software, students can simultaneously measure and graph such physics quantities as position, velocity, acceleration, force, temperature, light intensity, sound, and voltage.

In this hands-on workshop, participants will use MBL tools selected from force, motion, sound, heat and temperature, and light to explore methods of effectively teaching physics in universities and high schools. They will also examine curricula that have been successful in teaching basic physics concepts. The primary focus of the workshop will be on methods of teaching physics using MBL tools rather than on the means of interfacing computers to the physical world.

Use of Microcomputer in Introductory Laboratories at Ohio State

Gordon J. Aubrecht II
Department of Physics, Ohio State University, Columbus, OH 43210-1106 and Marion, OH 43302-5695

The availability of microcomputers is prompting a change in the sequence and selection of topics in physics instruction. Physicists who own and use microcomputers in their classes are more receptive to introducing topics not usually found in our current introductory physics textbooks—for example, nonlinearity or real air resistance. The very existence of this conference attests to the ferment the technological changes have caused in the physics community. This willingness to change should be channeled and used to assist in the reevaluation and restructuring of the entire introductory physics curriculum. The laboratory component of introductory courses is one area particularly amenable to change.

A 1986 Ohio State University physics department self-study reveals overwhelming discontent with the introductory laboratories. Students and graduate-assistant graders regard them as loathsome, and lecturers as irrelevant. Discontent with the state of the laboratories led the curriculum committee to recommend sweeping changes. Such changes take considerable thought and money, so it seemed prudent to try out various alternatives in small pilot programs.

I teach at the Marion campus of Ohio State. The enrollment is low—about 900 students overall—but the introductory courses are identical to those offered in Columbus. Because of its small size, the Marion campus physics program offers an ideal test site to pilot changes proposed in the curriculum committee report. As a member of the curriculum committee, I find it relatively simple to feed back experience into the department.

My report covers my experience with some of the pilot introductory laboratories. In particular, I discuss laboratories I have designed to give the students kinesthetic experiences—those that integrate the body's feedback mechanisms with physically measurable quantities. With the cooperation of Ron Thornton of Technical Education Research Centers, I have tested this software in college-level classes.

Measurement of the Tension in a String

J. Hellemans and G. Hautekiet
Afdeling Didaktische Fysika, Katholieke Universiteit Leuven, 3000 Leuven, Belgium

Students studying mechanics often do not understand that the tension of a string attached to a moving object changes with the acceleration of the object.[1] Measuring this tension is not possible with a classical forcemeter (i.e. a dynamometer), but can be done with an arrangement of strain gauges.

The experimental setup is given in Figure 1. The mass M can slide along a two-meter air track, and is connected with a string to the mass m, which can move vertically. A leaf spring, mounted close to m, measures the tension in the string. Two strain gauges are glued on each side of the leaf spring. The strain gauges are mounted in a bridge, the signal of the bridge is amplified, and using the appropriate interfacing and software, the value of this signal is read on a monitor. Changing the angle leads to a different acceleration of M and m. When the tension in the string changes, the reading on the monitor will change accordingly. The detail of the leaf spring s with strain gauge g glued on top is given in Figure 1(B).

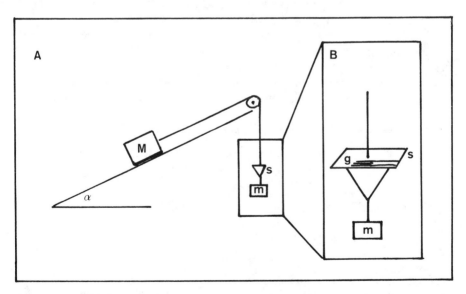

Figure 1. (A) The experiment set-up. (B) Detail of the leaf spring s with strain gauge g glued on top.

1. R. Gunstone, "Student Understanding in Mechanics: A Large Population Survey," Am. J. Phys. **55**, 691 (August 1987).

The Tension in the String of a Simple Pendulum

J. Hellemans and G. Hautekiet
Afdeling Didaktische Fysika, Katholieke Universiteit Leuven, 3000 Leuven, Belgium

We gave 25 third-year physics students who had taken an extensive mechanics course in their first year a few minutes to draw the acceleration of a swinging pendulum (see Figure 1). Only a few students succeeded in doing it correctly. Most of the students made the classical mistake[1] of saying that in A and A' the acceleration is zero, and that in B and B' it is tangent to OB and OB'. Many were convinced that the tension T in the string is always equal to $mg \sin \theta$. To correct these misconceptions, we asked students to determine experimentally the tension in the string of a pendulum as a function of time.

The easiest construction would be to use a pendulum attached to a dynamometer, but a dynamometer is unequal to the task because the reaction time of its spring is too long in comparison to the period of oscillation of the pendulum. In our experiment, the string is attached to one end of a leaf spring on which two strain gauges are attached on the upper and undersides. The other end of the leaf spring is firmly fixed. The strain gauges were mounted in a bridge, and the signal of the bridge is amplified. Both strain gauges and amplifier were purchased from R. S. Component, Ltd. The mounting of strain gauges and amplifier is very simple and can be done by any physics teacher. The output of the amplifier is introduced to a microcomputer via the appropriate interface, and the signal can be seen on the screen with the help of the standard storagescope program that comes with the interface.

From simple mechanical reasoning the following equations are obtained:

$$T - mg \cos \theta = ma_n = mv^2 / R = mR \, (d\theta/dt)^2$$

$$mg \sin \theta = ma_t = mR \, (d^2\theta/dt^2)$$

For small angles, the results of these equations are:

$$\theta = \theta_o \cos \omega t$$

$$T = mg \, (1 + \theta_o^2 \sin^2 \omega t)$$

From the last equation it is clear that the tension is a function of time, and will be minimal for $t = 0$; $T / 2$; . . . (which corresponds to $\theta = \pm \theta_m$); maximal for $t = T / 4$; $3T / 4$; ... (which corresponds to $\theta = 0$).

In our demonstration, we will show the results of a measurement of the tension

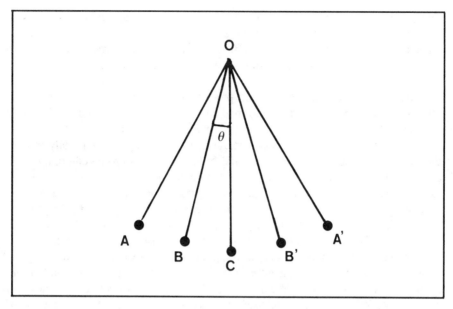

Figure 1.   The acceleration of a swinging pendulum.

In our demonstration, we will show the results of a measurement of the tension as a function of time.

1. F. Reif, "Scientific Approaches to Science Education," Phys. Today **39**, 48 (November 1986); R. Gunstone, "Student Understanding in Mechanics: A Large Population Survey," Am. J. Phys. **55**, 691 (August 1987).

Advanced Laboratory Computer Data Acquisition and Analysis

Don M. Sparlin
Physics Department, University of Missouri–Rolla, Rolla, MO 65401

Advanced laboratory projects generate lists of independent/dependent pairs of numbers representing measurements. Students are asked to extract values of fundamental constants, or to assume the constants and test the fit of one or more theoretical functions. There are now software packages to assist in such analysis. These packages allow professional quality acquisition, analysis, and graphical presentation at an affordable cost and minimal student programming time.

The physics department of the University of Missouri–Rolla has for many years supported an intensive junior-year project laboratory.[1] Working in groups of two or three, the students complete a coherent project during each of two successive semesters. This project laboratory requires extra resources in terms of space, apparatus, and faculty time, but we find it indispensable because it requires students to develop in performance areas not often found in encyclopedic, or series-scheduled laboratories.

Students must carefully select their project for the semester from a list of over 40 suggestions. After groups have formed and compromised on a project title, they must complete library research; develop a final formulation of the project; specify, assemble, and calibrate the needed apparatus; write a proposal and orally present their project at midsemester; develop and follow a laboratory procedure; take and analyze data; keep individual, functional laboratory notebooks; write a term paper to *AIP Style Manual* standards; and, finally, present their results in a sequence of individual, formal, ten-minute talks before a critical audience of their peers, former students in the course, and the physics department faculty. Armed with a loud timer, the graduate assistant for the course serves as master of ceremonies.

The goal of the laboratory is to develop laboratory process rather than specific skills, so students are encouraged to avoid writing programs already available in commercial software packages. Fortunately, there are now many software packages by scientists for the use of scientists, so it is no longer necessary to force the use of business software into the service of scientific analysis.[2] Until recently, adequate data acquisition and experiment control at UMR meant that students had to master programming an AIM 65 microcomputer using custom-built 8-bit I/O boards. This required considerably more time than was appropriate to the average project.

We have now acquired software that reduces programming to selection from menus. We acquired this software, *ASYSTANT+,* during an educational-discount opportunity. It is used with a 12-bit I/O board,[3] and offers menu-driven analysis, graphics, control, and data acquisition. We use *ASYSTANT+* with MS-DOS machines with 640K memory, Hercules graphics presentation, and floppy disk storage. The package provides the reliability and instant graphics presentation necessary to complete a project during a single semester. Students must still use and calibrate operational and instrumentation amplifiers for signal conditioning.

I am very satisfied with the level of effort, understanding, and accomplishment achieved by our students with such instrumentation. Since the project requires a term paper, publication-quality graphics (in-house plot routine on IBM plotter), and oral presentation of the project to the department, the students deserve the best available equipment and instrumentation. Using computers as versatile tools does not interfere with the full involvement of the student with the physics of the project. I do not have the same opinion of full computer simulations.

Among the projects completed using this software have been two-dimensional thermal diffusion, sub-Rayleigh criterion Zeeman data,[4] Cavendish balance displacements, surface plasmon absorption peaks, computer-generated holograms, and the value of the spin-spin relaxation time, $T2$, from NMR signals.[5]

1. L. Reinisch and D. M. Sparlin, "UMR Inc.: A Role Playing Experience in the Advanced Laboratory," Am. J. Phys. **46**, 3 (1978).

2. D. M. Sparlin, "*ASYSTANT*—How Software Changes Laboratory Goals," Physics Section Missouri Academy of Science, 1987 Annual Meeting, Lincoln University, Jefferson City, MO, 24–25 April 1987; *ASYST, ASYSTANT+* (Rochester, NY: *ASYST* Software Technologies, Inc., 1987).
3. IBM data acquisition and control board and terminal, Mendelson Electronic Co., Inc., 340 East First Street, Dayton, Ohio 45402, (800) 422-3525. Priced at $195 per set. Limited number.
4. D. M. Sparlin, "Beyond the Rayleigh Criterion with *ASYSTANT*," Paper HH6, Summer Meeting of the American Association of Physics Teachers, Bozeman, MT, 15–19 June 1987.
5. D. M. Sparlin, "Computer Analysis of Advanced Laboratory NMR Data," Paper ED8, Annual Meeting of the American Association of Physics Teachers, Washington, DC, 25–29 January 1988.

Advanced Electronics with Physics Applications

Michael J. Pechan
Department of Physics, Miami University, Oxford, OH 45056

Advanced Electronics with Physics Applications (AEPA) is a two-semester senior/graduate-level laboratory sequence that builds upon the experience our physics and engineering physics majors gain with computers throughout our curriculum. This experience includes using the computer as a "black box" tool in the three-semester, introductory physics sequence, interfacing physics experiments to a personal computer at the sophomore level, and computational physics in our mathematical physics sequence at the junior level. We will discuss these three other applications in other presentations at this conference.

AEPA introduces electronics techniques employed by experimental physicists. The microprocessor is given considerable attention and is used where appropriate. However, it is used only if it is the best tool for the job. Design is an important aspect of AEPA. Accordingly, the lab handouts are decidedly not "cookbook" in nature. Students are expected to present their results in a professional journal format.

Each experimental work area for AEPA is equipped with a Macintosh Plus, a Metaresearch BenchTop Instrument, a digital multimeter, oscilloscope, function generator, digital trainer, and frequency meter. The Macintosh was chosen for its ease of use and small "footprint" on the workbench. The BenchTop has several unique features that make it ideal for training in data-acquisition techniques. Its highly developed system of I/O along with a supplied training manual make it ideal for studies of microprocessor architecture. Sophisticated A/Ds (12-bit, 16-channel, 35 kHz, true differential, programmable gain, unipolar/bipolar switchable) and D/As (12-bit, up to 6 channels) combined with the base CPU provide a flexible, yet powerful, training system.

AEPA I is the first course in the two-semester sequence. AEPA I is a three-credit, 15-week course that meets for two, two-hour labs and a one-hour lecture each week. Twenty-three experiments in the course include introduction to tools (one experiment), digital techniques (four experiments and a student project), microprocessor architecture (six experiments and a student project), and applied physics (eight experiments and two student projects).

Digital techniques cover gates, flip-flops, monolithic counters, and wave-shaping circuits. Microprocessor architecture covers the basic instruction set of the Intel 8031 microprocessor including program control, arithmetic and logic, port manipulation, and input and output interfacing through A/D and D/A devices. Students learn to control and monitor the applied-physics experiments at the machine level, to control the machine routines and manipulate the output with higher-level languages, and to analyze their data in commercial spreadsheet programs. The applied-physics experiments include verification of Thevenin's analysis, passive-filter frequency response (manual data collection), higher-order passive and active filters (automated data collection), Fourier analysis, impedance matching, oscillators, FET characterization, voltage-controlled gain, digital resistance, determination of magnetic hysteresis using an analog trace,[1] and a digital-capture technique.

AEPA II, the second course in the sequence, meets once a week for a two-hour lab and uses the experiences gained in AEPA I to tackle more advanced projects. Topics include capturing the induced emf of a falling magnet,[2] measuring the speed of light, phase-locked loops and lock-in amplifiers, electron diffraction, and CAD production of a printed circuit board.

The students find AEPA to be challenging and overall response has been decidedly positive. The course content appeals to a wide range of interests and prepares our majors to pursue technical careers at a variety of levels. The ongoing goal for the course is to continually upgrade the physics content.

1. John W. Snider, "Magnetic Hysteresis Measurements with an Integrating Magnetometer," Am. J. Phys. **39**, 964 (1971).
2. R. C. Nicklin, "Faraday's Law-Quantitative Experiments," Am. J. Phys. **54**, 422 (1986).

Computer Interface for Measurement of Coefficient of Linear Expansion

Todd A. Hooks, David L. DuPuy, and Philip B. Peters
Department of Physics and Astronomy, Virginia Military Institute, Lexington, VA 24450

In order to control an experiment to measure the coefficient of linear expansion of copper, we have implemented a computer interface and the required software. We designed the interface to allow the minute changes in length upon cooling a copper tube to be measured by the variation in output voltage of a linear potentiometer. This voltage is amplified and scaled and subsequently converted to a digital number by an ADC.

The temperature at four different places on the metal tube is monitored by temperature transducers. Each of these transducer outputs is selected in sequence using a multiplexer. This voltage is amplified and scaled and converted to a digital number.

The I/O interface with the computer is an 8255-programmable peripheral interface chip. A total of 50 data points is taken, covering the temperature range from 370K to 320K. The measured value of the expansion coefficient is accurate to within 2.7 percent of the handbook value.

A Parallel Approach to Data Acquisition and Control

Charles D. Spencer
Department of Physics, Ithaca College, Ithaca, NY 14850

After considerable experience developing hardware and software for computer-based physics laboratories and instruments at Ithaca College, I have built, tested, and refined a parallel data collector (PDC).[1] My objective was to develop hardware and software that works as independently as possible from the host computer, supports a wide range of measurements, and is adaptable to new data-acquisition needs.

The best way for a computer to communicate with data-collection hardware is through parallel I/O ports designed for the computer's expansion slots. Once ports are installed, the remaining PDC hardware (including that which implements timing and control functions and makes measurements) and all software are independent of the host. Ports can be purchased for any computer with expansion slots. However, commercial devices are overly expensive and complex, and simple ports can be easily built. We have produced versions for Apple II, IBM PC/XT/AT, and old S-100 computers, and we are working on versions for IBM PS/2 and Macintosh II machines.

The PDC uses two 8-bit I/O ports: control-in, control-out; and data-in, data-out lines. The data-collection hardware is set up with sequences of output commands to control-out lines (such as programmable interval timer or PIT integrated circuits). Polling the control-in lines allows software to monitor the status of data acquisition (such as when data is ready). Data-in lines read measured values (from PITs, ADCs, and memory). Data-out lines send PIT programming codes, and initial count values to the hardware.

The PDC supports measurement and control circuits composed of combinations of up to six Intel 8253 PITs, up to six Analog Devices AD583 ADCs, and up to eight Hitachi 8263 8K by 8-bit static RAMs (as well as combinations of numerous other devices).

So far we have implemented three versions of the PDC: (1) a general-purpose system that, with appropriate software, can act as a logger for one or two voltages at rates selectable from 0.001 to 2,000 readings per second (with real-time graphics up to 1,000 readings per second using an 8 MHz PC-AT compatible with Microsoft *QuickBASIC 4.0*); as a pulse height analyzer with 9-bit resolution; as a single-channel real-time multiscaler with the counting interval selectable from 1 millisecond to 1,000 seconds, as a sequential event timer with the minimum time between adjacent events 1 millisecond; or as a combination timer and voltage logger, which can, for example, measure the velocity of an air-track glider, the impulse of its collision with an end bumper, and the return velocity; (2) a "fast" voltage system in which the hardware acquires and stores 8-bit ADC values in local memory at programmable rates up to 40K readings per second; (3) a 16-channel real-time multiscaler with the counting interval selectable and with 16-bit resolution per channel. After acquisition, the host computer inputs the values from PDC memory.

In addition to these capabilities, the PDC is adaptable to whatever new measurement objectives come along. Possibilities include output controls for stepper motors and temperature controllers, and data outputs for digital-to-analog converters. When it's time to upgrade to a faster, more powerful host computer (with enhanced languages, operating system, graphics, storage and communication capabilities), all that must be done to get the PDC up and running is to acquire "simple" parallel ports for the expansion slot, transfer high-level language software from the old system to the new, and probably make some software modifications. No change in operation principles or data-acquisition hardware is needed.

1. C. D. Spencer and P. Mueller, "Multichannel Analyzer Built from a Microcomputer," Am. J. Phys. **47**, 445 (1979); P. A. Smith, C. D. Spencer, and D. E. Jones, "Microcomputer Listens to the Coefficient of Restitution," Am. J. Phys. **49**, 136 (1981); D. A. Briotta, P. F. Seligmann, P. A. Smith, and C. D. Spencer, "The Appropriate Use of Microcomputers in Undergraduate Physics Labs," Am. J. Phys. **5**,

891 (1987); C. D. Spencer, P. Seligmann, and D. A. Briotta, "General-Purpose Measurement Interface for Physics Experiments," Comp. in Phys., 59 (November/December 1987).

EduTech's Interface Kit

John Elberfeld
EduTech, Rochester, NY 14609

Many teachers would like students to understand that computers can do more than Logo programming and word processing; computers are valuable tools used in industry and research. But how can a student with a strong interest in computers learn more about the actual circuits and electronics that make computers run? Some teachers feel that computer-literate students need an introduction to digital logic and integrated circuits.

To help teachers and students gain hands-on experience designing and constructing interfacing circuits, EduTech has developed the Interface Kit. The kit consists of a solderless breadboard and cable that attaches to the Apple II's 16-pin internal game port. Using solderless breadboards rather than hot and dangerous soldering guns to assemble the projects saves time and trouble, and means that parts can be used an unlimited number of times. Every required electronic component, including phototransistors, photoresistors, thermistors, capacitors, resistors, LEDs, potentiometers, integrated circuits (ICs), and connecting wires is included in the kit. The kit introduces digital electronics by having students actually wire and test integrated circuits and connect these circuits to Apple II computers through a game-paddle port. The IC used (7400) is used to demonstrate AND, OR, NOT, YES, NAND, and NOR logic gates. These logic gates are the basic building blocks of even the most complex computer circuits.

The Interface Kit contains all the electrical components, software programs, and step-by-step instructions needed by the most hesitant novices. The mystery of computer interfacing slowly fades away as students work through the exercises in the kit. It provides practical programs and interfacing circuits to measure temperature, relative light intensity, and extremely precise time intervals. Each circuit is explained in detail, and each program listing is explained line by line.

The Interface Kit was originally designed for teacher workshops. Teachers, who preferred to work at their own rates, required a detailed manual for their self-paced progress. Although the fundamental programs were listed and explained in the manual, the teachers required that the programs also be included on a disk so that valuable workshop time would not be spent on typing and proofreading. Teachers demanded projects that were practical and informative, not just flashy or amusing. Each time we held a teacher workshop we incorporated suggestions into revision and expansion. After four years of development, we introduced the kit into the classroom.

The circuits in this kit have been specially adapted to work on all current Apple II computers, including the II+, IIe, IIc, and IIGS. All programs have been tested successfully on the latest LASER 128 and on other compatibles. The modified game ports on the Apple IIc and the LASER eliminate the Annunciator Output experiments.

A Gentle Introduction to Apple Game-Port Interfacing

John Elberfeld
EduTech, Rochester, NY 14609

This miniworkshop is designed specifically for teachers who have hesitated to attend other technically oriented workshops, but who want a hands-on, low-pressure opportunity to create sensors and to understand elementary sensor-monitoring software. No computer background or electronic training is required.

The workshop uses Apple game-port interfacing to introduce computerized real-time data acquisition. Starting with the location of the Apple's on-off switch, the workshop moves in gradual steps until participants are building and monitoring temperature, light, and motion detectors.

All activities are based on EduTech's Interface Kit. The kit consists of a solderless breadboard and cable that attaches to the Apple II's 16-pin internal game port. Every required electronic component, including phototransistors, photoresistors, thermistors, capacitors, resistors, LEDs, potentiometers, integrated circuits (ICs), and connecting wires are included in these kits. The kits evolved over a four-year period based on feedback from dozens of workshops and courses. Each computer work station will be supplied with an Interface Kit, and kits will be offered at half price ($35) to participants at the end of the workshop.

The workshop is designed to give newcomers a feel for the simplicity of game-port interfacing, an overview of some BASIC programming tricks used in interface-monitoring programs, the confidence to pursue other interfacing topics in more depth, and the background needed to evaluate commercially available interfacing equipment. Digital inputs and analog inputs will be covered in depth with several examples of each actually created and tested in the workshop. The circuits created in the workshop are identical to commercially available interfacing equipment so participants will come to unravel the mystery of the magical black boxes. The scope of the workshop will be determined by the background and interest of the participants and the time allotted to the course.

Interfacing Physics Experiments to a Personal Computer

John W. Snider and Joseph Priest
Department of Physics, Miami University, Oxford, OH 45056

The laboratory part of the course consists of two, two-hour activities each week. Students complete an experiment during each laboratory period. A manual describing each experiment has been prepared for student use. About half the exercises are devoted to analog electronics with heavy emphasis on the use of operational amplifiers for tailoring electronic signals for the input to an ADC. A variety of transducers are introduced throughout the course. Experiments include the design, construction, and evaluation of an inverting amplifier, a noninverting amplifier, a two-stage thermocouple amplifier, a strain-gauge amplifier using a Wheatstone bridge, a current-to-voltage converter using a photovoltaic cell, and an integrating amplifier using a magnetometer.

We have developed software to assist the student in evaluating experimental results and in preparing computer-generated graphs.

Interfacing physics experiments to the computer begins with building and testing an 8-bit ADC for an Apple II computer. Students can use this converter in a variety of experiments that include measuring and analyzing the angular displacement-time spectrum of a damped physical pendulum, constructing and evaluating a thermistor thermometer and temperature controller, measuring and analyzing the large-amplitude period of a plane physical pendulum, measuring Planck's constant, measuring the characteristics of a junction transistor, measuring and analyzing the intensity distribution in a single-slit diffraction pattern, and measuring the coefficients of static and kinetic friction.

In most experiments the computer is used to record data from the experiment, to compare the data with a theoretical expectation, and to glean some meaningful physical quantity from the comparison. For example, the theoretical formula for the intensity distribution in a single-slit diffraction pattern is fitted to the experimental data. From the fit, the student obtains the width of the diffraction slit, which can be compared with an independent measurement.

The course capstone is an interfacing project. Each student designs an experiment, constructs the apparatus, writes and tests the software involved in performing the experiment, performs the experiment, and writes a report.

Emphasis on doing physics with the interfacing projects and using the computer as an analytical instrument are rewarding features of the course. Students find that the course prepares them well for future physics endeavors whether as a student or in the work force.

Laboratories in Sound Analysis Using Fast Fourier Transforms

Carl R. Nave
Department of Physics and Astronomy, Georgia State University, Atlanta, GA 30303

Darrell L. Bell
Benjamin E. Mays High School, Atlanta, GA 30331

Under the sponsorship of the REAP[1] program during the summer of 1987, we developed some short experiments making use of the Apple II computer for sound analysis. The experiments use some BASIC routines to demonstrate discrete Fourier analysis and fast Fourier analysis of time-varying signals. The main purpose was to develop convenient routines that could accept data from a VELA[2] laboratory interface by analog-to-digital conversion, from a mathematical formula, from hand-entered data, or from a text file. The fast Fourier transform routines are based on some machine-language FFT routines supplied to us by Vasilis Pagonis.[3]

The VELA is capable of A/D conversion with a minimum time interval of 30 microseconds and produces satisfactory wave forms for most musical instrument or voice sounds. The FFT routine developed for use with the VELA uses an accessory card to manage the transfer of data from the memory of the VELA to the Apple IIe memory. Another version is set up to read Apple text files in the format used by Vernier's *Graphical Analysis* program so that the FFT program can analyze data that has been collected by other means. A third version was configured to pause and prompt the user to enter BASIC program lines either in the form of "read" statements for direct entry of numerical data or a loop to evaluate a mathematical function.

The multiwindow display for the discrete Fourier analysis program consists of a plot of the original wave form or function, a bar graph of the first ten harmonic amplitudes, a plot of the synthesized wave form for comparison, and a scrollable list of the numerical values for the harmonic amplitudes. Any of the windows can be zoomed to fill the screen and dumped to an Epson printer using a Grappler card. The display for the FFT program consists of a plot of the original wave form, a bar graph of ten harmonic amplitudes, and a third half-screen window to display a user-chosen number of up to 127 harmonic amplitudes in bar-graph form.

During the past year we have used the routines successfully in an elementary acoustics course, an electricity and magnetism lab on Fourier analysis, and a senior-level acoustics laboratory. We have also used the function version in an optics course with various widths of single pulses to show that the Fourier transform of the single slit produces the single-slit diffraction pattern. We will show the display screens and examples of applications.

1. Research and Engineering Apprenticeship Program, Academy of Applied Science for the Army Research Office, 1 Maple Street, Concord, NH 03301.

2. Versatile Laboratory Aid (VELA) (TEL-Atomic, Inc., P.O. Box 924, Jackson, MI 49204).
3. Vasilis Pagonis, Allegheny College, Physics Department, Meadville, PA 16335.

Computerized Experiments in Physics Instruction

H. M. Staudenmaier

Interfakultatives Institut für Anwendungen der Informatik und Fakultät für Physik, Universität Karlsruhe, Federal Republic of Germany

Two computer-instruction projects have been installed in the physics department of the University of Karlsruhe. The first is a computer-aided instruction laboratory (CAIL),[1] and the second is the integration of computerized experiments into the laboratory exercises of a graduate course.

The central hardware is a VAX-750 computer with various peripheral devices and six terminals reserved for the CAIL participants. We use the CAMAC system as an interface between the computer and experiments because of its modular structure and its extensive use in nuclear and high-energy physics.

Computer-Aided Instruction Laboratory

The CAIL was started to familiarize students with real-time applications of computers in physics experiments. The CAIL offers typical topics of online computing, including data collection, data reduction, and process control; but these are combined with physics applications. Students carry out various experiments that are connected to a VAX minicomputer. The CAIL increases students' knowledge both of physics and computer science. Experience has shown that a knowledge of computer science improves job prospects for physics graduates.

Students take between five and seven hours per week or a total of about 80 hours to complete the CAIL. The CAIL can be taken as an optional course within the physics curriculum, and so far 360 physics students have passed successfully.

At the moment, the CAIL consists of nine experiments or exercises. Students start with exercises that concern the operating system of the computer, the compiler (mainly FORTRAN or Pascal), graphics, and two simple examples of data collection in assembly language. In my presentation, I will present three of the online experiments in detail.

Semiconductor Characteristics. With the help of the computer, students measure the characteristics of various semiconductor elements such as Si, Ge, tunnel, or zener diodes. They then plot the current-voltage characteristic and extract the parameters of the elements.

Sampling of Analog Signals. Methods of detection and analysis of noisy signals are important tools in experimental physics. Students sample and digitize analog signals of different shapes with various noise components using a fast 12-bit A/D converter. Students repeatedly sample one signal, leading to a data file with about 20,000 data points. They then analyze this file with various methods of signal processing, for instance, tests of the sampling theorem, the aliasing effect, or cumulative averaging. A second very important field is the transformation of a signal from the time domain to the complex frequency domain. Here, students use the date sampled in the experiment to illustrate Fourier transform (FT), discrete FT, fast FT, leakage effect, convolution, and correlation.

Elements of Picture Processing. Methods of picture processing have been used in physics for a long time. Pattern recognition has been the most effective procedure in high-energy physics to detect new particles. In this experiment, students digitize video pictures with a specially equipped IBM PC and then transfer the image file to the VAX computer. Students work on the following subjects of image processing: gray-scale histograms, pseudocolors, two-dimensional Fourier transform, pattern recognition, digital filtering, and edge sharpening. Finally, they determine the fractal (or Hausdorff) dimension of the coastline of Great Britain. For that purpose, they digitize a map and "survey" with a software "ruler" of variable length.

Computerized Experiments in Graduate Laboratory Courses

We have incorporated two computerized experiments into the regular graduate laboratory course. The idea is to reduce the time spent for data collection and analysis so that students can measure more test pieces or more interesting effects. The important point is that the computer remain a "black box" as much as possible, the emphasis being on teaching physics.

Cosmic Ray Experiment.[2] In physics research the measurement of "single events" is already a common method, but up to now this method has been practical only if a computer interfaces with the experiment. The cosmic-ray experiment is an example of single-event experiment. We use two large bars of scintillation material, 240 cm long, placed above each other. The student places photo multipliers at both ends of each bar, and a cosmic particle passing through generates scintillation light that is detected in the photo multipliers. The position of the particle is determined electronically and stored in the computer together with the resulting pulse height. The physical objectives are angular and energy distribution of cosmic rays, east-west effect, shower generation, etc. Students have not been totally positive about this experiment; they have trouble understanding and adjusting the electronics of the experiment.

Heat Capacity Measurement.[3] The measurement of heat capacity has been a traditional experiment of this laboratory. The experiment was computerized to speed up the measuring procedure and analysis of the data. The sample, for instance, copper, is placed in a cryostat and the specific heat is measured in the temperature range 80 to 300K. The sample is heated by a resistance heater controlled by the computer connected to the apparatus. An important feature is a thermal shield that is kept at the same temperature as the sample; the shield's temperature is controlled with a PID algorithm implemented in the computer. The analysis is done online and the result is displayed simultaneously on the graphics terminal. In the old experiment, students measured only one sample. They can now measure three at the same time, for instance, copper, lithium fluorid, and potassiumdihydrogenphosphate (KDP). KDP is very interesting material since a ferroelectric phase transition can be observed around 120K. The experiment has been a part of the laboratory course since the summer of 1987 and student acceptance is unqualified.

1. H. M. Staudenmaier, "Use of Computer in Science Education," J. Phys. **3**, 144 (1982).
2. S. Bauer, "Rechnergekoppelter Veruch zur Messung der Hoehenstrahlung," Diplomarbeit, Universität Karlsruhe (1986).
3. G. Mayer, "Apparatur zur Automatischen Messung der Spezifischen Waerme von Festkoerpern," Diplomarbeit, Universität Karlsruhe (1987).

Popcorn, Nuclear Decay, and Counting Statistics Using an MBL

Priscilla Laws and John Luetzelschwab
Department of Physics and Astronomy, Dickinson College, Carlisle, PA 17013

Courses in the Workshop Physics Project at Dickinson College replace, formal lectures with student observations and experiments enhanced by computer-based measurements, data analysis, graphing, and numerical problem-solving. This workshop introduces materials developed by the project to facilitate teaching of high school and college-level introductory physics.

In this workshop, each participant will use a Macintosh computer equipped with a serial interface and two transducers—the Geiger tube and a small microphone—to explore nuclear and popcorn decay phenomena. The statistics of repeated counting of a long-lived source will be investigated for a low average number of counts and a high average number of counts per interval, of counting time. Computer software capable of generating a real-time histogram of frequency of occurrence vs. number of counts per interval will be used to observe the development of both the Poisson and the Gaussian distributions in real time.

If time permits, participants will conduct repeated measurements on a mechanical system that can be characterized by a Gaussian distribution. In this case, Microsoft *Works* spreadsheet and *Cricket Graph* software will be used to analyze and display the data.

The workshop will conclude with a discussion of how counting statistics can be used at the beginning of an introductory physics course to help students obtain an understanding of the nature of statistical uncertainties that are an inevitable part of all quantitative physics experiments. Interested participants will be given information on how the software and hardware might be implemented on Apple II or IBM PC computers by those with the requisite technical skills in interfacing and software design.

The Workshop Physics Project was funded by a grant from the U.S. Department of Education Fund for Improvement of Postsecondary Education.

Visual Photogate Timing and Graphical Data Analysis

Priscilla Laws and John Luetzelschwab
Department of Physics and Astronomy, Dickinson College, Carlisle, PA 17013

Courses in the Workshop Physics Project at Dickinson College, replace formal lectures with student observations and experiments enhanced by computer-based measurements, data analysis, graphing, and numerical problem solving. This workshop introduces materials developed by the project to facilitate teaching of high school and college-level introductory physics.

Microcomputer-based photogate timing systems have been available for more than five years from several leading scientific supply companies. Some of these hardware and software packages are oriented toward specific standard experiments. In some cases students merely need to drop an object and press "return" on a computer keyboard, and the computer system then takes over to deliver a printout of data that has been analyzed and graphed. Participants in this workshop will explore appropriate and inappropriate ways to use microcomputer-based photogates as tools for physics education.

Each participant will use a Macintosh computer equipped with a serial interface. A photogate consisting of a phototransistor and an infrared LED will be used to explore linear and rotational motion. The hardware will be used in conjunction with highly flexible, pedagogically oriented software developed jointly by Dickinson College and Technical Education Resources Centers. Students can use the software to obtain a strip chart style of display that shows when the photogate is blocked and when it is unblocked. This display invites the use of operational definitions of position, velocity, and acceleration to analyze motion data. Workshop participants will analyze and display trial free-fall observations using the Microsoft *Works* spreadsheet and *Cricket Graph* software.

The workshop will conclude with a discussion of why it is important for novice students to use operational definitions for all physical quantities in preference to the more abstract but ultimately more powerful paraphrase definitions usually presented in introductory textbooks.[1]

The Workshop Physics Project was funded by a grant from the U.S. Department of Education Fund for Improvement of Postsecondary Education.

1. R. Karplus, "Educational Aspects of the Structure of Physics," Am. J. Phys. **49**, 238 (1981).

Linearization of Real Data

Priscilla Laws and John Luetzelschwab
Department of Physics and Astronomy, Dickinson College, Carlisle, PA 17013

Courses in the Workshop Physics Project at Dickinson College, replace formal lectures with student observations and experiments enhanced by computer-based measurements, data analysis, graphing, and numerical problem solving. This workshop introduces materials developed by the project to facilitate teaching of high school and college-level introductory physics.

Aside from the rather shallow process of memorizing equations, there are several ways that students can know something about a functional relationship in physics: they can employ qualitative reasoning based on some general principles that they are already familiar with, they can derive a relationship mathematically using physical laws and definitions, and they can take actual measurements and attempt to linearize the data to determine an empirical relationship.

The goal of this workshop is to give the participants some hands-on experience in collecting and linearizing different types of data sets. It will also give participants some insight into the difficulties students typically have with the concept of linearization. Participants will use low-cost apparatus along with the Microsoft *Works* spreadsheet and *Cricket Graph* software to measure, graph, and linearize several different functional relationships quickly and easily. These relationships include: the sliding friction force as a function of the mass of a block, the time of fall of a ball as a function of the distance through which it falls, and the flux of parallel nails through a loop (representing an electric, magnetic, or gravitational field) as a function of the angle between the nails and the normal to the loop.

The workshop will conclude with a discussion of some software packages that can be used with IBM PCs or Apple II–series machines to perform similar analyses.

The Workshop Physics Project was funded by a grant from the U.S. Department of Education Fund for Improvement of Postsecondary Education.

Coupling Numerical Integration with Real Experience

Priscilla Laws and John Luetzelschwab
Department of Physics and Astronomy, Dickinson College, Carlisle, PA 17013

Courses in the Workshop Physics Project at Dickinson College, replace formal lectures with student observations and experiments enhanced by computer-based measurements, data analysis, graphing, and numerical problem solving. This workshop introduces materials developed by the project to facilitate teaching of high school and college-level introductory physics.

Although students learn how to integrate a number of the standard closed-form functions prevalent in physics equations, they have a great deal of difficulty understanding the physical meaning of the integral expressions typically found in calculus-based introductory physics courses.

Participants in this workshop will measure quantities associated with several simple physical systems and enter these quantities into a Microsoft *Works* spreadsheet in a manner that lends itself to numerical integration using a finite sum. In addition, *Cricket Graph* software will be used to display the data graphically. We will consider three physical systems for which both analytic and numerical integration techniques can be used: the work done as a function of the displacement resulting from a force applied to a spring, the area under a $P–V$ curve for a thermodynamic system, and the potential as a function of distance from a continuous charge distribution.

The workshop will conclude with a discussion of other physical systems that might be used as opportunities to verify the physical significance of solving a closed form integral by using a finite sum of real data.

The Workshop Physics Project was funded by a grant from the U.S. Department of Education Fund for Improvement of Postsecondary Education.

Coupling Simulations with Real Experience

Priscilla W. Laws
Department of Physics and Astronomy, Dickinson College, Carlisle, PA 17013

Courses in the Workshop Physics Project at Dickinson College, replace formal lectures with student observations and experiments enhanced by computer-based measurements, data analysis, graphing, and numerical problem solving. This workshop introduces materials developed by the project to facilitate teaching of high school and college-level introductory physics.

Physics is an experimental science, and a primary focus of the Workshop Physics Project is to use direct hands-on engagement with real physical phenomena as a way of helping students learn about the process of doing physics. The project is strongly opposed to the use of computer simulations as a substitute for real experience for two reasons. First, students have grown up with simulations in the form of video games and cartoons that systematically violate the laws of physics. Second, simulations do not teach students to appreciate the real difficulties inherent in reducing real observations about the natural world to theoretical idealizations.

However, simulations can be very powerful aids to observation and reasoning when coupled with real experience. Participants in this workshop will use three computer simulations available for the Macintosh computer in conjunction with real apparatus to explore physical systems. These sample simulations include: *Graphs and Tracks*, developed by David Trowbridge at the University of Washington and Carnegie Mellon, a spreadsheet-based relaxation method solution of Laplace's equation for mapping the potential difference in the space between two line electrodes, and *Coulomb*, developed by Blas Cabrera at Stanford University, for displaying the electric field lines in two dimensions from charges lying in a plane.

The workshop will conclude with a discussion of other physical systems that might be coupled productively with computer simulations to consolidate student understanding of various phenomena.

The Workshop Physics Project was funded by a grant from the U.S. Department of Education Fund for Improvement of Postsecondary Education.

VideoGraph: A New Way to Study Kinematics

Robert J. Beichner
*Center for Learning and Technology, State University of New York at Buffalo, Buffalo,
NY 14260*

Michael J. DeMarco, David J. Ettestad, and Edward Gleason
Physics Department, State University of New York, College at Buffalo, Buffalo, NY 14222

The *VideoGraph* software package provides students with a new way of taking data on the motion of objects.[1] Targeted for any introductory high school or college-level physics course, this software not only makes it easier and faster to collect motion data, but also helps students make the cognitive link between the physical event and the mathematical graph representing it.

Students begin by videotaping an interesting motion. This motion could be something like the collision of air-track carts or the oscillation of a mass on a spring. It could also be a "real-world" landing of a jet airplane or the acceleration of an automobile. Students use a tripod-mounted videocamera to record the motion of the object as it moves across the field of view. These images are played back one at a time on a single-frame–advancing videocassette recorder. As the images are displayed by the VCR, they are digitized and sent into a Macintosh microcomputer and stored as "MacPaint" documents. We use Pixelogic's *MacViz* interface to do this, although others will work.

To run the *VideoGraph* software, the students select the "Open" choice under the "File" menu and load the video frames into the computer. By making the proper selections from the "Set-up" menu, they select an origin of coordinates, calibrate the image (by selecting an object of known size from a frame and then entering its size and units of measure), and mark a point of interest on each frame. This last task is usually very time consuming during traditional labs, because it involves measuring extremely small dots on a photograph.

The *VideoGraph* software presents each frame individually. The student moves a mouse-controlled cursor until it is on a readily discernible part of the moving object. Clicking the mouse button records the position of that point and automatically advances to the next frame. This process continues until all frames have been marked. It generally takes fewer than two seconds per frame.

Now the student selects a graph from the "Windows" menu. For example, if the student selects the "X-Position" graph, the software opens a new window and draws appropriate axes. Selecting "Animate" under the "View" menu then produces a movie showing the object as it goes across the screen. At the same time, the software generates a graph of the object's position along the x direction. It can also produce graphs of position, speed, and acceleration for both x and y directions or can display a list of coordinates for the graph.

We are currently conducting a thorough evaluation of the effectiveness of the software.[2] The research design permits us to compare the *VideoGraph* technique with the more traditional stroboscopic method of gathering motion data. We will also be able to see if viewing an object moving on the screen reduces the need for students actually to produce the motion. We certainly do not want to imply that using the software will completely remove the requirement for hands-on experimentation. But if students need not be involved in producing every motion, they could be assigned a series of previously produced motions for thorough analysis. This could be done as homework assignments or as part of laboratory exercises. Using *VideoGraph* will thus allow students to examine a broader variety of motions, both those they produce and those produced by others.

Since this technique of gathering data will work for a wide range of motions, we are developing an entire series of laboratory exercises using the *VideoGraph* package.[3] We are also designing supplementary software to allow students to "build" motion graphs by selecting line segments from a palette of choices. Distance, speed, and acceleration graphs will be linked so that adding a segment to one graph will automatically update the other two. Student creation of a simple object that moves according to the kinematics graphs will lead students from graph to motion. We hope this will complement the current *VideoGraph* software, which goes from motion to graph.

The*VideoGraph* project was partially supported by the National Science Foundation (grant no. CSI8750443). Mainstay, Inc. donated their software development environment, V.I.P.

1. The *VideoGraph* software is available from the Center for Learning and Technology. The current version requires at least two megabytes of memory in order to work with a reasonable number of images. The animation is smoothest on a Macintosh II.
2. The development and evaluation of the *VideoGraph* software is being done as part of Mr. Beichner's Ph.D. dissertation in science education at SUNY Buffalo. Statistical results of the research will be available at the conference.
3. This curriculum package will be made available by the department of physics at the SUNY College at Buffalo. Titles of the labs include: "Measurement and Error," "Vectors," "Acceleration," "Free Fall," "Projectiles," "Harmonic Motion," "Work," "Linear Collisions," "Two-Dimensional Collisions," and "Angular Momentum."

Computer Analysis of Physics Lab Data

John Elberfeld
EduTech, Rochester, NY 14609

Computerized data-acquisition devices can now record more data in a few seconds than students can analyze in a semester using standard techniques. Fortunately, the computer also provides students with the power and speed to discover patterns and find relationships in the masses of data they record. This miniworkshop introduces participants to computerized data analysis of information recorded in physics labs. No knowledge of computers, programming, or statistics is required or expected.

All activities are based on EduTech's *Data Analysis*, a versatile but straightforward program designed to help high school students process data recorded in labs. Data may be entered from the keyboard or read off disks used by EduTech's gameport interfacing programs. Students interact with various least-squares curve-fitting routines to determine the "best" relationship between two variables. Options are included for the study of polynomial, exponential, and logarithmic relationships. The program gives the students the ability to test hypotheses by predicting the dependent variable's value based on entered values for the independent variable and on the equation selected by the student as the best.

Data Analysis brings sophisticated mathematical techniques to students from the junior high level on up, making complex analysis as accessible as graphing.

Data Analysis provides a time-efficient alternative to traditional methods of graphing and analysis. Once students have learned the techniques of hand analysis, they can use *Data Analysis* to do more testing and graphing on other experimental data. This allows them to explore relationships between variables, form a hypothesis as to the best relationship, and test it.

In *Data Analysis*, students enter x and y values into the program on a keyboard. The program plots y against x and can fit a variety of student-selected curves to the data. Students can save and recall data from data disks. Polynomial, exponential, and log curves can be fit to the data. Graphical and numerical results can be printed out on most dot-matrix printers.

Data Analysis runs on the Apple II family of computers and Apple II–compatible computers. A single disk drive is required. A second disk drive and graphics-capable printer are helpful options. The software package comes complete with extensive manual and back-up disk.

IEEE-488 Interface for a Voltage Power Source and an Analog Multiplex Unit in a Microcomputer-Based Undergraduate Laboratory

C. W. Fischer and P. Sawatzky
Physics Department, University of Guelph, Guelph, Ontario N1G 2W1, Canada

Although undergraduates are increasingly using microcomputers in the laboratory to control experiments and collect data,[1] the instructor must be careful to choose the proper degree of computer interaction. Students must not become so enamored with operating the computer that they fail to understand the physics of the experiment.

This paper describes the microcomputer system and peripherals needed for a pedagogically sound data-collection and experiment-control system. Our system incorporates an IBM PC equipped with an IEEE-488 data bus. We describe in detail the interface of a programmable voltage source and an MUX (time division analog multiplex unit). These units plus a standard DVM permit "computerization" of most of the experiments required of students in our junior undergraduate laboratory. While satisfying our primary objective of introducing modern tools and techniques to the undergraduate laboratory, our system ensures that the computer and its peripherals are as easy to use as other standard laboratory tools.

We assume that the students have familiarity with a programming language such as BASIC, Pascal, or FORTRAN, and we provide them with the elementary subroutines needed to control bus activity. Since these subroutines are specific to the type of plug-in board chosen for the PC, they are not described here.

The work station is a portable desk carrying a PC equipped with a plug-in board interfacing the PC data bus to an IEEE-488 port. These boards are commercially available or can be constructed from reports in the literature.[2] The work station also carries a 4 1/2 digit DVM fitted with a IEEE-488 port. The student uses the DVM initially to test the experiment and later to measure voltages presented to the MUX. Because a single DVM may be used to read various voltages (eight in our case), the use of the MUX reduces system cost considerably. We have used the programmable power supply to control oscillator frequency, output of low-frequency wave forms, and DC voltages.

The MUX and programmable voltage source have features important for student use. We designed the MUX and programmable power supply so that the student can manually simulate the action of the computer during an experiment. This aspect is essential because it allows the student to separate the computer and its software from the physics. The DVM could be replaced by an ADC to save cost, but then the student would not be able to use the DVM to test the experiment before relinquishing control to the computer. The DVM is also used during data collection to provide a visual check on the data.

The IEEE-488 interface is a 16-line digital bus that permits parallel transfer of data and commands between instruments. Eight of the 16 lines are devoted to the interchange of data and commands, five are interface management lines, and the remaining three are hand-shaking lines. Each device connected to the data bus has a unique address. These devices may be configured as talkers, listeners, or controllers. The MUX and power supply are listeners only; the DVM is a talker or listener; and the PC functions as a talker, listener, and controller.

The heart of the MUX interface is an 8-to-1 analog switch based on Bi-FET technology. An analog signal in the range of $\pm$ 30 volts presented to an addressed input channel appears at the output. The selected channel is indicated by a seven-segment display. The user may override the computer and manually select an MUX channel. During the setup and testing phase of the experiment, the ability to select the addressed channel manually is useful.

The heart of the IEEE-488 power supply interface is the 12-bit DAC. The DAC accepts data over the bus after the interface has been addressed to listen. After the proper address is received, a comparator outputs a true signal and sets the NRFD (not ready for data) hand-shaking line false, i.e., interface ready to receive. The controller places the first data byte (8 MSB) on the bus. This byte is latched and presented to the DAC. The controller is then signaled to send the second byte containing the 4 LSB that are also latched. Following the second byte the interface is again ready to receive, or it may be placed in the idle state by having the controller issue the unlisted command. A complete schematic in ACAD format is available from the authors upon receipt of a DOS 3.X formatted disk.

1. D. A. Briotta, P. F. Seligman, P. A. Smith, and C. D. Spencer, "The Appropriate Use of Microcomputers in Undergraduate Physics Labs," Am. J. Phys. **55**, 891 (1987).
2. B. Hall, "Programmable Laboratory Interface to the IEEE-488 Bus," Rev. Sci. Instr. **57**, 695 (1986).

21X Microcomputer-Based Logger

Boris Starobinets
Department of Geophysics and Planetary Sciences, Tel Aviv University, Ramat-Aviv 69978, Israel

It is not easy to find sophisticated microcomputer-based instruments that interface readily with a wide variety of external devices and are also easy to use. The 21X micrologger (Campbell Scientific) combines both these features.

The microcomputer-based, battery-operated 21X logger was designed to measure, collect, and record meteorological data (wind direction and speed, temperature, air pressure, and humidity). This miniature precision instrument inputs data from the sensors, processes the information, and stores this processed data in its memory to be read later using a cassette recorder, a printer, or a remote computer.

But the 21X is far more than a simple sampling logger that does no more than periodically sample and record inputs. The 21X has programming capabilities that can convert it to a very useful, general-purpose instrument for measurement, control, and interface. The 16-digit keyboard on the 21X panel is used to enter programs and commands. A serial I/O connector connects the 21X to the printer, and can also be used for remote programming or data transfer through a RS232 interface.

The panel has two terminal strips. The upper strip connects analog inputs. On the lower terminal strip, the first four numbered terminals are the switched excitation channels, which provide precisely regulated excitation voltages for bridge measurements. Two constant analog output channels can provide analog output voltage under the program control to be used, for example, with *x-y* plotters, and strip charts. These outputs are very helpful for constructing proportional controllers, automatic curve tracers, and calibrators, etc. The six digital control ports can be set to either 0 or 5 volts. Thus, the 21X can be used to control external devices such as stepmotors or relays. The four-pulse count inputs can be used as pulse counters.

The 21X's almost 100 instructions are divided into four categories: (1) input/output instructions are used to make measurements and store the readings in input locations, transfer the content of these input locations to analog outputs, or to initiate digital output ports; (2) processing instructions perform numerical operations on data; (3) output processing instructions use data from input locations to generate time or event dependent values that are stored in the final storage. Data from final storage can be transferred to remote computer; and (4) program control instructions control the sequence of execution.

Using these helpful features of the 21X logger, we have built an automatic sun photometer. The sun photometer tracks the sun and measures the spectrum of the solar radiation in the visible region. Tracking is carried out with the help of the quadrant detector. An error signal arises when the image of the sun shifts out from the center of quadrant photodetector. The 21X processes these signals and uses them to control two step-motors, which correct the position of the sun photometer. The 21X simultaneously checks the position of the filter wheel (a wheel on which are fixed seven narrow-band interference filters). The 21X measures the solar radiation at the instant that the corresponding filter is positioned exactly over measurement sensor.

Any physics teacher working with the 21X will appreciate this remarkable microprocessor-based measurement and processing tool.

Using Vernier's Graphical Analysis in Physics Labs

Betty P. Preece
Melbourne High School, Melbourne, FL 32901

Participants will explore various ways to use Vernier's *Graphical Analysis* program with data from physics labs. Using actual data from a lab on the period of a simple pendulum, we will use *Graphical Analysis* to analyze and plot graphs so that we can search for possible relationships such as direct linear, to the second or to the one-half power, or independent. We will search both on screen and on hard copy. We will use data from demonstrations, lab group, and individual analyses.

I will also discuss applications of this *Graphical Analysis* to rectilinear motion and to centripetal motion labs. I have successfully carried out computer analysis of these labs in my physics lab even when only one Apple IIe and printer has been available for a class of 30. Even teachers of physics with limited experience in computers and physics can use *Graphical Analysis*. The software enables students to see relationships that are often obscured by the time-consuming plotting of data by hand.

ASYST, a Tool for Physicists

J. D. Thompson
Department of Physics, Augustana College, Sioux Falls, SD 57197

The use of computers has revolutioned the teaching of physics. The computer now serves as a powerful tool both in the laboratory and in the classroom. Part of the progress in using computers is due to powerful computer programs like *ASYST*.[1]

ASYST is an integrated package of modules that support data collection, data analysis, curve fitting, and graphical presentation of results. The *ASYST* package is versatile, convenient to use, fast enough to be useful for the laboratory, and also useful in the classroom, e.g. for displaying curves, wave form synthesis, fast Fourier transforms, inverse transforms, and various other mathematical operations. The system comes complete with several utilities that facilitate program and file preparation and management, including a text editor.

ASYST is a stack-oriented package similar in use to FORTH language and the reverse Polish notation of some calculators. A large number of primitive words (similar to commands such as "Print," "Sin," etc., of BASIC) are supplied with the package. User-defined words are easily created in a manner similar to creating a subroutine in BASIC or a procedure in Pascal.

Several useful mathematical tools are included in *ASYST*. First, *ASYST* has an extensive array-handling capability. It includes an array editor, and it allows the user to select subarrays and carry out array arithmetic, including inner and outer products. Using *ASYST*, the user can store arrays in the computer's expanded memory, if available.

ASYST also offers extensive graphics capabilities, including automatic scaling and graphing on the monitor in linear or exponential form. Numerous *ASYST* commands facilitate user selection of scale, annotation, color, and number of superimposed curves. Words are also available for doing contour and axonometric plots of data. All *ASYST* graphs may be dumped to an attached printer or directed to an *x-y* plotter.

ASYST also accommodates some limited statistical analyses and curve fitting. Statistical analyses include one-way, two-way, and regression analysis of variance.

Curve-fitting capabilities include smoothing, least-squares polynomial, exponential, and logarithmic fits, and fits to orthogonal polynomials such as Tchebyshev, Legendre, and Hermite polynomials.

Other mathematical operations in *ASYST* include polynomial integration, differentiation, and root extraction; matrix manipulations including inner and outer products, inversion, multiplication, and solution of simultaneous equations; diagonalization of self-adjoint matrices, and determination of eigenvalues; fast Fourier transformations, inverse transformations, and frequency-domain filtering.

Of particular interest to physicists is the ability of *ASYST* to do hardware-based data acquisition. With the addition of readily available boards, 12-bit A/D at conversion rates to 50 kHz and higher are attainable. The acquired data may be stored in arrays or transferred directly to data files on disk, including direct memory access to files in a RAM disk for the highest transfer rates. Data may also be acquired through an IEEE-488 interface board from equipment such as digital oscilloscopes, digital voltmeters, etc., equipped with IEEE-488 capabilities. Words to effect the transfer of data through several available boards are included in the package and the setup required for individual board installation is relatively easy. With suitable available hardware, the system may be used for digital to analog conversion as well as digital I/O.

The minimum hardware requirements for *ASYST* are: IBM PC/XT/AT or compatible, 512K RAM, two floppy disk drives or one floppy drive and a hard disk, IBM color graphics adapter (equivalent or better), and a math coprocessor (Intel 8087, 80287 or 80387). The ability to do hardware-based data acquisition requires an analog to digital (A/D) conversion board or an IEEE-488 board.

1. *ASYST* (Rochester, NY: ASYST Software Technologies).

Microcomputer-Based Integrated Statistics, Analysis, and Graphics Software for Introductory Physics Laboratories

J. D. Kimel
Department of Physics, Florida State University, Tallahassee, FL 32306

Students in introductory physics laboratories can use microcomputer software to analyze and graph the results of their experiments. Such powerful and convenient tools cut through the calculational tedium normally required of such analyses and get directly to the physics being investigated in the laboratory.

We have designed an integrated series of statistics, analysis, and graphics software for introductory physics laboratories. Our system requires mobile work stations consisting of an IBM PC–compatible microcomputer and a graphics-capable dot-matrix printer (set up for IBM graphics). Each work station is mounted as a unit on a rolling cart. Our software is in use in introductory physics laboratory sections of 20 students working in pairs, with ten mobile work stations.

The software is now used in the first semester of a two-semester laboratory associated with a calculus-based introductory physics course at Florida State University. After we acquire more work stations, we will expand the use of the software to both semesters of the calculus-based course, and eventually to our non-calculus-based course. These laboratories will serve approximately 300 students each semester.

The software gives the student a convenient means of calculating and visualizing the statistical uncertainties associated with any repeated measurement of a physical quantity. The student must take such uncertainties into account in determining what physical conclusions might be inferred from the measurements. An advantage of the software is that it gives the student almost-immediate analysis and graphics feedback on measurements. The immediacy of the system encourages the student to make hypotheses about the data and draw conclusions about the statistical relevance of the laboratory measurements to the hypotheses. This enhances the student's insight into the physical implications and limitations of physics laboratory experiments.

The software provides the student with the means of recording, analyzing, printing, graphing, and saving to diskette, the experimental results of the laboratory measurements. Statistical uncertainties in repeated measurements are calculated and graphically displayed. These uncertainties are fed into the analysis portion of the software, which provides a least-squares polynomial fit to the data on linear, semilog, and log-log scales. The parameters of this fit are assigned uncertainties propagated from those in the experimental data. The process allows the student to draw statistically significant conclusions about the physics being investigated in the experiment. When the data is saved to diskette, the student can quickly analyze the data in several different ways and can identify the measurements both analytically and graphically as exhibiting a linear, exponential, or power-law dependence on the independent variable. Thus, the student is led to hypothesize about the underlying physics of the experiment.

The software is written in *QuickBASIC* and consists of two parts. The first part, *SIGMA*, allows the student to enter and display on the screen a series of measurements of a single quantity in a worksheet-style format. The program calculates the mean, the standard deviation of the sample, and the standard deviation of the mean for the series of measurements. The software allows the student to graph the results to screen or printer and to save the worksheet on a diskette for later use.

The second part, *ANAGRAPH*, which is also in a worksheet-style format, allows entry and display on the screen of a list of measurements and their experimental uncertainties versus the independent variable points. The student uses a least-squares algorithm to fit the experimental data to a polynomial of degree up to six, and the software returns the expansion coefficients together with uncertainties resulting from uncertainties in the experimental measurements. The results can be saved to diskette, printed, or graphed to the screen or printer. The student can select three different types of polynomial fit:

$$1.\ y = a_0 + a_1\Sigma x + a_2\Sigma x^2 + ...$$

$$2.\ \log(y) = a_0 + a_1\Sigma x + a_2\Sigma x^2 + ...$$

$$3.\ \log(y) = a_0 + a_1\Sigma\log(x) + a_2\Sigma[\log(x)]^2 + ...$$

and can graph these on linear, semilog, or log-log scales. Thus the student can easily identify a linear, exponential, or power dependence of the laboratory measurements versus the independent variable points.

The software is self-booting and menu driven. Help screens appear where necessary to remind the student of the next step in applying the software. Although written instructions are provided, the software is intuitive, self-explanatory, and very easy to use.

The hardware requirements are an IBM PC or compatible with one 5.25-inch floppy-disk drive, 256K of memory, and a CGA card. The software can utilize an 8087 mathematical coprocessor, but the coprocessor is not required. For printer-graphics output, the software requires a dot-matrix printer with a parallel interface, set up for IBM graphics.

The software programs *Analysis and Graphics Package* are part of the collection *Computers in Physics Instruction: Software*, which can be ordered by using the form at the end of this book.

Application of a Commercial Data-Acquisition System in the Undergraduate Laboratory

William R. Cochran
Department of Physics and Astronomy, Youngstown State University, Youngstown, OH 44555

I use a commercial data-acquisition system (the Keithley 570) in conjunction with the IBM-PC in an upper-division undergraduate laboratory. In my presentation I will discuss the advantages and disadvantages of this system from the point of view of a small college physics department. I will also discuss example experiments from optics and thermodynamics.

In the optics experiment, the PC and Keithley 570 are used to drive a stepper motor connected to the movable mirror of a Michelson interferometer, to sample the interference signal incident on a photodiode, and to calculate and plot the resulting spectrum, using the Cooley-Tukey algorithm to obtain the transform.

In the thermodynamics experiment, the student calibrates a thermistor by determining the constants in the Steinhart-Hart equation, and then uses the PC and Keithley 570 to drive a transconductance amplifier that measures the voltage across the thermistor. The system also collects the data and calculates the Steinhart-Hart constants. The student then uses a second program, into which the constants for any thermistor can be entered, to determine unknown temperatures.

Physics Education Research and Computers

Computers and Research in Physics Education

Lillian C. McDermott
Department of Physics FM-15, University of Washington, Seattle, WA 98195

Computer-based materials for teaching physics are being produced at a rapid rate. The range of instructional programs already generated is broad, including exercises for drill and practice, tools for graphic display, simulations of phenomena, guidance for solving textbook problems, and tutorials for teaching specific material. Programs for solving complex problems numerically are making it feasible for students to consider real-world situations instead of being limited to idealizations. As more powerful programming environments are developed, the scope of what is technically possible continues to grow. Advances in technology regularly give rise to suggestions of new ways to use computers in physics education.

Although development has been fast, we have devoted relatively little attention to examining what students at a computer may be learning or how they may be thinking. To take full advantage of the current technology, we should pause to reflect on what computer-based instruction has contributed to student learning, on what we would like such instruction to accomplish, and on how we might best direct our efforts to bring about the desired results. Such considerations will provide a much more fruitful direction for our labors than will intuition or opinion.

In the fall of 1982, a special conference was held at the University of Pittsburgh to consider how research could help realize the educational potential of the computer.[1] Among the approximately 40 people who attended were cognitive scientists, computer scientists, psychologists, science educators, precollege teachers, and university faculty. Several physicists were present, including F. Reif, who was one of the two chairmen. The report of the conference, which was published in 1983, includes a research agenda. Two main categories are identified: basic cognitive research and prototype research. The first deals with teaching and learning at a fundamental level, an orientation that is characteristic of the cognitive scientist or psychologist. The second is of more immediate interest to physicists and is the emphasis of this paper. The point of view expressed is that of a physics instructor whose primary motivation for research is to understand better what students find difficult about physics and to use this information to help make instruction more effective.

This paper examines the relationship between the computer and research in physics education in the context of specific examples. Most of the examples presented here illustrate how research can be used as a guide for development, i.e., how research drives the computer program. The remaining examples show how the computer can be a tool for doing research, i.e., how the computer program drives research. Discussion of the examples gives rise to questions warranting further study. Reflection on the potential of the computer, both as a means of delivering instruction and as a means of investigation, leads to some suggestions for future research. The paper concludes with a summary of how research and curriculum development, both traditional and computer-based, can interact within an instructional environment to contribute to the improvement of physics education.

Research as a Guide for Development of Programs

Results from research can be used to guide the development not only of written materials but also of instructional programs on the computer. Empirical investigations can help identify common conceptual and reasoning difficulties that students encounter in learning physics. Careful examination of student thinking on a topic may also reveal useful intuitions that might be exploited in instruction. This information can be used to design computer-based instruction that addresses the difficulties identified or that develops new concepts or skills. The main emphasis in other research may be on the development and testing of specific instructional strategies, both those that can be used generally and those that are peculiar to the computer. Optimally, such research involves preliminary testing of the instructional procedures before they are incorporated into a computer program.

Research can provide a basis for the design of programs to teach problem solving. The focus is often on the development or the use of a theoretical model of cognition or instruction. The features of the model can be incorporated into a computer program for teaching students how to solve certain types of physics problems. The model may be prescriptive or descriptive. A hypothetical task analysis may be the starting point for constructing a prescriptive model of good problem-solving performance. In this case, the model may consist of the knowledge and procedures necessary for arriving at a correct solution to a problem. Some problem-solving research focuses on how individuals with different levels of expertise solve problems. Analysis of the data leads to the formulation of descriptive models for expert and novice problem-solving behavior. Ideas from artificial intelligence about how knowledge is acquired, stored, and accessed are sometimes incorporated into this theoretical framework. The models that result from this type of research may be used to develop computer programs that will lead students from novice to expert status.

Another way in which research can play an important role in program development is to ensure that the quality of interaction between the computer and the student is the very best that it can be. Matters of motivation, screen appearance, computer feedback to student response, ideal student group size, etc., all make a difference in how effective computer instruction will be. This important area for research is beyond the scope of this paper.

The computer programs discussed below are examples in which the results of research were used to guide program development. They were chosen to illustrate a variety of ways in which this has been done. A number of other programs could also have served as examples.[2]

Addressing Student Difficulties Identified through Research

The first two programs were deliberately designed to address specific difficulties identified through research. Both are based on empirical investigations by the Physics Education Group at the University of Washington.

Tutorial/Simulation: Distinguishing between Position and Velocity

Peter Hewson has applied the results of research to design a computer program that helps students distinguish between the concepts of position and velocity.[3] One manifestation of confusion between these two concepts is the use of a position criterion to determine relative velocity. This misconception, or alternate conception (a term Hewson prefers), was identified in an investigation of student understanding of the concept of velocity by D. Trowbridge and L. C. McDermott.[4] They reported that a sizable number of introductory physics students claimed that two objects had the same speed when one passed the other.

During individual demonstration interviews, students were shown a demonstration in which two balls rolled along adjacent tracks. One ball rolled along a horizontal track and maintained a constant speed, while the other rolled up an inclined track and slowed down. The ball that rolled uphill started behind the ball on the level track and passed it. Then as the ball on the incline continued to slow down, it was passed by the ball on the level track. The students were asked if and where the two balls ever had the same velocity. There is an instant when the ball that is rolling uphill and slowing down has the same velocity as the ball on the level track. However, a significant number of students claimed that the balls had the same velocity at the two passing points.

Hewson developed a microcomputer simulation in which two cars race against each other along parallel paths. The accompanying tutorial program consists of two phases: diagnostic and remedial. The first diagnostic phase identifies the inappropriate use of a position criterion. The student observes six car races and for each one pushes a button at the instant that the cars appear to have the same velocity.

In the remedial phase, students who do not distinguish position from velocity are led to confront and resolve their confusion. There are two races in this part of the program. The first is designed to create dissatisfaction with a position criterion by providing an extreme situation in which one car moves across the screen at constant speed, passing another that is not moving. The second race is designed to provoke the correct criterion. Both cars travel at the same constant speed across the screen with one maintaining a constant distance behind the other. The questions asked by the computer and the commentary made in the program focus attention on the difference between position and velocity. The program does not point out the correct criterion; the students must recognize themselves that when the two cars are at the same speed the distance between them appears approximately constant.

The program was tested with eighth-grade students at a high school in Johannesburg, South Africa, and at the University of the Witwatersrand, where it was used in a freshman physics course for students from educationally disadvantaged backgrounds. The program was successful with both groups, but more so with the university students. Most of the latter group were able to state a correct criterion several weeks after they had used the program and also to recognize that their ideas had changed as a result of using the program.

In designing the program, Hewson had in mind a model of learning as conceptual change. He considered the particular methodology (microcomputers) and the specific content (velocity) to be examples of a broad theoretical approach that can be applied to other methodologies and other concepts. It is interesting to note that his instructional strategy is similar to the one developed at the University of Washington to address the same conceptual difficulty. Although the Physics Education Group does not describe its approach to curriculum development in terms of a theory of instruction, the result is much the same.

Tutorial/Simulation: Connecting Graphs and Actual Motions

Graphs and Tracks, by D. Trowbridge, is designed to help students make connections between actual motions and the graphs that represent them.[5] In an investigation that extended over several years, L. C. McDermott, M. Rosenquist, and E. H. van Zee identified some specific difficulties students have in graphing motion.[6]

Before Trowbridge developed this tutorial program, the Physics Education Group had developed a laboratory-based curriculum to address commonly recurring errors identified through research.[7] We designed, tested, and refined the instructional strategies incorporated in this curriculum during several cycles of use in courses taught by members of the group. The curriculum includes several exercises and experiments that make use of steel balls rolling on aluminum tracks, equipment also used in some of the research. Students are asked to observe the balls rolling on various track arrangements and to construct graphs of position, velocity, and acceleration versus time. They are also asked to do the reverse operation, i.e., given various motion graphs, the students must produce corresponding track arrangements.

Drawing on both the research and instructional materials, Trowbridge designed a computer program to help students overcome specific difficulties in making connections between real motions and graphs of position, velocity, and acceleration versus time. The program is in two parts. Part 1 is entitled "From Graphs to Motion" and part 2 "From Motion to Graphs."

Part 1 uses the animation capabilities of the computer. The student is shown a set of example graphs together with a set of tracks and a ball. For each example, the student's task is to arrange the tracks to reproduce the graph. The display on the screen for one of the examples is shown in Figure 1. The tracks consist of five segments that can be raised or lowered by moving and clicking the mouse. The mouse can also be used to select initial values for position and velocity. After adjusting the tracks and choosing the initial conditions, the student tells the computer to roll the ball. As the ball rolls along the tracks, the corresponding motion graph is generated. The student can compare this graph with the original one, modify the arrangement, and try again as many times as desired. The student is responsible for detecting any errors and deciding when the match is good enough to go on to another example. The computer program itself is nonjudgmental; it is up to the student to determine if a response is correct. The student may choose to consult the help facility that is available to give step-by-step guidance toward the correct solution. The program also allows students to generate new problems by creating and saving their own graphs.

Part 2 uses the graphics (drawing) capabilities of the computer. The screen shows a series of motions on various track arrangements, one at a time. In each case the student is asked to produce the three corresponding motion graphs either by using a graphing palette containing basic shapes (straight line, parabola, etc.) or by drawing freehand. When the student asks for feedback, the program analyzes the student's graph and attempts to give step-by-step guidance for correcting a mistake. Thus Trowbridge had to program the computer to recognize many different errors. The effectiveness of the program depends heavily on knowledge obtained from previous research with students.

Graphs and Tracks has been used to extend the research on which it is based. As discussed later in the paper, preliminary results show that the computer makes it possible for us to identify and analyze some student errors in graphing, that otherwise might be undetected or not as well understood.[8] Computer-based interviews also allow observation of the strategies students use to perform the assigned tasks.

Developing and Testing Specific Instructional Strategies

The next four programs illustrate research in which the computer is used to develop or test a specific instructional strategy. In the first two programs, the computer itself is an integral part of the strategy. In the other two programs, the method that is being tested is not inherently restricted to the computer. However, the computer makes feasible the large amount of practice required in one of these and the high degree of individualization required in the other.

Microcomputer-Based Laboratory (MBL)

A special instructional strategy for addressing difficulties with kinematics and with motion graphs makes use of the microcomputer-based laboratory. R. Tinker and R. Thornton have designed a sonar probe that can serve as a motion detector.[9] When interfaced with a computer, this device allows students to generate and view graphs of their own motion in real time. The immediate feedback from the computer helps students relate the motions of their own bodies to various features of motion graphs.

Prompted by the results from the research cited above and by his own observations of college students, Thornton designed MBL materials for use in teaching introductory physics. He and instructors at several other institutions have found that use of these materials in sessions lasting from one to a few hours can markedly improve the ability of students to draw and interpret motion graphs.[10] Studies with both calculus and noncalculus physics students, demonstrate that those who have had experience with MBL do significantly better on examination questions involving motion graphs than those who have not had such experience. After using MBL materials, even students without a strong science background outperform calculus-level students who have not. Results from research at the precollege level indicate that instructional strategies based on MBL are also effective with younger students.[11]

Videodisc/Computer Graphics Overlay: Ray Diagrams and Optical Systems

F. Goldberg and S. Bendall used the computer in combination with a videodisc to help students understand the relationship between an optical system and the ray diagram that describes it.[12] This work is based on research by F. Goldberg and L. C. McDermott, who investigated student understanding of the real image formed by a converging lens.[13] One of the most common difficulties that Goldberg and McDermott identified was the inability of students to draw an appropriate ray diagram to make predictions about a simple system consisting of a light bulb, a lens, and a screen.

In developing an instructional strategy to address this particular difficulty, Goldberg and Bendall reasoned that if students were introduced to the idea of a ray diagram by seeing the diagram superimposed on a video picture of the real equipment, they would retain a strong impression of the relationship between the two. There is some evidence from psychological research to support this idea.

To investigate the effectiveness of this strategy, Goldberg and Bendall developed two versions of a lesson on image formation by a converging lens. Both presented the same textual information. In the "video" version, a video picture of the lens system and the corresponding ray diagram are presented simultaneously on the computer screen. The ray diagram drawn by using computer graphics appears as an overlay over the picture of the lens system. In the nonvideo version, either the video is presented without graphics or both the ray diagram and lens system are drawn by the computer.

Both versions have been tried in preliminary testing with future elementary school teachers who had no previous formal instruction in optics. Immediately following the lesson, the investigators assessed student understanding by asking questions with the actual physical apparatus present in front of the student. The students who had worked through the "video" version of the lesson were more successful in making correct predictions and drawing correct ray diagrams than those who had worked through the non-video version.

Analogy-Based Computer Tutor

Remediating Physics Misconceptions Using an Analogy-Based Computer Tutor is a computer program designed to help students overcome some common misconceptions in mechanics.[14] One of these is that a static, rigid body (e.g., a table) cannot exert a force on another body with which it is in contact (e.g., a book). The program attempts to make plausible the existence of a passive normal force through computer-generated Socratic dialogue in which a series of analogies is introduced. The *Analogy-Based Computer Tutor* implements an instructional strategy designed to bring about conceptual change. The tutor is based on earlier research by J. Clement and D. Brown, who developed and tested the procedures with students before any programming on the computer was undertaken.[15] This work built on research that had been conducted in a high school physics classroom by Jim Minstrell, who in turn drew on an instructional sequence used earlier by A. Arons.[16]

Like Minstrell, Clement and Brown did their research in the classroom. To help students accept the idea that a table exerts an upward force on a book resting on it, Clement and Brown suggest an anchoring situation that may be more intuitively acceptable to students than is the target situation (the book on the table). For example, the students might be asked to compare the book on the table with a book resting on a hand (a possible anchoring situation). Clement and Brown found that students may admit that the hand exerts a force but may not believe that an analogy between a hand and a table is valid. As other investigators also note, students are often reluctant to accept the idea that something inanimate can exert a force. At this point, the students are presented with one or more bridging analogies. In one sequence, for example, a book on a coiled spring serves as an intermediate analogy. Although students usually recognize that the spring can exert an upward force when compressed from above, they often do not see the table and spring as analogous. Other bridging analogies may then be proposed, such as a book on foam rubber that sags and a book on a thin board that bends slightly.

When this instructional strategy was tried in several high school classes, there was a significant difference in experimental over control groups in acceptance of the physicist's interpretation that the table exerts an upward force on the book. Clement and Brown found that sometimes many discussions were necessary. Often a long chain of bridging analogies had to be suggested to convince some students that the target and anchor systems were similar with respect to the feature under consideration.

Proceeding from this experience in the classroom, Clement and Brown worked with T. Murray and K. Schultz to design a prototypical computer program for an analogy-based tutor. They were interested in seeing whether the classroom strategy could be simulated in a computer environment and whether it would be effective under those circumstances. The computer program they designed contains many possible bridging analogies. Whenever they experience difficulty in seeing the proper connection between any two analogies in the chain, the students are offered a new analogy that attempts to narrow the conceptual gap. To replace the feedback that a human tutor gets from students through facial expressions and other clues, the computer asks students to rate their confidence in their answers at each stage of the program. With this information and knowledge of the analogies that have been tried, the computer can keep track of the student's state of learning. The tutor is intelligent, because it can tailor instruction at any point in the program according to individual need.

The analogy-based tutor was tried with a small group of students. After working through the program, almost all succeeded in giving the correct answer that the table exerts an upward force on the book. They indicated that this answer made sense to them at a high level of confidence, even though most had started by being convinced that the table could not exert an upward force.

Hierarchical Analysis Tool (HAT)

Gerace and Mestre have developed a computer program to help students improve their problem-solving performance.[17] The program is called the *Hierarchical Analysis Tool* (HAT). It is based on research that indicates that experts and novices organize knowledge and approach problem solving in very different ways.[18] Experts seem to think hierarchically, classify problems according to basic principles, and generally begin to solve problems with a qualitative analysis. Novices, on the other hand, seem to store information more homogenously, classify problems according to surface features, and approach problem solving by searching for equations that might lead to an answer. *HAT* is designed to help students make a transition from novice toward expert behavior.

HAT presents students with 25 problems in classical mechanics. In each case, the students select from a computer-generated list the appropriate principle that could be applied to obtain a solution to the problem. Gerace and Mestre have compared the effect of this type of instruction to that of two other treatments. In one, the students use the textbook from their physics course to solve the problems. In the other, the students use an equation database program (*Equation Sorting Tool*) to obtain the equations needed. The large number of equations in the data base can be greatly reduced by performing sequential sorts according to basic principles, surface feature attributes, or a particular variable.

Gerace and Mestre have compared the effectiveness of the three instructional strategies by examining student performance on a test designed to be similar to a traditional final examination. The results show that students who had used the *Hierarchical Analysis Tool* shifted their criteria for categorizing problem type away from surface features, toward basic principles. However, the *HAT* group was no more successful than the other two groups in being able to solve the problems. Thus the instructional strategy of having students classify problems, at least as it was implemented in this research, does not seem to be very useful for improving problem-solving ability.

Teaching Problem Solving

Following are two examples of programs designed to teach problem solving in two different areas of physics. The first draws on the results of research but is also based on the experience of the author in teaching physics. In the second, the approach to instructional design is from a more theoretical perspective and has a strong cognitive-science emphasis.

Tutor for Kinematics Problems

G. Oberem has designed an intelligent tutoring system called *Albert* to teach students how to solve one-dimensional kinematics problems with constant acceleration.[19] *Albert* can understand and solve problems presented to it in natural language taken directly from a standard textbook. The tutor acts as a coach, helping on request, usually by making suggestions rather than by giving direct answers to questions. Students can use any method to try to solve a problem. They may ask for help at any point. The actions or responses of students affect the direction in which the program proceeds. Multiple branching makes possible highly individualized feedback. *Albert* is an intelligent tutor because it can diagnose difficulties, monitor a student's progress, and provide help at a level that is appropriate for a particular student.

Given a problem, *Albert* translates from English to the mathematical symbolism in which the kinematical equations are expressed. *Albert* can find a solution by using these equations or, alternatively, by using the more fundamental relationships from which they are derived: the definition of average velocity, the definition of acceleration, and the expression for the average velocity when the acceleration is constant.

The tutorial-management system that drives *Albert* uses an instructional approach that is based on research by F. Reif and coworkers. In broad outline, the problem-solving strategy has three stages: problem description and analysis, construction of a solution, and assessment of the solution. Each of these stages has been analyzed, described in detail, and validated in research by Reif and J. Heller.[20] Oberem draws on this work explicitly, and on the work of other groups as well. In addition, in order to anticipate difficulties that might need to be addressed, Oberem analyzes written responses by students to sample problems. He also examines student/computer interaction by taking the place of the computer during exploratory testing. When students typed their comments on the keyboard they were unaware that they were communicating with Oberem and not with the computer.

Albert has been used with a limited number of students at Rhodes University in South Africa. In initial testing, *Albert* has been able to maintain a coherent dialogue with each student. Although the process has sometimes been time consuming, the dialogue has led in each case to a successful solution of the problem selected by the student. In its present state of development, *Albert* can help students on an individual basis but cannot give very deep explanations of its reasoning.

Tutor/Microworld for Electric Circuit Problems

The work of B. White and J. Frederiksen reflects the more theoretical approach to computer-based instruction that characterizes cognitive scientists.[21] Drawing on the results of empirical investigations, cognitive theory, and artificial intelligence, White and Frederiksen have designed a program to teach students to solve electric circuit problems. Their objective was to create a computer learning environment that utilizes a microworld (interactive simulation) within an intelligent tutoring system. The authors proceed from the proposition that progress toward expert status can be viewed as an evolution of successive causal models. Each of these mental models consists of a set of rules. A particular model may be adequate for solving a given class of circuit problems but inadequate for solving problems at the next level of complexity. For these, additional rules must be introduced. Simple models evolve into more complex ones. The transition from novice to expert status is viewed as a transformation of mental models.

White and Frederiksen study the procedures followed by an expert troubleshooter who was teaching at a technical high school. They analyze the procedures and put them into the form of rules that, in their entirety, constitute a mental model for an expert (expert model). The rules are arranged into sets (student models) to match the progression from lower to higher levels of expertise. In formulating the student models, the designers were guided by findings from previous research on student understanding of electric circuits.

The program is capable of simulating circuits on the computer screen and allowing students to perform experiments. The computer poses problems, explains the behavior of circuits, and presents rules for a causal model appropriate for the student's stage in learning. Instruction begins with simulations of simple circuits. The solution of problems based on these circuits requires only a simple model. As the students progress to more complicated circuits and problems, new models are introduced that contain additional rules. The students are described as moving from a novice to an expert state on the basis of their ability to solve more difficult problems.

Questions for Research in Context of Examples

Research plays a significant role in all the computer programs discussed above. However, the programs themselves suggest directions for further research. Some of the questions are broad in scope. Does learning that appears to result from a computer program extend beyond the medium itself? Can the student apply the concepts and skills learned to other instructional environments and to a real-world situation? Does success on computer tasks indicate that a student has fully understood the concepts and reasoning involved? Questions such as the foregoing apply, of course, to all forms of instruction, not only to materials that are computer based. Nevertheless, in reflecting on the use of computers in physics education, we should not dismiss questions of this kind simply because other types of teaching are equally at fault. After all, a major reason for incorporating computers in instruction is to improve upon what is already being done. The questions considered in the discussion below are not necessarily applicable to the particular examples that have been described. However, the programs provide a good context in which to raise some important issues. Often the better the program, the more interesting and sharply focused are the questions that can be extracted from it.

Experience has shown that students seem to enjoy working on judgment-free, goal-oriented tasks such as attempting to design a set of tracks to reproduce a graph shown on a computer screen. They also seem enthusiastic about trying to replicate a graph by their own motion. Could success on either type of task be analogous to becoming more proficient at a game? Is it possible for students to develop the necessary skill without being able to generalize the process to situations outside of the context of the computer program or MBL activity? Can students translate their experience with tracks displayed on the computer screen to the manipulation of real apparatus? Can students who have seen computer-generated graphs of their own motion, sketch a qualitatively correct graph of the motion of an object that they observe in the laboratory? Are students who have learned about the graphing of motion by manipulating balls and tracks on a computer screen, or by using motion detectors in the laboratory, also able to explain precisely how the shape, slope, and intercepts of a graph correspond to the various features of a particular real motion?

In addition to general questions about the efficacy of using analogies in introductory physics, there are questions that bear directly on computer-based instruction. How engaged is the student at every step? Rather than struggle with a particular bridging analogy and try to identify the connection between it and the anchor, students may find it considerably easier to call for another suggestion from the computer. Also, how can we be sure that the relevant common feature is the one on which the student is concentrating? When there are so many possible bridges available, is it possible that students will lose track of the critical element in the relationship between target and anchor? Can they learn to recognize the limitations of the bridging analogies or are they likely to develop new misconceptions by making correspondences that are not valid?

The ability to solve standard textbook problems is frequently taken as a measure of student understanding in physics. It is often assumed that successful problem solving involves all the important aspects of understanding. However, there has been a great deal of research that shows that the ability to solve standard textbook problems does not necessarily imply a functional understanding of the subject matter. How much does instruction in how to solve problems contribute to student understanding of concepts and representations? Does practice in solving standard textbook problems promote the development of scientific reasoning ability so that a student can reason successfully about new situations? What happens when problems are presented that cannot be solved by the patterns taught?

A computer can remember and quickly retrieve a series of long and extremely complicated rules. Will the student, when removed from the computer, be able to remember and apply a rule-based model that may not have a strong conceptual unity? Is it possible that the use of the computer to teach students how to solve problems may encourage routine application of formulas without enhancing understanding? When students attempt to follow prescribed procedures, are they thinking of the physics involved or is their attention mostly devoted to recalling and following directions? Is it realistic to expect that we can help novices become experts by teaching them a set of procedures? Perhaps expert performance cannot be adequately described in terms of the sum of its identifiable characteristics.

How does the passage of time affect retention of learning by computer-based instruction? For example, is it possible that the difference in performance between the students who used the video version of the geometrical optics lesson and those who did not, might disappear if testing were to be repeated at a later date instead of immediately after the lesson? Perhaps simultaneous presentation of a phenomenon and its scientific representation is much more effective for short-term learning than for long-term retention. Of course, the question of long-term learning should be asked with respect to any instructional strategy. But the same consideration applies to problem solving, whether taught by prescription, by a rule-based model, or by any other method. If problem solving really is, as many physicists believe, the most important skill developed during physics instruction, then it is important to determine how long the process learned, will be remembered.

Computer as Tool for Doing Research

The discussion in the previous section focuses on the use of research to provide guidance for instructional design. Here we look at the computer as a means for doing research, rather than as the beneficiary of research. The computer has been a research tool for many years in basic cognitive research, where it has been used to model human cognition. Recently the computer has also begun to be used as a tool to extend the scope of empirical investigations, to aspects of student understanding inaccessible through other means. The examples below illustrate how the computer program can drive research.

Construction of Models of Cognition

Cognitive scientists use computers to build models that can study human thought. For example, in studies on problem solving, the computer may be used to simulate differences between novice and expert behavior and to construct a dynamic performance model for the transition from one to the other. Insights obtained in this way, may be used to develop intelligent tutoring systems for facilitating this transition. This work is mentioned here only briefly since the main emphasis in this paper is on prototype research.

A specific illustration of this type of research is provided by the work of J. Larkin.[22] She has designed the program *Able* that can "learn" from solving successively more complicated problems in mechanics. Having solved a problem, the computer records the method used in memory. The next time it encounters a similar problem, it arrives at the solution more rapidly. Thus the program gradually develops into a more expert program: *More Able*. Larkin and several collaborators have also produced a more recent program *Fermi*, that incorporates more general problem-solving procedures.[23] These can be applied to problems in different topics in physics, such as fluid statics or electric circuits.

Investigation of Student Understanding

The Physics Education Group at the University of Washington has begun using materials designed for computer-based instruction as investigatory tools. In some preliminary studies, D. Grayson has utilized the computer to investigate student difficulties with some of the more subtle aspects of the graphical representation of real motions. A new computer program in the early stages of development is also currently being used to probe student understanding of wave motion.

Graphs of Simulated Moving Objects

During individual interviews, students who had studied kinematics in one of the laboratory-based courses taught by the group, were shown part 1 of an early version of *Graphs and Tracks*.[8] This version differs from the one discussed earlier in the paper in that it does not function as a tutorial. The program provides no assistance to students who have difficulty with the tasks.

Each student was shown three position-versus-time graphs and asked to arrange the tracks so that the motion of the ball generates the given graphs. Figure 1 shows the graph contained in the first example given to the students, together with the correct track configuration and initial conditions needed to produce the graph. Initially, the ball must have a negative velocity, which diminishes gradually until the ball reverses direction. The velocity then becomes positive and continues to increase.

Nearly half the students tried to make the ball move entirely in the positive direction. By inclining the first section of the track upward, as shown in Figure 2, they tried to make the ball slow down but not turn around. With the range of initial conditions possible, however, the ball cannot be given a high enough velocity to continue moving forward. The resulting curve generated by the ball is shown in Figure 2 at the lower left of the example graph. Some of the students who tried to make the ball move only in the forward direction, were able to give a correct verbal description of the motion depicted in the example graph. However, they did not seem to recognize that although the graph extends toward the right along the time axis, it reverses direction along the x-axis. Hence, the motion must reverse direction in space. A similar difficulty had emerged in earlier interviews in which real balls and tracks were used. It was the computer-based interviews, however, that helped elucidate how the students were thinking when they apparently ignored the reversal in direction depicted in the example graph.

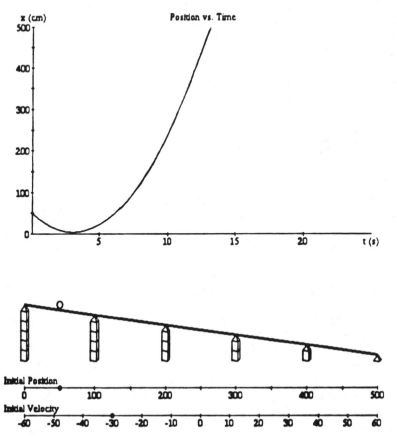

Figure 1. Graph presented to students during computer-based interviews and the correct track configuration for producing this graph.

The x-versus-t graph that was presented to the students in the second task, is a parabola that ends in a straight line. Curve A in Figure 3 represents this graph, and the accompanying track configuration is one that will produce that graph. Thus, curve A represents motion up and down a straight inclined track, ending on a level section. Many of the students were able to produce an incline that sloped upward to yield a parabola that was approximately the right shape but was either too broad or too narrow. The way in which the students attempted to alter the width of the parabola to match the original graph, indicated that they could not properly relate a change in the shape of the graph, to a change in the physical situation. For example, half of the students who wanted to make the curve less broad, tried to do so by increasing the initial velocity. However, as shown by curve B in Figure 3, increasing the velocity broadens the graph. The larger the initial velocity for a given incline, the further the ball will travel up the incline and the longer it will take to turn around.

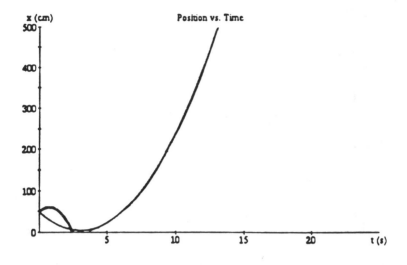

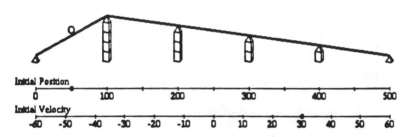

Figure 2. Graph presented to student during computer-based interviews. This incorrect track configuration was produced by students who tried to make the ball move only in the forward direction. Within the range of initial conditions possible, the ball always reverses direction and generates a small curve similar to the one at the lower left.

In the third task, the students were shown a symmetric curve. Almost half the students did not recognize that because the graph was symmetrical, the track arrangement that would produce it also had to be symmetrical. Thus we may speculate that symmetry arguments, which are used often by physicists, are not intuitive for students.

The semiquantitative types of difficulties noted during the interviews, could not have been observed without the computer. The instant feedback feature of the program enabled the interviewer to observe the strategies that the students used in trying to alter their incorrect hypotheses. In a laboratory situation, it would have been difficult, if not impossible, to make these observations. Moreover, use of the program as an investigatory probe served to point out where help sequences should be

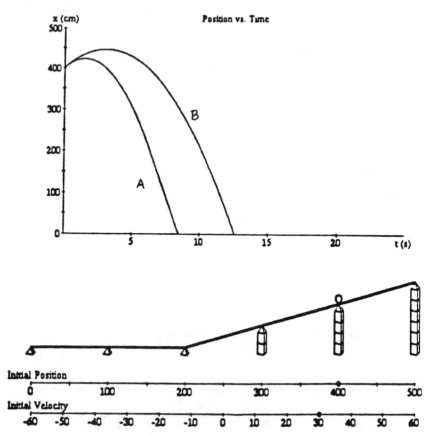

Figure 3.   Curve A corresponds to the correct track configuration needed to
reproduce the given graph. Curve B results from an initial velocity
that is too large.

inserted and what the content should be. These additions strengthened the program
by making it more responsive to student needs.

Graphs of Student-Generated Motions

In another set of interviews, 24 students who had experimented with balls and
tracks in the laboratory, were shown the four position-versus-time graphs in Figure
4.[8] They were asked first to describe how they would have to move their bodies to
reproduce the graphs and then to replicate them by moving in front of an MBL
motion detector. When they were satisfied that they had produced a good match to
the given graphs, they were to write a description of how they had actually moved.

In order to produce the first graph (Figure 4a), the student must stand still at a
position close to the detector, move away at a constant speed, and then stand still

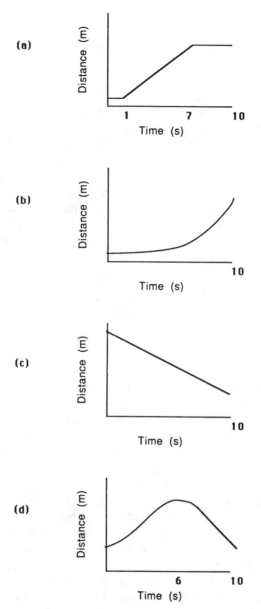

Figure 4. Using an MBL motion detector, students were asked to move their bodies to replicate the four graphs shown.

again. For the second graph (Figure 4b), the student should start fairly close to the detector, move away very slowly initially, and then continue moving away with increasing speed. The third graph (Figure 4c) represents motion toward the detector at constant speed. The fourth graph (Figure 4d) requires walking away from the detector at an increasing speed, slowing down, changing direction, speeding up a little, and then walking at constant speed.

Although eventually almost all the students were able to produce a correct set of four graphs, observation of their attempts revealed three general types of difficulties. One fourth of the students initially ignored their predictions and attempted to produce the desired graph by trial and error, almost as if they were playing a video game. Sometimes students appeared not to notice that the graphs they generated did not correspond in detail to the given graphs. The steepness of a line or shallowness of a curve were either not considered important or the changes needed to make the graphs match more closely could not be envisioned.

In their written comments, some of the students demonstrated a tendency to describe their motions in terms of their expectations rather than in terms of how they actually moved. For example, in trying to reproduce a particular graph, the students might state that they were speeding up as they had predicted would be necessary. However, they might actually be moving at constant speed. It is interesting to speculate whether perceiving one's own motion properly may require a different type of judgment than analyzing the motion of external objects. If this is the case, it might be especially important that MBL activities always involve two or more students so that they can help one another relate their internal kinesthetic sense to the motion that an observer would see.

Graphs of Transverse Pulses

Development of *Pulses on a String* was begun without an established research base. In designing the program, D. Grayson was guided by an instructional strategy already tested and in use in courses taught by the Physics Education Group. Drawing on her teaching experience with these materials, she produced a preliminary version of a program to help students learn to draw and interpret different graphical representations of a transverse pulse moving along a string. At A. Arons' suggestion, the program focuses on asymmetric pulses. If a pulse on a string is symmetric, a student need never distinguish between the graphical representation of the transverse displacement as a function of position or as a function of time. Since the transverse displacement depends on both variables, one needs to be held constant while the effect of varying the other is studied. Such experiments are very difficult to do in the laboratory.

The original incentive for designing the program was to use the computer to implement a written exercise that had proved to be instructive but tedious in the classroom. However, it was considered crucial that the computer not do the thinking for the student. Thus it was necessary to differentiate between activities that require thinking and that are essential for learning and those that are merely drudgery. In working through *Pulses on a String*, students must generate graphs point by point but they do not have to perform repetitive calculations. The computer plays the role of an accurate plotter.

The program asks the student to create a pulse on a string either by using the graphing palette or by choosing the example pulse, which is asymmetric. The graph of y (transverse displacement) versus x (position along the string) represents the shape of the string at a particular time. We may think of the graph as a snapshot of the string at that time. From the graph of y versus x, the student has the option of plotting three different graphs for a pulse moving to the right: y versus t, v (transverse velocity) versus t, and v versus x. Figure 5 shows a graph of the example

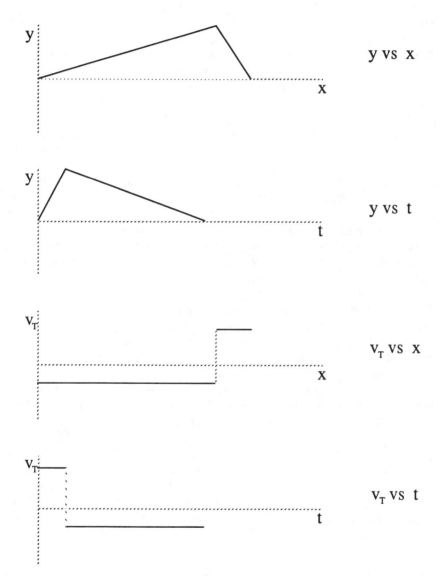

Figure 5. The example graph of transverse displacement, y, versus position along the string, x, is shown at the top. Below it are corresponding graphs of transverse displacement versus time for the point labeled A, transverse velocity versus position along the string, and transverse velocity versus time.

pulse together with the three derived graphs. As can be seen, there is a reversal in shape between each set of graphs: y versus x, and y versus t, and v versus x, and v versus t. Parts of the pulse at larger values of x pass a given point on the string at earlier times.

To plot the y-versus-t graph, the student focuses attention on one point on the string. When the student marks that point with the mouse, the y-coordinate is immediately plotted by the computer on the y-versus-t graph at the appropriate value of t. The student then moves the pulse forward and repeats the process until a complete pulse is generated on the y-versus-t graph. This is illustrated in Figure 6. The other two graphs are produced in a similar way except that it is the change in y, rather than y, that the student marks with the mouse. From the transverse displacement and the time between snapshots, the computer calculates the velocity and plots v at the appropriate value of x or t. Before plotting a graph, the student is asked to predict the shape and to draw a sketch that remains on the screen. Thus the student can make a direct comparison between the prediction and the actual graph that is generated point by point.

Pulses on a String was used with a class of 20 in-service middle and high school teachers after they had studied kinematics. The teachers were specifically asked to relate the concepts they had learned in kinematics to the concepts involved in the program. Although no formal research was conducted, observations made by the staff indicate that the program has promise. Many of the teachers claimed that this experience on the computer helped them reinforce and extend their understanding of the kinematical concepts. They also recognized and could explain the reversal in shape of the graph that results from switching the independent variable. If these preliminary results are borne out in subsequent research, an argument might be made that it is possible and perhaps even desirable in teaching introductory physics to introduce the study of wave motion immediately after kinematics. Hence, an important part of the background needed for the teaching of modern physics would be in place early in the course.

At the same time as it is being developed, *Pulses on a String* is being used as an investigatory probe to explore student understanding of wave motion. Its use with students thus serves the dual purpose of identifying student difficulties with the subject matter and providing formative evaluation for program development. The program may also be a useful tool for examining student understanding beyond the topic of waves. There are very few contexts in which introductory physics students encounter functions of two variables. Projectile motion is one such context; wave motion is another. Thus, use of the program with an appropriate questioning sequence may yield insights into difficulties with interpreting the effect of changing one variable when the other is kept constant.

Suggestions for Future Research

Recent advances in technology have greatly expanded the variety and quantity of instructional programs produced for the computer. Improvements such as enhanced

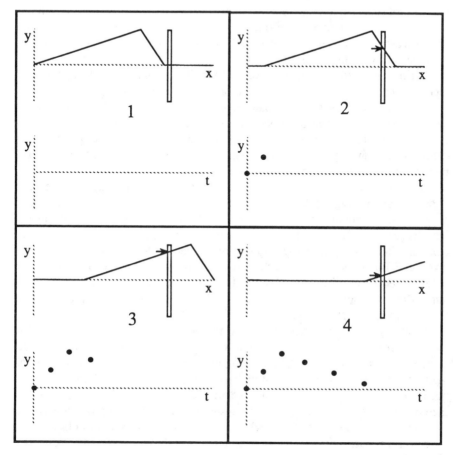

Figure 6. Illustration of the process by which the graph of y versus t is generated from the graph of y versus x.

graphics capability, increased speed of execution, simpler programming languages, more powerful work stations and microcomputers, new microcomputer-based laboratory tools, etc., have greatly increased the educational potential of the computer. These same advances have brought the computer into research in physics education and also expanded its scope. The technology available today has already had an effect on the type of research being done, and it has created new possibilities for the future.

Using a computer program to do research allows us to control our research so that we confront every student with exactly the same data. In a laboratory-centered interview, on the other hand, the possibility exists that data might not have been obtained under exactly the same set of circumstances for each student. Unlike a laboratory situation, in a computer-based presentation of tasks we can eliminate mechanical failures or influence by the investigator. We can thus direct our attention toward substantive issues and the more subtle aspects of a student's response. Instead of concentrating on a uniform presentation of the task, we can focus on the line of reasoning used by the student and explore the role of higher-order thinking skills that may be involved. Furthermore, once a computer task is perfected, we can extend testing to large numbers of students. Broadening the student population increases the level of confidence in the validity of any generalizations.

There are many issues that should be addressed by research in addition to the examples of research that have been described in the paper and to the questions raised within the context of the discussion. Some are general and pertain to an entire instructional approach or to a whole group of programs. Others are more narrowly focused and refer to specific subject matter.

Some general subjects for inquiry are related to the building of models of human problem-solving behavior: Is there sufficient similarity between a computer program and a human mind that computer modeling is likely to yield useful insights for instruction? Will success in building an expert problem solver on the computer help us significantly improve our understanding of how people solve problems? Having identified the skills that experts possess, the computer may be programmed to exercise those skills and to solve a certain class of physics problems. Does it follow that teaching these skills to students will lead to the same level of proficiency?

Another group of questions relates to the building of computer microworlds, models of real or artificial situations with which students can gain interactive experience. Does practice in a special computer environment (e.g., a Newtonian microworld) really improve student understanding, or does it only improve performance on a computer activity? Is the student engaged at a deep intellectual level or merely acquiring a skill that does not transfer to other situations? Will the experience have a lasting effect? Can what has been learned by manipulation in a microworld be articulated by the student in terms of the concepts and principles of physics?

Examples of questions of a more specific nature arise in the context of modern or contemporary physics. Many concepts require a level of abstraction that cannot be readily represented in traditional instruction. The computer can help students construct highly formal concepts by providing visual models that can serve as a foundation. The computer can also give students experience in applying these concepts to representations of inaccessible phenomena, both those that are real and those that are idealized. In turn, computer programs that have been developed for such instruction can be used to investigate student understanding of the concepts. There is a special need for such research at this time. If we plan to extend the curriculum of introductory physics to include special relativity, statistical physics, or quantum mechanics, we should consider carefully what students are likely to find

most difficult in studying those topics and design instructional materials accordingly.

Conclusion

It is an enormous intellectual challenge to build into a computer program the
knowledge of the subject matter, of the student, and of the strategies that are essential for effective teaching. There is a need for research to identify specific student
difficulties and for research to develop and test instructional strategies that address
them. When possible, implementation of an instructional procedure on the computer should take place after preliminary trials in the classroom or laboratory. It is critically important that testing with students continue throughout development of a
computer program. The long time required to produce computer-based materials
should be invested in only the most promising approaches.

Research and development are more likely to yield useful results when carried
out concurrently and continuously in an instructional setting. Of course, the interaction of these components is important not only for the design of good computer
programs but for producing effective printed materials as well. Moreover, when
both types of instructional materials are developed in the same environment, there
is the additional advantage that feedback from one can benefit both. A flow chart
that shows the iterative process appears in Figure 7.

The computer offers the distinct advantage of making interactive one-on-one
instruction feasible for large numbers of students. It makes possible certain types
of instruction not available in any other way. The opportunities for visualization
can be a great aid in teaching material that requires a high level of abstraction. For
some students, the computer provides an environment that is more comfortable and
nonthreatening than the traditional classroom or lecture hall. These advantages
provide reasons for optimism, but they do not guarantee improvement in the quality of instruction. There is a need for research to examine precisely what is occur-
ring intellectually while the student works at the computer. In the process of
looking critically at computer-based instruction, we may learn a great deal about
what we should and should not do in the teaching of physics, regardless of the
mode of instruction.

The author would like to thank Diane Grayson and Emily H. van Zee for their contributions
to this paper and also Arnold Arons, Peter Shaffer, Mark Somers, and David Trowbridge for
helpful conversations. Special appreciation is due Joan Valles for her invaluable assistance.
Support by the National Science Foundation for the work of the Physics Education Group at
the University of Washington is gratefully acknowledged.

1. A. M. Lesgold and F. Reif, *Computers in Education: Realizing the Potential. Report of
   a Research Conference.* (Washington, DC: U.S. Government Printing Office, 1983).
2. See, for example, M. Reiner and M. Finegold, "Changing Students' Explanatory
   Frameworks Concerning the Nature of Light Using Real-Time Computer Analysis of
   Laboratory Experiments and Computerized Explanatory Simulations of e.m.
   Radiation," in *Proceedings of Second International Seminar: Misconceptions and*

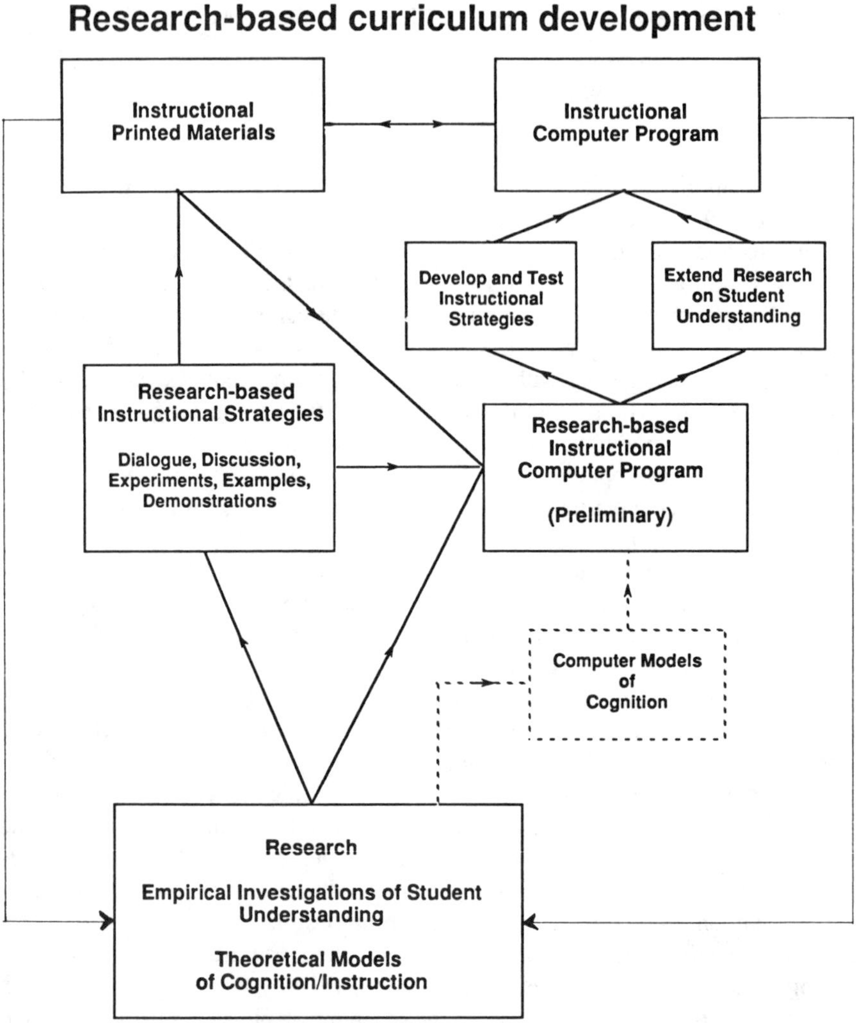

Research-based curriculum development

Figure 7. A flow chart that summarizes the iterative process in which
research and curriculum development, both traditional and computer-based, can interact within an instructional environment.

Educational Strategies in Science and Mathematics II, edited by J. Novak (Ithaca, NY:
Cornell University, 1987), pp. 368-77; M. Wiser, "The Differentiation of Heat and
Temperature: an Evaluation of the Effect of Microcomputer Teaching on Students'
Misconceptions," Educational Technology Center Technical Report (Cambridge, MA:
Harvard Graduate School of Education, 1986); J. Snir, "Making Waves: Software for
Studying Wave Phenomena," in these proceedings.
3. P. Hewson, "Diagnosis and Remediation of an Alternative Conception of Velocity
Using a Microcomputer Program," Am. J. Phys. **53**, 684 (1985).

4. D. E. Trowbridge and L. C. McDermott, "Investigation of Student Understanding of the Concept of Velocity in One Dimension," Am. J. Phys. **48**, 1020 (1980).
5. D. E. Trowbridge, "Applying Research Results to the Development of Computer-Assisted Instruction," in these proceedings.
6. See, for example, L. C. McDermott, M. L. Rosenquist, and E. H. van Zee, "Student Difficulties in Connecting Graphs and Physics: Examples from Kinematics," Am. J. Phys. **55**, 503 (1987); E. H. van Zee and L. C. McDermott, "Investigation of Student Difficulties with Graphical Representations in Physics," in *Proceedings of Second International Seminar: Misconceptions and Educational Strategies in Science and Mathematics III*, edited by J. Novak (Ithaca, NY: Cornell University, 1987).
7. M. L. Rosenquist and L. C. McDermott, "A Conceptual Approach to Teaching Kinematics," Am. J. Phys. **55**, 407 (1987); M. L. Rosenquist and L. C. McDermott, *Kinematics* (Seattle: University of Washington, 1982).
8. D. J. Grayson, E. H. van Zee, and L. C. McDermott, "Investigation of the Use of Computers to Address Student Difficulties with Graphs," AAPT Announcer **17**, 71 (1987); D. J. Grayson, E. H. van Zee, T. D. Gaily, and L. C. McDermott, "Investigating How Students Produce Motion Represented on a Graph," AAPT Announcer **16**, 92 (1986).
9. Further information on probes may be obtained from R. F. Tinker, Technical Education Research Center (TERC), 1696 Massachusetts Ave., Cambridge, MA 02138.
10. R. K. Thornton, "Learning Physics Concepts with Microcomputer-Based Laboratories," in these proceedings; R. Thornton, "Access to College Science: Microcomputer-Based Laboratories for the Naive Science Learner," Collegiate Microcomputer **5**, 100 (1987); Thornton, "Tools for Scientific Thinking—Microcomputer-Based Laboratories for Physics Teaching," Phys. Ed. **22**, 230 (1987).
11. H. Brasell, "The Effect of Real-Time Laboratory Graphing on Learning Graphic Representations of Distance and Velocity," J. Res. Sci. Teach. **24**, 385 (1987); J. R. Mokros and R. F. Tinker, "The Impact of Microcomputer-Based Labs on Children's Ability to Interpret Graphs," J. Res. Sci. Teach. **24**, 369 (1987).
12. S. Bendall and F. M. Goldberg, "Using a Videodisc to Facilitate Student Learning in Geometrical Optics" (Paper given at American Educational Research Association meeting, New Orleans, LA, 1988.).
13. F. M. Goldberg and L. C. McDermott, "An Investigation of Student Understanding of the Real Image Formed by a Converging Lens or Concave Mirror," Am. J. Phys. **55**, 108 (1986).
14. T. Murray, K. Schultz, D. Brown, and J. Clement, "Remediating Physics Misconceptions Using an Analogy-Based Computer Tutor," Artificial Intelligence and Education Journal (in press).
15. J. Clement, "Overcoming Students' Misconceptions in Physics: The Role of Anchoring Intuitions and Analogical Validity," *Proceedings of Second International Seminar: Misconceptions and Educational Strategies in Science and Mathematics III,* edited by J. Novak (Ithaca, NY: Cornell University, 1987), pp. 84–97.
16. J. Minstrell, "Explaining the "At Rest" Condition of an Object" Phys. Teach. **20**, 10 (1982).
17. R. Dufresne, W. Gerace, P. Hardiman, and J. Mestre, "Hierarchically Structured Problem Solving in Elementary Mechanics: Guiding Novices' Problem Analysis," in *Proceedings of Second International Seminar: Misconceptions and Educational Strategies in Science and Mathematics III*, edited by J. Novak (Ithaca, NY: Cornell University, 1987), pp. 16–130.
18. See, for example, M. T. H. Chi, P. J. Feltovich, and R. Glaser, "Categorization and Representation of Physics Problems by Experts and Novices," Cognitive Science **5**, 121 (1981).

19. G. Oberem, "*Albert*: A Physics Problem-Solving Monitor and Coach," in Proceedings of Conference on Learning in Future Education: Computers in Education. (Calgary, Canada: University of Calgary 1987); Oberem, "An Intelligent Computer-Based Tutor for Elementary Mechanics Problems" (Ph.D. dissertation, Rhodes University, Grahamstown, South Africa, 1986).
20. F. Reif and J. I. Heller, "Knowledge Structure and Problem Solving in Physics," Educational Psychologist **17**, 102 (1982).
21. B. White and J. Frederiksen, *Causal Model Progressions as a Foundation for Intelligent Learning Environments* Report No. 6686. (Cambridge, MA: BBN Laboratories Inc., 1987).
22. J. H. Larkin, "Cognition of Learning Physics," Am. J. Phys. **49**, 534 (1981).
23. J. H. Larkin, F. Reif, J. Carbonell, and A. Gugliotta. "FERMI: A Flexible Expert Reasoner with Multi-Domain Interfacing," Cognitive Science, in press.

Applying Research Results to the Development of Computer-Assisted Instruction

David E. Trowbridge
Department of Physics, FM-15, University of Washington, Seattle, WA 98195

Curriculum development has always been part art, part science, part trial and error, and part common sense. Development of computer-assisted instruction will, no doubt, require the same mix. The best examples of physics materials embody both inspiration and methodology. Typically, they have been nurtured in an instructional setting over a period of several years, have had extensive testing at numerous sites, and have undergone many cycles of improvement.

A relatively new approach in curriculum development has been to incorporate results from research in physics education. Using this approach, development of educational software can be guided by research just as traditional instruction can. The computer has made curriculum development more difficult, but has given us a new and potentially very powerful asset for physics education. The complexity of software development has made a grounding in research all the more important.

This paper illustrates one example of our effort to apply research results to the development of computer-assisted instruction. We refer the reader to some research into student difficulties with graphs of motion and then describe some instructional software that addresses those difficulties. The software package, *Graphs and Tracks*, teaches elementary concepts of graphing position, velocity, and acceleration versus time for simple one-dimensional motion. Use of this program in a laboratory-centered course has guided ongoing development and has suggested ways of integrating this and other computer materials into existing courses.

Focus

The goal of curriculum development is to improve the teaching of physics. Computers may be able to help, provided we have a clear idea of what we are trying to accomplish. To understand the instructional process better, one must examine both on what the teacher is doing and on what the student is doing. Likewise, when choosing an area in which to use computers in instruction, one may attempt to automate some of the functions of a teacher, or to create learning activities that enable the student, independently of the teacher, to engage in activities conducive to learning.

While others are focusing attention on what highly skilled teachers do in the classroom, we have chosen to focus on what students do when they learn physics. Current research on student difficulties suggests that it is useful to watch students carefully as they engage in learning activities or confront tasks during individual demonstration interviews. In doing so, one can sometimes infer what students are thinking. We can begin to answer questions such as: What beliefs do students have about the physical world? How do they apply their current conceptual framework to physical situations? In what situations do their reasoning skills work, and where do their skills break down? Where are the gaps in their understanding of physical concepts?

For the development project described in this paper, we have drawn from observations of what students do with activities in a laboratory-centered physics course, from reports of their responses to oral questions in individual demonstration interviews, and from studies of their responses to conceptually oriented written questions. From this, we have tried to identify critical learning experiences that are necessary to achieve deeper levels of understanding of a particular subject. We have developed a pair of highly interactive graphical programs, *Graphs and Tracks I* and *II*, that facilitate those experiences.

Research Basis

Researchers in physics and mathematics education have documented several common difficulties students have with drawing graphs of motion.[1] Certain kinds of misconceptions are widespread and highly predictable. A few of these difficulties are reported in a recent paper by McDermott, Rosenquist, and van Zee.[2] In one series of investigations, students observed a ball that rolled along an arrangement of level and inclined sections of track. They were then asked to sketch graphs of the motion. The apparatus is shown in Figure 1, along with a correct graph of position versus time.

Students displayed a variety of difficulties, including: not representing continuous motion by a continuous line; not distinguishing the shape of a graph from the path of the motion; and not representing the continuity of velocity (i.e., kinks in a position versus time plot). Figure 2 illustrates some common incorrect graphs.

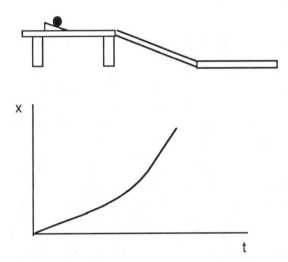

Figure 1. Apparatus used to probe student difficulties, with sketch of correct graph (not shown to students).

Some students did not fully understand what a point on an x-versus-t graph represents, or what the horizontal axis represents. Given the opportunity to take measurements using meter sticks and clocks, some students plotted distances traveled along each of the three track sections (rather than positions) versus clock readings, as in Figure 2a, without regard for the continuous nature of the motion. Others drew graphs that resembled the shape of the tracks, as in Figure 2b. Some believed that the first and third straight-line segments in their x versus t graph should be parallel, because the first and third track sections were parallel. Others thought that the middle segment of their graph should be straight, because the middle track section was straight. Their reasons were evident in questioning during interviews. One of the most common errors with x-versus-t graphs was to neglect trying to match slopes of adjoining segments, as in Figure 2c. A number of other difficulties concerned velocity and acceleration graphs and translation among different graph types.

Closely allied research on students' understanding of graphs of functions conducted among first-quarter calculus students shows that only about half can correctly interpret a graph of a simple monotonic function of time.[3] Even among honors physics students, after regular instruction in kinematics, only 30 percent can sketch a reasonably correct position-versus-time graph for a simple motion demonstrated on an inclined air track.[4]

An Example of Research-Based Software

Graphs and Tracks consists of two programs designed to address the difficulties described above: "From Graphs to Motion" and "From Motion to Graphs."

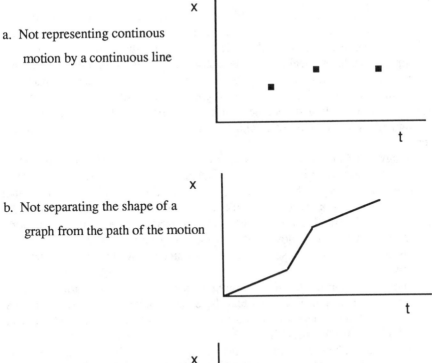

a. Not representing continous
   motion by a continuous line

b. Not separating the shape of a
   graph from the path of the motion

c. Not representing continuity
   of velocity

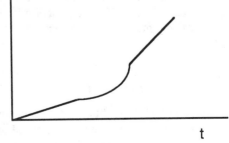

Figure 2.   Examples of incorrect student graphs.

Part 1: "From Graphs to Motion"

In *Graphs and Tracks 1* students interpret motion graphs by generating motions that correspond to a set of given graphs. The program displays graphs of position, velocity and acceleration for a number of examples of rectilinear motion (see Figure 3). Students are presented with an adjustable set of sloping tracks and with scales of initial position and initial velocity. The students' task is to set up an arrangement of tracks and suitable initial conditions so that when a ball rolls on those tracks, the graphs it generates match the given graphs. Students use a mouse to adjust the incline of the ramps and to set the initial position and initial velocity of the ball. Upon releasing the ball, students see a graph generated on the screen as the ball is moving. Students can see immediately whether their graphs differ from the given graph.

Design of this program was based on a laboratory exercise in which students were shown graphs on paper and asked to construct an apparatus consisting of steel balls rolling along aluminum U-channels that would illustrate the motion shown in the graphs. While this technique was instructive, it was very time consuming. Students could perform only two or three experiments during a class period, and the time available for thinking about the correspondence between real motions and graphs was limited. In addition, the number of experiments that students could do was limited by the number of teaching staff available to evaluate each student's work.

Graphs and Tracks I contains several examples of motion, from simple to more complex. In addition, it can to save any example of motion the student is able to generate using the adjustable tracks. Thus the student may create any example that he or she finds instructive or would like to use to challenge another student. We discovered during field testing that this mode of usage was especially motivating with those students who enjoyed the competition. Alternatively, an instructor can use this program to create a set of exercises for students for homework quizzes or classroom activities.

A help facility provides specific suggestions for correcting errors. It works equally well for any problem that can be entered using the adjustable tracks. Students may request help a little at a time, or continuously. Some students, who reached an impasse felt strongly that they wanted only one small hint, not a series of help messages, before returning to their own solution. Others liked the continuous feedback feature. The help facility is completely general, so that the program generates help messages that are specific to any attempted solution of any problem of this class.

Part 2: "From Motion to Graphs"

Graphs and Tracks 2 poses the complementary task: given a motion and a verbal description of that motion (generated automatically by the computer), sketch the corresponding graphs of position, velocity, and acceleration versus time (see Figure 3). The graphs need not be precise to be judged correct, but they must represent the most salient features of the motion correctly. For instance, an object traveling in the positive x-direction and speeding up should be represented by an upward curving x-versus-t graph. The program looks for common kinds of errors and gives appropriate feedback. The facility for giving feedback is completely general; specific help that refers to particular segments of the student's graphs can be automatically generated for any problem that can be created using *Graphs and Tracks 1*.

Students may ask for feedback on their graphs at any time. The program is able to evaluate both the overall qualitative correctness of the graphs and detailed attributes such as differentiability of x versus t continuity and correspondence of segments among different graph types. The evaluation facility of "From Motion to

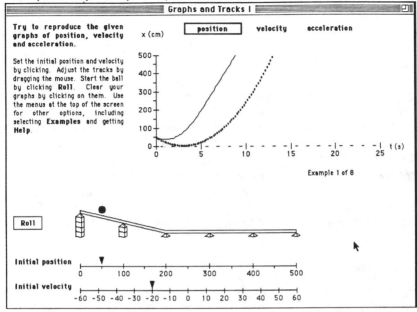

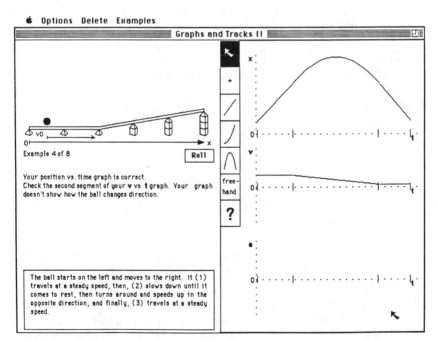

Figure 3.   Graphs and Tracks 1 and 2.

Graphs" was built using the specific research results on student difficulties with graphs.[3] For example, the program anticipates and gives feedback on student graphs that do not represent continuous motion by a continuous line, or do not distinguish the shape of a graph from the path of the motion, or do not represent the continuity of velocity. If the student were to sketch the graph shown in Figure 2b for the motion demonstrated in Figure 1, the program would respond with the following: "Check the second segment of your x-versus-t graph. Your curve shows the ball maintaining a constant speed."

The program uses a strategy of feedback in which, it finds errors, directs the students' attention to a particular part of their graph, and then tells what their graph would indicate, in contrast with what actually occurred in the motion. Just as teachers often do when working individually with students, the program usually avoids telling students directly that they are wrong, but simply focuses their attention on the dubious parts of their solution. We prefer to tell or show students the implications of their own actions and let them draw their own conclusions.

Design Characteristics

Graphs and Tracks has a hands-on feel to it; the student uses the mouse to adjust a simulated apparatus and to sketch curves. We designed the program to be used by students with limited computer experience. No particular knowledge of commands or special keys is assumed. Students can learn to use the mouse for pointing and making selections from menu items simply by using the program. Interactive instructions on how to use the program are contained in the program itself, so no user manual is required. A help facility enables the first-time user (or anyone wanting a review) to exercise each function of the program. All available options are described in words and indicated on the screen as icons or menu items. No supervision of students is required, but some desirable strategies for usage (e.g., having students work in pairs) can be implemented by an instructor.

Integration into the Curriculum

The skills of graph interpretation that this software addresses, are skills that have been taught in introductory physics courses in other ways for a long time. What is new is the level of engagement and the amount of practice that is possible with interactive computer programs. Laboratory experiences using actual equipment are essential for developing a solid understanding of physical concepts. Unfortunately, in most instructional labs, students have a limited opportunity to think of experiments and try them out. A convenient, easy-to-use simulation program, frees students from some of the tedium of data collection and graphing and enables them to generate and test more hypotheses. In the area of graphs of motion, the computer can plot graphs as quickly as the motion itself evolves.

Graphs and Tracks is used at the University of Washington as an adjunct to the kinematics curriculum developed by the Physics Education Group.[5] It is used concurrently with activities involving the TERC motion detector.[6] It seems to work particularly well in conjuction with the microcomputer-based laboratory exercises.

The *Graphs and Tracks* software package fills a gap between the student's everyday experience with motion, and the mathematical treatment of kinematics presented in typical physics courses. It provides a convenient, easy-to-use environment for learning the qualitative relationships between motion and graphs.

Summary

Many of the concepts of introductory physics are far from self-evident to students. In general, these concepts are not learned merely by listening to explanations, no matter how clearly articulated these instructions may be. Students seem to need a variety of experiences, some repeated over an extended period of time, to acquire the deep understanding characteristic of the trained physicist. The program described herein offers one way for students to deepen their understanding by having a direct experience making connections between motion and its graphical representation.

We have found that a multifaceted approach to curriculum development is crucial to designing instructional software. An ongoing research program for delving into student difficulties, an instructional setting in which to inspire and guide the creation of learning materials, and a powerful programming environment[7] can be brought together to create good materials for introductory physics courses. In addition, we have found this and other programs, designed originally for instructional use, to be useful as research tools for further investigation of student conceptual development.[8]

Many people have contributed to the development of *Graphs and Tracks*. Original ideas for these programs sprang from classroom activities developed by the Physics Education Group at the University of Washington. Some preliminary design was conducted at the Educational Technology Center at the University of California, Irvine. Wilfred Hansen of the Information Technology Center at Carnegie Mellon University implemented the first mouse-oriented version of part 1 in the C language. The current programs were developed using the cT programming environment (formerly CMU Tutor) from Carnegie Mellon University. They currently run on the Macintosh family of microcomputers (Mac Plus, SE, II) and UNIX workstations equipped with CMU's Andrew system. They will soon be available on the IBM-PC family as well. Bruce Sherwood of the Center for Design of Educational Computing at Carnegie Mellon has given me generous assistance and support in developing this and other programs in the cT language. This work was supported in part by NSF grants DPE 84-70081 and MDR 84-70166.

1. E. H. van Zee and L. C. McDermott, "Investigation of Student Difficulties with Graphical Representations in Physics," *Proceedings of the Second International Seminar on Misconceptions and Educational Strategies in Science and Mathematics,* (Ithaca, NY: Cornell University, 1987); J. Clement, "Misconceptions in Graphing," *Proceedings of the Ninth Conference of the Netherlands* (1985); D. J. Grayson, E. H. van Zee, T. D. Gaily, and L. C. McDermott, "Investigating How Students Produce Motion Represented on a Graph," AAPT Announcer **16**, 4 (December 1986).

2. L. C. McDermott, M. Rosenquist, and E. H. van Zee, "Student Difficulties in Connecting Graphs and Physics: Examples from Kinematics," Am. J. Phys. **55**, 6 (1987).
3. G. S. Monk, "Students' Understanding of Functions in Calculus Courses" (unpublished paper, Mathematics Department, University of Washington, May 1987).
4. P. C. Peters, "Even Honors Students Have Conceptual Difficulties with Physics," Am. J. Phys. **50**, 6 (1982).
5. M. Rosenquist and L. C. McDermott, "Module 2: Kinematics," (3rd trial ed.) (Seattle, WA: ASUW Publishing, University of Washington, 1982).
6. R. K. Thornton, "Tools for Scientific Thinking: Microcomputer-Based Laboratories for the Naive Science Learner," *Proceedings of the Seventh National Educational Computing Conference,* June 1986.
7. B. A. Sherwood and J. N. Sherwood, "CMU Tutor: An Integrated Authoring Environment for Advanced-Function Workstations," *Proceedings IBM Academic Information Systems AEP Conference*, San Diego, CA, April 1986, 4:29–37.
10. D. J. Grayson, D. E. Trowbridge and L. C. McDermott, "Use of the Computer to Identify Student Conceptual Difficulties," AAPT Announcer **17**, 4 (December 1987).

Computer Tutors: Implications of Basic Research on Learning and Teaching

Ruth W. Chabay

Center for Design of Educational Computing, Carnegie Mellon University, Pittsburgh, PA 15213-3890

The vision of computers as sophisticated tutors is not new. More than 20 years ago, in a *Scientific American* article entitled "The Uses of Computers in Education," Patrick Suppes said: "One can predict that in a few more years millions of school children will have access to what Phillip of Macedon's son Alexander enjoyed as a royal prerogative: the personal services of a tutor as well-informed and responsive as Aristotle."[1]

More than a few years have passed since that article was written, but despite impressive advances in computer technology we still cannot claim to have populated the schoolrooms of the nation with a million silicon Aristotles. Why not? In part, the lack of spectacular success is due not simply to limitations in the capabilities of computers, but to limitations in our understanding of how successful students learn and how good teachers teach.

Because of the difficulty of providing sophisticated, effective direct instruction with computers, many developers of educational software have chosen not to attempt this at all. Instead, they have developed other kinds of programs, such as simulations and laboratory aids, which can be used with guidance from a teacher. Many people have come to believe that it is not possible to create a truly effective computer tutor that can deliver direct instruction.

However, during the two decades since Suppes' paper appeared, there have evolved a number of exciting strands of basic research on the processes of learning and teaching. This paper will explore the impact of a few of these developments on the vision and the reality of direct instruction by sophisticated computer tutors.

Research on Processes

This paper discusses detailed research on the processes of human learning and problem solving in real educational domains, including but not restricted to physics. Some researchers have studied novices and experts solving mechanics problems. Other work has focused on geometry, reading comprehension, elementary mathematics, and computer programming. Instead of focusing primarily on outcomes of different sorts of instruction, this research analyzes, at a very fine grain size, the processes of solving problems, learning, and teaching. Though the underlying aim of this research is to construct a fundamental theory of human thinking, it is not just an abstract exercise. Experiments have already demonstrated large, significant gains in student performance due to instruction based on such detailed analyses.[2]

Components of a Computer Tutor

In considering the potential impact of such research on the design of instructional computer programs, we need to ask what goals would shape the design of a good computer tutor. Three major components are shown in Figure 1.

COMPONENTS OF A COMPUTER TUTOR

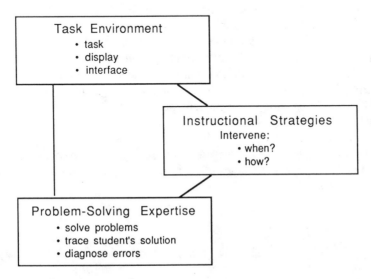

Figure 1.   Components of a computer tutor.

The task environment itself—the sequence of problems, the displays, the interface, the way in which the student interacts with the computer—establishes the framework for a tutorial session. The task environment is a strong determinant of instructional outcomes. In a poorly designed task environment, students may end up spending most of their effort on things peripheral to the skills and concepts we really want them to learn. For example, a multiple-choice activity is not appropriate if the goal is to teach students to construct and not simply to recognize solutions. An awkward interface may force students to spend more effort on communicating with the tutor than on solving problems. On the other hand, a well-designed activity can minimize demands on a student for less important processing, such as evaluating arithmetic expressions, and free the student to concentrate on principles and strategies for solving problems.

To support and guide students solving nontrivial problems, a tutor must have enough problem-solving expertise to be able to generate solutions itself. It must be able to recognize and track multiple correct-solution paths, and must be able to identify simple errors and poor strategies.

What should a tutor do when it detects an error? Some sort of principles for delivering guidance and feedback to a student are an important component of an instructional computer activity. Instructional principles for a computer tutor must be specified explicitly, and will not necessarily be identical with instructional strategies appropriate to tutoring by humans.

Problem-Solving Expertise

Research on scientific problem solving has focused on describing the behavior of experts and novices as they solve problems. This research is qualitatively different from traditional educational studies that try to measure the outcome of a particular kind of instructional approach because it focuses on the details of the actual thinking and learning process itself. Such studies are not easy, since frequently even experts who can solve complex problems quickly and easily cannot explain in detail all the decisions and choices, blind alleys and backtracking, insights and shortcuts, that went into the solution process. Researchers in cognitive science have therefore developed experimental techniques designed to help reveal the details of experts' behavior. One common approach is to ask problem solvers to "think aloud" as they work on problems—to voice each thought as it comes to them, without worrying about where it may lead or whether it is correct or not. These "think aloud" sessions are recorded, and the resulting "protocols" analyzed in careful detail to reveal strategies, thought patterns, plans, and changes in the level of detail at which problem solvers work.[3]

Problem Spaces

The model of problem solving that has emerged from such studies depicts the solution process as a search through a problem space, beginning at an initial state and progressing through successive states of partial solution until finally the goal state—an acceptable solution—is reached (Figure 2). The terminology of problem

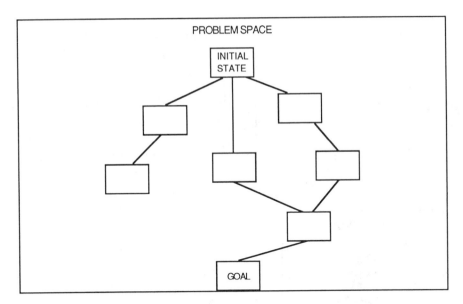

Figure 2.   Problem space.

solving sounds very quantum mechanical: transitions between different problem states are effected by applying appropriate "operators" (rules or solution steps). Global and interim goals—ideas about what the solution or the next step should look like—are explicitly associated with the search for appropriate operators or solution steps. Unproductive paths and dead ends are usually a part of the problem space, and not all operations will lead to a correct solution. There may be more than one path that does lead to a solution, because in many problems more than one approach can succeed.

As an example, consider the solution of a simple mechanics problem depicted in Figure 3. Assume, for instance, that we wish to find the velocity of a block sliding down a frictionless incline at some time, given that the block started at rest. Figure 3 depicts two states out of many in the problem space: the initial state, a picture of the objects involved; and a subsequent state, where the application of the operator "draw a force diagram" has produced an augmented picture of the problem with force vectors.

For an expert, drawing a force diagram is a simple and straightforward operation. As we all know, however, it can be a complex and challenging problem in its own right for a student. Thus, to understand in detail what a student must go through to solve the problem, we must describe the problem space in much greater detail. In the problem-space diagram in Figure 4, the operator "draw a force diagram" has become a subgoal in the solution process. The first operation necessary to achieve this goal is to pick a subsystem, so this is now shown as the first operation in the search for a solution. Of course, poor choices are possible; picking the plane as a subsystem is a legal choice, but not a useful one if our goal is to compute the speed of the block; it will eventually lead to a dead end, and the solver will have to backtrack and choose a different path.

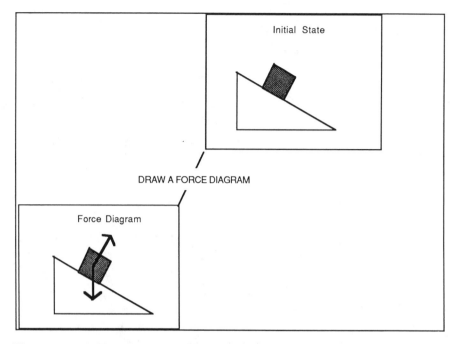

Figure 3.   Initial state and subsequent state in problem space.

We could, and perhaps should, add even more detail to the analysis; for example, we could include the operations necessary to select the center of mass of the block as the point at which to draw our vectors, and so on.

Why must the model be so detailed? One compelling reason is that by describing the model in full detail, we make it possible to test it. Just as mathematical expressions and relations provide a natural notation for models in physics, computer programs provide a natural way to formulate models of problem solving. A computer program that has the knowledge necessary to solve problems, should itself be able to solve a variety of such problems, more or less in the way a human expert would. In attempting to design a program that can really solve problems, one finds that it is necessary to descend to a finer and finer grain size in order for the program to be able to execute all the necessary substeps of each operation. Similar results are obtained for students. Heller and Reif show that students' performance on physics problems improves dramatically if they follow a very detailed analysis procedure; leaving some of the detail out of the procedure results in a substantially inferior performance.[4]

Intelligent Tutoring Systems

A second consequence of such a fine-grained analysis of the process of solving problems is the opportunity it provides to track and model the thinking process of a

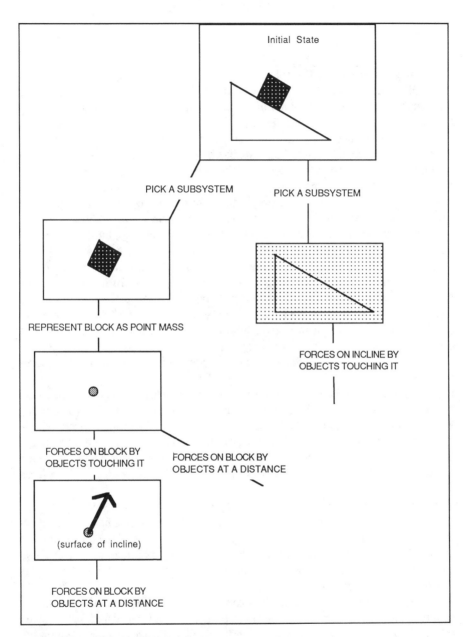

Figure 4.   More detailed view of problem space.

student as he or she works to solve a problem. A new generation of instructional computer programs, called *Intelligent Tutoring Systems* (ITS), do exactly this: they track the student's progress through the problem space and identify points of difficulty—incorrect rules, errors in applying rules, and so on.[5]

Task Environment

Displays

Displays often play a key role in solving problems, especially physics problems. Often the first step taken by an expert solving a physics problem is to draw a diagram.[6] The diagram is an important focus in the solution process, and much of the work involves the diagram directly. This is evident to physics teachers, who continually urge students to draw diagrams as the first step in solving a problem. Recent work by cognitive psychologists has begun to elucidate why diagrams are so important and how they are useful in solving problems.[7]

Studies of human memory suggest that there are two kinds of memory: long-term memory and short-term memory. Things remembered over a long period of time are stored more or less permanently in long-term memory. Short-term, or working, memory, in contrast, does not endure for a long time, but provides very quick access to information.

Problem solvers must keep information about the problem and the solution process active in working memory for immediate access. The difficulty is that the capacity of working memory is quite limited, and only a few pieces of information can be stored there at once. We have all seen beginning students who can't seem to keep the pieces of a problem in their heads, or who lose track of what they have done so far; their short-term memory capacity may be overloaded.

Experts often get around this limitation by combining several associated pieces of information into one "chunk" in memory. Another important strategy, however, is to store much of this important information in a display or diagram, rather than trying to keep it all in memory. As a display such as the one in Figure 4 is built up by the problem solver, more and more information about the problem itself and the progress of the solution is encoded in the display, and therefore need not be held in working memory all the time. For example, the fact that the plane exerts a force on the block in a direction normal to the surface is encoded in the display by a force vector.

Diagrams also provide a way of keeping track of goals in the solution process; they indicate what the next step should be. In Figure 5, the solver has indicated that the next step is to write equations for the x and y components of the forces, but has not yet done so. Even if interrupted in the middle of the problem, the solver could easily reconstruct what the next step should be simply by looking at the display and seeing that the equations are not complete.

In addition, well-organized displays can decrease the time needed to search for different pieces of information that must be used together, because such related information can be grouped physically in the display. Besides decreasing memory load, the use of pictures or diagrams makes it possible to substitute relatively easy perceptual inferences (about positions or directions, for example) for more difficult logical inferences.

CAI Expertise

Displays in dynamic, interactive computer programs must satisfy criteria quite different from those for static or passive media. There is not yet a solid theoretical

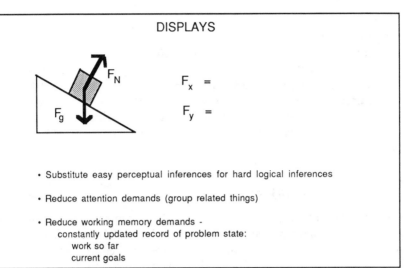

Figure 5. Role of displays in the solution process.

basis or set of principles to guide designers. Nonetheless, there are experienced developers of instructional computer activities who have, through experience with thousands of hours of student use of their programs, learned how to develop robust, easy-to-use, effective instructional programs. Some of them have begun to try to articulate empirical rules of thumb for designers of interactive educational computer activities.[8]

A good interface and good displays are not simply desirable but essential. Poor choices may actually increase the load on the student, making the problem solving much harder by requiring the student to devote considerable effort to controlling the program or understanding the displays. Cluttered, disorganized displays or complex or awkward modes of interaction demand attention and effort that could much better be devoted to the problem-solving process itself.

Instructional Strategies

In working with a student, a tutor must make detailed decisions at a number of levels about how best to interact with the student. Human tutors make many of these decisions automatically, on the basis of experience and intuition, without consciously thinking about them. However, a computer tutor has no intuition or social experience; it must have explicitly encoded, detailed rules and guidelines for every interaction with a student.

For example, an important class of decisions concerns tutorial interventions: when to intervene, and how. Who should control interventions, the student or the tutor? Should the tutor intervene only when asked for help? Should the tutor always provide help when requested? When should help be given? Immediately after an error? After several errors? Before the student makes an error? What kind of help should the tutor give? Hints? Answers? Diagnosis of errors? Directions? A complete solution?

Until recently, there has been little research on the actual process of tutoring. At best, tutorial strategies in instructional computer programs have been derived by an empirical, trial-and-error process; at worst, they reflect only the untested intuitions of the program designers. However, recent studies of the strategies of expert human tutors, combined with attempts to articulate empirically derived guidelines, begin to provide a principled basis for designing computer-based tutorial interactions.

Studies of Expert Human Tutors

Protocol studies of expert human tutors working with students solving problems[9] have begun to uncover patterns and strategies in tutorial interactions. Some of the results are surprising, and even run counter to established wisdom.

Perhaps the most striking observation is that each tutorial episode (one problem in a sequence of problems, for example) consists of three distinct phases, as illustrated in Figure 6. The tutor's goals are both cognitive and motivational, and active goals and strategies vary from phase to phase.

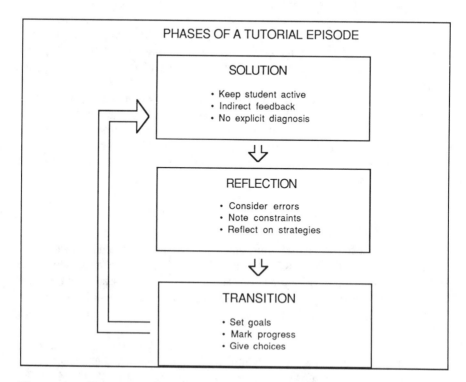

Figure 6. Phases of a tutorial episode.

In the solution phase, the tutor's primary goal is to work through to a correct solution of the problem. The student must be kept active and interested, and must work as independently as possible. Thus the tutor gives just enough help and guidance to allow the student to take the next step in the solution. Feedback from tutors is almost always indirect in this phase; tutors rarely say "no" explicitly, and though they may well recognize exactly the conceptual or procedural difficulties a student is having, they almost never deliver this explicit diagnosis to the student. Errors are corrected, but not dwelt on. The main goal is to get to a solution.

After the student obtains a correct solution, the tutor helps the student reflect on the process of solving the problem. Usually the reflection phase is short. A tutor may point out to a student, for example, that he or she correctly extended a rule to a new situation, or that an error was due to an attempt to apply a rule or procedure under conditions where it was not applicable. The tutor and student may review the student's overall strategy and discuss possible improvements. The main emphasis in the reflection phase is in getting the student to consider the solution process itself, rather than on applying particular rules or procedures.

The transition phase appears to be a crucial part of a tutorial episode. The tutor's goals in this phase are primarily motivational: to set for the student a comfortable level of challenge, to give the student a sense of control, and to stimulate the student's interest in the problems. In this phase, tutors explicitly mark the progress made by the student so far and help the student set new performance goals. They may give the student a choice of type of problem or level of difficulty, and may try to stimulate curiosity about new features of the problem.

Some instructional computer materials have been used heavily by thousands of students over a period of several years. Perhaps it should not be surprising that conventions evolved in many such programs reflect some of the goals of human tutors. For example, most good tutorial CAI programs continually mark progress and set goals by displaying information about the level of the problems, the number correct, and the number remaining to work. Students are given control over which problems to work and what level of difficulty to attempt. Students are not allowed to get permanently stuck on a problem; after a few tries the program provides help or a way out.

Necessary Differences?

The major differences between typical computer tutors (from both the ITS and CAI traditions) and experienced human tutors are, first, that computer tutors give much more direct feedback, usually saying "no" very explicitly, and often delivering a detailed error diagnosis; second, that there is usually no reflection phase in computer-based tutorial interactions.

It may in fact be necessary for computer tutors to provide much more explicit feedback than human tutors. Because the bandwidth for human communication is very high, a human tutor can convey much information in vocal inflection and timing, so a response of "well," "hmmm," or even "ok" can easily convey the information that an answer is not correct. With a more limited verbal bandwidth, computer tutors' responses are open to misunderstanding, and explicitness is often essential to clarity. However, it is sometimes possible to compensate for limited verbal communication with enhanced visual communication; in some activities the only feedback necessary may be the display generated as the computer works out the consequences of the student's input.

The absence of the reflection phase appears less functional, and is probably due to a lack of understanding of its importance. In fact, a computer tutor that traced the details of a student's solution could easily point out important features of the solution process; perhaps this will come to be a standard feature of future computer tutors.

Summary

Recent results from research in cognitive science provide insights that make a significant difference in the effectiveness of instruction. The importance of detailed models of the knowledge necessary to solve problems has been clearly demonstrated in experimental studies involving both paper-and-pencil tasks and computer tutors. An understanding of the role of diagrams and displays in the problem-solving process and of principles for structuring tutorial interactions can potentially have equally dramatic effects on instruction. Combining these insights with much empirical experience in educational computing may in fact allow us to design sophisticated computer tutors capable of delivering effective direct instruction to students on a one-to-one basis.

1. P. Suppes, "The Uses of Computers in Education," Scientific American 215, 206 (1966).
2. J. R. Anderson, F. Boyle, and B. Reiser, "Intelligent Tutoring Systems," Science 228, 456 (1985); J. I. Heller and F. Reif, "Prescribing Effective Human Problem-Solving Processes: Problem Description in Physics," Cognition & Instruction 1, 177 (1984); A. S. Palincsar and A. L. Brown, "Reciprocal Teaching of Comprehension-Fostering and Comprehension-Monitoring Activities," Cognition & Instruction 1, 117 (1984).
3. K. A. Ericson and H. A. Simon, *Protocol Analysis: Verbal Reports as Data* (Cambridge, MA: MIT Press, 1984).
4. Heller and Reif, "Prescribing Effective Human Problem-Solving Processes."
5. Anderson, Boyle, and Reiser, "Intelligent Tutoring Systems" J. H. Larkin and R. W. Chabay editors *Computer-Assisted Instruction and Intelligent Tutoring Systems: Shared Issues and Complementary Approaches*, (Hillsdale, NJ: Lawrence Erlbaum Associates, forthcoming); Sleeman and Brown, editors *Intelligent Tutoring Systems* (New York: Academic Press, 1982).
6. Larkin, J. McDermott, D. P. Simon, and H. A. Simon, "Expert and Novice Performance in Solving Physics Problems," Science 208, 1335 (1980).
7. J. H. Larkin, "Display-Based Problem Solving," In *Complex Information Processing: Essays in Honor of Herbert A. Simon*, edited by D. Klahr and K. Kotovsky (Hillsdale, NJ: Lawrence Erlbaum Associates, 1988); J. H. Larkin and H. A. Simon, "Why a Picture is (Sometimes) Worth 10,100 Words," Cognitive Science 11, 65 (1987).
8. S. M. Alessi and S. R. Trollip, "Computer-Based Instruction: Methods and Development," (Englewood Cliffs, NJ: Prentice-Hall, 1985); R. W. Chabay and B. A. Sherwood "A Practical Guide for the Creation of Educational Software," *Computer-*

Assisted Instruction and Intelligent Tutoring Systems: Shared Issues and Complementary Approaches, edited by J. H. Larkin & R.W. Chabay (Hillsdale, NJ: Lawrence Erlbaum Associates, forthcoming); E. R. Steinberg, *Teaching Computers to Teach*, (Hillsdale, NJ: Lawrence Erlbaum Associates, 1984).

9. R. W. Chabay, M. R. Lepper, and J. H. Larkin, "Cognitive and Motivational Strategies of Human and Computer Tutors," *Computer-Assisted Instruction and Intelligent Tutoring Systems: Shared Issues and Complementary Approaches*, edited by J. H. Larkin & R. W. Chabay (Hillsdale, NJ: Lawrence Erlbaum Associates, forthcoming); R. T. Putnam, "Structuring and Adjusting Content for Students: A Study of Live and Simulated Tutoring of Addition," American Educational Research Journal **24**, 13 (1987).

Addressing Students' Conceptual and Cognitive Needs

A. B. Arons
Department of Physics, University of Washington, Seattle, WA 98195

Research in physics education continues to develop compelling evidence that many students enter introductory physics courses with primitive, strongly rooted preconceptions that remain unaltered by exposure to the usual course materials.[5] On emerging from the courses, many students exhibit residual misconceptions and inadequate mastery of some of the most basic concepts and modes of reasoning. They acquire passing grades largely through partial credit for memorization of bits and pieces of vocabulary and erratically employed problem-solving procedures. Many of the investigations lend credence to the idea that one-on-one instruction and exercises of the highly interactive kind possible in computer-based instruction (exploiting graphics) can contribute significantly in helping students attain more effective grasp of the abstract concepts and modes of reasoning.[5] This is especially true when emphasis is placed on operational definition of technical terms, on qualitative observation and interpretation of physical phenomena, and on verbal explanation of lines of reasoning. This emphasis supplements the usual use of formulas and numerical solution of typical end-of-chapter problems. Many of the interview questions and situations that reveal student difficulties and misconceptions are, in themselves, helpful in remediating the very difficulties they initially uncover.

Despite the seeming promise of strong cognitive assistance from highly interactive computer-based materials, very little has happened so far to fulfill this promise. Much of the software that has been written is of very low instructional quality; it tends to attack peripheral rather than fundamental cognitive problems and provides very little interactive feedback that would lead students to recognize error or inconsistency or revise their thinking of their own volition (instead of simply being told the "right" answer). Furthermore, we have not had a sufficient volume of appropriate material to allow statistically convincing evaluation of impact and effectiveness over an entire introductory physics course (or even a major portion).

Nagging questions still remain regarding cost effectiveness. The writing, coding, and debugging of such materials is exceedingly time consuming and therefore exceedingly costly. Will the effectiveness of such instructional materials justify the cost? We still do not have even a preliminary, much less a convincing, answer to this question.

Nevertheless, in the hope that cost effectiveness might be demonstrated, I shall list some of the specific items of computer-based materials that are badly needed, that seem to be neglected, and that could, in combination, have a significant impact on student mastery of concepts and lines of reasoning in our introductory courses. I draw these specific items from my own observations and from the research literature.

Acute readers may note that certain obviously important items do not appear on the following list of suggestions. Material on kinematical concepts of position, clock reading, velocity, and acceleration is in the class of "badly needed," but since a great deal of effort is currently being invested in generating very promising materials, I do not include them on the present list. The same is to be said about materials on the law of inertia and the concept of "force." These areas are excluded not because of lack of importance but because they are the subject of intense and fruitful effort. Other readers may discern that I do include other items that are in fact currently being worked on or that are available in good versions. Concerning such, I can say only that I can lay no claim to being up to date on all work being conducted in this huge field.

Examples of Desirable Drill

One very primitive impediment to learning in introductory physics courses is lack of drill to help students imbed a definition, a procedure, or a physical phenomenon in the memory to such a degree that it becomes second nature and no longer stands in the way of the next level of utilization, concept formation, or reasoning. Space in textbooks is costly and limited, and adequate drill with immediate feedback and reinforcement can rarely be provided. Such drill, however, can readily be provided on the computer; although the mode may seem pedestrian and unexciting to the writer of software, the results for the learner can be very important. Specific areas in which availability of such drill seems currently inadequate are as follows:

1. Drill on significant figures in numerical calculations.

2. Drill on propagation of error in numerical calculations.

3. Drill on the definitions of sine, cosine, and tangent as ratios of sides of simple right triangles. Setting up each ratio when the triangle is presented in various different orientations. Reversing the above procedure (i.e., given labeled sides of triangles in various orientations, recognizing the name of a given ratio).

Recognizing directly the meaning of products such as $c \sin\theta$ and $c \sin\theta$ (where c denotes the length of the hypotenuse) without going through algebraic manipulation of the original defining ratios.

4. Drill on addition and subtraction of vectors in *one* dimension both algebraically and graphically. Many students never master the meaning of purely numerical operations with plus and minus signs along the number line without also interpreting these operations in terms of arrows. The recommended drill helps them understand the full meaning of the rectilinear kinematic equations. Such drill connects with similar drill on addition and subtraction of vectors in two dimensions.

5. Drill on calculating numerical values of torque (or setting up algebraic expressions). Forces should be presented in random orientations on the screen, differing from any original figure in the text. Different angles should be invoked (i.e. sometimes the angle between the force and the normal to the radius arm, sometimes the angle between the force and the radius arm). Forces should occasionally be directed through the axis of rotation. Net torques should be calculated when several are imposed simultaneously.

6. Drill on basic electromagnetic phenomena. Magnetic field direction around various configurations of current carrying conductors. Forces and torques on current carrying conductors in a magnetic field. Exercises with the Lorentz force on moving charged particles in a magnetic field. Exercises with direction of induced emf.

Underpinnings

In addition to drill of the kind suggested in the preceding section, there are a number of other concepts and modes of reasoning that underpin the learning of physics and with which many students have substantial difficulty. A significant fraction of students in calculus-physics courses have trouble with the following ideas:

1. Verbal interpretation of ratios.[1,4] What is the meaning of the number obtained when one length is divided by another length or one mass by another mass? (Here one is comparing two numbers by finding how many times one is contained in the other. Many students fail to recognize division as counting how many times one number can be subtracted from the other because they have never had to say something to this effect in their own words.) What is the meaning of the number obtained when the mass of an object (in grams) is divided by its volume (in cm^3)? (Here one is finding how much of the property in the numerator is present in or goes with, or corresponds to *one* unit in the denominator. Many students fail to articulate the necessary *one* of the denominator and hence fail to comprehend the meaning of the new number. Thus, they fail to use it successfully in subsequent steps of arithmetical reasoning.) It is also neces-

sary to reverse the interpretation and deal explicitly with the reciprocals, e.g. with volume divided by the mass.

2. Meaning of π in simple geometrical terms, i.e., as the ratio of circumference to diameter of circles. Few students have carried out the necessary simple measurements and fewer still can articulate the insight that π is a common property of all circles.

3. Arithmetical reasoning involving division. Given 800 g of material that has 2.3 g in each cm^3, what must be the volume occupied by the sample? Students should not be allowed to use the density formula (they simply memorize the rearrangement of symbols). They should be led to argue that, since the "package" of 2.3 g is associated with just *one* cubic centimeter, we find the total number of cubic centimeters if we find how many such "packages" are contained in 800 g.

4. Coupling arithmetical reasoning with graphical representation in cases of simple linear relationship, e.g., associating density with the slope of a graph of total mass M versus total volume V; interpreting an arithmetical calculation by representing it on a graph and vice versa; interpreting graphically the connection between finding the total volume given the total mass and density on the one hand and finding the change in volume when the same mass is added to a given initial mass. (Most students see these as two entirely different calculations and subtract the final and initial total volumes without perceiving that they can calculate changes directly. Parallel exercises are usually necessary with other contexts such as concentration of solutions, circles and π, etc.)

5. Distinction between general linear relationship and direct proportionality. Many students fail to register the distinction in earlier experience and must be helped to articulate it in their own words.

6. Ratio reasoning involving geometrical scaling from linear dimensions to areas and volumes and vice versa. Ratio reasoning involving scaling of various physical effects when functional relationships are known (e.g., variation of centripetal force with angular velocity, variation of gravitational and electrostatic forces with distance of separation when the inverse square law is applicable, etc).

7. Radian measure. Very few students have acquired any genuine understanding of the reasons for introduction of radian measure in place of degrees; they have simply memorized a ritual if they use radian measure at all. They should be led to see why radian measure is introduced, how it differs from measure in degrees, and in what sense it is to be regarded as "natural." They should be led to see that a quantity can be dimensionless without being unitless. The examination should not be confined to the relation among angle, arc length, and radius, but should extend to the observation that the sine and tangent of an angle become more and more nearly equal to the numerical value of the angle provided the measure is in radians rather than in degrees. For students in calcu-

lus-physics courses, this examination should extend to consideration of the limits of the relevant ratios underlying the derivation of the derivatives of the sine and cosine functions.

Examples Involving Various Basic Physical Concepts

Following are examples of subjects in which many students encounter serious conceptual difficulty and thus fall irretrievably behind. Dialogues helping them surmount the difficulties as early as possible would help retain many students who otherwise drop out. Dialogues would also serve to improve the grasp and course performance of many of those who survive with marginal understanding of the material.

1. Showing velocity, acceleration, and force vectors (in separate diagrams, never different vectors on the same diagram) for:

    a. an object thrown vertically upward while rising, falling, and at the top of its flight;

    b. various points in a trajectory of projectile motion;

    c. a car at various points in a roller coaster ride;

    d. a pendulum at various points in its swing, including the end points;

    e. a bob on a spring in simple harmonic motion;

    f. a bob in circular motion in a horizontal plane and at various points in a vertical plane;

    g. a pendulum bob suspended from the roof of an accelerating car.

2. Making operational distinctions between gravitational and inertial mass.

3. Understanding frictional force. Many students tend to memorize the formula $f = \mu N$ and regard this as giving the magnitude of the frictional force acting on a stationary block under all circumstances. They begin to grasp the concept when they are led through a sequence in which they must articulate the insight that the frictional force increases from zero to f as the external applied force on the block increases. Many students acquire the misapprehension that frictional forces always oppose the motion of the entire body on which they act. Although it is true that frictional forces always oppose the sliding of one surface over another, they do not always oppose the motion of the entire body on which they act (witness the frictional force exerted by the road on the wheels of an accelerating car).

4. Understanding tension in strings. Many texts introduce the term "tension" as though it were a primitive concept with obvious meaning to everyone. Since "tension" is actually a subtle concept requiring careful and explicit operational definition, many students become seriously confused.

5. Understanding "massless" strings. Most texts introduce the "massless string" as though it were an obvious concept that requires no special introduction or discussion. An effective dialogue would lead students to consider first a simple case in which a string with appreciable mass is being accelerated (e.g., a block being accelerated along a horizontal surface by a pull exerted on a massive string). One would draw separate free body diagrams of both the block and the string and establish the fact that the forces on opposite ends of the string would not be equal in magnitude. One would then examine what happens to the magnitude of the difference between the forces as the mass of the string is made smaller and smaller relative to the mass of the block.

6. Using free-body force diagrams. Many students (in fact, a sizable majority even in calculus-physics courses) fail to develop the capacity to draw correct force diagrams when homework is limited to conventional end-of-chapter textbook problems. They tend to fiddle with the algebraic manipulations until they get the "right" answer in the back of the book, and they rarely check the correctness of the force diagrams. Students also tend to follow the space-saving pattern set in the majority of textbooks and do not draw separate force diagrams of interacting objects. Hence many students fail to distinguish clearly between the forces exerted on a car by the road and the forces exerted on the road by the car. Effective dialogues would lead students to draw numerous force diagrams for situations familiar in everyday experience. These might include:

    a. one's own body and the ground while walking or running (including speeding up and slowing down);

    b. one's own body, the seat of the car, the body of the car, and the road surface when the car is speeding up, slowing down, or going around a curve;

    c. one's own body sitting on a box that rests on a merry-go-round;

    d. an electrically charged balloon sticking to a wall;

    e. two unequal point charges attracting or repelling each other (many students show unequal electrical forces acting on the two objects even though they can repeat the words of Newton's third law).

7. Considering the following two rectilinear cases:

    a. A ball rests on a cart; the cart is accelerated horizontally.

    b. A bob hangs suspended from a string (pendulum); the upper end of the string is accelerated horizontally. Many students performing such experiments will say that both the ball and the bob are "thrown backward." They need help in straightening out the confusion of frames of reference and recognizing explicitly that neither object is "thrown backward" in the inertial frame established by the ground.

8. Understanding centripetal force. Many students labor under the misapprehension that any centrally directed force is a "centripetal" force regardless of magnitude, whereas the term "centripetal force" is usually meant to apply to the net force that imparts the acceleration v^2/R. Students must be led to recognize explicitly that the object will spiral inward if the centrally directed force is greater than the quantity mv^2/R and will spiral outward if it is smaller. Because the discussion of centripetal force usually starts with the bob on a string moving in a horizontal plane, many students acquire the misapprehension that the tension in the string is invariably equal to the centripetal force. They will, for example, say that the tension in the string is equal to the centripetal force at both the top and the bottom when the bob revolves in a vertical circle.

9. Sketching the shapes of asymmetric transverse and longitudinal wave pulses reflected at (and transmitted through) various free and rigid boundaries.

10. Performing exercises in hypothetico-deductive reasoning.

   a. Given some mechanical situation arising in an end-of-chapter problem, what will happen if the angle of application of this force is changed, if this mass is increased or decreased without limit, or if the angle between this surface and the horizontal is increased or decreased?

   b. Given a DC circuit consisting of batteries and bulbs, how do the bulb brightnesses compare in the initial configuration? What will happen to the brightness of each remaining bulb if a given bulb is removed? What will happen to the brightness of each bulb if a wire is connected between two arbitrarily chosen points?

   c. Given simple R and C configurations, simple R and L configurations and the simplest R, L, C configurations, students must be able to answer some purely phenomenological questions.

11. Visualizing effects that transcend direct sense experience. At the most primitive level, a large majority of students have great difficulty accepting the proposition that the (inanimate and seemingly rigid) table exerts an upward force on an object that rests on it.[22] They need to be led to the perception that the upward force is made possible by the springlike action associated with deformation of the table, a deformation that is not zero even when a sheet of paper rather than a book is being supported. Students must be led to visualize the deformations that take place when bodies collide with each other; very few students do so spontaneously. A long wooden or metal rod is laid on a table, and a force is applied to one end. The rod is displaced. If asked whether the far end of the rod moves at exactly the same instant as the end to which the force was applied, most students say "yes." They do not visualize the passage of the elastic wave that precedes motion of the object as a whole. Visualizing such effects consciously and explicitly, however, is an important prelude to the introduction of time delays and propagation

effects in electromagnetic phenomena. It helps students comprehend the very basic phenomenological questions which Faraday and Maxwell were asking and prepares them for recognition of the failure of action at a distance (Newton's third law) and the motivation behind the invention of field theory.

Conclusion

The preceding examples are meant to be illustrative and are neither prescriptive nor exhaustive. Many such dialogues, without constituting entire courses, would be very helpful to many students within our existing instructional framework. Availability of such tutorial assistance would free teachers to deal more effectively and successfully with the development of more sophisticated levels of knowledge and understanding. Not only would this enhance the development of future scientific and engineering professionals, it would significantly improve the science competence of school teachers in precisely those areas of subject matter in which we now observe their greatest deficiencies. It would thus make a significant contribution to achievement of wider public understanding of science. The immediate problem, however, is to resolve the question of cost effectiveness.

1. A. B. Arons, "Cultivating the Capacity for Formal Reasoning," Am. J. Phys. **44**, 834 (1976).
2. A. B. Arons, "Thinking, Reasoning, and Understanding in Introductory Physics Courses," Phys. Teach. **19**, 166 (1981).
3. A. B. Arons, "Phenomenology and Logical Reasoning in Introductory Physics Courses," Am. J. Phys. **50**, 13 (1982).
4. A. B. Arons, "Student Patterns of Thinking and Reasoning," Parts One, Two, and Three. Phys. Teach. **21**, 576 (1983); **22**, 21 (1984); **22**, 88 (1984).
5. A. B. Arons, "Computer-Based Instructional Dialogs in Science Courses," Science **224**, 1051 (1984).
6. A. B. Champagne, L. E. Klopfer, and J. H. Anderson, "Factors Influencing the Learning of Classical Mechanics," Am. J. Phys. **48**, 1074 (1980).
7. J. Clement, " Students' Preconceptions in Introductory Mechanics," Am. J. Phys. **50**, 66 (1982).
8. J. Clement, J. Lochhead, and G. S. Monk, "Translation Difficulties in Learning Mathematics," Am. Math. Month. **88**, 286 (1981).
9. G. Erickson, and J. Aguirre, "Student Conceptions About the Vector Characteristics of Three Physics Concepts," J. Res. Sci. Teach. **21**, 5 (1984).
10. M. H. Fredette and J. Clement, "Student Misconceptions of an Electric Circuit: What Do They Mean?," J. Coll. Sci. Teach. **10**, 280 (1981).
11. F. M. Goldberg and L. C. McDermott, "Student Difficulties in Understanding Image Formation by a Plane Mirror," Phys. Teach. **24**, 472 (1986).
12. F. M. Goldberg and L. C. McDermott, "An Investigation of Student Understanding of the Real Image Formed by a Converging Lens or Concave Mirror," Am. J. Phys. **55**, 108 (1987).
13. R. F. Gunstone, "Circular Motion: Some Pre-Instructional Alternative Frameworks," Res. in Sci. Ed. **14**, 125 (1984).
14. R. F. Gunstone, "Student Understanding in Mechanics: A Large Population Survey," Am. J. Phys. **55**, 691 (1987).
15. R. F. Gunstone, and R. White, "Understanding Gravity," Sci. Ed. **65**, 291 (1981).

16. Halloun, and Hestenes, D., "The Initial State of College Physics Students," Am. J. Phys. **53**, 1043 (1985) and "Common Sense Concepts About Motion," Am J. Phys. **53**, 1056 (1985).
17. R. A. Lawson and L. C. McDermott, "Student Understanding of the Work-Energy and Impulse-Momentum Theorems," Am. J. Phys. **55**, 811 (1987).
18. M. McCloskey, A. Camarazza, and B. Green, "Curvilinear Motion in the Absence of External Forces," Science **210**, 1139 (1980).
19. L. C. McDermott, "Research on Conceptual Understanding in Mechanics," Phys. Today **37**, 24 (1984).
20. L. C. McDermott, M. L. Rosenquist, and E. H. van Zee, "Student Difficulties in Connecting Graphs and Physics: Examples from Kinematics," Am. J. Phys. **55**, 503 (1987).
21. J. W. McKinnon and J. W. Renner, "Are Colleges Concerned with Intellectual Development?," Am. J. Phys. **39**, 1047 (1971).
22. J. Minstrell, "Explaining the 'At Rest' Condition of an Object," Phys. Teach. **20**, 10 (1982).
23. P. C. Peters, "Even Honors Students Have Conceptual Difficulties with Physics," Am. J. Phys. **50**, 501 (1982).
24. D. E. Trowbridge and L. C. McDermott, "Investigation of Student Understanding of the Concept of Velocity in One Dimension," Am. J. Phys. **48**, 1020 (1980) and "Investigation of Student Understanding of the Concept of Acceleration in One Dimension," Am. J. Phys. **49**, 242 (1981).
25. L. Viennot, "Le Raisonnement Spontane en Dynamique Elementaire," Eur. J. Sci. Ed. **1**, 205 (1979).
26. B. White, "Sources of Difficulty in Understanding Newtonian Dynamics," Cog. Sci. **7**, 41 (1983).

Can Computers Individualize Instruction?

Tryg A. Ager
Institute for Mathematical Studies in the Social Sciences, Stanford University, Stanford, CA 94305

The Spectre of Skepticism

To understand why many have been arguing that it is impossible to establish the educational effectiveness of computers, we must glance back on 20 years of discussion of the effectiveness of computers in instruction.

By 1968 Bitzer, Bunderson, Suppes, Bork and others had developed a variety of CAI materials and compared them with traditional instruction. These early developers differed in their assessments, visions, and approaches to instructional uses of computing, but their work showed that computer implementations of programmed learning methods demonstrated modest learning gains in controlled experiments.

Still, when the instructional methods were such that teachers and computers could perform them equally well (drill and practice in arithmetic), the experiments tended toward no significant differences.[1] During the same period, there was deliberate exploration of the computer as a *sui generis* interactive and graphics medium. Happily, fundamental ideas of physics and mathematics were often the context of these explorations. In this first decade, three main questions were on the CAI agenda:

1. Are computers acceptable media for instructional methods such as drill and practice in elementary mathematics?

2. Are computers as effective as teachers at some instructional tasks?

3. Do computers have unique instructional properties or capacities?

By 1978 scores of academics were involved in CAI, and funding agencies for materials development and instructional improvement experienced exponential growth in proposals. These agencies had an interesting problem: deluged with proposals for instructional computing, how were they to filter the good from the bad? There were very few physics instructors with track records in the field. Alderman's landmark study of TICCIT had just come out,[2] but new proposals went beyond the programmed learning and drill-and-practice methods that had been studied previously in controlled experiments. Proposals dealt with complete college courses by computer, microcomputer simulations, computer-driven interactive video, and language instruction, to name just a few. Effectiveness was a reasonable criterion to apply to these new instructional ideas. But the dismantling of NSF and the Department of Education that began in 1981 prevented leading-edge, major evaluative projects on the scale of the previous PLATO and TICCIT evaluations. Metaanalysis and literature reviews replaced large-scale evaluations for most of the 1980s.[3] Interpretations of these studies of studies have become the arena of evaluation debate—an arena several times removed from the trenches where actual work on CAI is done. Furthermore, inexpensive microcomputers democratized CAI and more energy was put forth on innovation than on understanding issues of evaluation and effectiveness.

Historically, demonstration of effectiveness has been central to vindication of computer-assisted instruction. Research and debate about the effectiveness of CAI is conducted on a wide scale. It is not a side show in either educational research or individual disciplines. It is not over; and it is not especially free from ambiguity and confusion. Two strategies for vindicating CAI can be distinguished: "vindication by research" and "vindication by the revolution." The research is conventional educational research, and its results are mixed. The revolution is the computer revolution, and its results are unavoidable even if not systematic or univocal. We find similar conflict between establishment science and innovation in other instances of major scientific, technical, or social change where there is pressure to revise existing evaluative criteria, descriptive categories, or theories.

Both the CAI researcher and the CAI revolutionary want to demonstrate effectiveness, but they use different methods. The researcher measures quantifiable variables defined independently of computers and embedded in general learning theories. Effectiveness, defined as change in achievement, is then tested in the context of existing theory. Achievement is conceptually independent of the uses of computers for instruction, so controlled experiments to establish empirically the relationship of computer use to achievement are apparently in order. As the literature shows, these experiments have been done by the score.

The revolutionary usually uses pragmatic criteria and less entrenched methods. The revolutionary is more inclined to start from the premise that computer-assisted instruction is *sui generis* and requires new categories and methods to define and test effectiveness. For example, many write-ups present qualitative data, formative evaluations, or anecdotal evidence as demonstrations of effectiveness. Vindications of CAI by such methodologies are often more notable for vigor than rigor.

Clark's Critique of the Research Vindication

Richard Clark has argued that instructional methods are related to learning outcomes independently of media.[4] So if computers are media rather than methods, a research vindication of CAI (or any other delivery medium) is impossible. Clark contends that most attempts to vindicate CAI fall into a 50-year-old trap of confusing instructional methods with media. Besides computers, he considers textbooks, video, mechanical models, workbooks, and sometimes teachers (as opposed to teaching) to be media. The crux of Clark's claim is that for matched studies where media are the only experimental variables, the null hypothesis is the expected outcome. Clark says: "Consistent evidence is found for the generalization that there are no learning benefits to be gained from employing any specific medium to deliver instruction,"[5] and adds that "studies comparing the relative achievement advantages of one medium over another will inevitably confound medium with method of instruction." I will refer to this as the principle of parity of media. The issue that faces Clark is clarification of whether the principle of parity of media is true by definition or a testable empirical claim. His writings indicate he believes both, and that is not good science.

Two years later, after an exchange with Petkovich and Tennyson,[6] Clark asserted even more pointedly about experiments to compare the effectiveness of CAI with traditional instruction: "There seems to be ample evidence in existing studies that no theoretical reason exists to ask such a question. Whenever adequately designed studies have asked the question, no differences in achievement have been found that may be unambiguously attributed to 'computers.'"[7]

We have to take the principle of parity as being true in order to make sense of that claim. Unambiguous relationships between effectiveness (achievement gains) and computer-assisted instruction cannot be established because there is always an alternative implementation of the instruction. This is true regardless of media or methods being considered. If we can always exchange media, then no particular medium is decisively a cause of achievement. So the cause of achievement must lie in method, not medium. It is then a corollary that educational methods can be tested for effectiveness, but educational media cannot.

We can interpret some episodes in the history of CAI by taking the principle of parity of media as an empirical, testable claim. The principle is illustrated in the history of programmed learning. Originally, the medium was printed in fill-in-the-blanks workbooks that branched users to different pages depending on self-scored performance. In the 1960s computers were found to be another suitable medium. Similarly, drill-and-practice teaching methods are effectively implemented in hand-out sheets, workbooks, radio programs, and computer programs, each with demonstrable effectiveness compared to non–drill-and-practice methods. Effectiveness of either method in one medium leads one to expect, and on Clark's theory actually predicts, effectiveness in another. In fact, these expectations/predictions were used to justify undertaking the transposition of programmed instruction and drill and practice to other media. Further supporting this idea, one early study showed that increased achievement from computer-based drill and practice stimulated noncomputer classes to do more offline drill and practice, which also resulted in increased achievement.[8] The effectiveness of drill and practice, not computers per se, had been validated. At best, computers catalyzed an improvement in instructional methods, which caused an increase in achievement. Many similar catalytic effects of computers are reported at this conference.

In the area of computer-assisted instruction in physics, the evolution of Alfred Bork's work reflects the idea that methods are independent of media. Bork explicitly draws on the theory of self-paced mastery learning to guide the design of computer tutorials in physics. His assumption must have been that the methodology was essentially media independent and hence would transfer to the new computer medium. I think it is fair to say that Bork also implies that because of its interactivity, the computer medium could enhance learning more than any other available instructional technology such as television or hand calculators. I think Clark would say the first step is correct and the second fallacious because it assumes that for a given method there are more effective media. Remember that in all of this, "effectiveness" means measurable changes in achievement.

Experiments to Test Clark's Theory

We only have time to do a *gedanken* experiment to test Clark's theory, but consider the following: The *Valid* instructional program teaches an entire semester of symbolic logic without scheduled lectures, textbooks, or conventional homework.[9] Everything in the course is done online except the final examination. The instructional methodology used by *Valid* is composed of three main parts:

1. Interactive exposition of the concepts and principles of logic (e.g., explanation of the concept and a formal definition of 'valid argument').

2. Interactive theorem proving (i.e., students construct proofs that are checked step by step so feedback about errors is given at the earliest possible moment).

3. Self-paced progress and on-demand instruction (i.e., students work when they want to, for as long as they wish). Let's call this instructional methodology

"method V." Method V is generic and has been applied to other subjects in the family of axiomatic mathematics and formal systems.

Clark's theory predicts that the effectiveness of method V, which is currently implemented on computers, is independent of media. I predict that a human implementation of method V would differ in effectiveness for reasons intrinsic to the alternative medium, viz., intense interpersonal relationships, high error rates on the human proof checkers, unreliable implementation of the real-time responsiveness requirements of method V, but most of all, the almost moral certainty that each human implementation of method V would differ more one from the other than each computer implementation. The human implementations lack the desirable properties of stability and replicability.

But I would not dispute that there probably are media independent methods other than method V that are just as effective for logic instruction. This, however, seems to catch Clark at his own game, since his argument based on parity of media can be applied to his own position, if transposed to a principle of the plurality of equally effective methods. Since achievement cannot be unambiguously attributed to method, method is not the causal factor in achievement. This leaves students as the only active items not factored out of Clark's learning model. Could students be the causes of achievement?

This *gedanken* experiment is symptomatic of the clash between the research and revolutionary programs to vindicate computers. Method V requires immediate feedback, self-directed pace, and zero tolerance of errors in construction of proofs. It requires time-critical and impersonal interactions with students that can be met by some media but not by others. Therefore, for method V, medium makes a difference and effectiveness of method V will vary depending on medium. Method V is a revolutionary instructional method, borrowing here and there from mastery learning and programmed instruction, but it includes in its specification some requirements that were impractical 30 years ago. So the revolutionary argues that computers generate a new class of media-specific methods, and hence the 50-year-old conceptual distinction between media and methods no longer applies. The categories that define instructional methods need revision.

A Physics Analogy to the Parity of Media Principle

Further doubts about the parity of media principle arise when we draw an analogy from high school physics. Pressure, temperature, and volume are related by the Charles-Boyle gas law, $PV = nRT$. Say a physicist, we'll call him Charles Boyle Clark, tells us that there are various media in which volumes exist, like jugs, boxes, tubes, and balloons, but that his law of gases holds independently of how we "implement" volume or what "medium" we use to enclose a volume. We can live with that until we try to build a real-world device like an internal combustion engine that gets 26 miles per gallon of gas. It is not clear that we can improve mileage ("effectiveness") without taking cylinder geometry and materials ("media") into account in implementing the volume variable in the gas law and the other relevant thermodynamic principles.

To argue that the geometry of cylinders is irrelevant to the design of effective engines because cylinders are only one of any number of logically possible implementations of the volume variable, reflects some profound confusion about the methodology of applied science (technology) as opposed to the establishment of abstract theoretical principles. Real-world engineering is a jungle of trade-offs to take account of interfering factors that either are absorbed into physical constants or packed into experimental error at the abstract theoretical level. No eighth grader doing classical Boyle-Charles Law experiments needs to bother with Van der Waals forces or chemical reactions of the gas with the sides of the vessel. However, in applying that theory, the engine designer must take those things into account.

Clark's idea that computers belong to a set of interchangeable media, depends on a simplistic view of the relationship between pure and applied theory. The presence of complex media, which have objective and affective properties (e.g., reliability and enjoyment) known to influence achievement, must be taken into account just as the materials and shape of cylinders are important in the internal combustion engine applications of thermodynamics and the ideal gas law. The practical effectiveness of any design always depends on both the underlying theory and the implementation.

Clark's Confounding Claims

Finally, let's interpret parity of media as authorizing a redescription of experiments, because by definition instructional methods are generic and allow multiple alternative implementations or media. Interpreted this way, parity of media is a built-in component of what we mean by an instructional medium. Therefore the causal impotence of computers, for example, is entailed by the meaning of the term "media," but of no empirical significance since it is an invariant structural feature of the descriptive categories, not the facts. This way of categorizing things results in all interesting differences between experimental and control groups being assimilated to method. In certain cases this assimilation makes sense, and in others it is arguable. There is an example of each below. But when Clark says the assimilation or confounding is inevitable, his view approaches the "true by definition" interpretation, thereby making the empirical content of his skepticism about computers vanishingly small.[10]

In his examination of 42 studies of the effectiveness of CAI, Clark claims the experiments are pervasively flawed either because of media-method or media-media confusion, both of which result from not understanding the generality of method as opposed to the specificity of media.[11]

1. *Media-Method Confounding.* By failing to understand the general meaning of method variables, the experimenter does not control for instructional method, so effects attributed to media are confused with effects of the instructional method.

Consider a comparison between a lecture-discussion presentation of the laws of orbital motion and a self-paced, interactive, exploratory computer tutorial on the same subject. Suppose the latter turns out to be more effective. We confound media and method if we say computers are better than people as instructional media for orbital motion. We should say that the self-paced, exploratory method is better than lecture-discussion.

2. *Media-Media Confounding.* The experiment is flawed because it does not control for medium. Even if instructional method is controlled, media may confound. An interesting limiting case arises if the teacher is classified as a medium. When the same person teaches the control group and is teacher and/or developer of the CAI group, effect sizes tend to be smaller or negative compared with similar experiments where the teacher differs. Why does this happen and what does it mean? The research vindication of CAI can argue either that there is a "John Henry" effect: the teacher works harder and more conscientiously in the control group, adjusting his or her performance to the level of CAI, or there is a "Peter Principle": the CAI design or control group adjusts to the teacher's level of competence. Either way, disappearance of the predicted variance is explained by reduction of the experiment to the null hypothesis by a Peter Principle or a John Henry effect.

Contrarily, Clark argues that the disappearance of the variance doesn't need to be explained away, because it is predicted by his theory where teacher simpliciter or teacher-cum-computer are just "media" and media cannot make a difference. An alternative hypothesis Clark doesn't discuss is that because teachers can reprogram themselves, they can become a dominant medium. A simple way of stating this is that teachers (*qua* media) can retard or advance achievement.

Summary of the Critique of Clark

Clark's skepticism about research vindications of CAI uses a principle of parity of media with respect to method to assimilate all interesting differences to "method" variables. By definition, computer components are classified as "media" variables. But media variables cannot affect achievement, so computers cannot affect achievement.

I make two criticisms:

1. The principle of parity of media seems eminently testable in extreme cases or for highly specific instructional methods such as method V. It's just that nobody wants to take credit for showing that Spanish as an instructional medium, is better than Chinese for children in Honduras, for some fixed subject matter and instructional method.

2. More important, though, Clark's critique misconstrues the long-term process of technological research and development. His strongest arguments come from studies that use well-established instructional methods such as programmed

learning or drill and practice as method variables in experiments to vindicate CAI. As a development strategy, it was correct for early CAI to bootstrap from the programmed learning and drill-and-practice methods, using established achievement levels as development targets. It is to be expected that new technology will make some contributions of its own to the repertoire of educational methods. At first, these new methods will be implementation- specific, like method V. But there will be an attempt to generalize new methods and make them less media specific. I think it is extremely interesting that there are instructional methods that are robust over different media, subject matters, and students. It is not, however, a trivial or inevitable theoretical consequence that this is so, but rather that the practical objectives of the development of new educational media or methods, require meeting existing achievement standards empirically established by existing methods.

Finally, because of the complexities of CAI development, we have to content ourselves to work in subjects that are well understood from an instructional standpoint and consequently for which we are asymptotically approaching the maximum achievement gains possible for large numbers of randomly selected students, who arrive with and continuously reveal great individual differences. As Clark suggests, we should not expect dramatic achievement gains in subjects we already teach rather well by a variety of media-method combinations. But by the same token, it is only reasonable to insist that we avoid spending time and money on media that are not competitive in effectiveness.

Issues of Individualization in CAI

The purpose of this part of the paper is to discuss strategies for making CAI effective by individualizing instruction. Because this is an emerging area of CAI, I want to assume for the sake of the argument, that for subjects where incoming ability and achievement vary yet instructional goals are uniform, individualization is a plausible method to reach a common learning goal by different routes or strategies. Adaptive teaching methods and individualized test-and-branch regimens for elementary mathematics instruction indicate individualization is effective in many cases. The question explored here is whether to extend individualization to complex subjects with strong formal components at the advanced secondary and undergraduate levels.

Why Computer-Based Instruction?

Should logic, mathematics, or physics be taught on an individualized basis instead of more impersonally to groups? In fact, why use computers in any direct instructional role for these subjects? Computers have an important place already as tools for calculation and simulation, not to mention their usefulness in writing up class assignments or preparing handouts.

1. *Access*. Delivery of curriculum by computers can increase access to subjects. We find that although calculus appears to be taught to about 600,000 American high school students, about three fourths of the nation's high schools are not represented among the approximately 60,000 students who take the advanced placement calculus examination. The underrepresentation of schools in the AP tests is largely because most American school districts are small and frequently cannot allocate resources to teaching advanced placement courses.

2. *Serving Special Needs at the Margins*. CAI technologies can be designed so that time, place, level, and rate of learning can be variables in the overall system. Classroom instruction tends to fix all four.

3. *Resource Allocation*. If stable parts of a curriculum are offloaded to computer-based instruction, possibilities for reallocation of teaching resources occur. Resource reallocation is good if you don't have to grade 300 exercise sets on the Boyle-Charles law every year, bad if it leaves you with nothing to do.

4. *Standardization and Quality Control*. Calculus is a good example. Most calculus students are taught by relatively inexperienced graduate students. Nationally about 50 percent of calculus students fail. Regardless of whether these two facts are related, clients of mathematics departments would benefit if they could count on a quality-controlled, reasonably standardized knowledge of calculus in their students. Similarly, the elementary physics courses serve clients in the university who would enjoy similar guarantees.

5. *Effectiveness and Side Effects*. Even if effectiveness is merely maintained and not dramatically improved, there are ancillary, desirable side effects of undertaking CAI development work in new areas with new methods. Our work on calculus and logic instruction is directly related to research in computer algebra and theorem proving.

Why Individualized Instruction?

Individualization is an essential ingredient of most established CAI methodologies including dialogues, programmed learning, adaptive drill and practice, and self-paced learning. It is widely accepted that these methods can be individualized for many subjects, but most successfully for skill- or fact-oriented subjects.

For the subjects we are considering (logic, elementary calculus, elementary physics, etc.), we have evidence that however narrowly we specify nontrivial problem-solving tasks, individual students solve the same problem differently. We have additional evidence that student preferences affect achievement, and that when time is taken into account, the same solutions to problems can exhibit remarkably different latencies. As students are given more problem-solving resources and assigned more complex problems, differences among student solutions increase. Some responsiveness to these differences on the instructional side is necessary.

Here are several examples of the kind of variance we find in data collected from the *Valid* and *Excheck* programs at Stanford, both of which require extensive interactive theorem proving from students.[12] Most students complete all work for the *Valid* program in 60 to 80 hours, but the low is about 35 hours and the high is over 110 hours. For a given proof in logic, we found in a spot check of two students, that one did the problems in about 60 interactions with the computer; the other used over 350 interactions to produce a virtually identical proof. For problems with a fixed set of correct solutions, some solutions are favored, but all are represented. In proofs of the more complicated theorems required in set theory, there is much greater variance in length of proof. For a sample of 1,500 proofs of 75 different theorems, the average length of the shortest proof of each theorem was 3.5 steps; the average length of the longest proof of each was 54.7 steps. Further analysis of these student proofs fails to reveal patterns of steps that would describe strategies or predict patterns in future proofs by the same student.

The fundamental reason for individualizing instruction in mathematics and mathematical sciences is that students solve complicated problems differently; good instruction should be able to accommodate and be responsive to these differences. In the *Valid* and *Excheck* programs we have tried to individualize instruction insofar as we can, constrained on the one hand by the limitations of the computer medium and on the other by the nature of the subject matter.

Media Constraints

In the *Valid* course, which deals with elementary symbolic logic, we do not encounter serious media constraints on the individualization of logic instruction, except for the following humorous case. A student had exhausted all available computer memory with a proof of a simple theorem that in his case had swelled to something over 600 steps. However, in the *Excheck* system to teach axiomatic set theory, computational complexity is a practical constraint of the medium. Certain routines depending on resolution-theorem proving have to be terminated after exhausting a certain cpu allocation. (This allocation was 30 cpu seconds for the machine on which the initial studies of student proofs was done; the current machine will decrease that allocation by an order of magnitude.) It will be interesting to compare student use of the resolution-based tools on a faster medium, since students previously complained about slowness.

Method and Content Constraints

Both of these systems implement decision procedures for fragments of the theoretical domain they cover. But in most parts of advanced logic and mathematics the best we can hope for is partial decision procedures. Our goals have been to implement systems that are consistent (every step they certify in a proof is a valid step.) but not complete (they cannot certify every valid step.) Turning off a resolution-theorem prover after exhausting some cpu allocation is a media constraint. The general inability to decide mechanically the correctness of every proposed inference is a theoretical constraint due to the content of the courses. Within such constraints of the medium, students are free to construct any valid proof as a solution to a theorem-proving problem.

Individualized Evaluation of Student Solutions

I have already mentioned that in complex subjects there is great variance among student solutions to nontrivial problems. Our approach, which I call "interactive reasoning," has been to make the computer smart enough (within the inherent limitations of media and subject matter) to recognize correct answers when it sees them. How is this done? Problems in these subjects tend to have stepwise solutions. We check each step as it is proposed and certify or reject it, based on whether it follows from steps already in the proof. In this way a connected, valid segment of reasoning is built up step by step. Once a natural and convenient granularity of such steps is found, students are free to find any correct solution to a problem. The design cost of this freedom is that essentially the entire logical and mathematical underpinnings of the subject have to be present in the instructional system, and there must be a representation of arguments or derivations as connected wholes, not just as isolated calculations.

A second approach, often called "intelligent tutoring," is to make the machine smart enough so that it can solve the problem itself and compare its internal solution to the student's. Taken together, such approaches are often called intelligent computer-assisted instruction (ICAI). (Wenger[13] provides a review of the field and Kearsley[14] and Sleeman and Brown[15] provide recent compendia of essential ICAI papers.) The trouble is that for the types of courses being considered here, ICAI approaches do not, as a practical matter, now accommodate the variety of solutions that students actually produce, and there are no well worked out logical or mathematical theories of how to construct the relevant set of correct proofs for hard problems in serious subject matters like calculus. Whereas the interactive reasoning approach gives the students maximal freedom to construct correct solutions, intelligent tutoring tends to converge students on solutions the machine is able to generate, which, given the state of the art, is not a significant set in any but the most elementary logical theories that have constructive decision procedures.

A third ICAI approach, which can be called "diagnostic tutoring," would be to have the computer figure out what students are up to when they need help. Anyone who has ever tried such an approach for even a simple subject, will tell you that in the 1980s such a dream is premature. Data from students using the *Excheck* set-theory program show that detection of a strategic or tactical pattern from initial segments of a proof is not possible because random choices occur in the construction of solutions to hard problems by nonexpert students.[16]

We are facing these issues as we try to build a system to teach the first year of calculus.[17] And our approach is interactive reasoning: we will build a system in which consistency of mathematical inference can be maintained, but that will not be able to solve complex calculus problems on its own or infer what a student might be doing strategically. Interactive reasoning works because of an explicit partnership between the student and the computer. On the other hand, calculus (unlike logic or set theory, which are more austere subjects) supports the use of handy subroutines such as finding critical points, finding zeroes, taking a limit at a point, taking a derivative, and establishing continuity that fit together in standard ways in solutions to more complex problems. Such building blocks, if they are identified and used as such by students, could map more local structure onto diverse student solutions, possibly allowing a cautious step in the direction of diagnostic tutoring.

Adjudicating Design Issues in Complex CAI Systems

I have explained some of the factors that need to be taken into account to design CAI systems for individualized instruction in logic or calculus, and by obvious analogies, physics or elementary real analysis. I want to conclude by indicating some practical difficulties in making good design decisions for these very complicated research and development efforts.

In designing advanced CAI systems, we are constrained by the possible. In addressing individualization issues such as those described above, we have two complementary ways to try to increase individualization:

1. By maximizing the capacity for students to articulate and construct connected chains of correct reasoning within the particular theory or subject.

2. By maximally understanding student behavior.

The former requires massive efforts in subject matter refinement and implementation. The latter requires precise, implementable theories of cognition and subject-matter–specific theories of individual differences in learning. The ICAI program would place its bets on the ability to construct acceptable student models and individualize instruction by reference to the characteristics of such models. The approach we are taking at Stanford is more conservative. We place our bets on the ability to do an acceptable implementation of the subject matter we are trying to teach. To teach calculus, implement calculus. To teach kinematics, implement kinematics. To teach logic, implement logic. This gives students freedom to learn the theory and its tools by freely operating within an honest and (insofar as possible) complete theoretical framework. Both approaches are formidable; both ought to be tried, but not at once. Our judgment is that by focusing on subject matter rather than cognition as the core of the system and interactive reasoning as a method, we will be able to attain acceptable levels of effectiveness for a computer-based instructional system for the whole of elementary calculus.

In the first part of this paper I discussed one of Clark's skeptical arguments about CAI evaluation at great length. I am now at a point where if his lessons made sense, I could apply them as follows: It is an enormously complicated problem to state the criteria for valid inference in a subject like physics or calculus where the notation used does not explicitly or uniformly represent the entire semantic content of the statements it is used to make. Nobody really knows how to do this in 1988. However, it is not too hard to find people who can recognize correct proofs in calculus and reliably grade them. Grading the AP calculus test brings hundreds of them to New Jersey each year. Since the problem of representing this knowledge is a consequence of choosing computers as a medium, using Clark's reasoning, I might more profitably focus my attention on an instructional method for calculus, leaving the media details aside, since in a properly categorized world, computers cannot make a difference.

Research on the *VALID* and *EXCHECK* systems was partially supported by National Science Foundation Grants EPP-74-15016-A01, SED-74-15016-A03 and SED-77-096998 and by the Fund for Improvement of Postsecondary Education Grant G00-780-3800. Research on the calculus instructional system is supported by NSF Grants MDR-85-50596 and MDR-87-51523.

1. D. Jameison, P. Suppes, and S. Welles, "The Effectiveness of Alternative Instructional Media: A Survey," Review of Educational Research **44**, 1 (1974).
2. D. Alderman, "Evaluation of the TICCIT Computer-Assisted Instructional System in the Community College," ETS PR 78-10 (Princeton, NJ: Educational Testing Service, 1978).
3. J. Kulik, R. Bangert, and G. Williams, "Effects of Computer-Based Education on Secondary School Students," Journal of Educational Psychology **75**, 19 (1983); C. Kulik, J. Kulik, and R. Bangert-Drowns, "Effectiveness of Computer-Based College Teaching: A Meta-Analysis of Findings," Review of Educational Research **50**, 525 (1984); R. Niemiec and H. Walberg, "Comparative Effects of Computer-Assisted Instruction: A Synthesis of Reviews," Journal of Educational Computing Research **31**, 19 (1987).
4. R. E. Clark, "Reconsidering Research on Learning from Media," Review of Educational Research **53**, 445 (1983); R. E. Clark, "A Reply to Petkovich and Tennyson," Education Communication and Technology Journal **32**, 238 (1984); R. E. Clark, "Evidence for Confounding in Computer-Based Instruction Studies: Analyzing the Meta-Analyses," Education Communication and Technology Journal **33**, 249 (1985).
5. R. E. Clark, "Reconsidering Research or Learning from Media."
6. M. Petkovich and R. Tennyson, "Clark's 'Learning from Media': a Critique," Education Communication and Technology Journal **32**, 233 (1984).
7. R. E. Clark, "Evidence for Confounding in Computer-Based Instruction Studies."
8. P. Suppes and M. Morningstar, "Computer-Assisted Instruction," Science, **166**, 343 (1969).
9. T. Ager, "Computation in the Philosophy Curriculum," Computers and the Humanities **18**, 145-156 (1984); P. Suppes and J. Sheehan, "CAI Course in Axiomatic Set Theory," *University-Level Computer-Assisted Instruction at Stanford: 1968-1980*, ed. P. Suppes, pp. 3-80 (Stanford, CA: Stanford University, Institute for Mathematical Studies in the Social Sciences, 1981).
10. R. E. Clark, "Reconsidering Research on Learning from Media," Review of Educational Research **53**, 451 (1983).

11. R. E. Clark, "Evidence for confounding in computer-based instruction studies."

12. P. Suppes and J. Sheehan, "CAI Course in Logic," ed. P. Suppes, *University-level computer-assisted instruction at Stanford: 1968-1980*, pp. 193-226.

13. E. Wenger, "Artificial Intelligence and Tutoring Systems," (Los Altos, CA: Morgan Kaufmann Publishers, Inc., 1987). T. Ager, "From Interactive Instruction to Interactive Testing" (in press).

14. G. P. Kearsley, ed. *Artificial Intelligence and Instruction* (Reading, MA: Addison Wesley Publishing Co. 1987).

15. D. Sleeman and J.S. Brown, ed., *Intelligent Tutoring Systems*, (New York: Academic Press, 1982).

16. P. Suppes and J. Sheehan, "CAI course in axiometric set theory."

17. P. Suppes, T. Ager, P. Berg, R. Chauqui, W. Graham, R. Maas, and S. Takahashi, "Applications of Computer Technology to Precollege Calculus: First Annual Report," Tech. Rep. 310, Psych. and Educ. Series. (Stanford, CA: Institute for Mathematical Studies in the Social Sciences, 1987); P. Suppes, Ed. *University-Level Computer-Assisted Instruction at Stanford: 1968–1980* (Stanford, CA: Stanford University, Institute for Mathematical Studies in the Social Sciences, 1981).

The Effects of Microcomputer-Based Laboratories on Exploration of the Reflection and Refraction of Light

John C. Park
Department of Mathematics and Science Education, North Carolina State University, Raleigh, NC 27695-7801

New technology has contributed to a new strategy for teaching concepts through laboratory science. This strategy is the use of the microcomputer-based laboratory. MBL applications have removed much of the drudgery that is often part of laboratory investigation, allowing students to focus more clearly on the scientific phenomena under investigation and to perform original investigations.

Four questions emerge from the literature on the effect of microcomputer-based laboratories on scientific exploration. (1) Do students who interact with MBL equipment and curriculum generate more experimental trials than without such materials? (2) Does composition of the lab groups by gender influence the number of unique trials? (3) What is the relationship between general achievement in science and the number of experimental trials? (4) What is the relationship between the number of unique trials and the level of understanding of the concept?

We are conducting a study of the effects of the microcomputer-based laboratory on the scientific exploration of the reflection and refraction of light by seventh-grade students. The study consists of a 2 by 2 by 3 factorial design with laboratory mode, standardized achievement score, and lab pairings as the independent variables. The dependent variables are the number of unique experimental trials and concept achievement score. Concept achievement score is defined as the adjusted post-test achievement score on the reflection and refraction of light with the pretest as the covariate.

Students from three middle schools were involved in the study. The children in each class were stratified and paired by achievement on the science subtest of the California Achievement Test (CAT). These pairings consisted of three categories: girl-girl pairing, boy-boy pairing, and girl-boy pairing. The lab pairings were randomly assigned to one of two laboratory modes stratified by pairing category and by standardized achievement score.

Half the students participated in a laboratory activity that used an Apple IIe microcomputer to monitor the angles of incidence, reflection, and refraction of an infrared light source. The other half used a He-Ne laser for the light source and measured the angles of incidence, reflection, and refraction. Both groups measured the angles as the light interacted with a rotated jar half-filled with water.

Students in each laboratory mode recorded the data on prepared data sheets. The students were asked to determine the relationship among angles of incidence, reflection, and refraction. Observers recorded experimenting time and number of trials. All students were briefly interviewed at the end of the exploration session.

A post-test on reflection and refraction was given to the participants three days after the conclusion of the laboratory activity. We are using a three-way analysis of variance to analyze the data. We will present the analysis of the data, which was collected in March and April 1988, during this session.

Kinesthetic Experience and Computers in Introductory Laboratories

Gordon J. Aubrecht II
Department of Physics, Ohio State University, Columbus, OH 43210-01106 and Marion, OH 43302-5695

In the last decade many papers have been published on the way students learn.[1] Many students learn best by doing.[2] When one learns to ride a bicycle, drive a stick-shift car, ice skate, or ski, at first mastery seems impossible. A single successful trial, however, produces an immediate physiological recognition of success, and progress thereafter is rapid. Such experience is called *kinesthetic*.

Students have informal remembrances of past real-life experiences, and sometimes physics teachers will try to translate between these experiences and the textbook. In lecture demonstrations teachers supply "canned" vicarious kinesthetic experiences. In physics laboratories teachers supply direct kinesthetic physics experiences. Of these three methods, only physics laboratories provide the *active* participation, either in problem solving or in understanding theoretical concepts, that most reliably leads to student intuition.

It is relatively easy to construct kinesthetic laboratory experiences involving force. Other concepts are more difficult, though they may seem no less fundamental to physicists. The computer makes it easier for students to experience such concepts kinesthetically.

The ideas of displacement and velocity, for example, are often difficult for students to grasp. Several versions of laboratory apparatus using sonic range finders to measure distance (displacement), speed (velocity), and acceleration and to display these data on the screen of a microcomputer in real time are now available. Such apparatus allows students to learn about graphing and gives them the sort of kinesthetic experience of motion they need to really understand it.[5]

Such computer-assisted laboratory experiences are becoming freely available. The enriched laboratory experience they provide allows students to focus on their interaction with nature.

Physics teachers often regard the laboratory as a place where students come to understand physical concepts and the limitations of the measurement process. It is true that the apparent accuracy of interfaced computer measurements will subvert this aim. But the laboratory is more than a place for measuring quantities. The use of computer tools should be welcomed because they allow students to *feel* their measurements during the measurement process. This sort of kinesthetic "instant acculturation" in the labs frees the lecturer from ineffective "show and tell" and allows more useful employment of the lecture hour.

1. The way students learn: A. Arons, *Development of Concepts in Physics* (Reading, MA: Addison-Wesley, 1965); A. Arons, "Phenomenology and Logical Reasoning," in *Proceedings of the International Conference on Education for Physics Teaching*, edited by P. J. Kennedy and A. Loria (Edinburgh, Scotland: International Commission on Physics Education, 1980); A. Arons, "Student Patterns of Thinking and Learning (I, II, & III)," Phys. Teach. **21**, 576 (1983) and Phys. Teach. **22**, 21 and 88 (1984); F. Reif and M. St. John, "Teaching Physicists' Thinking Skills in the Laboratory," Am. J. Phys. **47**, 950 (1979); P. J. Peters, "Even Honors Students have Conceptual Difficulties with Physics," Am. J. Phys. **50**, 501 (1982); J. A. Rowell and C. J. Dawson, "Laboratory Counterexamples and the Growth of Understanding in Science," Eur. J. Sci. Ed. **5**, 203 (1983); J. Solomon, "Learning about Energy: How Pupils Think in Two Domains," Eur. J. Sci. Ed. **5**, 49 (1983); E. Guesne, "Children's Ideas about Light," in *New Trends in Physics Teaching*, edited by J. Solomon (Paris: UNESCO, 1984), pp. 179–92.
2. J. Spears and D. Zollman, "The Influence of Structured versus Unstructured Laboratory on Students' Understanding the Process of Science," J. Res. Sci. Teach. **14**, 33 (1977); R. F. Tinker and G. A. Stringer, "Microcomputers! Applications to Physics Teaching," Phys. Teach. **16**, 436 (1978); R. F. Tinker, "Microcomputers in the Teaching Lab," Phys. Teach. **19**, 94 (1981); A. Hofstein and V. N. Lunetta, "The Role of the Laboratory in Science Teaching: Neglected Aspects of Research," Rev. Ed. Res. **52**, 201 (1982); M. L. de Jong and J. W. Layman, "Using the Apple II as a Laboratory Instrument," Phys. Teach. **22**, 291 (1984); H. Brassel, "Effectiveness of a Microcomputer-Based Laboratory in Learning Distance and Velocity Graphs," Ph.D. dissertation, University of Florida (1987).
3. S. Tobias, "Peer Perspectives on the Teaching of Science," Change **18**(2), 36 (1986), "Peer Perspectives on Physics," Phys. Teach. **26**, 77 (1988).

4. Y. Kakiuchi, "Research on the Process of Learning and its Application: A Case Study," in *Trends in Physics Education*, edited by T. Ryu (Tokyo: KTK Scientific Publishers, 1986).

5. R. Thornton, "Access to College Science: Microcomputer-Based Laboratories for the Naive Science Learner," Col. Microcomputer **5**, 1 (1987); R. Thornton, "Tools for Scientific Thinking: Microcomputer-Based Laboratories for Physics Teaching," Phys. Ed. **22**, 230 (1987).

Students' Construction of Concept Maps Using Learning Tool

Robert B. Kozma
NCRIPTAL, University of Michigan, Ann Arbor, MI 48109

Computer-based tools are software programs that use the capabilities of the computer to amplify, extend, or enhance human cognition. Their impact has not been empirically established, but these software packages are designed to provide an external representation of internal cognitive processes. By acting as a mirror of the learner's thought process, the computer may not only facilitate the learning of the particular domain to which it is applied, but more important, it may aid the development of general learning skills and strategies.

To be effective, a tool for learning must closely parallel the learning process and address both the limitations and capabilities of human cognition. The aspects of the learning process that most inhibit learning are limitations on short-term memory, difficulty in retrieving needed information from long-term memory, and the ineffective or inefficient use of cognitive strategies to obtain, manipulate, and restructure information. Important capabilities that students bring to the learning situation are previously learned knowledge, concepts, and skills.

To compensate for limitations and build on strengths, a computer-based tool should do one or more of the following: supplement limited short-term memory by making large amounts of information immediately available for the learner's use; make relevant, previously learned information available simultaneously with new information; prompt the learner to structure, integrate, and interconnect new ideas with previous ones; provide for self-testing and practice, thus increasing the retrievability of information; provide for the easy consolidation and restructuring of information as the student's knowledge base grows.

Learning Tool is a Macintosh program that uses principles of cognitive psychology to help students learn any subject, from philosophy to physics. There is no specific information in *Learning Tool*; rather, students learn by entering, organizing, and using information.

Learning Tool operates at three coordinated levels. The "Master List" is an outliner; it allows the student to enter and order key concepts, such as "atom," "electron," and "nucleus." Each entry automatically creates a labeled "note card" icon at the second level, the "Concept Map." In the "Concept Map," the students can spatially organize and link the note cards to display user-defined relationships in hypertext fashion. Thus, for example, the user can graphically show the structure of the atom by linking the above terms in "made of" relationships. Note cards can also be stacked to create submaps, so "neutron," "proton," and "quark" can be embedded in the "nucleus" note card. They will automatically be indented under this term in the "Master List." At the third level, the student can enter information for each note card. Both text and graphic information can be entered.

Among the other tools designed to facilitate the students' use of their notes are multiple-term, Boolian search function (which, for example, permits the student to look for all the cards that contain information on "electromagnetism" and "gravity"); and the capability to create and take self-tests to practice the retrieval of factual details and conceptual relationships.

Computational Physics and Spreadsheets

Teaching Computational Physics

Steven E. Koonin

Department of Physics, California Institute of Technology, Pasadena, CA 91125

Modern physics research is concerned increasingly with complex systems composed of many interacting components: the atoms in a solid, the stars in a galaxy, or the values of a field in space-time that describes an elementary particle. In most such cases, previously unknown phenomena can arise solely from the complexity of the system. Although we might know the general laws that govern interactions between the components, it is difficult to predict or to understand the new phenomena qualitatively. General insights in this regard are difficult to envision, and we quickly reach the limits of the analytical pencil-and-paper approach that has served physics so well in the past. Because numerical simulations are essential to further understanding, computers and computing play a central role in much of modern physics research.

Using a computer to model physical systems is, at its best, more art than science. Through a mix of numerical analysis, analytical modeling, and programming, the computational physicist exploits the power of the computer to solve otherwise intractable problems. Computational physics is a skill that can be acquired and refined: knowing how to set up the simulation, what numerical methods to employ, how to implement them efficiently, and when to trust the accuracy of the results.

A Neglected Discipline

Despite its importance, computational physics has largely been neglected in the standard university curriculum. In part this is because it requires balanced integration of three commonly disjoint disciplines: physics, numerical analysis, and computer programming. Another factor is the lack of computing hardware suitable for teaching. Students usually acquire what skills they have by working on a specific thesis problem; as a result, their exposure is often far from complete.

This situation and my professional background in large-scale numerical simulations motivated me to begin teaching an advanced computational physics laboratory course at Caltech in the winter of 1983. My goal was to provide students with direct experience in modeling nontrivial physical systems and to impart to them the minimal set of techniques for dealing with the most common problems encountered in such work. The computer was to be viewed neither as a "black box" nor as an end in itself, but rather as a tool for getting at the physics.

A factor motivating the decision to develop a computational physics curriculum at this time, was the ready availability of hardware that could provide each student with an individual computing environment. Personal computers can be used easily and interactively through a variety of high-level languages, and they offer numerical power sufficient for illustrating many research-level calculations. Moreover, the graphics facilities commonly found on such systems allow an easy but often startling insight into many problems. In short, I (and my students) could concentrate on the strategy of a calculation and the analysis of its results rather than on the mechanics of using a computer.

A New Course

How, then, was I to teach the art of computational physics? It was apparent that the traditional lecture-*cum*-assignment approach was not optimal, as there is no teacher better than direct experience, and computing is a very personal activity for most physicists. I therefore planned a curriculum in which the computer would teach by example and exercise. This format had the added benefit that students could work largely on their own, at their own pace, and at times of their own choosing.

It was relatively easy to identify the broadly applicable numerical methods that should be covered in the course, but choosing the physical situations in which to demonstrate those methods was more difficult. I attempted to satisfy simultaneously the following criteria:

1. That the physics discussed be an "interesting" extension or enrichment of the usual quantum, statistical, or classical mechanics material.

2. That the scale of the computation be appropriate to the numerical power of the hardware.

3. That the problem not be soluble analytically.

In the end, I found 16 such "case studies." In more than half of them, the student compares calculated results with experiment or observations.

Developed during the 1983–1984 academic year, the curriculum consists of eight units. The student begins each unit by reading a text that provides a heuristic discussion of several related techniques for accomplishing a particular numerical task. Intuitive derivations of simple, general methods are emphasized, with appropriate references cited for rigorous proofs or more specialized techniques. Short, mathematically oriented exercises involving only a small amount of programming reinforce this material.

After reading the text section, the student works through the remaining two sections of the unit: an example and a project. Each is a brief exposition of a particular physical situation and how it is to be modeled using the numerical techniques taught in that unit, together with a set of exercises for exploiting and understanding the associated program. The example and the project differ only in that the students are expected to use a "canned" program for the former and perhaps write their own programs for the latter. The units and their accompanying examples and projects are listed in Table 1.

Table 1.

Computational Physics Curriculum

Unit	Numerical Methods	Example	Project
1	Differentiation, quadrature, finding roots	Semiclassical quantization of molecular vibrations	Scattering by a central potential
2	Ordinary differential equations	Order and chaos in two-dimensinal motion	Structure of white-dwarf stars
3	Boundary value and eigenvalue problems	Stationary solutions of the one-dimensional Schrödinger equation	Atomic structure in the Hartree-Fock approximation
4	Special functions and Gaussian quadratures	Born and Eikonal approximations to quantum scattering	Partial wave solution of quantum scattering
5	Matrix inversion and diagonalization	Determinining nuclear charge densities	A schematic shell model
6	Elliptic partial differential equations	Laplace's equation in two dimensions	Steady-state hydrodynamics in two dimensions
7	Parabolic partial differential equations	The time-dependent Schrödinger equation	Self-organization in chemical reactions
8	Monte Carlo methods	The Ising model in two dimensions	Quantum Monte Carlo simulation of the hydrogen molecule

Each of the programs I developed functions in several capacities: as an easy-to-use demonstration of the physics; as a laboratory for exploring changes in the numerical algorithms or parameters; and, in the projects, as a model for the student's own program. Thus the programs not only had to work correctly, they had to be easily readable and understandable. Simple organization, full documentation, and a structured programming style were essential. Because much of each program involves input/output (I/O) and "bookkeeping," the few important numerical sections had to be called out clearly. "Elegance" in coding and speed of execution often had to be sacrificed for the sake of intelligibility. More significant, my ambitions were frequently restrained by a desire to keep the calculations and the graphics displays "simple." With some thought and care, however, I was able to work to my satisfaction within these constraints.

The choice of language invariably invokes strong feelings among scientists who use computers. Any language is, after all, only a means of expressing the concepts that underlie a program. The important ideas in the curriculum would remain relevant no matter what language was used. FORTRAN is the computer language used most widely for scientific computation. However, its I/O is notoriously cumbersome. More importantly, FORTRAN programs must go through the time-consuming step of compilation before they can be run. This removes the rapid feedback that is so desirable when a student is developing a program.

I finally settled on BASIC in spite of its acknowledged deficiencies, such as lack of local subroutine variables and its awkwardness in expressing structured code. Those deficiencies, I felt, were more than offset by the simplicity and widespread use of BASIC, its ready availability on the microcomputers we were using, its powerful graphics and I/O statements, the existence of interpreters that are convenient for writing and debugging BASIC programs, and the existence of compilers for producing rapidly executing finished programs. Virtually all of the students are familiar with some other high-level language and so can learn BASIC "on the fly" while taking the course. Nevertheless, several students have elected to write their projects in other languages and have had no problems in doing so.

I teach the course in a laboratory format for junior and senior physics majors. All the students have taken (or are taking) the conventional courses in classical, statistical, and quantum mechanics, so they are familiar with many of the physics concepts involved. Moreover, there is enough of a "computer culture" among Caltech undergraduates so that most of them are familiar with the hardware before beginning the course; those who are not, become proficient after a few hours of individual instruction. As mentioned above, students work through the material largely on their own, although a teaching assistant and I hold weekly office hours during which we are available for help and consultation. Individual half-hour interviews upon completion of each unit serve to monitor the students' progress and assess their understanding. The typical student, working six or seven hours a week, can complete three or four units in a ten-week term, perhaps writing his or her own code for two of the projects and using my code for the others.

Brief discussion of some of the examples and projects will serve to give a feeling for the level of the material and the style in which it is presented to the students. The complete curriculum, together with a diskette containing the BASIC source code for the examples and projects, is published in my text *Computational Physics*.

Example 7, the time-dependent Schrödinger equation, illustrates techniques for solving parabolic partial differential equations and also provides the student with a laboratory for exploring the evolution of wave packets in various potential fields (the function specifying the forces acting on the particle), a much discussed (but difficult to illustrate) situation in elementary quantum mechanics. The basic problem is to solve the Schrödinger equation,

$$i\,\hbar\,\frac{\partial\psi}{\partial t} = -\frac{\hbar^2}{2m}\frac{\partial^2\psi}{\partial x^2} + V(x)\psi\,,$$

for the complex wave function $\psi(x,t)$, which describes a quantum particle (an electron, for example) of mass m moving in a potential field $V(x)$. Here, x is the particle's location, t is the time, i is the unit imaginary number, and is Planck's constant divided by 2π. Given the initial wave packet, $\psi(x,t=0)$, the problem is to find ψ for all subsequent times.

This general problem cannot be solved by analytical methods, so numerical methods are essential. The most efficient approach is to describe the space and time coordinates with small but finite "steps" and thereby replace the continuous partial differential equation by a large number of algebraic equations that can be solved on the computer.

The program I wrote for this situation allows the student to specify the potential field by "drawing" it with the computer's cursor. The initial wave packet is then specified, and the evolution begins. The results are displayed as a "movie" of $|\psi(x,t)|^2$ as shown in Figure 1. This quantity, the squared modulus of the complex wave function, gives the probability that the particle can be found at position x at time t. After the student stops the evolution, the potential or the initial wave packet can be changed for another run.

Students working with this program check the accuracy of the numerical methods used, to program and explore several alternative evolution algorithms, and investigate various potentials and wave packets. The third activity demonstrates vividly such intrinsically quantal phenomena as tunneling and resonance—concepts difficult to convey in a conventional lecture.

Project 6 illustrates techniques for solving elliptic partial differential equations by considering the steady-state (time-independent) flow of a viscous fluid about an obstacle—for example, the flow of a stream around a rock. The mathematical description of the flow is based on continuity (fluid is neither created nor destroyed) and the response of each bit of fluid to the pressure and viscous forces acting on it. When the flow is two-dimensional (coordinates x and y), these two physical principles can be embodied in the coupled nonlinear elliptic equations

$$\left[\frac{\partial^2}{\partial x^2} + \frac{\partial^2}{\partial y^2}\right] \psi(x,y) = \zeta(x,y), \text{ and}$$

$$\nu \left[\frac{\partial^2}{\partial x^2} + \frac{\partial^2}{\partial y^2}\right] \zeta(x,y) = \left[\frac{\partial \psi}{\partial y}\frac{\partial \zeta}{\partial x} - \frac{\partial \psi}{\partial x}\frac{\partial \zeta}{\partial y}\right].$$

Here, ψ is the stream function that specifies the direction of flow at each point, ζ is the vorticity of the fluid, and ν is the kinematic viscosity. Specification of the problem is completed by imposing boundary conditions on ψ and ζ (for example, the fluid cannot flow into the surfaces of the object).

Like the Schrödinger equation, these equations cannot be solved analytically but are amenable to numerical treatment using small, discrete steps. The resulting large number of nonlinear algebraic equations can be solved by a relaxation pro-

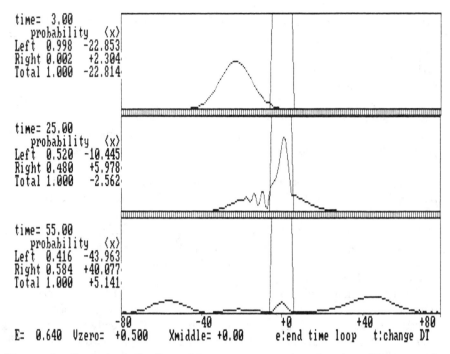

Figure 1. Evolution of a Gaussian quantum wave packet describing a particle approaching a rectangular potential barrier. The probability
density lψ(x,t)l^2 is plotted as a function of the particle's location at
several different times; the barrier is indicated by the vertical lines
in the center of each frame. The mean energy of the packet is
slightly greater than the height of the barrier so part of the packet
is transmitted and part reflected. The top frame shows the packet
approaching the barrier. In the middle frame it penetrates the barrier, the rapid oscillations in the probability density caused by
interference between incident and reflected waves. Somewhat
broadened transmitted and reflected packets are seen at a later
time in the lower frame, together with a small, slowly decaying
part of the wave function trapped within the barrier.

cess in which initial guesses for the stream function and vorticity are refined iteratively.

After deriving the equations and briefly discussing them, the students are guided through a series of steps culminating in writing a program for solving the flow
about a rectangular plate at various speeds. Results can be displayed as character
plots, as shown in Figure 2, and the drag and pressure forces acting on the plate
can be calculated from them. At low velocities, the flow is smooth and easy to

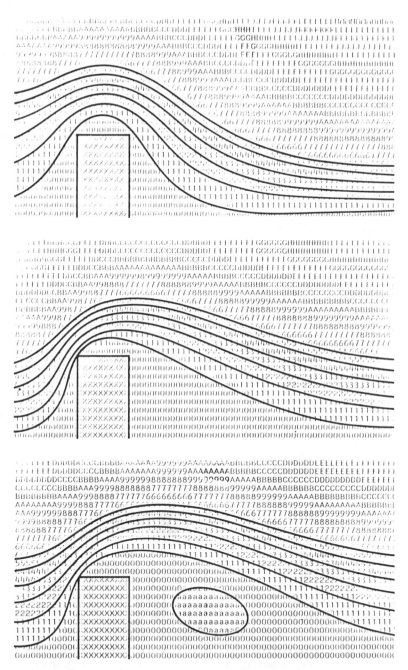

Figure 2. Streamlines for a viscous fluid flowing around a rectangular plate
at various velocities. The direction of flow is from the left, and
each pattern is reflection-symmetric about the lower edge. Note
particularly the hydraulic "jump" above the plate, the laminar flow
at low velocity (top pattern), the increasing separation of the flow
from the rear of the plate at higher velocities (middle pattern), and
the vortex behind the plate at the highest velocity (bottom pattern).

understand. As the velocity increases, the fluid takes a "jump" as it passes over the plate, a phenomenon familiar to anyone who has rafted over river rapids. At the highest velocities, the flow separates from the back face of the plate and a vortex (eddy) forms behind it. Here again, the computer is used to simulate situations for which analytical solutions are impossible and for which intuition is difficult to develop.

Example 5 illustrates the use of a computer in a different way—the analysis of experimental data by least-squares fitting. The physical situation here is the scatter-

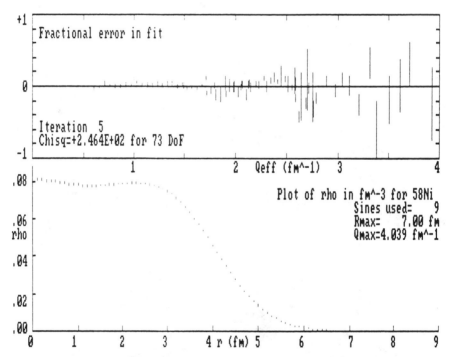

Figure 3. Least-squares fit to the elastic scattering of electrons from the nucleus ^{58}Ni (nickel of 58 protons and neutrons). The upper panel shows the fractional error of the fit to the experimental cross section at each momentum transfer (scattering angle). The vertical lines indicate the uncertainties in each of the experimental measurements; most of the points are fitted within their uncertainty. The lower panel shows the corresponding charge density of the nucleus deduced as a function of distance from the nuclear center, r, measured in femtometers (fm, or 10^{-15} meters). The small oscillations in the interior density, at left in the plot, are a "shell effect" caused by the particular motion of a few nucleons within the nucleus.

ing of high-energy electrons (approximately 200 million electron volts or more) from atomic nuclei, currently a topic of high current research interest in nuclear physics. Because the electrons interact with the nucleus through the Coulomb force, the cross sections for such scattering are sensitive to the distribution of electrical charge (that is, the structure) of the nucleus. Indeed, the measured cross sections can be "inverted" in a model-independent way to obtain the nuclear charge density.

My program for this problem analyzes actual experimental data to infer the charge densities for nuclei of calcium, nickel, and lead. An iterative, nonlinear least-squares fit procedure is used to adjust the charge density to the measured cross sections. The density and fit to the data are displayed as the iterations proceed; typical results are shown in Figure 3. In running and understanding this program, the student checks the accuracy of the fits obtained, extracts information about the nuclei from them and compare with simple nuclear models, and explores alternative fitting strategies.

Conclusion

While developing and teaching this curriculum, I have been impressed by several unexpected advantages. To write a program for simulating a given physical situation, the student must understand the physics in way that is different from (and complementary to) that required for an analytical approach. Then too computer programs bring a flexibility and vividness of presentation that is difficult to obtain otherwise. Moreover, the simulating of systems brings a sense of exploration and surprise to the learning process, since an understanding of the results of changing parameters or algorithms often leads to greater insights. Finally, because complex situations can be presented, students are exposed in detail to research-level problems at an earlier stage of their education. In these ways, I expect computer-based education in physics to supplement, rather than supplant, the traditional mode of lecture instruction.

1. S. E. Koonin, *Computational Physics*, (Menlo Park, CA: Benjamin/Cummings, 1985).

Mathematics by Computer

Stephen Wolfram
Center for Complex Systems Research, Department of Physics, University of Illinois, Urbana, IL 61801

I have been involved in a practical problem that is going to have important consequences for the future of physics. This practical problem is the development of a computer program called *Mathematica*. *Mathematica* is a comprehensive system for doing mathematical computation. It runs on a variety of computers and does a whole range of mathematics. *Mathematica* is useful for many kinds of users, ranging from math professors or physics professors to engineers and down to college and even high school students. Our idea in developing *Mathematica* is to produce a mathematical tool that is as generic as the calculator but provides much broader coverage of mathematics.

The most elementary way to think about *Mathematica* is as an enhanced calculator—a calculator that does not only numerical computation but also algebraic computation and graphics. *Mathematica* can function much like a standard calculator: you type in a question, you get back an answer. But *Mathematica* goes further than an ordinary calculator. You can type in questions that require answers that are longer than a calculator can handle. For example, *Mathematica* can give you the numerical value of π to a hundred decimal places, or the exact result for a numerical calculation as complicated as the result of 3^{1000}, (see Fig. 1).

Mathematica has a big collection of mathematical functions built into it. Even a standard calculator has sines, cosines, and logarithms built in, but *Mathematica* allows you to ask the value of a function like a Bessel function $J_0(10.5)$ or the Reimann zeta function $\zeta(1/2 + 13i)$ to 40 decimal places, (see Fig. 2). Such special

In[10]:=
  N[Pi,100]

Out[10]=
  3.1415926535897932384626433832795028841971693993751058209749445\

  9230781640628620899862803482534211 7068

In[11]:=
  N[3^1000,100]

Out[11]=
  1.3220708194808066368904552597521443659654220327521481676649203\

                                                                477
  6822682859734670489954077831385060806196 39 10

Figure 1.   Mathematica as a calculator. π and 3^{1000} to 100 significant digits.

```
In[12]:=
  BesselJ[0,10.5]
Out[12]=
  -0.236648
In[13]:=

  N[Zeta[1/2 + 13I],40]
Out[13]=
  0.4430047825053681891978974413328849126 -

    0.6554830983211689430513696491913355506 I
```

Figure 2. Numerical results for the Bessel function $J_0(10.5)$ and the Riemann zeta function $\zeta(1/2 + 13i)$.

functions in mathematical physics are built into *Mathematica*. In fact, *Mathematica* has built in virtually all of the functions found in standard handbooks of mathematical functions. Our goal was to make obsolete all the tables of mathematical functions that you find filling up a lot of space in physics libraries.

But *Mathematica* does much more than numerical computation. Another major kind of computation handled by *Mathematica* is symbolic algebraic computation. If we type in the algebraic expression $(1 + x)^5$, at first *Mathematica* just echoes back the same expression, (see Fig. 3). But we can tell *Mathematica* to perform various mathematical operations on the expression. For example, we can tell it to expand the expression out, or to factor it. *Mathematica* will recognize that the expression of that standard form can be reduced back to the original expression. And *Mathematica* can do much more complicated operations than that. It will remember that an expression that goes on for several screens actually simplifies back down to the original product of a few terms.

As an example of an algebraic computation, if you try to solve a simple quadratic equation like

$$x^2 + 2x - 1 = 0$$

```
In[24]:=
  Expand[ (1+x)^5]
Out[24]=
                  2        3       4     5
    1 + 5 x + 10 x  + 10 x  + 5 x  + x
In[25]:=
  Factor[%]
Out[25]=
           5
    (1 + x)
```

Figure 3. Algebraic expansion and factoring of $(1 + x)^5$.

```
In[34]:=
   x^2 + 2x - 1 ==0

Out[34]=
                    2
   -1 + 2 x + x   == 0

In[35]:=
   Solve[%,x]

Out[35]=
           -2 + 2 Sqrt[2]              -2 - 2 Sqrt[2]
   {{x -> --------------}, {x -> --------------}}
                 2                          2
```

Figure 4. Algebraic solution of $x^2 + 2x - 1 = 0$.

for x, you will get a symbolic result of the solution of this equation in terms of the symbolic parameter x, (see Fig. 4). If you ask *Mathematica* to solve an equation like

$$x^5 + 3x + 1 = 0$$

for which there is no analytical solution, *Mathematica* will give you the symbolic representation of the results that cannot be found. In addition, *Mathematica* can give you the numerical result. Even though it may not be mathematically possible to obtain a result in algebraic form, *Mathematica* will give you a numerical result in terms of complex numbers, (see Fig. 5).

```
In[31]:=
   x^5 + 3x + 1 == 0

Out[31]=
                 5
   1 + 3 x + x   == 0

In[32]:=
   Solve[ %,x]

Out[32]=
                            5
   {ToRules[Roots[3 x + x   == -1, x]]}

In[33]:=
   N[%]

Out[33]=
   {{x -> -0.839072 - 0.943852 I}, {x -> -0.839072 + 0.943852 I},

    {x -> -0.331989}, {x -> 1.00507 - 0.937259 I},

    {x -> 1.00507 + 0.937259 I}}
```

Figure 5. Numerical solution of $x^5 + 3x + 1 = 0$.

```
In[41]:=
   Integrate[x/(1-x^3),x]
Out[41]=
                 1 + 2 x
   -(Sqrt[3] ArcTan[-------])
                 Sqrt[3]      Log[1 - x]   Log[1 + x + x ]
   --------------------------  - ---------- + ----------------
                3                    3              6
```

```
In[42]:=
   D[%,x]
Out[42]=
       1                  2              1 + 2 x
   ---------  -  ------------------ + ---------------
   3 (1 - x)            2                    2
                   (1 + 2 x)         6 (1 + x + x )
            3 (1 + ----------)
                       3
```

```
In[43]:=
   Simplify[%]
Out[43]=
     x
   ------
       3
   1 - x
```

Figure 6. Analytical solution of $\int x(1-x^3)^{-1}\,dx$ and subsequent differentiation of the answer.

Another thing *Mathematica* can do is symbolic integration. Let's start off with a very simple one. If you ask *Mathematica* to integrate x^m with respect to x, i.e. you get the result

But *Mathematica* is capable of more complicated integrals. If you ask *Mathematica* to integrate

$$\int x^m dx = \frac{x^{m+1}}{m}$$

it will produce an expression involving arc tangents and some logarithms, and so

$$\int \frac{x}{1-x^3}\,dx,$$

on. We can check the result by asking *Mathematica* to differentiate. (See Figure 6).

In addition to numerical and algebraic calculations, *Mathematica* is capable of sophisticated graphics. *Mathematica* does graphics by using a "postscript page description language." This is the kind of graphics control language that's used, for example, by the Apple Laserwriter, and is being increasingly used on many different kinds of computers. Using this language, *Mathematica* allows you to take a picture and actually see the Postscript form of the picture. This Postscript form lies behind the original picture that *Mathematica* produced. Postscript is a much more obscure language than *Mathematica*, but if you know Postscript well enough, you can edit the Postscript that *Mathematica* actually produced. For example, when you originally plot a curve you can tell *Mathematica* to make it blue, but by editing the Postscript, you can change the color of the curve to red.

An advantage of Postscript is that it provides a very portable description of the graphics. You can, for example, take a graphical image that you produce in *Mathematica* and paste it into a document that you're making up with Pagemaker or any other desktop publishing system. The graphics that you produce will be rendered in the highest resolution that's available on the kind of printer that you have.

Mathematica will also allow you to do three-dimensional graphics. You can ask it to plot the sin (xy) as a function of x and y, with x running from 0 to 3 and y running from 0 to 3. The result is shown in Figure 7. If you want to, you can, for example, ask *Mathematica* to show you the plot as it would appear if you simulated shining a light onto the surface from one side, (see Figure 8). You can also alter the viewpoint, or display the surface with different parameters.

That's a capsule summary of *Mathematica's* capabilities in numerical computation, algebraic computation, and graphics. I would like to go on to describe how to program *Mathematica*, and how to create documents and things like live textbooks that make use of *Mathematica*.

Mathematica is a rather complete and powerful programming language that allows you to build on top of the large number—about 700—built-in functions. You can build in your own functions for a particular application using any of several different styles of programming. One such style of programming is writing a function in a standard structured programming fashion, as you would in a language like C or Pascal. It's somewhat easier to program in *Mathematica* than it would be to program directly in C or Pascal because *Mathematica* is an interactive system. *Mathematica* allows you to see exactly what your program does as soon as you've typed in pieces of it.

Because *Mathematica* is a symbolic system you don't have to worry about creating all the kinds of data structures that you might need. *Mathematica's* general-expression data structures will hold anything that you have.

Another style of programming is perhaps more interesting. The idea of this second style is to use transformation rules to take textbook formulas and transcribe them almost directly into *Mathematica*. This is the way that *Mathematica* gets taught lots of mathematical information: you just take tables of rules like those for logarithm functions and enter them into *Mathematica*. Afterward *Mathematica* will automatically use them.

Another style of programming in *Mathematica* is to add to *Mathematica's* knowledge about graphics. For example, to teach *Mathematica* about polyhedras, you begin by reading definitions of polyhedra into the file.

Let's see what happens if we want to show a tetrahedron as a three-dimensional graphical object. *Mathematica* takes the symbolic description of the graphics and actually renders it as a three-dimensional object, (see Figure 9).

In[5]:=

```
Plot3D[ Sin[x y],{x,0,3},{y,0,3}]
```

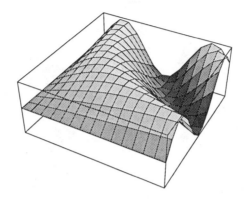

Out[5]=
  -SurfaceGraphics-

Figure 7.   A 3-dimensional plot of sin (*xy*).

In[6]:=
```
Show[%, Lighting->True]
```

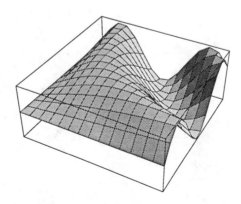

Out[6]=
  -SurfaceGraphics-

Figure 8.   A 3-dimensional plot of sin (*xy*) with simulated illumination.

In[50]:=
```
Show[Graphics3D[Tetrahedron[]]]
```

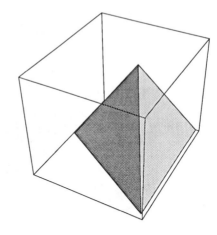

Out[50]=
```
-Graphics3D-
```

Figure 9. A 3-dimensional plot of a tetrahedron.

In[52]:=
```
Show[Graphics3D[Stellate[Dodecahedron[]]]]
```

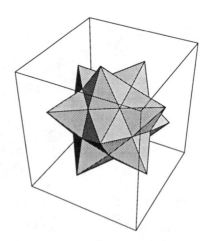

Out[52]=
```
-Graphics3D-
```

Figure 10. A 3-dimensional plot of a stellated dodecahedron.

If you ask *Mathematica* to render even more complicated graphical objects, it will follow the same procedure. Say, for example, that you ask *Mathematica* to show you a stellated dodecahedron. It will take the symbolic description of the dodecahedron, apply the operation of stellation, and then show the results, (see Figure 10). Because *Mathematica* uses Postscript graphics, you can expand the results to any size you want. In fact, you can even drag it out to eight feet wide, print it out, and glue it together to make a poster.

The *Mathematica* program is broken into two pieces. The first piece is a kernel that actually does computations. This kernel runs exactly the same on many different kinds of computers. The kernel of *Mathematica* exists right now for the Macintosh and for work stations that run similar types of graphics—IBM RTs, various kinds of super computers, and other such things. Most of the operations that I've shown you today make use of the kernel of *Mathematica*. The kernel understands input expressions and produces either output expressions or Postscript graphics.

The second piece of the *Mathematica* program, the front end of the program, is responsible for interaction with the user. This piece takes advantage of whatever user capabilities exist on a particular kind of computer. It is possible for you to run the front end of the program on a Macintosh, and the kernel on a remote computer. In such a case, the calculations that we've been doing here would look essentially the same. However, we would see the results come back a bit more quickly.

You can use the front end of *Mathematica* to prepare documents based on *Mathematica*. You can, for example, make the examples we've illustrated today into a section of a book. When you type in the word "polyhedra," you will find under a menu called "styles" a list of possible styles in which you can show that text. You can type in information about polyhedra to annotate the work that you're doing.

You can also use *Mathematica* to prepare your own textbooks. On a machine like the Macintosh, you can use the front end of *Mathematica* as something that acts as a live textbook. You can provide texts to be read on the screen, imbed graphics into that text, and even input commands that can be executed to do computations.

There are a number of efforts underway to write textbooks in the kind of live form that *Mathematica* makes possible. Such efforts typically accompany a printed textbook with a *Mathematica* notebook. One of the more ambitious projects along these lines is a calculus book to be published by Addison-Wesley.

One thing you can do in a *Mathematica* textbook that you can't do in an ordinary textbook is take the formulas in the book, make changes to the examples that are given, and then reevaluate them and see what the results would be. This capability gives you a way to explore what the formula in the book really means.

You can also use *Mathematica* to produce animation in your textbook. First you tell *Mathematica* to compute a function that could be a collection of sine curves in two dimensions. Then you tell it to produce a sequence of these pictures. By running through the whole sequence of pictures, you'll produce a sort of animated movie, and after it's read into memory it will run reasonably quickly. The actual code necessary to generate such a sequence of frames is often quite short—only about 20 lines.

In the past we have distributed *Mathematica* by giving it away, but giving it away in a slightly unusual fashion. Our favorite method is to bundle it with hardware that computer manufacturers produce. We like *Mathematica* to be included as part of the standard system software that comes with the machine. This is the way we have distributed *Mathematica* on the NEXT Computer, and *Mathematica* will also be bundled as part of the standard system software on some other computers that are coming out later this year.

For other machines *Mathematica* is typically being sold by the hardware manufacturers who produce those machines. *Mathematica* is also being sold by IBM for the RT and by Sony Graphics.

To find out more about *Mathematica*, I refer you to my book, *Mathematica*,[1] which provides a complete documentation and description of what *Mathematica* does.

1. Stephen Wolfram, *Mathematica*: A System for Doing Mathematics by Computer (Redwood City, CA: Addison-Wesley Publishing Company, Inc., 1988) 749 pp.

Concurrent Supercomputers in Science

Geoffrey C. Fox and David Walker
Caltech Concurrent Computation Program, California Institute of Technology, Pasadena, CA 91125

The importance of supercomputing is illustrated by the establishment of the five NSF centers at Cornell, Illinois, Pittsburgh, Princeton, and San Diego. Many articles, government reports, and two recent books[1] have described the rationale and use of these centers, and we do not intend to repeat that discussion here. Rather we would like to look to the future and describe the nature and performance of the computers that will be at such centers around the year 2000.

In the next section, we will review the twin driving forces from the technological (computer science) and scientific (computational science) sides. We also review trends in supercomputers and why we need parallel processing. Then we describe why parallel computing works, with general principles drawn from using prototype machines over the last five years. In the final section we use some computational science activities at Caltech to illustrate the impact of concurrent supercomputers on science.

Our recent book[2] gives more information on most of this material, and there are several excellent reviews of parallel and high performance computer architectures.[3]

The Technological and Scientific Motivation

Parallel supercomputing is driven by a confluence of two complementary forces. First, many computational fields need huge increases in computer performance to perform realistic calculations. Second, it is technically possible to build much faster concurrent computers. The first point can be illustrated by computations in scientific fields such as aerodynamics (design of new aircraft), astrophysics (evolution of galaxies, stellar and black hole dynamics), biology (modeling new genetic compounds, mapping genome, neural networks), computer science (simulation of chips/circuits), chemistry (predication of reaction rates), engineering (structural analysis, combustion calculations), geology (seismic exploration, modeling earth), high-energy physics (proton properties), material science (simulation of new materials), meterology (accurate weather prediction), nuclear physics (weapon simulation), plasma physics (fusion reaction simulation). These fields are both academically and commercially interesting. There are also other major fields of interest to industry and the military: business (transaction analysis, spreadsheets), film industry (graphics), robotics (machine/person), space (real time control of sensors), signal processing (analysis of data), defense (control nation's defense, codes breaking). In all of these cases, we anticipate much higher performance supercomputers will lead to major progress.[4]

Advances in computer technology offer two things: (1) *a disappointment.* Sequential (Von Neuman) machines are approaching fundamental performance limits due to the speed of light and heat dissipation constraints. We can expect perhaps a factor of ten increase from component improvement by the year 2000; (2) *an opportunity.* Advances in VLSI allow parallel computation to give one at least another factor of 100 in performance.

The concept of parallel processing is well-known. Large problems are solved in the real world by joining many individuals together into teams (villages, nations) to work on a single task. This task is divided into parts, with each part assigned to a single person. This analogy is explored in our book and article, where we contrast parallel computing with the use of a team of masons on a large construction project. Here we will content ourselves with noting that a parallel computer currently consists of up to 1,024 individual computers (more for specialized applications such as with the connection machine[5]) working together to solve some problem. Such a 1,024-node NCUBE hypercube was used by a group from SANDIA to illustrate speedups of over a thousand on problems quite typical of major scientific computations.[6] We can call this a "village of computers." In the next 15 years, we can expect up to 105 such complete computers in a single machine—a "city of computers." There are many choices for the way the computers communicate with each other and coordinate work.[7] This is computer architecture and analogous to the "government or management of the village." Caltech pioneered the development of one particular parallel computer, the multicomputer or hypercube, where individual computers communicate by sending and receiving messages.

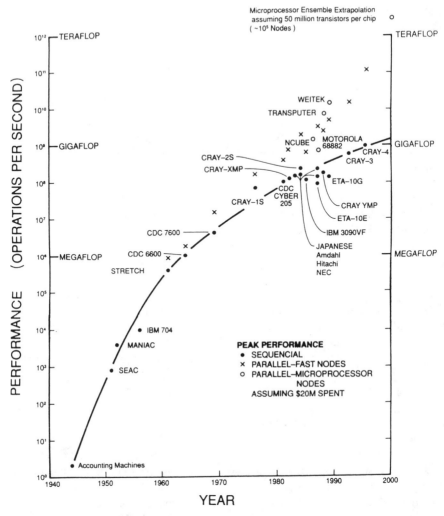

Figure 1. The performance of sequential and parallel computers from 1940 to 2000.

The performance in sequential and parallel computers is illustrated in Figure 1, which was adapted by Fox and Messina from an earlier graph by Buzbee. We see sequential machines leveling off at around a gigaflop performance (10^9 floating point operations per second), corresponding to a one nanosecond clock cycle time. Parallel machines are expected to reach a performance of about 1,000–10,000 times this by the year 2000. This trend in supercomputing is shown in Figure 2, which emphasizes that we have chosen to define a supercomputer by a cost of $20 million. Such a definition is necessary because once you admit parallel machines, a given design can be made with more or less nodes (individual computers) at more or less cost.

SUPERCOMPUTERS TODAY and TOMORROW

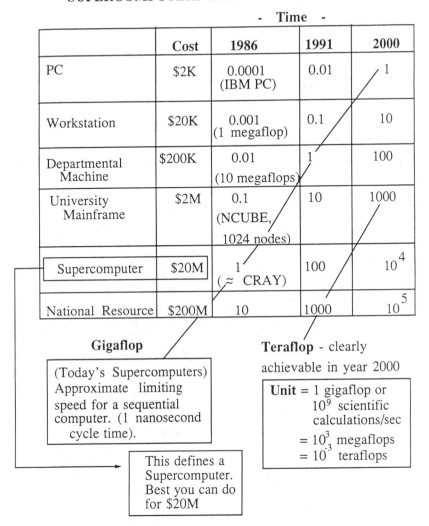

	Cost	- Time - 1986	1991	2000
PC	$2K	0.0001 (IBM PC)	0.01	1
Workstation	$20K	0.001 (1 megaflop)	0.1	10
Departmental Machine	$200K	0.01 (10 megaflops)	1	100
University Mainframe	$2M	0.1 (NCUBE, 1024 nodes)	10	1000
Supercomputer	$20M	1 ($\approx$ CRAY)	100	10^4
National Resource	$200M	10	1000	10^5

Gigaflop

(Today's Supercomputers) Approximate limiting speed for a sequential computer. (1 nanosecond cycle time).

This defines a Supercomputer. Best you can do for $20M

Teraflop - clearly achievable in year 2000

Unit = 1 gigaflop or 10^9 scientific calculations/sec
= 10^3 megaflops
= 10^{-3} teraflops

Figure 2. Computers of today and tomorrow ranging in power from the PC up to top-of-the-line supercomputers. Our working definition of a supercomputer is the best machine that can be purchased for $20 million.

A single board of the NCUBE hypercube contains 64 individual computers, each with a performance that is about twice a PC/AT and each with half a megabyte of memory. One can also buy similar systems built around the transputer chip from INMOS. Such nodes cost around $2,000 each and use ten chips in all—the NCUBE node has six one-megabit memory chips and one custom VLSI chip with 11 communication channels and arithmetic and floating point units.

Year	1980	1988	2000
Transistors per complex chip	5×10^4 -10-	5×10^5 -100-	5×10^7
Examples: Intel	8086	80386	?
Motorola	68000	68030	

Figure 3. Transistor budget extrapolated to the year 2000.

In the next 10 to 15 years, we can expect to see a factor of increase of 100 in the number of transistors contained in a chip of given size. As emphasized by Gordon Moore from INTEL (private communication) one can expect a transistor "budget" of $ fifty million per chip by the year 2000.

The performance increase projected in figures 1 and 2 assumes that one can combine this factor of 100 with another order of magnitude coming from architectural improvements, e.g., RISC, and decreasing clock cycle time.

A chip with fifty million transistors has far too much capability to produce just a single sequential computer. It requires some 10 to 50 individual nodes per chip to use the transistor budget. Thus if we assume that computer circuitry has an approximate fixed cost per unit area, then we can contrast personal computing economics, which dominates in the 1980s, with parallel computing economics, which will dominate the 1990s. The 1980s saw the increase in transistor density converted to

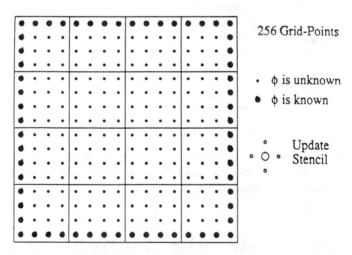

Figure 4. A finite difference grid decomposed among the nodes of a concurrent processor. Each processor is responsible for a 16×16 subgrid. Laplace's equation is to be solved for φ using a simple relaxation technique (equation 1). The five-point update stencil is also shown.

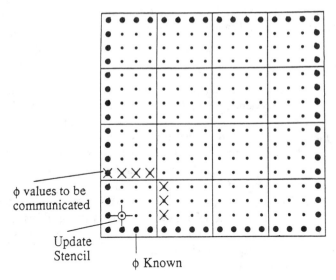

φ values to be communicated

Update Stencil

φ Known

Figure 5. The role of processor 0 (in the bottom left corner). The update stencil equation 1 is applied to each point at which φ is unknown. In order to update the grid points on the upper and righthand boundaries, processor 0 must communicate with the processors above and to the right of it.

smaller, cheaper machines at approximately unchanged performance levels, whereas in the 1990s the increase of transistor density will be converted to increased performance for machines of fixed physical size (cost).

Why Does Parallel Computing Work?

The requirements for the success of parallel processing are quite simple. Crudely, one can say that the problem must be large.[8] To understand this, note that all successful parallel machines have obtained their parallelism from "data parallelism" or "domain decomposition." We can consider a problem as an algorithm applied to a data set. We obtain concurrency by dividing the data between the nodes and applying the given algorithm concurrently to the part of the data for which each node is responsible. This is illustrated in Figures 4 and 5, which show a 16-node concurrent computer used to solve Laplace's equation in two dimensions.

The sample problem has 256 grid points in a 16×16 mesh at which the potential $\phi(i,j)$ is to be determined. As shown in Figure 4, we suppose that a simple iteration (relaxation) algorithm is to be used to solve this problem. This is not the best way of solving this particular algorithm. However, more sophisticated iterative techniques are probably the best approach to large three-dimensional finite-difference or finite-element calculations. Thus, although the simple example in Figure 4 is not "real" or interesting itself, it does illustrate important issues. The computational solution consists of a simple algorithm,

$$\phi(i,j) = 1/4 \left[\phi(i+1,j) + \phi(i-1,j) + \phi(i,j+1) + \phi(i,j-1) \right] \tag{1}$$

applied to each point of the data set. As stated above, in general we get concurrency by applying Eq. (1) simultaneously in different parts of the underlying data domain. We achieve concurrency by leaving the algorithm unchanged and executed sequentially within each node. Rather, we divide up the underlying data set. One attractive feature of this method is that it can be extended to very large machines. A $100 \times 100 \times 100$ mesh with 10^6 points is not an atypical problem. Nowadays, we divide this domain into up to 1,024 parts—clearly such a problem can be divided into many more parts, and use future machines with very many nodes.

Returning to the "toy" example in Figure 4, we associate a 4×4 subdomain with each processor. Let us examine what any one node is doing; this is illustrated in Figure 5. We see that, in this case, an individual processor is solving the "same" problem (i.e., Laplace's equation with an iterative algorithm) as a sequential computer. There are two important differences.

1. The concurrent algorithm involves different geometry—the code should not address the full domain but rather a subset of it.

2. The boundary conditions are changed.

Referring to Figure 5, one finds conventional boundary conditions, ϕ is known, on the left and bottom edges of the square subdomain. However, on the top and right edges, one finds the unusual constraint, "Please communicate with your neighboring nodes to update points on the edge."

For this class of problem, the hypercube and sequential codes can be quite similar; they differ "only" in the geometry and boundary value modules. We have found it relatively easy to develop codes that run either on concurrent or sequential machines depending on input data. Typically, we have developed such code from scratch and not modified existing sequential code; because existing sequential code is not usually structured in an appropriately modular fashion to allow direct conversion to concurrent form.

Figure 5 and the novel boundary condition cited above make it clear that communication is associated with the edge of the region stored in each node. We can quantify the effect of this on the performance of a hypercube by introducing two parameters, t_{calc} and t_{comm}, to describe the hypercube hardware.

t_{calc}: The typical time required to perform a generic calculation. For scientific problems, this can be taken as a floating point calculation $a = b * c$ or $a = b + c$.

t_{comm}: The typical time taken to communicate a single word between two nodes connected in the hardware topology.

t_{calc} and t_{comm} are not precisely defined and depend on many parameters, such as size of message length for t_{comm} and use of memory or registers for t_{calc}. The overhead due to communication depends on the ratio τ given by

$$\tau = \frac{t_{comm}}{t_{calc}} = 1.5 \rightarrow 3 \tag{2}$$

where we have quoted the value for our initial cosmic cube and Mark II hypercubes built at Caltech. For problems similar to those in Figure 5, we have implemented code and measured the performance of the hypercube. We can express the observed speedup S as

$$S = \frac{N}{1+f_c}, \tag{3}$$

where the problem runs S times faster than a single node on a hypercube with N nodes. f_c is the fractional concurrent overhead, which on this problem class is due to communication. Because the latter is an edge effect, one finds that

$$f_c = \frac{0.5}{n^{1/2}} \frac{t_{comm}}{t_{calc}}, \tag{4}$$

where one stores n grid points in each node. $n = 16$ in the example of Figure 4 and 5. The ratio of edge to area in two dimensions is $4\ n^{1/2}$. We see that f_c will be ≤ 0.1 in this example, i.e., the speedup will be ≤ 90 percent of optimal, as long as one stores at least 100 grid points in each node. This is an example of how one can quantify the importance of the problem being large. On a machine with N nodes, the hypercube performs well on two-dimensional problems with at least 100 grid points in each node.

One can generalize equation 4 by introducing the fractional system dimension d associated with any problem.[9] In terms of grain size n and problem dimension d,

$$f_c = \frac{\text{constant}}{n^{1/d}} \frac{t_{comm}}{t_{calc}} \tag{5}$$

for a concurrent computer with

$$d_c = \text{dimension of (the topology of) the computer} \geq d. \tag{6}$$

For the hypercube, $d_c = \log_2 N$ is large and equation 6 is satisfied for most problems. This explains why a rich topology allows the concurrent computer to perform well. Some experimental results verifying equation 5 are summarized in Figure 6. We might have thought that one obtained small f_c only for local (nearest neighbor) problems such as that shown in Figure 4. This is not true as shown in Figure 7 which points out that f_c decreases as one increases the "range" of the algorithm. Indeed, long-range force problems have some of the lowest overhead known for parallel machines. What counts is not the amount of communication (minimized by a local algorithm) but the ratio of communication to calculation.

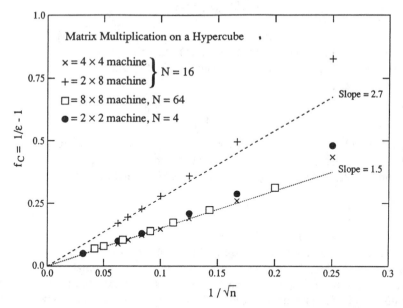

Figure 6. Experimentally measured overhead f_c for the multiplication of two MxM matrices on a hypercube. The number of matrix elements per node n is given by n = {M^2}/N. For this problem the system dimension d is 2. The dashed lines show the asymptotic behavior for large n. Note that f_c is independent of N for a square decomposition.

From equation 5, we see that the overhead f_c depends only on the grain size and that the speedup S is linear with N for extrapolation at constant grain size. This is contrasted in Figure 8 with an extrapolation of a fixed problem to machines with an increasing number N of nodes. In the latter case, $n \propto 1/N$ decreases and the importance of control and communication overheads increase; the speedup when plotted against N eventually levels off. It is our impression that one typically builds larger machines to solve larger problems and that the approximately fixed grain-size extrapolation, that is, S linear with N, is most appropriate.

In many problems, there are other degradations in the performance of current computers. For example, load imbalance is often a significant issue.[10] One needs to parcel out work to the nodes so that each has approximately the same amount of computation. This was trivially achieved in Figure 4 by ensuring that each node processes an equal number of grid points. We have shown that one can view load balancing as an optimization problem and apply a variety of techniques, of which simulated annealing[11] and neural networks[12] are the most attractive. These methods are quite interesting because they involve a deep analogy between general problems and a physics system.

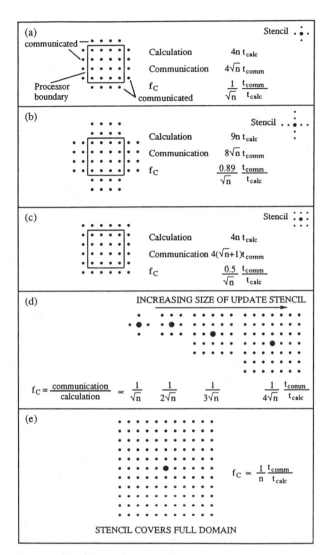

Figure 7. The effect of differing stencils on the communication overhead f_C. For the stencils in panels a, b, and c, f_C is proportional to , where n is the number of grid points per processor. As the stencil size increases (see panel d), f_C decreases, until when the stencil covers the full domain of the problem (panel e), f_C is proportional to 1/n. This corresponds to a long-range force problem.

Uses of Concurrent Supercomputers

Overview

At Caltech, we have set up the Caltech Concurrent Supercomputer Facility (CCSF) led by Dr. Paul Messina. This currently includes high-performance hypercubes from JPL (internal), NCUBE, INTEL, and AMETEK, as well as a half share in a 16K node connection machine CM-2 from Thinking Machines. We expect to keep

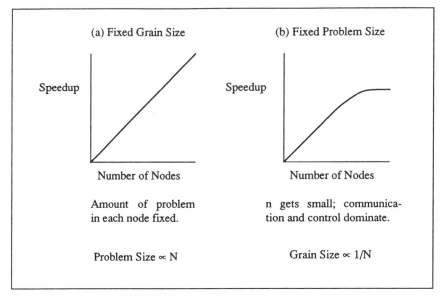

Figure 8. The relationship between speedup S and number of nodes N for (a) fixed grain size and (b) fixed problem size.

this facility up-to-date with new parallel machines as they become available. We have studied the use of such machines in various computational science and engineering fields and use local Caltech examples to highlight current and future uses of concurrent supercomputers.

Let us start with general remarks. Current large computations consist at best of some 1,000 hours on a CRAY/XMP, perhaps running for well-vectorized code at 100 megaflops. Taking a "year" as 5,000 hours, we can state that large computations → an integrated level of 20 megaflop years while more typically one might use one megaflop year. Even the latter would require a VAX11/780 dedicated for five years, and this machine has only just become obsolete. However, we should soon see gigaflop years as possible for large problems, while by the year 2,000, a teraflop year on a single problem seems quite possible. The impact of such huge increases in performance is not easy to predict, but we know that the impact will be large. We need to use present machines to learn how to harness the power of those in the future. In many fields current supercomputers are allowing scientists to solve the "real" problem for the first time, e.g., to study true three-dimensional cases rather than model two-dimensional test cases possible in the past.

We will now look at a few examples.

Astronomical Data Analysis

At Caltech, Tom Prince's group is using the NCUBE to analyze radio astronomy data to look for pulsars.[13] Similar uses are possible in optical astronomy with the advent of new detection technology with direct photon counting. We can also expect optical interferometers involving satellites, and perhaps the new Keck telescope needs. At a rough estimate, the current Mount Palomar 200-inch telescope needs 0.1 gigaflop for real-time processing, and the Keck telescope needs perhaps 10 gigaflops. We will need to develop novel (perhaps neural-net) parallel-image-clean-up and feature-extraction algorithms to exploit the new telescopes, and we will need to use novel architecture computers to perform the analysis.

Cortex Simulations

Caltech has just started a new Ph.D. program called Computation and Neural Systems (CNS) to study the issues involving biology, computer science and physics. Central to this research is the three-way interplay among computer simulation, experimentation, and theoretical neural network in deepening our understanding of the brain. We have only just started cortex simulations which are naturally suited to parallel simulations.[14] It is simply not known how large a simulation, and with what accuracy in neural modeling, is needed to reproduce essential functions of the cortex.

Vision

A more direct use of parallel computers in CNS is to implement high performance vision.[15] Real time vision requires 30 pictures/second, and teraflop performance should allow an excellent vision system with around 10^6 pixels. Of course there are many unsolved issues in vision, but the crucial point is that we can expect computers with the necessary performance; this should give researchers a major motivation to find effective algorithms.

Quantum Chemical Reactions

Kuppermann has been investigating the theoretical simulation of chemical reactions involving three or four atoms.[16] Even collisions between large proteins and enzymes only involve this number of active atoms. A typical problem is the chemical laser shown in Figure 9. Kuppermann's group is using a hyperspherical coordinate method involving the solution of a multichannel Schrödinger equation preceded by a finite element determination of the eigenfunctions. The current gigaflop computers allow the study of reactions like $O + H_2 \rightarrow OH + H$ and $OH + H_2 \rightarrow H_2O + H$, while the more complex states in $F + H_2 \rightarrow FH + H$ could need teraflop performance.

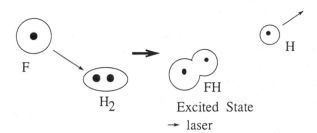

Figure 9. The reactions involved in the fluorine-based chemical laser are typical of those requiring computer performance approaching the teraflop range.

Geophysics

The hypercube has been used as a routine tool by the Caltech geophysics group for many years,[17] and Clayton is proposing to develop on interactive system around the hypercube for the interpretation of oil exploration data. Shell has been using the NCUBE hypercube for some time for some aspects of seismic data analysis.

Oil reservoir simulations are crucial in secondary and tertiary extractions from reservoirs, and these finite element computations are well suited to parallel machines. Hager has been simulating the long-term plate movement and convection within the earth over time periods of $O(10^7)$ years with such finite element methods.[18] In each case megaflop performance allows some two-dimensional simulations, while gigaflop machines will lead to some of the first good three-dimensional computations.

High Energy Physics

We have used the hypercube and perhaps 10,000 hours of hypercube computer time to produce 16 published papers on lattice gauge theory calculations, i.e., to predict properties of fundamental particles. These are Monte Carlo calculations in four dimensions and so are particularly time consuming.[19] Currently a 20×20×20×20 lattice needs on the order of 1000 CRAY/XMP hours, i.e., corresponds to 0.02 gigaflop years. More reasonably sized systems, such as 100^4, clearly need teraflop performance. New string theories linking all forces will be even more computationally demanding. Further, we have little idea how to calculate scattering amplitudes.

Space Science

Caltech's Jet Propulsion Laboratory (JPL) is known for its unmanned probes of the solar system. An interesting project planned is the Mars Rover, which will

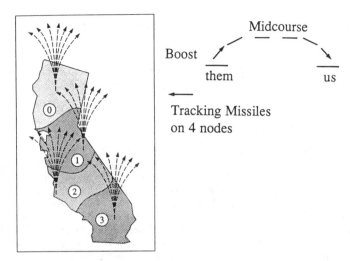

Figure 10. An imaginary missile launch scenario decomposed for a concurrent computer with four nodes. These types of target-tracking problems are well suited to concurrent supercomputers.

place a robot on Mars. This will need a small fault-tolerant on-board controller with gigaflop performance, and the same technology described earlier as giving us a teraflop "mainframe" should allow this goal.

Another major computationally intensive task at JPL involves satellite data processing. Fields like synthetic aperture radar and spectrometry need gigaflop computers for real-time analysis. This estimate ignores time needed for difficult feature extraction, which is currently in its infancy, due partly to the lack of high-performance general-purpose computers.

Defense of the Nation

To defend against the threat sketched in Figure 10, it has been estimated that one must track and aim countermeasures against some 2,000 missiles in boost and 100,000 objects in mid course. This is a large problem and quite well suited to concurrent supercomputers.[20] So we will have the computer power to address the strategic defense initiative! However, will the software work in practice? Who knows? We certainly don't, but simulation of such battle scenarios is a major computational task probably requiring much greater performance to cover the many possible scenarios.

Conclusions

Highly parallel supercomputers built from 1,000 to 100,000 individual nodes will surely be the standard in the next decade. They offer a challenge to the user but the opportunity for a revolution in the computational approach to science. The question is not "Will they work?" or "Will such systems be built?" but rather "How can we use them?"

1. S. Karin and N. P. Smith, *The Supercomputer Era* (New York: Academic Press, 1987); C. Lazou, *Supercomputers and Their Use* (Oxford, England: Oxford University Press, 1987).

2. G. C. Fox, M. A. Johnson, G. A. Lyzenga, S. W. Otto, J. K. Salmon, and D. W. Walker, *Solving Problems on Concurrent Processors* (Englewood Cliffs, NJ: Prentice Hall, 1988).

3. C. G. Bell, "Multis: A New Class of Multiprocessor Computers," Science **228**, 462 (1985); *Experimental Parallel Computing Architectures*, J. J. Dongarra, ed., (Amsterdam: North-Holland, 1987); G. Fox and S. Otto, "Algorithms for Concurrent Processors," Physics Today (May 1984); W. D. Hillis, *The Connection Machine* (Cambridge, MA: MIT Press, 1985); W. D. Hillis, "The Connection Machine," Scientific American **108** (June 1987); K. A. Frenkel, ed., Special Issue on Parallelism; Connection Machine and Shared Memory," Communications of the ACM **29**, 1168 (1986); L. S. Haynes, R. L. Lau, D. P. Siewiorek, and D. W. Mizell, "A Survey of Highly Parallel Computing," Computer **9** (January 1982); "Multiprocessors and Array Processors," ed. W. J. Karplus, *Proceedings of the Third Conference on Multiprocessors and Array Processors, Simulation Series, 18,* (San Diego: Society for Computer Simulation 1987); G. Fox and P. Messina, "Advanced Computer Architectures," Scientific American, (October 1987).

4. K. C. Bowler, A. D. Bruce, R. D. Kenway, G. S. Pawley, and D. J. Wallace, "Exploiting Highly Concurrent Computers for Physics," Physics Today (October 1987); B. Buzbee, "Supercomputers: Values and Trends," International Journal of Supercomputer Applications **1**, 100 (1987); W. A. Goddard, III, "Theoretical Chemistry Comes Alive: Full Partner with Experiment," Science **227**, 917 (1985); V. L. Peterson, "Use of Supercomputers in Computational Aerodynamics," *Proceedings of the 1985 Science and Engineering Symposium,* (Minneapolis: Cray Research Inc. 1985); D. J. Wallace, "Numerical Simulation on the ICL Distributed Array Processor," Phys. Rep. **103**, 191 (1984).

5. G. C. Fox and W. Furmanski, "Matrix," in *Proceedings of the Third Conference on Hypercube Concurrent Computers and Applications*, ed. by G. C. Fox, (New York: ACM, 1988).

6. J. L. Gustafson, G. R. Montry, and R. E. Benner, "Development of Parallel Methods for a 1024-Processor Hypercube," *SIAM Journal on Scientific and Statistical Computing,* (1988).

7. G. C. Fox, "Domain Decomposition in Distributed and Shared Memory Environment-I: A Uniform Decomposition and Performance Analysis for the NCUBE and JPL Mark IIIfp Hypercubes," Invited Paper at ICS 87, International Conference on Supercomputing, June 8–12, (1987), Athens, Greece.

8. G. Fox, "The Performance of the Caltech Hypercube in Scientific Calculations: A Preliminary Analysis," in *Supercomputers-Algorithms, Architectures,* and *Scientific Computation*, ed. F. A. Matsen and T. Tajima (Austin: University of Texas Press, 1987); G. C. Fox, "The Hypercube and the Caltech Concurrent Computation Program: A Microcosm of Parallel Computing," Caltech Report C^3P-422 (1987); G. C. Fox and W. Furmanski, "The Physical Structure of Concurrent Problems and Concurrent Computers," invited paper at Royal Society Discussion Meeting, December 9-10, (1987).

9. G. C. Fox and S. W. Otto, "Concurrent Computation and the Theory of Complex Systems," Caltech Report C^3P-455 (1986).

10. G. C. Fox, "A Review of Automatic Load Balancing and Decomposition Methods for the Hypercube," in *Numerical Algorithms for Modern Parallel Computer Architectures*, ed. M. Schultz, (New York: Springer-Verlag).

11. G. C. Fox, S. W. Otto, and E. A. Umland, "Monte Carlo Physics on a Concurrent Processor" in *Proceedings of the Conference on Frontiers of Quantum Monte Carlo*, September, 3–6, 1985 at Los Alamos, ed. J. E. Gubernatis, Journal of Statistical Physics **43**, numbers 5/6 (June 1986); J. W. Flower, S. W. Otto, and M. Salama, "Optimal Mapping of Irregular Finite Element Domains to Parallel Processors," 1986.
12. G. C. Fox and W. Furmanski, "Load Balancing Loosely Synchronous Problems with a Neural Network," in *Proceedings of the Third Conference on Hypercube Concurrent Computers and Applications*, ed. G. C. Gox, (New York: ACM, 1988).
13. P. W. Gorham and T. A. Prince, "Hypercube Data Analysis in Astronomy: Optical Interferometry and Millisecond Pulsar Searches," in *Proceedings of the Third Conference on Hypercube Concurrent Computers and Applications*.
14. J. M. Bower, M. C. Nelson, M. A. Wilson, G. C. Fox, and W. Furmanski, "Piriform (Olfactory) Cortex Model on the Hypercube," in *Proceedings of the Third Conference on Hypercube Concurrent Computers and Applications*.
15. W. Furmanski and G.C. Fox, "Integrated Vision on a Network of Computers," in *Biological and Artificial Intelligence Systems* (ESCOM Science Publishers, 1988).
16. P. G. Hipes and A. Kuppermann, "Gauss-Jordan Inversion with Pivoting on the Caltech Mark II Hypercube," and P. G. Hipes, T. G. Mattson, and A. Kuppermann, "Integration of Coupled Sets of Ordinary Differential Equations on the Caltech Hypercubes: Matrix Inversion," in *Proceedings of the Third Conference on Hypercube Concurrent Computers and Applications*.
17. R. Clayton, B. Hager, and T. Tanimoto, "Applications of Concurrent Processors in Geophysics," in *Proceedings of the Second International Conference on Supercomputing* (St. Petersburg, Fla: International Supercomputing Institute Inc., 1987).
18. M. Gurnis, A. Raefsky, G. A. Lyzenga, and B. H. Hager, "Finite Element Solution of Thermal Convection on a Hypercube Concurrent Computer," in *Proceedings of the Third Conference on Hypercube Concurrent Computers and Applications*.
19. J. W. Flower, J. Apostolakis, H. Ding, and C. Baillie, "Lattice Gauge Theory on the Hypercube," in *Proceedings of the Third Conference on Hypercube Concurrent Computers and Applications*, edited by G.C. Fox, (New York, N.Y.: ACM, 1988).
20. T. D. Gottschalk, "Multiple-Target Track Initiation on a Hypercube," in *Proceedings of the Second International Conference on Supercomputing*; T. D. Gottschalk, "Concurrent Multiple Target Tracking," in *Proceedings of the Third Conference on Hypercube Concurrent Computers and Applications*.

Wondering about Physics...Using Computers and Spreadsheet Software in Physics Instruction

Dewey I. Dykstra, Jr.
Department of Physics, Boise State University, Boise, ID 83725

This paper deals with two issues in the use of computers in physics instruction. The first issue is whether the use of computers can, in fact, be justified in physics instruction and, if it can, under what conditions. The second issue is the use of spreadsheet software as a positive example of conclusions reached from the first issue. The first part of the paper wrestles with the justifiability issue and proposes some conclusions. The second part presents the application of spreadsheet software to physics instruction as an exemplar. The third part presents examples of an approach to using spreadsheets in physics classes that takes advantage of the instructional possibilities of the software.

Computers Are Not an Inherently Superior Instructional Medium

In a report summarizing the findings of over 150 studies of the efficiency and/or efficacy of computer-based instruction Ted Shlechter claims that

> Consistent empirical evidence does not exist to support or deny claimed advantages of CBI (computer-based instruction) over other instructional media for (a) reducing training time; (b) reducing life-cycle costs; (c) facilitating students' mastery of instructional materials; (d) accommodating individual learning differences; and (e) motivating students' learning. The lack of empirical support for these issues is not totally explained by problematic courseware.
>
> Problematic research procedures were also found throughout the CBI literature. Most noticeable of these research problems were (a) confoundings due to differences in instructional content; (b) making comparisons with inappropriate media; (c) confoundings due to "program novelty effects"; and (d) findings that were not replicated.[1]

In a review of research on learning from media, Dick Clark suggests that

> The best current evidence is that media are mere vehicles that deliver instruction but do not influence student achievement any more than the truck that delivers our groceries causes changes in our nutrition. Basically, the choice of vehicle might influence the cost or extent of distributing instruction, but only the content of the vehicle can influence achievement.[2]

Ted Shlechter and Dick Clark are respected for their work in the field of instructional technology: Schlechter, because of his extensive recent work summarizing research in the effectiveness of CBI recently, and Clark because of his significant work in the field for decades. In his review, Clark says that those reports that support claim to the superiority of computer-based instruction attribute to such instruction effects that can just as easily be explained by the confounding factor of the instructional approach. The computer treatments in the studies he cites are not just computer deliveries of the same instruction received in a competing treatment. The computer-based treatments often personalize instruction, which is not done in the other treatment. This mastery-based approach can be shown to produce a significant effect independent of the use of computers.[3] Because the effect observed in the studies of computer-based instruction versus other media is no larger than that which can be induced by a mastery-based approach, one cannot solely attribute the observed effect to the computer-based training. Orlansky and String have pointed out that programmed instruction on the computer cannot be demonstrated to be any more efficient or effective than the same instruction from a programmed instruction book of the sort popular in the late 1960s.[4]

So, where does this leave us? Are we being swept away by another fad based on an expensive high tech toy? Are we wasting time and money? Why do we continue to use computers in instruction? What lessons can we learn from these summaries of the research?

The statements above, stopped me cold for a while. For me, responsible answers to the questions did not come easily. I was afraid that I might, in fact, be advocating the use of a fun toy, a fad. Yet at the time I was first exposed to these summaries of the research, many people for whom I have great respect were also advocating the use of computers to deliver instruction.

Why Should We Use Computers in Instruction?

In the light of the above, here are some thoughts that justify the expense of computers for instruction. These justifications are also the criteria that should be used for the design, selection, or evaluation of computer-based instruction.

It is not hard for us to believe that no medium for the delivery of instruction is inherently superior to another. We have all seen plenty of badly botched attempts at instruction. As Clark points out, how one uses the medium is more important than which medium is used. From this, it is easy to conclude that continued efforts at "horse race" experiments comparing computers with other media are largely a waste of time. But this does not justify the use of computers over other media for instruction.

Instruction that we do attempt, via the computer (or any other medium), should include as much of a mastery-based approach as possible. This means that if we are dealing with typical computer-assisted instruction (CAI), then the software should pretest the students on each objective. If a student passes the pretest successfully, that person should be moved on to the next objective. Those objectives that a student does not meet initially should be brought up and tested again in some way, until the objective is met. If the software you are evaluating or developing does not use this strategy, then you could probably do just as well, and do it less expensively, with some other medium. But just including mastery-based approaches still does not, by itself, justify the use of the computer as more effective or efficient.

The fact that the investigators so frequently fail to make the experiments "fair" sticks in the mind. Why have these people spoiled their own experiments? It must be easier, in fact, maybe more natural, to include things like a mastery-based approach when designing computer-based instruction, because the mode of interaction is usually one-on-one in the typical CAI. It is so natural and so easy, in fact, that it is hard not to use this approach. The resulting experiments were unfair because they did not hold all things the same, other than the medium being used.

Let Us Take This "Unfairness of the Tests" One Step Further

What if there are learning objectives that can be accomplished with a computer that no other medium can accomplish? Can one do a "fair" experiment to compare media in this context? Obviously not! What if these learning objectives are important for students to acquire? Clearly, our only choice is to use the computer to help them achieve that objective. Now, finally, we have a reason why computers should be used in instruction: to accomplish something that can be done only with a computer.

This may seem to be an extreme position, but there are a number of examples where this criterion is easily met.

Course Logistics. In cases where an instructor cannot be spread thin enough to handle all of the courses needed by students, CBI can handle much of a course, including monitoring progress and managing grades, etc., freeing the professor for other duties, most importantly the out-of-the-ordinary needs of students in the computer-based course that can be handled only by human intelligence.

Learning Objectives. The work with the ultrasonic motion transducer by Bob Tinker of the Technical Education Resource Center and Ron Thornton of Tufts University is an example in which the computer makes possible an understanding of graphs of motion not possible before.[5] It seems that the ability to plot graphs of student-controlled motion in real time is instrumental in learning results from this technology. No other technology available can generate graphs of student-controlled motion in real time.

There is one more reason justifying the use of computers in physics instruction: to support the development of skills and quantitative habits of mind that are a part of the physicist's way of understanding the world. An essential feature of the physicist's attitudes about the world is that the outcome of an experiment should be predictable in some quantitative fashion. If something happens that we would not have predicted, we work at it until we can quantitatively describe the outcome. Computers bring large amounts of computing power into the hands of the physicist. If physicists-in-training do not come to view the computer and computing software as second nature in their efforts to understand the physical world, then we, the teachers, are derelict in our duties. Those students who are not studying to be physicists are likely in need of similar computational skills. If they are taking a physics course as part of the requirements for graduation, but are not majoring in a quantitative field, then one contribution that their experience in physics can make is for them to see, if only briefly, through the quantitative "eyes" of the physicist; this is made possible by accessible, easy-to-use computers and software. Obviously, the only way to develop the requisite skills is to use the computer.

Using Spreadsheet Software in Physics Instruction

Rather than explain in detail what spreadsheet software is, I urge you to gain access to one of the packages mentioned below and work through some tutorial material for it, and then try some physics investigations of the sort described below or in the literature. Spreadsheets cannot be fully understood from static displays on paper.

For the sake of completeness, I shall describe spreadsheet software. Briefly, spreadsheet software displays on the screen, rows and columns of information stored in cells. Into each cell one can enter a label (text material), a number, or a formula. A cell containing a formula will display the value that is the result of the calculation indicated by the formula. The formula can include references to the values displayed in other cells. Changing the value displayed in a cell in some way will result in a change in the value displayed by any other cell that refers to the original cell in its formula. Most spreadsheet packages can now graph data generated in the spreadsheet portion of the package. This paper assumes that spreadsheet software appropriate for physics instruction includes features that will graph the data in the spreadsheet. This feature is highly desirable and will be assumed to be a part of the software package whenever the term "spreadsheet" is used in this paper.

Useful Properties of Spreadsheet Software

1. *Ubiquity*. Spreadsheet software adequate for the investigation of physical models or computational chores in much of the physics curriculum is available on most of the common microcomputers. A nonexhaustive list includes Apple II (*SuperCalc*), Commodore (*PowerPlan*), Macintosh (*Excel, Quattro*), and MS DOS (*Lotus1-2-3, Quattro*).

2. *Similarity*. Because the features of spreadsheet software are basically common to all of the software packages, we can speak in the same general terms to each

other about the teaching of physics in this context without being lost in the details of differences in programming languages. Because of marketing competition, many of the producers of this software have worked out schemes for porting files from one software package to another.

3. *Gentility.* To become as popular as it has, the software must have a powerful but friendly interface with the user. The user can easily edit input. Output formatting is already arranged for. Calculations are easy to input and propagate through the rows and columns. Spreadsheets have the power to carry out lots of calculations quickly and most can iterate calculations. Tools called "macros" customize files to facilitate initial student use. The numbers generated can be easily displayed in graphical form.

The appeal to undergraduates of spreadsheets for physics calculations is vividly illustrated by an anecdote told by a colleague at a well-known university. This physics professor, an Apple user and, originally, a non–MS DOS user, gave assignments that he thought would require his students to write short programs in BASIC or Pascal. The students found that their campus had microcomputer labs that gave them access to *Lotus Symphony* (a spreadsheet package) on MS DOS machines. The students chose to use *Symphony* over a traditional programming language. The professor has become a *Symphony* expert to assist his students, although he uses *Excel* with the Macintosh for his own work.

With spreadsheet software there is little mental overhead required for making things happen. One's working memory is not filled with the mechanics of operating the spreadsheet software. Both students and instructors can concentrate on learning and reasoning about the physical world without being unduly distracted by the mechanism of the software. Because one is only adding the "physics" to a well-developed software package, when things go awry, it is usually with the physics that has been entered and not with some obscure misunderstanding by the student about programming thus both instructor and student can focus on the physics. The features of the software make it possible to ease students into using it. Starting with spreadsheets that are almost completely built, with each subsequent investigation the student is asked to do one more thing, until the student is building the whole spreadsheet. We can realistically expect that students eventually build their own spreadsheets. At the Air Force Academy, for example, spreadsheet investigations have become an integral part of some physics classes.[6] With the students exploring physics using spreadsheets, it is easy to imagine them coming to view spreadsheets as an obvious tool for this purpose.

How many students have resisted taking or really learning physics because of the fear or sheer burden of the mechanics of computation? What might we have lost from the physics community because we never received the benefit of their imaginations? How many might we bring into the fold if we reduce the barriers to experiences in which students can see the excitement of matching nature quantitatively?

Spreadsheets provide a low-threshold entry into explorations of physical phenomena which students do not yet possess mathematical tools to handle analytically in closed form. For that matter, students can handle situations for which there is no closed analytic form. What has held us back from these types of situations before is that the numerical methods involved were too tedious computationally. We can now get something for our students not possible before. Instead of being limited to free fall in a vacuum, students can include the effects of air drag. Many of today's real problems are being solved numerically on computers. Computations using spreadsheets can introduce students to the basic processes involved in the computations from the beginning. What better introduction to computational physics than to acquire the habit of applying basic principles of physics and computational power to model what might happen in a given situation? Isn't this the essence of computational physics?

One additional idea deserves mention. While I have already indicated some of the advantages of spreadsheets over programming languages for learning physics, one other important reason in favor of spreadsheets exists: spreadsheet software is the most efficient or effective way to accomplish many tasks, and every well-trained professional should be able to use it. The only valid reason for writing a program to do such things as developing budgets for projects or keeping grades, is to learn the language. Writing a program to do these jobs will never be the most efficient or effective way to accomplish these tasks. This is not to say that programming languages have no place, but they get in the way of introductory students' learning physics. They can and should be used later to do the things that programming languages can do well.

Using Spreadsheets to Find Out about Physics

At the introductory level it is best to focus instruction on helping students understand the basic points of view in physics. Students come to class with a set of working beliefs that do not reflect scientists' view of the world, and our task is to change these beliefs. This is not an easy task. Just telling them the "truth" as physicists see it, is not enough to change students' beliefs. We have to help them construct new ideas about the world for themselves. Activities aimed at getting the right answer generally do not accomplish this.

Since the technology available to students in the past did not support much more than "single-answer" problem-solving, most of us have developed skills generating this sort of problem. Spreadsheets and microcomputers make much more possible. We have to shift gears to dream up activities that go beyond the confines of the old technologies. If we do not ask the computer to give us more than we were getting from the old technologies, then we are wasting our time and money.

The following are examples that contrast the single-answer approach to designing spreadsheet-based assignments to a more expansive investigation approach.

	A	B	C	D
1	Clocks on Other Planets			
2				
3	0.99	length of earth pendulum (m)		
4				
5			Period	Correct
6			(earth clock	Length
7	Planet	"g"	on planet)	for Planet
8		(m/s^2)	(s)	(m)
9	Earth	9.8	2.00	0.99
10	Moon	1.6	4.94	0.16
11	Mars	3.8	3.21	0.39
12	Jupiter	25	1.25	2.53
13	Sun	270	0.38	27.36

Figure 1. Single-answer approach: SHM spreadsheet.

Physics of Simple Harmonic Motion (SHM)

Single-Answer Approach. Teachers can write an interesting handout that leads the students to set up a spreadsheet to calculate the period of a pendulum on various planets. Students might put the problem in the context of colonists from earth going to another planet. Should they take along the family heirloom grandfather clock? In one column they might list g on various planets, and in another column calculate the corresponding ideal simple-pendulum period for a pendulum that gives the right period on earth. They might calculate in a third column the correct length for the proper period on each planet.

Investigation Approach. Starting with the same motivation, teachers can have students set up a spreadsheet that calculates the time to various positions (starting with the initial position) and velocity, calculating from the forces involved to the resulting accelerations, to velocity, and to time at each position. Then students can explore to find the best length by having the length of the pendulum as an input along with g on the planet in question.

This approach offers several unique experiences to students: How often are students asked to work out the time as a function of position? How often are students asked to calculate the physics of pendulums from first principles instead of derived equations? Because the calculations in this version are not limited to small angles, one can have the students explore the discrepancies between the spreadsheet calculation and the ideal equation. Since the calculations average over discrete intervals, one can also have students explore the effects of adjusting interval sizes. When they have finished this investigation, they might write a short report summarizing their findings. This will help them bring their ideas into focus and give the instructor an insight into their understandings.

This spreadsheet calculates the time to position from first principles and also displays the SHM small-angle approximation value, both allowing the student to vary the pendulum length. (In this figure and all subsequent spreadsheet images, numbers in bold that also have labels in bold are inputs that can be changed, forcing the spreadsheet to recalculate other values that it displays.)

	A	B	C	D	E	F	G	H
1			TWO SECOND PENDULUM					
2								
3	1.0	Mass (kg)						
4								
5	Accelerations due to gravity (m/s^2)							
6	Earth	Moon	Mars	Jupiter	Sun			
7	9.8	1.6	3.8	25	270			
8								
9								Ideal SHM
10	9.8	acceleration due to gravity (m/s^2)						1/2 period
11	0.9	Trial length (m)						(s)
12								0.952
13	theta	theta	Force	alpha	omega	time		
14	(°)	(radians)	(N)	(rad/s^2)	(rad/s)	(s)		
15	5	0.087	-0.85	-0.95	0.00	0.000		
16	4.5	0.079	-0.77	-0.85	-0.13	0.139		
17	4	0.070	-0.68	-0.76	-0.17	0.198		
18	3.5	0.061	-0.60	-0.66	-0.21	0.244		
19	3	0.052	-0.51	-0.57	-0.23	0.284		
20	2.5	0.044	-0.43	-0.47	-0.25	0.320		
21	2	0.035	-0.34	-0.38	-0.26	0.354		
22	1.5	0.026	-0.26	-0.29	-0.27	0.387		
23	1	0.017	-0.17	-0.19	-0.28	0.418		
24	0.5	0.009	-0.09	-0.10	-0.29	0.449		
25	0	0.000	0.00	0.00	-0.29	0.479		
26	-0.5	-0.009	0.09	0.10	-0.29	0.510		
27	-1	-0.017	0.17	0.19	-0.28	0.540		
28	-1.5	-0.026	0.26	0.29	-0.27	0.572		
29	-2	-0.035	0.34	0.38	-0.26	0.604		
30	-2.5	-0.044	0.43	0.47	-0.25	0.638		
31	-3	-0.052	0.51	0.57	-0.23	0.674		
32	-3.5	-0.061	0.60	0.66	-0.21	0.715		
33	-4	-0.070	0.68	0.76	-0.17	0.761		
34	-4.5	-0.079	0.77	0.85	-0.13	0.819		
35	-5	-0.087	0.85	0.95	0.00	0.959		

Figure 2. Investigation approach: SHM spreadsheet.

Relativistic Electrons in a Linear Accelerator

Single-Answer Approach. Teachers can introduce the idea that the electrons go so fast in the Stanford linear accelerator that the two-mile-long accelerator is Lorentz contracted to a length of two feet. Students can build a spreadsheet that has in one column a list of possible Lorentz-contracted lengths and in the next column the speed that the electrons would have to go for the accelerator to be contracted to those lengths.

Investigation Approach. Using the same introduction, teachers can have students set up a spreadsheet that simulates an electron accelerator. The input should be the accelerating voltage. From the accelerating voltage they can generate columns to calculate the kinetic energy of the electron, the electron speed, the relativistic length of the accelerator, the relativistic mass of the accelerated electron, and the ratio of the relativistic mass to the rest mass of the accelerated electron at each 100-meter interval for the full 3,200-meter length of the accelerator. They can

	A	B	C
1	LORENTZ CONTRACTION		
2			
3			
4			
5			
6	Lorentz		
7	Contracted	Required	
8	Length	Velocity	
9	(m)	(m/s)	
10	3200	0.00000E+00	
11	3000	1.04396E+08	
12	2000	2.34187E+08	
13	1000	2.84975E+08	
14	500	2.96315E+08	
15	100	2.99853E+08	
16	50	2.99963E+08	
17	10	2.99999E+08	
18	5	3.00000E+08	
19	2	3.00000E+08	

Figure 3. Single-answer approach: Lorentz-contraction spreadsheet.

	A	B	C	D	E	F	G	H
1	LINEAR ELECTRON ACCELERATOR							
2								
3	750	: accelerating voltage (kV/m)			9E+16	c^2	9E-31	m(e)
4								
5								
6	Dist	E	v	Rel. L.	Rel. m	mass		
7	(m)	(J)	(m/s)	(m)	(kg)	ratio		
8	100	1.20E-11	2.9999E+08	21.84	1.33E-28	147		
9	200	2.40E-11	3.0000E+08	10.92	2.67E-28	293		
10	300	3.60E-11	3.0000E+08	7.28	4.00E-28	440		
11	400	4.80E-11	3.0000E+08	5.46	5.34E-28	586		
12	500	6.00E-11	3.0000E+08	4.37	6.67E-28	733		
13	600	7.20E-11	3.0000E+08	3.64	8.01E-28	879		
14	700	8.40E-11	3.0000E+08	3.12	9.34E-28	1026		
15	800	9.60E-11	3.0000E+08	2.73	1.07E-27	1172		
16	900	1.08E-10	3.0000E+08	2.43	1.20E-27	1319		
17	1000	1.20E-10	3.0000E+08	2.18	1.33E-27	1465		
18	1100	1.32E-10	3.0000E+08	1.99	1.47E-27	1612		
19	1200	1.44E-10	3.0000E+08	1.82	1.60E-27	1758		
20	1300	1.56E-10	3.0000E+08	1.68	1.74E-27	1905		
21	1400	1.68E-10	3.0000E+08	1.56	1.87E-27	2051		
22	1500	1.80E-10	3.0000E+08	1.46	2.00E-27	2198		
23	1600	1.92E-10	3.0000E+08	1.36	2.14E-27	2344		
24	1700	2.04E-10	3.0000E+08	1.28	2.27E-27	2491		
25	1800	2.16E-10	3.0000E+08	1.21	2.40E-27	2637		
26	1900	2.28E-10	3.0000E+08	1.15	2.54E-27	2784		
27	2000	2.40E-10	3.0000E+08	1.09	2.67E-27	2930		
28	2100	2.52E-10	3.0000E+08	1.04	2.80E-27	3077		
29	2200	2.64E-10	3.0000E+08	0.99	2.94E-27	3223		
30	2300	2.76E-10	3.0000E+08	0.95	3.07E-27	3370		
31	2400	2.88E-10	3.0000E+08	0.91	3.20E-27	3516		
32	2500	3.00E-10	3.0000E+08	0.87	3.34E-27	3663		
33	2600	3.12E-10	3.0000E+08	0.84	3.47E-27	3810		
34	2700	3.24E-10	3.0000E+08	0.81	3.60E-27	3956		
35	2800	3.36E-10	3.0000E+08	0.78	3.74E-27	4103		
36	2900	3.48E-10	3.0000E+08	0.75	3.87E-27	4249'		
37	3000	3.60E-10	3.0000E+08	0.73	4.00E-27	4396		
38	3100	3.72E-10	3.0000E+08	0.70	4.14E-27	4542		
39	3200	3.84E-10	3.0000E+08	0.68	4.27E-27	4689		

Figure 4. Investigation approach: electron accelerator spreadsheet.

	A	B	C
1	DOPPLER SHIFT		
2			
3	3000	source frequency (Hz)	
4			
5			
6	Source	Observed	
7	Velocity	Frequency	
8	(m/s)	(Hz)	
9	1	2991	
10	2	2982	
11	5	2955	
12	10	2909	
13	20	2818	
14	50	2545	
15	100	2091	
16	200	1182	

Figure 5.   Single-answer approach: Doppler shift spreadsheet.

experiment with accelerating voltages until at the end of the accelerator, its relativistic length is about 2/3 meter.

After they have built the simulator in spreadsheet form, students should write a short report describing such things as (1) how the velocity at the end of the first 100 meters compares with that at the end for the voltage that they have found and at several different fractions of that voltage, and (2) how the relativistic mass and the rest mass compare at the end of the track.

Graphs of the various calculated quantities versus distance along the track for different accelerating voltages can be quite instructive.

Doppler Shift

Single-Answer Approach. Referring to ambulance sirens and car horns, teachers can ask students to calculate the shifted frequency for various source velocities. If they have the source frequency as an input; in column one they can list a range of velocities and in column two calculate the shifted frequency.

Investigation Approach. Referring to the same phenomenon, teachers can lead students to calculate the frequency as a function of time as the source passes by. Students should assume that the observer is offset from the linear path of the source, which moves with constant velocity. Thinking of the angular position of the source along its path with respect to the observer, students can calculate the observed frequency from the component of the velocity toward the observer. From angular position they can calculate linear position along the path of motion and from that, time at position. Only one more step remains: to calculate the propagation time for the sound from the point of emission. Finally, students must consider frequency versus time of arrival of the sound at the observer. With source velocity, offset distance, and source frequency as input variables, students can explore the nature of the changes in frequency as the source goes by. Some very interesting results can occur when the source is faster than the sound. Graphs of this situation

	A	B	C	D	E	F	G
1	DOPPLER SHIFT						
2							
3	100	source velocity (m/s)			0.0175	° to rad conv	
4	10	offset (m)					
5	200	source frequency (Hz)					
6	330	sound velocity (m/s)					
7							
8					Sound	Sound	
9			Time at	Approach	Travel	Arrival	Observed
10	Angle	Position	Position	velocity	Time	Time	freq
11	(°)	(m)	(s)	(m/s)	(s)	(s)	(Hz)
12	80	56.71	0.000	98.48	0.175	0.175	259.7
13	70	27.47	0.292	93.97	0.089	0.381	257.0
14	60	17.32	0.394	86.60	0.061	0.455	252.5
15	50	11.92	0.448	76.60	0.047	0.495	246.4
16	40	8.39	0.483	64.28	0.040	0.523	239.0
17	30	5.77	0.509	50.00	0.035	0.544	230.3
18	20	3.64	0.531	34.20	0.032	0.563	220.7
19	10	1.76	0.549	17.36	0.031	0.580	210.5
20	0	0.00	0.567	0.00	0.030	0.597	200.0
21	-10	-1.76	0.585	-17.36	0.031	0.616	189.5
22	-20	-3.64	0.604	-34.20	0.032	0.636	179.3
23	-30	-5.77	0.625	-50.00	0.035	0.660	169.7
24	-40	-8.39	0.651	-64.28	0.040	0.691	161.0
25	-50	-11.92	0.686	-76.60	0.047	0.733	153.6
26	-60	-17.32	0.740	-86.60	0.061	0.801	147.5
27	-70	-27.47	0.842	-93.97	0.089	0.930	143.0
28	-80	-56.71	1.134	-98.48	0.175	1.309	140.3

Figure 6. Investigation approach: Doppler shift spreadsheet.

can be quite surprising. When students have explored, teachers can have them report their findings in a short report.

Falling in Air

Single-Answer Approach. Teachers can ask students to calculate distance fallen as a function of time for a sphere falling through the air. Students can look in the text for the equation for distance fallen as a function of time for the critical parameters.

Investigation Approach. Teachers can suggest a simulation of the Leaning Tower of Pisa thought experiment posed by Galileo. If the teacher can provide the height of the tower and drag factor and explain how the drag can be calculated, students can generate a spreadsheet that allows the comparison of the distance fallen versus time for a lead sphere, a wooden sphere, and a sphere in the absence of air. Students should generate separate columns for the drag force on each sphere and the ratio of the drag force to the weight of each object. Then they can generate graphs of these quantities.

When they have finished the Leaning Tower of Pisa spreadsheet, students should write a short report discussing the sizes of the drag force on the two objects and the reason why the wood sphere gets to the ground later than the lead sphere.

	A	B	C	D	E	F	G	H	I	J	K	L	M
1	47.8	Ht of Tower in m			Tower of Pisa Data								
2	2E-3	c in air	Earth										
3			No Air	lead ball			oak ball						
4			9.81	4.31	m in kg		0.391	m in kg		drag F	drag F	drag F	drag F
5			dist	a	v	h	a	v	h	(N)	(N)	Weight	Weight
6	No.	Time (s)	(m)	(m/s^2)	(m/s)	(m)	(m/s^2)	(m/s)	(m)	lead	oak	lead	oak
7	0	0.00E+00	47.80	9.81	0.00	47.80	9.81	0.00	47.80	0.000	0.000	0.000	0.000
8	1	3.33E-02	47.79	9.81	0.33	47.79	9.81	0.33	47.79	0.000	0.000	0.000	0.000
9	2	6.67E-02	47.78	9.81	0.65	47.78	9.81	0.65	47.78	0.000	0.000	0.000	0.000
10	3	1.00E-01	47.75	9.81	0.98	47.75	9.81	0.98	47.75	0.001	0.001	0.000	0.000
11	4	1.33E-01	47.71	9.81	1.31	47.71	9.80	1.31	47.71	0.002	0.002	0.000	0.001
12	5	1.67E-01	47.66	9.81	1.63	47.66	9.80	1.63	47.66	0.004	0.004	0.000	0.001
13	6	2.00E-01	47.60	9.81	1.96	47.60	9.79	1.96	47.60	0.006	0.006	0.000	0.002
14	7	2.33E-01	47.53	9.81	2.29	47.53	9.79	2.29	47.53	0.009	0.009	0.000	0.002
15	8	2.67E-01	47.45	9.81	2.62	47.45	9.78	2.61	47.45	0.011	0.011	0.000	0.003
16	9	3.00E-01	47.36	9.81	2.94	47.36	9.77	2.94	47.36	0.015	0.015	0.000	0.004
17	10	3.33E-01	47.26	9.81	3.27	47.26	9.76	3.26	47.26	0.018	0.018	0.000	0.005
18	11	3.67E-01	47.14	9.80	3.60	47.14	9.75	3.59	47.14	0.022	0.022	0.001	0.006
19	12	4.00E-01	47.02	9.80	3.92	47.02	9.74	3.91	47.02	0.027	0.027	0.001	0.007
20	13	4.33E-01	46.88	9.80	4.25	46.88	9.73	4.24	46.88	0.031	0.031	0.001	0.008
21	14	4.67E-01	46.73	9.80	4.58	46.73	9.72	4.56	46.73	0.037	0.036	0.001	0.010
22	15	5.00E-01	46.57	9.80	4.90	46.57	9.70	4.89	46.58	0.042	0.042	0.001	0.011
23	16	5.33E-01	46.40	9.80	5.23	46.41	9.69	5.21	46.41	0.048	0.048	0.001	0.013
24	17	5.67E-01	46.22	9.80	5.56	46.23	9.67	5.53	46.23	0.055	0.054	0.001	0.014
25	18	6.00E-01	46.03	9.80	5.88	46.03	9.65	5.85	46.04	0.062	0.061	0.001	0.016
26	19	6.33E-01	45.83	9.79	6.21	45.83	9.64	6.17	45.84	0.069	0.068	0.002	0.018
27	20	6.67E-01	45.62	9.79	6.54	45.62	9.62	6.49	45.63	0.077	0.076	0.002	0.020
28	21	7.00E-01	45.40	9.79	6.86	45.40	9.60	6.81	45.41	0.085	0.083	0.002	0.022
29	22	7.33E-01	45.16	9.79	7.19	45.16	9.58	7.13	45.17	0.093	0.092	0.002	0.024
30	23	7.67E-01	44.92	9.79	7.51	44.92	9.55	7.45	44.93	0.102	0.100	0.002	0.026
31	24	8.00E-01	44.66	9.78	7.84	44.66	9.53	7.77	44.68	0.111	0.109	0.003	0.028
32	25	8.33E-01	44.39	9.78	8.17	44.40	9.51	8.09	44.41	0.121	0.118	0.003	0.031
33	26	8.67E-01	44.12	9.78	8.49	44.12	9.48	8.40	44.14	0.131	0.128	0.003	0.033
34	27	9.00E-01	43.83	9.78	8.82	43.83	9.46	8.72	43.85	0.141	0.138	0.003	0.036

Figure 7. Investigation approach: Fall with drag force spreadsheet.

Propagation of Errors in Lab

An additional use of spreadsheets is to illustrate the consequences of variations in measured values that are inputs to a calculation in lab. The teacher sets up a spreadsheet that does the calculation of some derived value from measurements actually made in the lab. The teacher can have students explore changes in the value due to variations in the inputs. He/she could even use the absolute or percentage variation in the measured values as inputs to the spreadsheet. The teacher could have students compare the results from the spreadsheet with the results of the formulas taught to estimate the propagation of error in lab calculations. Additional ideas for use of spreadsheets in the lab can be found in a recent book by Ouchi.[7]

Clearly, the examples of the investigative approach shown here are a departure from the typical physics problems that we have all created in the past. Creating these investigations can lead to interesting surprises, even for relatively jaded physics professors. To aid in these efforts, here are some guidelines for creating investigations in physics using spreadsheets.

Guidelines for Developing Physics Spreadsheet Investigations

1. It often helps to think of the spreadsheet task as a simulator, not a calculator. As with the SHM and relativistic electron examples above, teachers should try to make the desired result an input. Students should try various values and let the spreadsheet calculate the response of the system.

2. An element of whimsy is useful. Sections on the Doppler shift in textbooks refer to the frequency change as a source goes by an observer; yet they calculate the difference only between the source frequency and the observed frequency, something not directly observable. What does the actual change in pitch look like on a graph?

3. Students should break down calculations into terms or factors that are calculated in separate columns or rows. The net result can be calculated in another column or row. This makes visible the interplay of contributions of various terms or factors that previously have gone unseen. The reduction in complexity of an equation in any given cell will reduce the frustrations that result from the strange hierarchy spreadsheets often use to evaluate the expression in the cell. The confusion everyone has with multiple layers of parentheses in long, single-line equations is also reduced.

4. For introductory students, teachers should keep the focus in spreadsheet investigations on exploring the physics in spreadsheet investigations. Teachers should give students useful calculational techniques, but avoid giving several to calculate the same or similar things. The spreadsheet should be a tool to uncover new ideas, not another layer of complexity to learn. Introducing the official names of these techniques is not necessary. Later, when students have become comfortable with the idea of using the spreadsheet to explore physics, teachers can introduce them to the methods of computational physics by giving them a topic to explore that their normal method fails to handle. Teachers can then suggest a method that works and give names to both this new method and the old one as the class explores strong and weak points of each.

Dykstra and Fuller have a number of additional examples of investigations that make use of these guidelines and can be used in introductory physics classes.[8]

This investigative approach to using spreadsheets in physics instruction has several good points. The focus can be maintained on the physics with less obscuring approximation and more emphasis on basic relationships. If students acquire the habits portrayed here, they will approach future problems by starting with first principles and let the simulation play out its course to discover what will happen. This is not a bad method. Once the students have learned this calculational habit of mind, we can subtly introduce the methods of computational physics where appropriate. A recent book by Orvis contains a number of calculation methods in science and engineering for spreadsheets.[9]

Finally, I add a WORD OF CAUTION. Teachers should not let the technology seduce them into throwing too much at students at one time. Very little learning goes on unless students have a stable platform from which to learn. If too much distraction comes from the mechanics of using the technology or it is unpredictable or unreliable for any reason, then they do not have a stable platform. After all, the most important thing is for students to come to a good understanding of physics. Computers are not the fundamental issue. They should hold only the status of tools to develop an understanding of physics. We must guard against training our students to use today's computers and software to learn physics. Instead we should help students to understand physics and to use technology (the computer, in particular) as a tool with which to think.

1. T. M. Shlechter, "An Examination of the Research Evidence for Computer-Based Instruction in Military Training," Technical Report No. 722, U.S. Army Research Institute for the Behavioral and Social Sciences, (Alexandria, VA: U. S. Government Printing Office, 1986).
2. R. E. Clark, "Reconsidering Research on Learning From Media," Review of Educational Research 53 445 (1983).
3. B. S. Bloom, "The 2 Sigma Problem: The Search for Methods of Group Instruction as Effective as One-to-One Tutoring," Educational Researcher 13, 3 (1984).
4. J. Orlansky, and J. String, "Computer-Based Instruction for Military Training," Defense Management Journal, 46 (1981).
5. R. K. Thornton, "Tools for Scientific Thinking—Microcomputer-Based Laboratories for Physics Teaching," Phys. Ed. 22, 230 (1987).
6. G. Hept, "Spreadsheets for Scientists and Everyone," (Paper given at the AAPT -APS Winter meeting in Crystal City, VA, January 1988) abstract in AAPT Announcer (December 1987).
7. G. I. Ouchi, *Lotus in the Lab: Spreadsheet Applications for Scientists and Engineers* (Reading, MA: Addison-Wesley, 1988).
8. D. I. Dykstra, Jr. and R. G. Fuller, *Wondering about Physics...Using Spreadsheets to Find Out* (New York: John Wiley, 1988).
9. William J. Orvis, *1-2-3 for Scientists and Engineers* (Alameda, CA: SYBEX, Inc., 1987).

Introducing Computation to Physics Students

William J. Thompson
Department of Physics and Astronomy, University of North Carolina, Chapel Hill, NC 27599-3255

Computation is now so much a part of physicists' work that it is important for physics students to become familiar with computing early in their training. We haven't made computing instruction straightforward for our students; we often present a haphazard potpourri of computing topics that we were involved in as graduate students or that are drawn from our current research. It is unusual to give much thought to the pedagogical merits of the materials used or to their order of presentation. We also seldom consider the balance between physics, mathematics, numerical analysis, and programming. How can we remedy this deficiency in the physics curriculum and provide relevant training? Computational physics is a discipline that melds physics, numerical analysis, and computer programming. Unfortunately, there are few published materials that integrate these disciplines in a balanced way. If you wish to develop an introductory computational physics course, I recommend the following:

1. Be prepared. Your training is in physics; don't wing it on the programming.

2. Provide convenient and reliable access to computers for both teachers and students.

3. Do a little computing well, rather than attempt much and fail.

The course developed at the University of North Carolina at Chapel Hill over the past decade provides understanding of basic numerical methods in the physical sciences, teaches efficient programming styles, and gives practice with real-world problems within students' comprehension. Students who take this course, which is required for undergraduate physics majors, improve their understanding of how mathematics integrates with physics and of how theoretical physics relates to data.

A Multicourse Menu

What should be the prerequisites for students taking an introductory computational physics course? The level of the physics and mathematics included in our course is that of college sophomores and juniors for whom the numerics required to understand physics is becoming sufficiently complicated and laborious so that students are motivated to learn how to compute. Some of the topics are also suitable for seniors and beginning graduate students, because of the mathematical expertise required. Previous exposure to computer use through "computer literacy" courses and some introductory programming experience are desirable prerequisites so that students approach the novel technology without trepidation.

The science must be well integrated into the course, and this is not a trivial problem. A big hindrance to starting a computational physics course is the scarcity of resource materials integrating physics, numerical analysis, and programming. Therefore I have developed a broad menu of topics from which teachers can organize appropriate courses (Table 1). A variety of topics suitable for the course goals and level of student experience is shown in the menu. The order down the menu is roughly the order in which physics students acquire mathematics and physics sophistication, and in which the numerical analysis and programming skills required also generally increase. In the course menu, the variety of topics and the range of physical sciences covered, provide enough materials for at least two semesters, although our course usually runs for one semester.

How do the three columns in the menu interrelate for teaching computational physics? In applied mathematics we develop elements of mathematics underpinning the numerical analysis required in computational physics. I don't assume that

Table 1

Menu for Introducing Computation to Physics Students

Applicable Math	Application/Development	Computing Lab Examples
Complex numbers and exponentials	Complex plane; basis of FFT; phase angles and vibrations	Conversion between polar and Cartesians
Power series expansions	Practical uses of expansions; binomial approximation	Realistic convergence; numerical noise
Numerical derivatives and integrals	Richardson extrapolation; roundoff and truncation; trapezoid and Simpson rules	Electric potentials; computer graphics
Curve fitting	Cubic-spline fits	Interpolation
Least squares principle		Radiocarbon dating
Monte Carlo simulation	Random number generation; Monte Carlo integration	Round off; entropy; radioactivity
Fourier expansions	Discrete transforms, series, integrals; windows; FFT	Fourier analysis of EEG
First-order differential equations	Logs and decibels; logistic-growth curve	World-record sprints; war games
Second-order differential equations	Catenaries; resonance; solving stiff DEs	Ammonia resonance; Kepler orbits

these topics have been covered in mathematics courses because such an expectation is contrary to my experience and because the physics context provides a different viewpoint. Practicing physicists are often tempted to forego the mathematical and numerical analysis developments and to get to the heart of the physics by using numerical recipes. Having seen many woeful examples in research, I am convinced that early training that fails to develop an understanding of the analysis and an intuitive awareness of the limitations of recipes is inappropriate. There is a tendency for computational physics courses to overreach the student's conceptual level and to confuse an ability to program calculations with the maturity to understand results.

Physicists often stress high-level and theoretical physics when selecting course topics in computational physics because we still have the notion that computers are research instruments rather than tools in a technological society. Few students in undergraduate physics courses will become research physicists, so let's provide appropriate learning experiences.

Matching materials to the students' level can be straightforward. For example, at the introductory level, the conversion between plane polar and Cartesian coordinates has applications to the complex plane and to graphics, but the programming is tricky because of the quadrant ambiguity in going from Cartesian to polar coordinates. At the intermediate level, least squares is not only useful for fitting data but is the basis of Fourier expansions. At a more advanced level, efficient algorithms for numerical solution of differential equations have many applications.

Many of the developments and computing projects in the menu relate to data analysis. Examples are spline fitting and least-squares fits of radiocarbon data from Egyptian antiquities. Such an emphasis is worthwhile in an introduction to computation in physics for several reasons:

1. It is useful for students to understand computer-based data-analysis methods that they should be starting to apply in experimental labs.

2. Data analysis is a primary activity of both pure and applied physics, so it is important to illustrate completely new data-manipulation modalities that computing speed and logical complexity make possible. Examples are cubic splines and the FFT, both of which would be tricky and laborious if attempted without computers.

3. Analysis of data provides many opportunities to connect formalism, numerics, and physics. For example, the Fourier transform of an ammonia inversion spectrum between frequency and time domains helps develop the concept of complementarity before it is encountered in quantum mechanics.

One way to introduce computation in an undergraduate physics curriculum is to include it in a laboratory course, perhaps by paring away some of those nineteenth-century mechanics and electromagnetism experiments that are of doubtful pedagogical value in the late twentieth century.

Computing, Programming, and Coding

Since one goal of our course is to teach clear programming methods, it is important for instructors to maintain a clear distinction among computing, programming, and coding. For physics these should form a nested heirarchy. The overall activity is computing, which begins with physical, mathematical, and numerical analysis of a problem, from which an algorithm for its solution is developed. Then comes the programming needed to specify this algorithm for computer calculation. What output should the program produce, and what input is needed to do this? To link input to output requires a programming language to implement this program plan. Finally comes coding, which includes documenting and testing the program.

One aim of an introductory computing course should be to train students so that they will not attempt to jump straight from the physics to the coding: "Think for a minute, code for an hour, debug for a day." If we emphasize appropriate numerical analysis, algorithms, and program specification, the coding steps of building and testing become more direct and less painful. Program organizational aids, such as pseudocode outlines and templates, modularity through subroutines or procedures, and consistent code layout and documentation, are all-important.

How much should an introduction to computation in physics rely on software packages and how much sweatware should the student expend? I suggest that the balance should depend on the depth of the course, with once-over-lightly courses being package oriented, and courses primarily for hard-core physics majors emphasizing quite detailed program writing. For computer graphics, I am convinced that those who will later make serious use of graphics should go through the effort of understanding a simple graphics program, even if they don't write it themselves. My example is usually printer-character graphics in a display matrix, allowing only very low resolution but conceptually the same as bit-map graphics. We provide a simple source program, then suggest improvements for increased elegance. Matrix-manipulation algorithms and packages are topics I have not attempted, because the best algorithms (especially for matrix inversion and eigenvalues) are usually beyond the mathematics sophistication of undergraduate physics students. If we overemphasize the use of packages in teaching computational physics, we will rapidly produce young physicists with the ability to use programs but no ability to compute.

Practical Course Design

The prerequisites for designing an introductory course in computational physics are hardware, software, knowledge, and time. In our classes at Chapel Hill we usually have about 25 sophomores and juniors, so it is feasible to provide open access to a half-dozen microcomputers. We have also run the course with terminals connected to the mainframe, using both batch and interactive computation. The computer-memory requirements are modest, with about 5,000 words being sufficient. Input-output needs are also minimal (keyboard, monitor, and printer). The software required is very simple, needing an efficient editor and a good compiler of the languages used. Compiler speed is not always an advantage, since speed encourages the student to correct program errors a line at a time, whereas what students often need to do is to redesign the program or to rethink the physics.

What knowledge should a teacher of computational physics have? A familiarity with numerical analysis and computation in physics, at least through the level of topics in undergraduate physics, is necessary. The teacher should also have at least one year's experience in the programming language, as well as experience with the computer system used. A graduate teaching assistant can be a help, but should not be a substitute for a well prepared faculty member.

The time devoted to the course is an important consideration, since computational physics does not yet have its own niche. In our full-scale one-semester course, which also introduces some mathematical physics topics, about two hours a week are used for applicable mathematics, applications, and developments. Another two hours a week are spent in a computing-lab discussion session to discuss algorithms, programming, and practical debugging strategies. None of this time is spent sitting at computers, since the hands-on work is done as homework.

Flexible office hours, about 15 minutes per student each week, are needed to help answer physics and coding questions. A low-scale computation course could run in two hours a week of formal instruction if the emphasis were on algorithms and programming and if extensive use was made of reliable, well-documented packages.

Course credit should not emphasize formal correctness of programs. Credit should also be given for clarity, originality, and exploration. A major power of exploratory computer experiments is that they seldom go out of control. Students should be encouraged to learn physics from their correctly executing programs, constrained by the models used. For this reason, it is useful at this level to choose computing problems with limiting analytical solutions and for students to include these in their programs as test cases.

In my heirarchy of computing, programming, and coding, I place choosing a language at a fairly low level. Over the decade of our course development we have run the gamut from interpreted BASIC to FORTRAN and Pascal. I think that BASIC is suitable only to establish a student's rapport with computers, because it encourages undisciplined programming habits and is therefore a poor choice for scientists who will have to write or modify programs. Pascal demands mental rigor, so that an executable program is often the program you intended, but Pascal experts can write dense and obscure code that is difficult for others (or themselves six months later) to understand. In FORTRAN it's easy to be undisciplined and executable programs are easy to produce, but correct programs are much more difficult. I sometimes teach Pascal and FORTRAN in parallel, especially if many in the class have used Pascal in an introductory programming course. Because I don't have a strong preference between the two languages at this level, I provided sample programs in both Pascal and FORTRAN when I wrote an introductory textbook for computational physics, *Computing in Applied Science* (John Wiley, 1984). I have not tried C language, but its features may not be much used at this level. In a course emphasizing data analysis, it might be interesting to use the date structures available in Pascal or C, but not in FORTRAN.

Sample Projects

I now briefly describe sample projects from our introductory computational physics course. In the first project radiocarbon dates of Egyptian antiquities are compared with their historical age. The radiocarbon dating scale is subject to variations because of the fluctuating $^{14}C/^{12}C$ ratio in atmospheric CO_2. A straight-line least-squares fit of the radiocarbon age, r, against the historical age, h, provides a calibration for the former. The algorithm derived for this project is unusual because it allows for errors in both variables. For example, a fair estimate is $\Delta h = 10$ years (half a lifespan in ancient Egypt) and $\Delta r = 45$ years (the precision of radiocarbon dates for artifacts about 4000 years old). Under the approximation that the ratio of errors at each datum is fixed, it is simple to derive the formula for the least-squares line and it has the intuitive appeal that, in the space $(h/\Delta h, r/\Delta r)$, one minimizes the perpendicular distances between data and line rather than the vertical distances of conventional least squares. Also, the slope of r on h is the reciprocal of the slope of h on r. The least-squares line, $r = -140 + 0.80\ h$, indicates that in the Nile Valley radiocarbon time seems to pass about 20 percent slower than historical time. The classes' attempts to understand this result have lead to discussions of fluctuations in atmospheric CO_2 and of the origins of atmospheric radioactivity.

The second sample project illustrates our goal of using computation to give insight into physics. The concepts underlying transformations between representations, such as frequency and time, are important to develop as background to quantum physics. Fourier transform spectroscopy of the ammonia inversion resonance is a pedagogically rich example, because it also relates to the mechanics of forced oscillations and resonance. A Lorentzian resonance line shape in the frequency domain transforms into an exponential decay in the time domain. Use of the FFT algorithm to transform discrete data forms, the computation part of this project. I avoided introducing the FFT until I devised a clear way of visualizing it, so I developed a complex-plane representation to show the basis of subdivision strategy, which lies at the core of the FFT algorithm. The programming in this project stresses the importance of an efficient and clear algorithm, but one with tricky logic that is more suitable for computer chips than for human wits.

Rewards

A course introducing physics students to computation can be very rewarding for students and teachers, though it is often time consuming and laborious. I have found that the students relate the formal mathematics to physics, and they learn how to program efficiently. They also appreciate the pitfalls of computing and the important distinction between algorithms, programming, and coding. The integration of mathematics and numerics with physics greatly enhances their understanding of each field. Above all, in an introductory computational physics course it is important for both teachers and students to be aware that the purpose of computing is insight, not numbers.

Spreadsheets in Research and Instruction

Charles W. Misner

Department of Physics and Astronomy, University of Maryland, College Park, MD 20742

The combined computation and graphics power of spreadsheets like Lotus 1-2-3 and Microsoft Excel allows for quick and comfortable exploration of solutions to ordinary differential equations, and more. This tool finds application at many levels from research through nonspecialist physics instruction.

Spreadsheet Examples

Before describing the advantages, disadvantages, and applicability of spreadsheets in general, I will try to build a bit of common background by presenting three examples. These examples are called worksheets. That is the common usage to distinguish the application or worksheet (the user program) from the commercial program (compiler or interpreter) that makes it run, which is here a spreadsheet. The best known spreadsheets are *Lotus 1-2-3*, the current standard, and *VisiCalc*, the ground-breaking invention. The worksheet files whose operation is described here, may be downloaded from the AAPT computer bulletin board (301-454-2086, 1200 and 2400 baud) as the file "CPI_MSNR.ARC" which will expand (using "ARCE.COM" that you also download) into the three "*.WK1" files I now describe.

Radioactive Decay

The first example is a worksheet with the filename "CHANCE.WK1" that provides a simple simulation of the decay of a single radioactive nucleus. Constructing the main part of this worksheet is a suitable assignment for nonscience students in the first weeks of an introductory course while they are learning to use a spreadsheet. Screen 1 shows the principal part of this worksheet. The theory it incorporates is that at each discrete time step (counted in column A) the nucleus has a fixed probability (chance = 10 percent) of decaying. A random number is produced in column B, and the first time this number is less than 10 percent, the state of the nucleus (column C) changes from 1 (undecayed) to 0 (decayed). A formula in cell B4 (column B, row4) calculates the sum of all the numbers in column C (after the first1 in cell C10) to find the number of time steps elapsed before the decay. Pressing the Calc key ([F9] in *Lotus 1-2-3*) produces a new set of random numbers to repeat the numerical experiment. After constructing this worksheet, then, students can easily collect data on dozens of repetitions of this theoretical model of radioactivity. They

```
C11:  @IF(C10=0,0,@IF(B11>$CHANCE,1,0))                              EDIT
@IF(C10=0,0,@IF(B11>$A$6,1,0))

          A        B        C                    D
1                 Simulated Radioactive Decay
2     Phys 499F
3
4     lifetime:        5
5
6          0.1 = chance (probability of decay)
7            2 = run number
8
9     time step rand num     state
10        0                    1
11        1      0.541         1
12        2      0.563         1
13        3      0.699         1
14        4      0.644         1
15        5      0.753         1
16        6      0.031         0
17        7      0.944         0
18        8      0.767         0
19        9      0.377         0
20       10      0.098         0
04-Aug-88   12:05 PM                        CIRC
```

Screen 1. The main screen from "CHANCE.WK1" which models the probabilistic decay of a single nucleus in a number of discrete time steps. The top line of the screen shows the contents of the current cell (where a highlight would show on an actual screen). Because the Edit key has been pressed ([F2] in *Lotus 1-2-3*), the second line redisplays this information for possible modification, but translates named cell references into their row and column designation.

can see that lifetimes very much different from the average are not at all uncommon according to this theory, and they may suspect that experiments could tell whether this theory is better than a more classical decay theory that might govern, for instance, how long a watch battery lasts. Thus, in addition to learning how to use a spreadsheet, students a couple of weeks into their first physics course have a chance to see physicists' ways of seeking evidence for a remarkable assertion—that pure probability, free of all memory of antecedent provocations, may govern some aspects of the microscopic world.

This worksheet is simple enough that it can be constructed (programmed) during a short lecture, or reconstructed by the reader who has a spreadsheet program available. A worksheet not very different from screen 1 can be constructed by entering formulas or labels into nine cells and completing one /Copy command as codified in listing 1, which we now explain. The step-by-step procedure that this listing abbreviates is as follows:

```
A6: 0.1    B6:   "= chance
           B9:   " rand num   C9:    "state
                              C10:   1
           B11:  @RAND        C11:   @IF(B11>$A$6,1,0)
           Bnn:  ...          Cnn:   ...
           B70:  @RAND        C70:   @IF(B70>$A$6,1,0)
A4: "lifetime:
           B4:   @SUM(C11..C70)
```

Listing 1. A beginning in the construction of the radioactive decay worksheet. The cells in the range B12..C70 are filled by copying the pair
of cells B11..C11, but the other nine cells are filled by simply moving the cell pointer to each cell in succession and typing the
appropriate content that follows the colon after the cell address in
the list above.

A6: Using the cursor keys (or mouse) move the cell pointer (highlight) to cell A6
and then type the three characters "0.1" without the surrounding quotation
marks. As you type, the characters will appear on the top line of the screen (in
Lotus 1-2-3; bottom line in some clones) and you can correct typing errors
using the [Backspace] key. When you've got it right, press [←] and the number
will appear on the worksheet in cell A6 as desired. This number can be changed
at any time by repeating this procedure. Be sure that you do not include any
leading or trailing spaces in numeric cell entries.

B6: Move the cell pointer to cell B6 and enter the characters ""=chance" there. (To
enter something means to type it, followed by a press of the [←] or [Return].)
The """ double-quote character is insurance against your spreadsheet interpreting the "=" equal character as the beginning of a formula that would confuse
the spreadsheet's calculator capabilities. It means that the following characters
are just text to be displayed at this location.

B9: Enter the characters "" rand num" in this cell.

C9: Enter the characters ""state" here. A further effect of the """" double-quote
character is seen here. It not only identifies the entry as a label (text), but also
causes it to be right justified within its cell when there is room to spare.

C10: Enter the single digit "1" here. This is the initial condition that says we are
starting out with a nucleus that has not yet decayed.

B11: Here enter the characters "@rand". This instructs the spreadsheet to assign to
this cell a numerical value that is a (pseudo-)random number between 0 and 1.
The initial character "@" is the spreadsheet's clue to search its library of built-
in functions when evaluating this formula. In our simulation, this cell represents
the initial "throw of the dice" that calls upon pure chance to decide whether the
nucleus is to decay or not during this time step.

C11: Here enter "@IF(B11>A6,1,0)". As always, no spaces are allowed anywhere in a formula or number, but lowercase is equivalent to uppercase. The built-in @IF function checks the condition (here B11>A6, which tests whether the random number in cell B11 on this row is greater than the odds we specified in cell A6) and yields the following argument (here 1, the undecayed state) if the condition was true, otherwise it yields the final argument (here 0, the decayed state). The dollar signs in A6 have no effect here, but are important for the copy step that follows.

B12..C70: Everything preceding this could be done with a hand calculator or a typewriter, except for keeping it together on a single active page. But this copy step really saves time. Move the cell pointer to cell B11 and call up the spreadsheet's main menu (in *Lotus 1-2-3* and most other *VisiCalc* progeny this is done by pressing the solidus or fraction-slash key " / "). Select the /Copy command by either pressing C or by moving the highlight to Copy and pressing Enter. You will be asked where to copy from; respond by moving the cell pointer so that it covers both cells B11 and C11, then press Enter. (That procedure is called pointing to a range of cells.) You are now asked where to copy to; respond this time by typing the edge of the target range which is B12..B70. (To do this second part by pointing, press the Help key, F1 in *Lotus 1-2-3*, and ask about pointing to ranges to learn the further clues you need.) When you press the Enter key to complete the specification, columns B and C will be filled with the appropriate formulae.

A4, B4: To summarize the computations, not all of which are visible on the screen at one time, enter the contents shown in the listing for these cells. The built-in @SUM function adds all the numbers in any list of cells and blocks of cells.

Note that much of the bulk in this detailed prescription is first-time learning of some basic spreadsheet skills such as entering text and formulas and using the /Copy command. After just a small amount of practice, students find a listing just as useful and much briefer than such a key-by-key prescription. In listing the use of the /Copy command is indicated by the row of ellipses "...".

Now that we have a worksheet model, does it do what we intended? Concurrent debugging is one of the important attractions of spreadsheet programming. Depending on the luck of the dice, the first screen display from this worksheet may have shown anomalies, or the user's suspicions may not have been aroused until half a dozen successive lifetimes—the Calc key, [F9], gives a fresh run—were reported as over 40 when an average 10 was expected. There is no need to learn how to use a separate debugging program; just pressing [PgDn] and inspecting the later time steps of almost any run of this program will show the problem. The logic in cell C11 that does the first step is not adequate for later steps: a nucleus cannot undecay. The cure is to change the formula in cell C11 to that shown at the top of screen, namely C11:@IF(C10=0,0,@IF(B11>A6,1,0)).

The logic for this in Pascallike pseudocode would be

```
IF  dead
    THEN  stay dead
    ELSE   IF beat the odds
           THEN alive
           ELSE dead.
```

After correcting this formula it is, of course, necessary to copy it down the remainder of the column.

By inspecting the column of copied formulas, you will see that the /Copy command by default treats cell addresses as relative addresses, so that if the model line uses the random number in the same line to evaluate the @IF statement, every copy also uses the random number in the cell to its left. This is the appropriate treatment for variables that change in every row to keep current. For constants, however, one wants to reference the same cell in every copy of the formula. To mark a reference as a constant, *Lotus 1-2-3* requires that one prevent adjustment of the row and column addresses in copies by marking them with $-sign prefixes as in $CHANCE ∫ A6.

By completing the preceding instructions, students learn enough to make a spreadsheet a practical tool, and see how little programming overhead is required to give themselves enough computational power to solve differential equations before they study calculus. A spreadsheet offers many other conveniences that good students will learn in the course of a semester, such as naming variables, inserting new rows and columns, or moving blocks of working code, and printing formula listings. The only essential we have not described is graphing, which amounts to calling up the /Graph menu and pointing out which column is to be plotted against which as y versus x.

The distributed file "CHANCE.WK1" contains additional computations that would be used as lecture demonstrations rather than student homework assignments in early weeks of an introductory course. Lotus *1-2-3* and its clones contain a /Data Table command that will run the worksheet many times (and when appropriate, with a specified list of input data) and report desired results. Here this command will make a table showing the lifetimes realized in a few hundred runs of the decay simulation, providing data for a statistical analysis. The /Data Distribution command will sort this data into bins for a histogram plot, which can be then graphed as shown in Figure 1 along with an easily computed continuum-limit theory for the expected average behavior.

Simple Pendulum

This next example is the worksheet "SIMPLPEN.WK1" that solves for the motion of a simple pendulum (point mass on the end of a light, rigid, pivoted rod). It is in the form that would be appropriate for a homework assignment in a calculus-based introductory physics course for engineers and scientists a couple of months into the first term after several dynamics problems have been solved using spreadsheets in previous assignments. One assumes that, from previous assignments, students have

Simulated Radioactive Decay

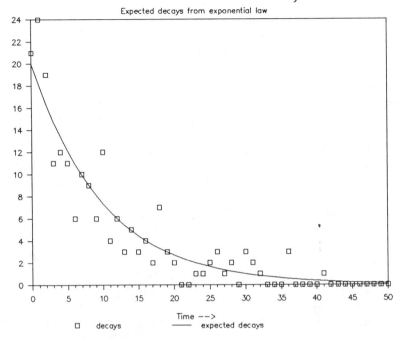

Figure 1. Histogram of the lifetimes realized in 200 runs of the radioactive decay simulation, compared to the average behavior expected in large numbers of runs.

```
C23: +ANGLE+ANG_VEL*$TIME_STEP                                    EDIT
+C22+D22*$E$17

          B         C         D      E       F       G        H       I
 9   Construction parameters:
10           Mass of bob:             0.1 kg      Gravity:    9.8 m/s^2
11           Length of shank:          2 m        Other:      4.9 = g/L
12                                                            19.6 =g*L
13   Initial conditions:
14           Initial Angle:           3.1 rad     Ang Veloc   0 rad/sec
15
16   Approximation:
17           Time step:         0.133333 sec
18
19   Resulting Period:          9.599999 sec     Nominal: 2.838453 sec
20
21   Time     Angle   Ang_Vel Ang_Accel         Plot_vel Energy
22       0      3.1  -0.01358 ?                     0     ?
23  0.133333 3.098188 -0.04193 ?                 -0.02775 ?
24  0.266666 3.092598 -0.07392 ?                 -0.05792 ?
25  0.399999 3.082741 -0.11235 ?                 -0.09314 ?
26  0.533333 3.067760 -0.16054 ?                 -0.13645 ?
27  0.666666 3.046353 -0.22267 ?                 -0.19161 ?
28  0.799999 3.016663 -0.30408 ?                 -0.26338 ?
04-Aug-88  12:12 PM                              CALC
```

Screen 2. A simple pendulum model as presented to students. The familiar formulas are given, as well as column (and graph) labels and an I/O area where the controlling parameters are displayed in an orderly way. The new physics in the torque and energy columns remains to be supplied by the student.

moderate skills in using spreadsheets, and that they have previously constructed complete worksheets to solve for one-dimensional motions of masses subject to forces. Thus they are given a nearly complete worksheet as shown in screen 2 with only the new physics left for them to complete.

This pendulum example illustrates that the use of spreadsheets (or other programming languages) by students may reduce the calculus skills required but does not replace analytical physics reasoning. To complete this assignment, the student must still have learned to sketch the geometrical (kinematic) description of the apparatus and to relate it to a free body diagram of the forces in play. Here the two forces acting on the point mass are a gravitational force Mg acting downward and an unknown tension force T acting toward the pivot point. The strategy of using $\tau = I\alpha$ instead of $F = Ma$ to obviate an evaluation of the tension force is implicit in the given spreadsheet outline, but should be discussed explicitly. Students will have to use paper and pencil to find the torque formula

$$\tau = \text{`--}(L \sin \theta)Mg$$

and the moment of inertia $I = ML^2$ and to solve for the angular acceleration $\alpha = -(g/l)\sin \theta$, as well as the corresponding energy formula. When these are entered into the spreadsheet and copied down their appropriate columns, the completed spreadsheet described by Listing 2 will result.

Students must be taught not only to construct spreadsheets to solve physics problems, but also to use them to explore how well they model the real world. Here they encounter an entirely different (and better) idea of what it means to "solve a problem." The solution is no longer a simple number or algebraic expres-

Time	Angle	Ang_Vel	Ang_Accel
t_0:	θ_0:	$\Omega_{1/2}$:	α_0:
0	θ_{init}	$\Omega_{init} + \alpha_0 \, dt/2$	$-(\mathbf{g/L}) \sin \theta_0$
t_1:	θ_1:	$\Omega_{3/2}$:	α_1:
$t_0 + dt$	$\theta_0 + \Omega_{1/2} \, dt$	$\Omega_{1/2} + \alpha_1 \, dt$	$-(\mathbf{g/L}) \sin \theta_1$

Listing 2. This shows the essential formulae for solving the differential equation for the pendulum motion. It corresponds to rows 21 through 23 in Screen 2, columns B through E. (The contents of input cells and their labels are evident from screen 2.) The format of this listing, called a "schema," differs from listing 1 in that it uses mathematical notation and expects students to translate to the detailed syntax of their spreadsheets. Each cell that is not just a label shows first the mathematical quantity stored there and then, after a colon, the formula by which it is computed (which students must enter into the worksheet). The last row is copied down many times to solve the differential equation, and constants that must be marked for absolute addressing during this copy process are shown in bold.

sion that could evaluate to a number. *The solution is a working theoretical model of the target physical system.*

Physics teachers used to imagine that conceptual models were being constructed inside students' heads, similar to the ones good physicists have inside their heads. End-of-chapter exercises were debugging tools to test this hypothesis—if students produced the same answer physicists would, that was taken as evidence that everyone was working from the same internal model. But students develop strategies to manufacture the evidence instead of the model, defeating teachers' attempts to convey insight. Now the model can be more external. It can be a Pascal program or a Lotus worksheet. The teacher can inspect it directly, and the student can see it and direct his or her energies toward it instead of toward its peripheral manifestations in a number or two.

Exploring a model must be taught. The simple pendulum worksheet, once completed, will at the press of the Calc and Graph keys ([F9]and [F10]), produce the graph shown in Figure 2. The student must be questioned about it. What (in a sen-

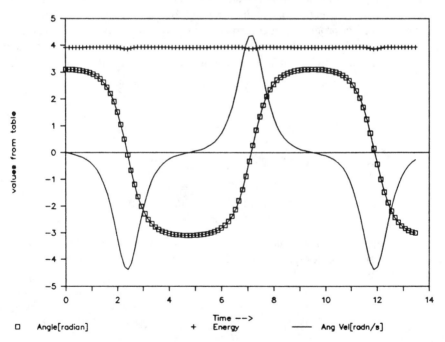

SIMPLE PENDULUM

Figure 2. A simple pendulum graph produced from the data shown in screen 2. Students cannot make sense of this graph until they learn to read the axis scales and legends, and until they recognize an angle of 3.1 radians as being near π and orienting the pendulum nearly upside down.

tence or two of English) is going on here? Why doesn't it look like a simple harmonic oscillator? Can someone make the real apparatus on the lecture demonstration table move that way? Other questions are needed to get students to change some of the input data. Can the model reproduce the simple harmonic motion one sees in a grandfather clock pendulum? How many other qualitatively different motions can the spreadsheet model produce?

The ease of making graphs also allows one to introduce the phase plane momentum versus position, in the introductory course. This gives students a tool that is useful for statistical mechanics, for Bohr-Sommerfeld quantization, and for the qualitative treatment of nonlinear differential equations.

White Dwarf Equation of State

This final example represents the use of a spreadsheet for research purposes. It is a curve-fitting problem that only has to be done once, so that it is inappropriate to use a hard-to-write program. The interactive nature of a spreadsheet allows the investigator to explore variations in the approximate formula that is being used to fit the data, and then to choose best-fit parameters.

This example problem is to find a direct formula ρ (P) for the equation of state of a degenerate ideal Fermi gas, including relativistic effects. The intended use is to simplify another program that constructs models of white dwarf stars where this equation of state describes the pressure of the degenerate electrons that support the star against its own gravity. An implicit formula can be found in the textbooks; it is

$$\rho = \rho_0 x^3$$

$$P = P_0 [x(2x^2 - 3)\sqrt{x^2 + 1} + 3\ln (x + \sqrt{x^2 + 1}\,)]. \tag{1}$$

From the asymptotic forms that these equations give,

$$\rho \propto P^{3/5}, P \text{ low}$$

$$\rho \propto P^{3/4}, P \text{ high}, \tag{2}$$

one can conjecture an approximate formula that will work at both low (nonrelativistic) pressure and at high (relativistic) pressure. It simply combines the two asymptotic forms:

$$\rho = \rho_0 \left[\left(\frac{5P}{8P_0}\right)^{3/5} + \left(\frac{P}{2P_0}\right)^{3/4} \right]. \tag{3}$$

The first step in the spreadsheet use is to test this proposal and to see how much difference there is between equation (1) and equation (3).

A simple spreadsheet can be constructed by making columns for x, P, and ρ using equation (1), and then making a fourth column in which ρ_{apprx} is computed from P using equation (3). To compare the two formulae one might then show the two ρ's on a graph against P, but since a white dwarf will have a very wide range of density and pressure from center to surface, it is more reasonable to make a log-log plot. This requires three more columns in the spreadsheet, log P, log ρ, and log ρ_{apprx}. They take hardly longer to program than writing the labels at the tops of the columns, since the spreadsheet has a built-in @LOG function that can be typed into one cell and then copied in one operation to the three columns. A reasonable range for the independent variable x is found to be 10^{-3} to 10^5, which could be done in 41 rows by setting $x_{i+1} = 10^{1/5}x_i$. Over such a wide range the errors don't look too bad on a log-log plot, but inspecting the tabulation shows errors as high as 50 percent around $x = 1$. (The variable $x = p_F/m_e c$ is the Fermi momentum of the electrons in relativistic units.) Also, "around $x = 1$" means for several decades in P around $x = 1$, so these really are substantial errors that one could hope to improve upon.

As an improved formula, one could think of blending the two asymptotic formulae in a root-mean-square sense instead of just adding them. But why a square here, rather than a cube? So one decides to try an arbitrary power:

$$\rho = \rho_0 \left[\left(\frac{5P}{8P_0} \right)^{3n/5} + \left(\frac{P}{2P_0} \right)^{3n/4} \right]^{1/n}. \tag{4}$$

To choose n we need to compare the results of several different cases, and to see the differences we need a numerical measure of the error of fit. I take this measure to be proportional to the area between the exact and approximate curves on the log-log plot. To do the integral that measures this area requires two more columns on the worksheet. One column is the difference

$$| \Delta \log\rho | = | \log\rho - \log\rho_{apprx} |;$$

the other accumulates this multiplied by $\Delta \log P$ using a trapezoidal rule to evaluate the integral

$$\int | \Delta \log \rho | \, d(\log P) = \int \left(\frac{| \Delta \ln \rho |}{\ln 10} \right) d(\log P) \approx \frac{1}{\ln 10} \int \frac{| \Delta \rho |}{\rho} \, d(\log P).$$

To give our error measure a simple interpretation as the fractional error in ρ times the number of decades of P with that error, we therefore define

$$\text{Fit\_Error} = (\ln 10) \int | \Delta \log \rho | \, d(\log P) \tag{5}$$

and report this quantity in the I/O area of our spreadsheet. With this measure, the first trial fit from equation (3) gives an error of 416 decade-percent, i.e., an average 40 percent error throughout 10 decades of P, using $n = 1$. Similarly for $n = 2, 3$, and 4 one finds errors, respectively, of 60, 6, and 29 decade-percent, so that substantial improvement can be found near $n = 3$.

The spreadsheet allows more systematic searching for the minimum error by using the /Data Table command. In an empty part of the worksheet we use /Data Fill to make a column of numbers from 2 to 4 at 0.1 intervals. Then the /Data Table command will on request, substitute these numbers successively for n and report in a parallel column the corresponding fit errors. (The computation time on an ancient PC is 30 seconds, the programming only a few times that.) From these one can then quickly make the graph shown as Figure 3. Repeating the /Data Fill command with a narrower range, e.g. 2.8 to 2.9 at 0.005 intervals, followed by a press of the Table key F8 to repeat the data-table construction, gives a redone graph with more precision near the minimum error. Gaining an order of magnitude precision every minute or two manually this way, is much more practical than designing a program to find a minimum automatically. In addition, one might learn something unexpected from looking at the numbers and graphs as they skitter by.

One point I noticed while finding a good value of $n = 2.856$ is that there were errors at small values of x where one expects the asymptotic formula to be good. These turned out to be errors in evaluating the "exact" formula for P from equation (1). You can discover by deriving its asymptotic form for small x that the coefficients of *four* terms in the Taylor series (x^1 through x^4) cancel between the logarithm and the other term. Even with 15 decimal digits of precision, round-off errors invalidate this formula for numeric computation for small x, and only the asymptotic formula is accurate.

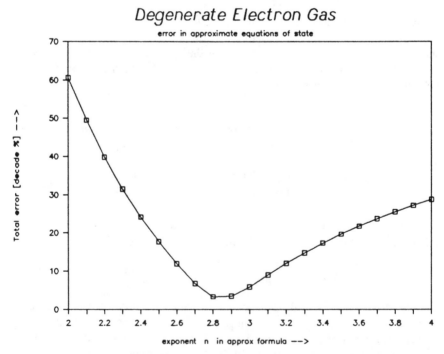

Figure 3. The error in fitting an approximate simple formula to data from an implicit equation for $\rho(P)$ is shown as a function of a parameter n in the approximate formula.

Spreadsheet Survey

Now that we have seen a few examples of spreadsheet applications, I will try to sketch out the areas in which I am aware of their usefulness to physicists, and some of the reasons that make them valuable in particular applications. Dewey has also addressed these questions in this session and there is substantial overlap in our assessments.[2]

A first consideration is accessibility, so it is important to realize that spreadsheets do not need to cost a lot apart from the computer, and that they do not take long to learn if one has an appropriate application on which to try them out. The best-known spreadsheet, *Lotus 1-2-3* has a list price of $495 and is widely available for $300. This is high for widespread student use, but there are good alternatives. One is the Borland *Quattro* spreadsheet, which has a special academic price to students and faculty near $45 (call Borland educational sales department at (408) 438-8400 for information). Another is *Joe Spreadsheet* from Holt, Rinehart, and Winston (call director of professional sales, (212) 614-3360) at a special academic price of $20. It runs complicated *Lotus 1-2-3* worksheets smoothly and has an excellent manual. The only serious disadvantage of the better *Lotus 1-2-3* clones is that they occupy more memory than the *Lotus* product. (*Lotus* has an inexpensive student edition without this problem, but its files are not compatible with the regular *Lotus 1-2-3* that may be found on students' parents' machines, or in the instructor's office, or in computer workrooms elsewhere on campus.)

Learning to use a spreadsheet requires about four hours for a faculty member who is comfortable with a personal computer in other applications, e.g., word processing. Of this time, about half should be spent working through the *Tutor* program that is supplied with version 1A of *Lotus 1-2-3*, and the other half working small examples such as a travel expenses statement or a solution of a one-dimensional conservative force motion modeled on listing 2. Tutorials are available in many varieties but use primarily business examples. The important parts to learn are the introductions (cursor control, file handling, etc.), basics (copying, pointing, built-in functions), and graphing. Data base and macro capabilities can be deferred until you acquire expertise.

Teaching Uses

There are appropriate uses for spreadsheets in teaching physics at all levels from graduate school through high school.[3] The most likely use among graduate teaching assistants would be to maintain a gradebook for the professor in a large, multi-section class. This use is, of course, just the kind of business format for which the programs were designed. After learning spreadsheets in this context, the graduate student would likely find uses for it in his other research as well.

Research students, by which I mean either graduate students or upperclass undergraduates oriented toward graduate school, should give Pascal and FOR-TRAN priority as they develop professional computer skills. Many, however, will find a spreadsheet easy to learn and very useful as described below under "Research Uses." It can also be used in mainstream courses[4] or to analyse the moderate amounts of data that are typically collected in experiments in laboratory courses at this level.

Students in introductory physics courses for physics majors, engineers, and other scientists, can benefit from learning to use spreadsheets because this takes less time than learning a conventional programming language. The simple pendulum model described earlier illustrates how the content of an introductory course can be modified to use student programming skills to achieve a better understanding of physics. Dykstra and Fuller have many useful suggestions.[5] Together with Pat Cooney, I am writing a book[8] to supplement standard texts that will offer more complete suggestions in this area and will provide the written support that students need.[6]

Nonscience undergraduates can be taught (as I have done in two courses) to use spreadsheet skills as a substitute for calculus skills. They may be attracted to a physics course that uses spreadsheets, because many of them are aware that spreadsheets skills are valuable in a large variety of nonscientific fields. The radioactive decay example presented earlier illustrates how teaching the use of a spreadsheet and teaching physics can proceed concurrently. Although a calculus prerequisite is desirable (to assure some competence in algebra) and is required in many business school programs, one should not anticipate significant skill or understanding of calculus. Besides, calculus is not necessary. Newton's second law can be phrased "Force causes a change in velocity," and programmed that way using small finite steps dt in a spreadsheet. Planetary motion can be solved in rectangular coordinates in the same style. You can emphasize the basic point that a fundamental law like Newton's, can lead to a variety of very detailed predictions without tempting students to memorize a hundred formulae that they will justifiably forget.

But the population that can benefit most from the introduction of spreadsheets is high school students. Spreadsheets are being used in high schools now. I expect that one could at least double the fraction of high school students who imagine themselves using mathematics in a future career if spreadsheets were widely used in high school science, mathematics, and economics. My basis for this nonscientific conjecture is the astounding fact that in the last five years, several million business men and women have learned to program computers productively using spreadsheets. Most of them do not regard constructing worksheets as computer programming, and very few use any calculus or other higher mathematics. Yet they are happily optimizing and projecting mathematical models of complex business situations. No ordinary differential equation on a spreadsheet is qualitatively different from compound interest, so we should regard differential equations as accessible at the high school level. Using finite difference forms, students should learn differential equations before they learn calculus. The application of differential equations to biological populations, chemical reactions, mechanics, electricity, and radioactivity are much more interesting than the usual calculus and precalculus problems, so motivation to learn should be improved. Later, calculus will give a deeper insight into an area with which the students are already familiar, just as grammar is normally taught to students who already know how to speak.

Research Uses

Some research uses of spreadsheets correspond to the purposes for which they were designed. These include the preparation and maintenance of grant budgets, cost estimation for experiments, and routine reports such as travel expense statements. The major advantages of spreadsheets are their ability to mix text and numbers, their correspondence to familiar pencil-and-paper preparation of the same sort of documents, and their ability to print out reports in familiar formats.

An analogous use that Professor J. P. Richard has shown me, is the preparation of error budgets for experiments as they are being designed. Here the noise characteristics of various blocks in the signal-detection and processing chain are typically computed separately and combined as functions of the various design parameters that can be adjusted. The controlling input parameters and various output noise and sensitivity figures can be arranged in a conveniently viewed and printable format, with the detailed calculations done in scratch areas when they don't fit easily into the report cells. Then the experimenter can interactively explore a variety of trade-offs by varying the controllable parameters to search for an optimal design.

Physics extends the use of spreadsheets far beyond the applications contemplated by the original designers. One such use is for calculations and graphs auxiliary to a larger project. An example is the white dwarf curve-fitting worksheet described earlier. Orvis includes a variety of other examples.[9] Others are small projects of any kind, theoretical or experimental, or first-try explorations to survey the landscape where a large project is contemplated. The ideal of a "back of the envelope" calculation can be extended by a couple of orders of magnitude in complexity, while retaining the same spontaneity and immediateness by using a spreadsheet instead of a piece of scrap paper.

Two characteristics make a calculation appropriate for a spreadsheet: a high ratio of design time to run time, and the need for small amounts of data. One should be reluctant to spend several person-months writing a program that will be run only 50 times at 30 seconds each on a personal computer. Better to spend a couple of person-days designing a worksheet that will take five minutes for each run. Small amounts of data (less than about 10^4 numbers) can also be explored interactively on spreadsheets. One would not want to type in more than a few dozen measured numbers, but many instruments can be programmed to produce ASCII files of measured data. Any file that would produce a tabular listing when copied to a printer can be read by a spreadsheet and, with the /Data Parse command, converted to one datum per cell in a spreadsheet. With the data thus organized by columns in the spreadsheet, they can be graphed, combined, fit, and otherwise manipulated interactively with great convenience. Built-in facilities such as least-squares fitting and matrix multiplication are valuable tools in such exploratory analyses.

Spreadsheet Qualities

Spreadsheets have many advantages over standard computer languages and some serious limitations. Among their most important advantages is convenience. They are easy to modify, edit, and reorganize. They are also reasonably powerful. They can solve systems of ordinary differential equations, or simple two-dimensional partial differential equations. Other important conveniences are their self-reporting capabilities and their concurrent debugging proclivities.

By self-reporting I mean that text and numbers can be intermixed on the same screen and in adjacent cells, so that the programmed storage locations for important data can (appear to) be the entries on the printable pages or viewable screens where data input and output is organized for convenience of comprehension and use. In most other computer languages, one could design and code a complete computation while leaving 90 percent of the job (I/O and user control) still undone. With a spreadsheet, these scientifically noncentral parts of the program are seldom more than 20 percent of the job.

The "CHANCE" worksheet above illustrates concurrent debugging. It refers to the fact that one normally supplies sample data before beginning to specify the computations when constructing a worksheet, and thus sees numerical results at every step during the programming. If one has a reasonable understanding of the example one uses during this construction process, errors in logic often produce recognizably spurious answers at intermediate steps. In this way it is likely that many errors will be caught, one at a time, as the programming (worksheet construction) goes on. With conventional languages it is more likely that, when the first test case is run, multiple interacting errors will make the debugging effort rather difficult. (There are practices in software engineering that mostly avoid this problem, but these are in even less common use by physicists than spreadsheets.)

Spreadsheets also have the advantage of being easy to learn, widely used, and widely applicable, all points that have occurred earlier in this discussion. The reasons for these qualities are several. One is the paper-and-pencil metaphor that is at the heart of spreadsheet design. This appearance makes the computer accessible to nontechnical people who are afraid they have no idea what goes on inside a computer. The spreadsheet converts the computer from a mysterious black box that reacts in nearly unpredictable ways, to a very concrete device that shows on screen every number it is calculating. One seems to be doing on screen exactly what one was accustomed to do on paper, except with a number of very powerful conveniences.

The most widely advertised convenience is the "what if" capability. If one input number is changed, one need not rewrite the entire worksheet; every dependent calculation will be redone automatically (or upon request if preferred). In effect, the spreadsheet programs while you calculate. Your effort is comparable to (but easier than) doing a computation with a hand calculator. But after you have done it once, the spreadsheet has learned how to repeat it for you with new data; it contains a program ready to run. I like to call this capability of the spreadsheet the service of a "invisible assistant." Early computational physicists such as Chandrasekhar and Hartree moved into positions where they could recruit persons called computors who would, once given the scheme and an example or two, do repetitive calculations using mechanical desk calculators and standard bookkeeping columnar paper (paper worksheets).[8] (Hylleraas would appear to have done all his own calculation throughout a long and distinguished career.[9]) The electronic spreadsheet provides the same services as these assistants (or an author's personal labor). It will repeat a complete calculation with new data, or (via the /Copy command) it will repeat a single step many times with evolving data. A further important service of the "invisible assistant" in current spreadsheets, is its ability to make graphs quickly from any columns of data that you point out.

In spite of its many advantages, a spreadsheet is not appropriate for every job. It occupies a particularly important niche in the ecology of computation. It is adapted to computations that involve limited amounts of data (e.g., 10^4 items), and that need limited logical control. It also demands relatively powerful computing facilities. Compared to a FORTRAN program doing the same principal computation, it "wastes" up to 95 percent of the computer's CPU and memory on the job of providing a convenient user interface, but that is usually considered an advantage now that programmers are so much more expensive than hardware.

Some remaining disadvantages of spreadsheets might be reduced if future computational environments are created with sufficient ingenuity. One major disadvantage of a spreadsheet, is that its logical structure is hard to display. Whatever degree of logic display is achieved is entirely the responsibility of the programmer; it is not demanded or even encouraged by the spreadsheet environment itself. As a consequence, programming errors that escape the first concurrent debugging phase are hard to detect. There is little systematic help built into the spreadsheet, so a heavy responsibility falls on the user to design a stringent suite of test cases to check that the worksheet does what it is expected to do. These are the same skills we have always taught graduate students who, however, need them as much to check for arithmetic and algebra errors as for errors in logical design.

Supported in part by a grant from the Fund for the Improvement of Post-Secondary Education (U.S.D.E.) with the further assistance of an equipment grant from IBM under its AEP program.

1. C. W. Misner, "Spreadsheets Tackle Physics Problems," Computers in Physics **2**, 37 (May/June 1988).
2. D. J. Dykstra Jr., "Wondering about Physics...Using Computers and Spreadsheet Software in Physics Instruction," in this proceeding.
3. J. R. Christman, "Invited and Contributed Papers on Spreadsheet Physics," AAPT Announcer **18** (May 1988); F. Griffin, D. Dykstra, and L. Turner, "AAPT Workshop—Exploring Physics with Spreadsheets," AAPT Announcer **18** 42 (May 1988); C. W. Misner and P. J. Cooney, "AAPT Workshop—Introductory Physics Using *Lotus 1-2-3* (and clones) on PCs," AAPT Announcer **17**, 39 (December 1987).

4. J. V. Kinderman, "A Computing Laboratory for Introductory Quantum Mechanics," in this proceedings.
5. D. I. Dykstra Jr. and R. G. Fuller, "Wondering About Physics…Using Spreadsheets to Find Out" (New York: John Wiley, 1988).
6. C. W. Misner and P. J. Cooney, "Spreadsheet Physics" (Reading, MA: Addision-Wesley, 1989).
7. W. J. Orvis, *1-2-3 for Scientists and Engineers* (San Francisco: Sybex, 1987).
8. S. Chandrasekhar, "The Highly Collapsed Configurations of a Stellar Mass," Mon. Not. Roy. Astr. Soc. **95**, 207 (January 1935); D. R. Hartree and W. Hartree, "Results of Calculations of Atomic Wave Functions III—Results for BE, CA and HG," Proc. Roy. Soc. **149**, 210 (1935).
9. Egil A. Hylleraas, "Uber den Grundzustand des Heliumatoms," Zeitschr. f. Phys. **48**, 469 (1928).

Teaching Computational Physics

Paul L. DeVries
Department of Physics, Miami University, Oxford, OH 45056

Mathematical physics has been the backbone of the undergraduate physics curriculum at Miami University since the days of George Arfken and his definitive text on the subject. Mathematical physics at Miami is taken in a two-semester sequence, normally during a student's junior year. In recent years, an increasingly important component of that sequence has been computational physics. In this talk, we will outline the history and motivation behind computational physics, its philosophical biases, and some of the actual content of the computational physics course as it is now being taught at Miami University.

There is, of course, some question of what computational physics is, or what it should be, and what (if any) are its limits. Certainly computational physics involves far more than evaluating integrals numerically or solving differential equations—although these are necessary tasks. We always try to convey to our students the sense of Hamming's admonition: "The purpose of computing is insight, not numbers."

Ideally, computational physics should be a synthesis of many topics: numerical analysis, computer programming, and most important, physics. Although our students are often competent in each of these areas in isolation, they may not be ready or able to pull it all together. Sometimes the curriculum itself adds to the problem; overcompartmentalization has sometimes meant that there has been little or no carry-over from one topic to another. Thus we find that much of our time is spent in illustrating the connections among topics. Sometimes, too, the student has an exaggerated notion of the "status" of the computer. We must constantly remind students that the computer is but a tool that we use in our work in physics—an assistant in our endeavors. We must never surrender the physics and ourselves to the machine.

Using Computer Experiments in Standard Physics Courses

Jan Tobochnik
Department of Physics, Kalamazoo College, Kalamazoo, MI 49007

Students need to learn how to use computers in the way that physicists use computers in their research. As a side-effect, these computational methods frequently provide educational benefits not found in traditional modes of instruction. I will present three ways of using computational physics in undergraduate physics courses: (1) replacement for two, three-hour labs in the standard calculus-based introductory physics sequence, (2) part of homework problems for sophomore-level mechanics, and (3) a weekly laboratory component for a thermal or statistical physics course. Pedagogical issues will be stressed, but some attention will be paid to programming details.

I will present examples of programs used in courses and the written material accompanying the programs. In some cases my text, *An Introduction to Computer Simulation: Applications to Physical Systems*[1], was used. The programs presented are written in *True BASIC*, a structured dialect of BASIC that is very similar to FORTRAN 77 and is identical on both the Macintosh and MS-DOS machines. The source code for the programs are available to the student and the student is expected to make minor modifications in the program and occasionally write their own programs based on model programs they have already seen.

I use the term "computer experiments" in the title of this talk because my use of the computer is different from the standard use of computer simulations. When using computer simulations, the student typically has no access to the source code. The emphasis is on interpreting the graphs or animation seen on the computer screen. In addition to this kind of interpretation, my goal is to show the connection between physics and simple numerical methods, to show how the simulation is produced by the computer, and to introduce the student to programming. Thus, these exercises allow the student to develop or extend programming skills while solving actual problems in physics.

Sample exercises to be discussed will include: (1) one-dimensional motion (introductory physics); (2) two-dimensional motion (introductory physics); (3) approach to chaos for a nonlinear oscillator (sophomore mechanics); (4) ideal gas in the microcanonical ensemble (thermal physics); and (5) magnetism (Ising model) in the canonical ensemble (thermal physics).

In the first two exercises the students learn how different functional forms for the force lead to a variety of types of motion for the particle. The Euler algorithm modified by Cromer[2] is used to solve the equations of motion. In the one-dimensional case, the position, velocity, and acceleration are plotted as a function of time. In the two-dimensional case, the trajectory in position, velocity, and acceleration space is plotted. Students make a hard copy of what they see on the screen, make measurements on the graphs, and interpret the various features of the graphs. Where possible, the students compare the numerical results to analytical results. Examples used in class include free fall with and without friction, projectile motion with and without friction, harmonic motion with and without damping and external forcing, and planetary motion.

In the third exercise a nonlinear damped forced harmonic oscillator displays the period-doubling approach to chaos. The computer simulation produces a Poincaré map.

The last two exercises use Monte Carlo methods to illustrate ideas in statistical mechanics and to introduce the student to numerical techniques used in current research. From the microcanonical ensemble, one can empirically discover the Boltzmann distribution, which is the basis for the canonical ensemble. Ideas such as the equation of state, phase transitions, and hysteresis can be illustrated using these methods.

I will show and briefly explain the programs to carry out these simulations, emphasizing the simplicity of the programming. As I go through the examples I will point out what actually happened when they were used at Kalamazoo College, what ideas appeared to be better learned using the computer versus traditional methods, and technical details to watch out for.

1. H. Gould and J. Tobochnik, *An Introduction to Computer Simulation: Applications to Physical Systems*, parts 1 and 2 (Reading, MA: Addison-Wesley, 1987).
2. Alan Cromer, "Stable Solutions Using the Euler Approximation," Am. J. Phys. **49**, 455 (1981).

An Example of "Task Management" in Constructing a Computer Program

Edward H. Carlson
Department of Physics and Astronomy, Michigan State University, East Lansing, MI 48824

Whether laboratory or computational, physics tasks ideally follow a common profile: analyzing the problem, devising modes of solution, implementing these methods while tracking progress, testing the results for correctness, and reporting or delivering the results in a clear package. However, while doing research, one commonly hacks out new paths through the jungle, and organized progress along the above profile is rarely completely achieved. The formal method of "task management" can serve as a foundation skill for well-organized work habits in laboratory work, computer programming, and possibly paper-and-pencil problem-solving by physicists.[1]

The new discipline of software engineering[2] applies task management concepts to large problems (constructing commercial computer programs of 30,000 to 10,000,000 lines of code). Suitably truncated and modified for the smaller programs (30 to 1,000 lines) that the typical physicist may need to construct, task management methods can improve the productivity and quality of programming, and serve as one mode (probably the most transparent) of teaching generalized task-solving skills.[3]

Compared to a typical laboratory measurement task, software construction is filled with detail and endowed with near-infinite tractability.[4] The programmer is enticed to jump back and forth along the task profile, becoming quite muddled in the process. The cure is to modularize both the process of constructing the code and the code itself. In fact, hierarchically arranged modules are a nearly universal human solution to the famous "seven, plus or minus two" capacity of our minds to hold thoughts.[5]

By employing a formal method with definite criteria for closure of each stage, a programmer can produce a structured, modular program with maximum clarity and correctness, and in minimum time. The formal method cannot guarantee that the programmer will not retrace steps along the task profile, but it does tend to minimize the risk; and by isolating details into clusters or modules, it helps prevent error propagation when items in the program are changed.

The experiment described in this paper was undertaken with students in a freshman-level "computing for physics" course. The students constructed a screen robot that simulates the mechanics of an organism hopping—similar to a human on a pogo stick. A team approach to this problem was employed. The students divided the problem into pieces, each done independently by several students so that a final working whole would result even if individual students had trouble meeting their goals.

The experiment was run for two successive years. In the first year, task management was not part of the curriculum; in the second, it was. Because the project was a simulation rather than a calculation, the appropriate steps of the method are: (1) analyze the task, (2) do the physics, (3) design the program, (4) construct the program, (5) test the simulation.

The problem itself is simple. All motion is in a plane, and the robot has only two massive parts: a body, and a leg that swings on an axle through the body. The leg length is variable because a foot-to-leg spring is present and a muscle (under control of the user of the simulation) exerts a torque between body and leg. The center of mass of the body and of the leg are placed on the body-leg axle. The problem is simple, rich, and interesting for the students. The trick is to control the robot's forward and backward hopping, without allowing it to fall.

Motion occurs in two phases: free fall when the foot is not in contact with the ground, and constrained motion—foot fixed—otherwise. (In the later condition, the leg length changes during motion, and the leg-foot spring exerts appropriate forces.) The transition between the two motions is a somewhat tricky point to code.

The software had to be driven by the hardware (PCs) and language choice available. To allow easy coding of screen graphics, a template was used,[6] and a canned Runge Kutta module was used to solve the differential equations. A test bed was designed to test the components of the module.

1. Edward H. Carlson, "Laboratory Skills and 'Task Management,'" submitted to the AAPT summer meeting, (1988).
2. Roger S. Pressman, *Software Engineering: A Practitioner's Approach* (New York: McGraw-Hill, 1982).
3. Edward H. Carlson, "Teaching Software Construction to Physicists," submitted to the Am. J. Phys.
4. Frederick P. Brooks, Jr., *The Mythical Man-Month* (Reading MA: Addison-Wesley, 1975), pp. 7–8.
5. George A. Miller, "The Magical Number Seven, Plus or Minus Two: Some Limits on Our Capacity for Processing Information," The Psychological Review **63**, 81 (1956).
6. Edward H. Carlson, "A Template for Writing Programs," Comp. in Phys. **1**, 65 (1987).

Solution of the Thomas-Fermi Equation for Molecules by an Efficient Relaxation Method

G. W. Parker
Department of Physics, North Carolina State University, Raleigh, NC 27695-8202

The Thomas-Fermi method provides a simple and elegant solution to the problem of finding an approximation to the electron density $n(x,y,z)$ in the ground state of an atom, molecule, or solid. In addition, it provides the initial member of a possible sequence of density-functional approximation schemes that lead to increasingly accurate descriptions of the ground-state density and related one-electron properties. The deficiencies of the Thomas-Fermi model are primarily consequences of the semiclassical approximation and the neglect of exchange, both of which require $n \gg 1$ for their validity. In spite of these basic limitations, the Thomas-Fermi equation becomes exact in the limit of large atomic numbers,[1] gives agreement with actual atomic densities in the region of large density $(n \gg 1)$,[2] and exhibits the correct dissociation behavior for molecules.[3] From a more general point of view, the deficiencies of the Thomas-Fermi model are targets for subsequent refinements and not reasons for discouragement with the approach, which avoids explicit reference to the underlying electronic wave functions.

I have developed a program that solves the multicentered Thomas-Fermi partial differential equation for an arbitrary molecule. Required data are the nuclear charges Z_a and their locations within a simple cubic computational grid that places the molecule more or less at its center. Boundary values are determined from the asymptotic solution, which corresponds to an atom located at the center of nuclear charge having total charge equal to the sum of the atomic charges.

Previous methods have been directed toward specific molecules, such as N_2, and the largest molecules treated have been H_2O and SO_2. This specificity is a consequence of the choice of dependent variable, which is related to the effective T-F potential energy V. At each nucleus, V must approach $-Z_a / r_a$, in atomic units, and this condition has been met, in diatomics for example, by writing

$$V = - (Z_a / r_a + Z_b / r_b) S_m,$$

where the molecular screening function S_m is required to approach unity at each nucleus. The latter condition has been met by constraining the nuclei to be at grid points.[4]

In my approach,[5] the dependent variable is chosen to be DV, the difference between the molecular V and the sum of atomic potential energies V_a. This difference is found to be bounded and smoothly varying through the nuclei and elsewhere, and it automatically incorporates the correct limiting behavior near nuclei through the exact atomic solutions $V_a = - Z_a S_a / r_a$, where S_a, the atomic screening function, goes to unity as r_a goes to zero. I have determined the universal T-F atomic screening function $S(x)$ to about 1 part in 10^5 by solving its differential equation as a two-point boundary-value problem using the well-developed and accurately known asymptotic solution.[6] Values of V_a at any point are then obtained from a cubic spline fit to the solution $S(x)$, or the asymptotic solution.

The corresponding equation for DV, a nonlinear Poisson equation, is solved on the three-dimensional grid by a relaxation method[7] that uses a sequence of grids starting with level 1, with cubic grid size H, and extending to finer grids, with level k having the spacing $h_k = H / 2^{(k-1)}$, where $k = 1, 2,..., M$. An approximate solution on level 1 is efficiently obtained by several Gauss-Seidel relaxation sweeps with one Newton iteration at each grid point to solve the nonlinear equation. The solution on level 1 is then interpolated to level 2 and similar relaxation there smooths errors on the scale of h_2. A correction cycle is then carried out to efficiently reduce errors on the scale of h_1. This involves a transfer of residual errors from level 2 back to level 1, where relaxation again gives a solution. Finally, the solution on 1 is linearly interpolated back to 2, followed by one relaxation sweep on 2 to smooth interpolation errors on the scale of h_2. The process continues with the interpolation to level 3 and so on until eventually terminating on level M, when the algebraic error is reduced below the inherent truncation error, which is $O(h^2)$. In terms of computational work, the final solution on level M is obtained in roughly the time required for six relaxation sweeps on level M.

I will present results for various molecules and compare them to predictions of molecular orbital calculations.

1. E. Lieb, "The Stability of Matter," Rev. Mod. Phys. **48**, 553 (1976).
2. P. Politzer, "Electrostatic Potential-Electronic Density Relationships in Atoms," J. Chem. Phys. **72**, 3027 (1980).
3. K. Yonei, "Extended Thomas-Fermi Theory for Diatomic Molecules," J. Phys. Soc. Japan **31**, 882 (1971).

4. B. Jacob, E. K. U. Gross, and R. M. Dreizler, "Solution of the Thomas-Fermi Equation for Triatomic Systems," J. Phys. B. **11**, 3795 (1978).

5. G.W. Parker, "Numerical Solution of the Thomas-Fermi Equation for Molecules," Bull. Am. Phys. Soc. **32**, 1084 (1987).

6. C. A. Coulson and N. H. March, "Thomas-Fermi Fields for Molecules with Tetrahedral and Octahedral Symmetry," Proc. Camb. Phil. Soc. **48**, 665 (1952); S. Kobayashi, T. Matsukuma, S. Nagai, and K. Umeda, "Accurate Value of the Initial Slope of the Ordinary TF Function," J. Phys. Soc. Japan **10**, 759 (1955).

7. A. Brandt, *Lecture Notes in Mathematics*, vol. 960 (Berlin: Springer, 1982), pp. 220–312.

Numerical Solution of Poisson's Equation with Applications in Electrostatics and Magnetostatics

G. W. Parker

Department of Physics, North Carolina State University, Raleigh, NC 27695-8202

The solution of Poisson's equation with Dirichlet boundary conditions is a fundamental problem in electrostatics and magnetostatics. A general approach to numerical solution begins with discretization on a lattice followed by a relaxation process to drive an arbitrary initial solution toward the exact discrete solution. Of course, any singular source terms must be transferred to the left-hand side of the equation where they combine with the potential to form a new unknown suitable for discretization (Taylor's theorem).

Different ways of accomplishing these steps have been discussed in the physics literature and at least two complete programs are available.[1] One approach makes use of existing spreadsheet programs to solve the discrete equations. This method focuses attention on the discrete form of the equation and the boundary values, and the resulting solution is obtained without any programming being required. This method is limited, in practice, to relatively coarse two-dimensional grids. The two published programs solve the same type of problems with essentially the same relaxation method as the spreadsheet approach, but they provide graphic output that, for example, shows the solution developing in "machine time," thereby illustrating the relation between relaxation and diffusion. Contour plots or "density plots" are also available. Koonin's open program could be modified to include additional features. Both these programs are generally used with relatively coarse grids (maximum resolutions are 24×32 and 23×79 points). In addition, there is no provision for singular sources or dielectric boundaries.

I have developed a program that provides a more detailed analysis than those just described for more elaborate systems with the same two-dimensional boundaries. A variable number of plates, with specified potentials, can be placed on a square grid of spacing H in fairly flexible fashion. For example, in addition to the usual parallel-plate configuration, they can intersect or can be merged into the boundaries to provide for a variable potential along any of the four boundary walls. Line charges, variable in number, strength, and position, are provided as well as a uniform charge density of some chosen strength. In addition, a dielectric boundary can be introduced along one of the grid lines, dividing the rectangle into two regions.

The solution is obtained using Gauss-Seidel relaxation on a sequence of square grids[2] from the coarsest spacing H, level 1, down to level M with the finest spacing $h = H / 2^{(M-1)}$, typically, $H = 2$ and $M = 4$ or 5. Multiple grids give accelerated convergence to the final solution on level M by calculating corrections on the coarser grids by relaxation, which are then transferred by interpolation back to finer grids. The sequence of grids making one cycle is M, $M - 1$, ..., 2, 1, 2, ..., $M - 1$, M. At least three cycles are used, which is equivalent in terms of computational work to about 15 relaxation sweeps on the finest grid. This reduces the algebraic error below the inherent truncation error, which is $O(h^2)$.

Program output includes data needed for equipotential contour plots, plots of induced charge density on the boundary and on both sides of each plate (except at their ends, which are singular points), and plots of the bound charge density along the dielectric boundary.

I will describe four examples of problems that may be treated. (1) A line of equally spaced charges within a grounded box, as suggested by a section in the Feynman lectures.[3] The rapid approach to a uniform field is easily demonstrated in a contour plot. (2) Parallel plates joined to one wall of the boundary at a common potential to form a cavity open at one end. The screening effect is shown by a equipotential plot. (3) The standard parallel plate problem with the additional detail provided by plots of the charge density on the plates. As the plate separation is decreased, the density on the plate's inner surfaces approaches uniformity except near the ends. (4) A superconducting boundary with the potential replaced by the vector potential $A_z(x,y)$, which must vanish on the boundaries. Line charges become line currents and the equipotential plots are seen to now give field lines of the magnetic induction.

1. S. E. Koonin, *Computational Physics* (Menlo Park: Benjamin/Cummings, 1986), example 6, p. 290; B. Cabrera, "Electromagnetism," *Physics Simulations*, vol. 2 (Santa Barbara: Kinko's Academic Courseware Exchange, 1986), program LAPLACE.
2. A. Brandt, *Lecture Notes in Mathematics,* vol. 960 (Berlin: Springer,1982), pp. 220–312.
3. R. P. Feynman, R. B. Leighton, and M. Sands, *The Feynman Lectures on Physics*, vol. 2 (Reading: Addison-Wesley, 1964), section 7–5.

A Computing Laboratory for Introductory Quantum Mechanics

Jesusa V. Kinderman
Physics Department, University of California at Los Angeles, Los Angeles, CA 90024

We have developed a set of 12 computing problems to be solved on a personal computer for our two-quarter junior-level quantum mechanics class at UCLA. These problems do not teach any new concepts, but they reinforce concepts taught in the lectures and develop the physical intuition needed to help students think in a new way. The problems duplicate a laboratory, but experiments are done using computers. Since this was the first time computing problems were assigned in this class, this laboratory was designated as an optional, extra-credit component.

Except for a package for three-dimensional graphics, no special software was developed for the course. The computing problems could be solved by writing a program or using a software package like *Lotus 1-2-3* or *SuperCalc4*. These spreadsheet packages were chosen because they can be learned easily and they easily generate graphics—an important consideration because 11 of the 12 computing problems have results displayed in graphs. Some of these results will be presented in this paper.

The topics of the computing problems are as follows:

- Properties of probability distributions; the Gaussian distribution; superposition of two wave functions.

- The uncertainty principle; determining probabilities by calculating areas under the probability-density curves.

- Time dependence of a state that is a linear combination of two eigenstates of a particle in a one-dimensional infinite-square well.

- Eigenvalues and eigenfunctions of a finite one-dimensional square well.

- Numerical solution of Schrödinger's equation for a potential of the form:

$$V = -V_0 / (1 - \exp(-b \text{ abs}(x) / a))$$

(V_0, b, and a were individualized for each student.)

- Angular distributions of particles moving under the influence of a central potential.

- Eigenvalues and eigenfunctions of an infinitely deep, spherically symmetric well.

- Numerical solution of Schrödinger's equation of the central potential:

$$V = -V_0 / (1 - \exp(-br / a))$$

(V_0, b and a were individualized for each student.)

- Probability densities of the electron in the hydrogen atom; three-dimensional plots of the probability densities in the xz plane.

- Numerical calculation of the eigenvalues and eigenvectors of a 3 by 3 real-operator matrix.

- The deuteron in a Yukawa potential; comparing the results of the variational method and the numerical solution of the Schrödinger equation.

- The anharmonic oscillator; comparing the result of time independent perturbation theory with the numerical solution of the Schrödinger equation; evaluation of the accuracy of the numerical solution.

Students had one or two weeks to complete each computing problem. A laboratory with 12 personal computers was open to students during weekdays and a teaching assistant was available for consultation six hours a week. Because the scope of each problem was limited, we hoped that each problem could be done in under two hours per week. However, students reported spending two or three times this amount of time.

The computing laboratory is interesting and unique in two ways. First, it covers the variety of topics that are introduced in an undergraduate class in quantum mechanics. Using a laboratory format, it reinforces topics discussed in lectures by assigning related experiments on computers. Second, since the students do all the work on the computer, they learn numerical analysis and computing skills, in addition to quantum mechanics. Students report using their newfound skills for other courses. After they have learned to use the computer to graph functions, they use this skill in other courses. They also use the computer for the data analysis and graphics needed in upper-division laboratories.

The Use of an Electronic Spreadsheet in Physics

Thurman R. Kremser
Department of Physics, Albright College, Reading, PA 19612-5234

Electronic spreadsheets have many applications in teaching physics, both in the classroom and the laboratory. They can be used to collect and analyze data from laboratory experiments, to solve numerical problems, and to graph and plot functions. The purpose of this workshop is to introduce the novice to the use of a spreadsheet in teaching physics. Participants will learn how to create a spreadsheet template and how to plot graphs from experimental data or mathematical functions. The emphasis will be on graph plotting. The workshop will be based on *SuperCalc3* or *SuperCalc4*, using the IBM-PC or compatible. Participants are urged to bring blank formatted disks to copy templates for their own use.

The workshop will include the following topics:

Organization and modes of operation of a spreadsheet. The spreadsheet is a rectangular array of cells, each of which can contain text, numerical values, or formulas for calculating values. The user can view the contents of the cells in the spreadsheet mode, and can enter text, formulas, and values in the data-entry mode. The command mode allows editing contents, printing output, graphing, and many more functions.

Data and formula entry. The data-entry mode is entered automatically whenever numerical values or formulas are entered. The cursor movement keys remain active so that data can be entered anywhere on the spreadsheet. Cells can be write-protected to prevent the accidental loss of formulas. Formulas can include arithmetic, trigonometric, logarithmic, and exponential functions; relational operators, logical functions, and statistical functions; as well as financial, calendar, and several special functions pertaining to spreadsheet operations.

Creating a spreadsheet template. A spreadsheet template consists of all the instructions, headings, constants, formulas, and column widths necessary to create a form for easy entry of data and calculation of results. A template can be created and saved to disk for future use.

Spreadsheet commands and functions. Of particular interest to physicists is the use of the spreadsheet to obtain graphs of experimental data or mathematical functions. The "/View" command is used to display data in the form of graphs on the monitor, the dot-matrix printer, or the pen plotter. This command allows the user to choose the range of data, the graph type, scaling, and labeling. The "/Replicate" (or "/Copy") command, which copies a formula from one cell to many cells, is indispensable in using the spreadsheet to create graphs of functions.

Applications. Participants in this workshop will use a spreadsheet template to analyze the data from the standard Atwood Machine experiment. They will use the data to obtain a graph relating the accelerating force and the acceleration. From this graph, they will determine the mass of the system. Participants will also use a spreadsheet template to calculate the position, velocity, and acceleration of a body falling in a resisting medium, with the resisting force proportional to velocity. Using the graphic capabilities of the spreadsheet, they will plot graphs of position, velocity, and acceleration as functions of time. Finally, participants will use the spreadsheet to calculate up to 25 terms of the Fourier series for a square wave function. Plots will be obtained for any number of terms in the series up to the maximum.

Interference Phenomena Using Spreadsheet Programs

Robert Novak

Physics Department, Iona College, New Rochelle, NY 10801

Commercial spreadsheet programs can be used to analyze experimental data[1] and to model theoretical problems.[2] It is simpler to use a spreadsheet package to perform such computational and graphical operations than to use a high-level language. Faster microcomputers with larger memories and spreadsheet software are becoming more accessible to introductory-level students, and the use of these can be a powerful tool in learning physics. This paper uses the graphics package of a spreadsheet program to present standing waves, beats, and Young's slits.

Wave phenomena can be described in terms of a sinusoidal wave form such as

$$F(x, t) = A \sin (kx - \omega t + \phi)$$

where $k = 2\pi$, and ω is the angular frequency. For two waves that interfere with each other, the amplitude can be written as

$$F_{tot}(x, t) = F_1(x, t) + F_2(x, t).$$

The intensity of the wave pattern is proportional to the square of the amplitude. This equation can be solved very quickly using a spreadsheet program.

We use a Compaq III computer (PC compatible with 640K of RAM, CGA graphics monitor, 12 MHz clock) with *Lotus 1-2-3* (Release 2) software and an MS-DOS Version 3 operating system. After the boot-up procedure, 572K is available. The spreadsheet program requires 215K and the remainder is available for the spreadsheet file. Figure 1 contains a screen printout used for the standing wave presentation.

Information is written into the spreadsheet as labels, numerical data, or formulas; use of a screen cursor indicates the location of the information. The labels in Figure 1 give the directions for entering data such as wave velocity, length of medium to be shown, wavelength, and amplitude. The frequency and the position values are then calculated from the data supplied. Time and position values are also calculated and presented in a matrix that consists of 500 position values and six time values. Graphics commands plot the displacements as a function of position.

The variables are written in column/row format (such as A12 or G32); constants use a "$" symbol with this format. The initial position value (A22) is taken as zero. The formula + A22 + \$D\$9/500 is placed into A23 and use of the copy function generates a list of position values. Similarly, a formula for the wave form is placed into the C23 position and it is copied for all the 500 x 6 time-vs-position values. The graphing commands are able to plot the function for the six times.

```
        A      B      C      D      E      F      G      H
1   TWO ONE DIMENSIONAL WAVES INTERFERING IN A LINEAR MEDIUM
2
3   Enter wave velocity  and length for the medium at D8 and D9.
4   For each wave, enter wavelength (C13 and G13) and amplitude(C14 and G14)
5   Enter a negative wavelength for the wave with a negative velocity.
6   Frequencies will be calculated automatically.
7
8   Wave velocity (m/sec)=      100.0000
9   Length of medium(m)=         50.0000
10
11  Wave#1                       Wave#2
12
13  Wavelength(m)=     20.0000   Wavelength(m)=   -20.0000
14  Amplitude(m)=       2.0000   Amplitude(m)=      2.0000
15  Frequency(Hz)=      5.0000   Frequency(Hz)=     5.0000
16
17  Enter time values from C19 through H19.
18
19          Time(sec) 0.0300  0.0600  0.0900  0.1200  0.1500  0.1900
20  Position
21  (meters)
22     0.0000         3.2361  3.8042  1.2361 -2.3511 -4.0000 -1.2361
23     0.1000         3.2345  3.8023  1.2355 -2.3500 -3.9980 -1.2355
24     0.2000         3.2297  3.7967  1.2336 -2.3465 -3.9921 -1.2336
25     0.3000         3.2217  3.7873  1.2306 -2.3407 -3.9822 -1.2306
26     0.4000         3.2106  3.7742  1.2263 -2.3326 -3.9685 -1.2263
```

Figure 1. Screen printout of part of the spreadsheet to illustrate standing waves. The entire spreadsheet contains 522 rows and 8 columns.

Figure 2 shows the screen output. This plot shows six time configurations of a standing wave.

Once this is set up, it is convenient to change data. Changing a wavelength or a velocity automatically changes the values throughout the program. These new val-

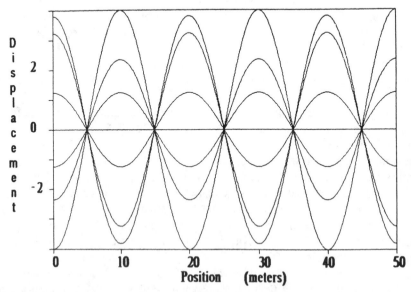

Figure 2. Graphical presentation of standing wave patterns for the data given in Figure 1.

ues are viewed graphically by pressing three keys. The program, as set up, uses standard rather than generalized units for the data. Using scientific format, magnitudes appropriate to visible light wavelengths can be used. After a change in data, the program recalculates all the values in less than 15 seconds; the plotting routine takes another two seconds to present the graph on the screen.

The plot in Figure 3 illustrates interference that produces beats at a point. In this case, the spreadsheet is set up so that the user can input the frequencies and amplitude of the two interfering waves, along with the time range. Both the beat and the average frequencies can be measured from the plot.

In the plot that illustrates Young's slits (Figure 4), the user inputs values from the slit separation, the distance from the slit to the screen, and the width of the screen. The image screen is divided into 1,000 points, and the distance between the point and each slit is calculated for each point. The two interfering functions each have a different optical path. The plot is taken at one time, and the envelope of this function, which is proportional to the time average, is also plotted. From this presentation, actual minimum and maximum positions can be obtained. These can be easily compared to experimental or theoretical values.

Spreadsheets by R. Novak for *Lotus 1-2-3* are part of the collection *Computers in Physics Instruction: Software*, which can be ordered by using the form at the end of this book.

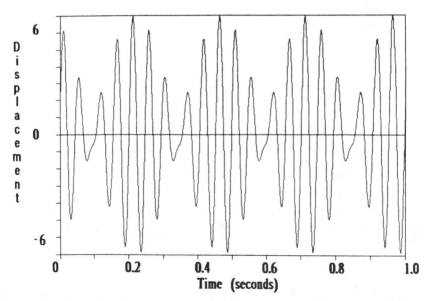

Figure 3. Superposition of two waves whose frequencies are 20 and 24 Hz. Their corresponding amplitudes are 4 and 3.

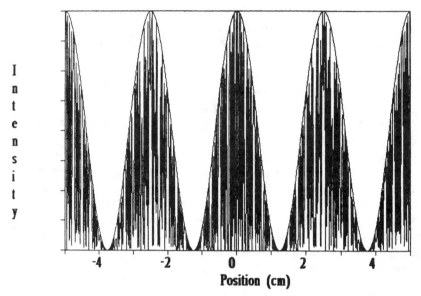

Figure 4. Intensity pattern from a Young's slits computation. The distance
between the slits is 20 μ; wavelength is 500 nm; and the distance
from slit to screen is 1 m. The pattern is shown for one time. The
envelope function that reflects the time average is also plotted.

1. R. Feinberg and M. Knittel, "Microcomputer Spreadsheet Programs in the Physics
   Laboratory," Am. J. Phys. **53**, 631 (1985).
2. T. T. Crow, "Solutions to Laplace's Equations Using Spreadsheets on a Personal
   Computer," Am. J. Phys. **55**, 817 (1987).

Wondering about Physics . . .
Using Spreadsheets to Find Out

Dewey I. Dykstra, Jr.
Department of Physics, Boise State University, Boise, ID 83725

Robert G. Fuller
*Department of Physics, U.S. Air Force Academy, Colorado Springs, CO 80840, on leave
from the Department of Physics and Astronomy, University of Nebraska, Lincoln,
NE 68588-0111*

This miniworkshop is designed to introduce physics teachers to the variety of ways
that spreadsheets can be used for student learning in physics. The participants will
undertake investigations drawn from the 53 examples in our book, *Wondering
about Physics....Using Spreadsheets To Find Out.* The investigations are appropri-

ate to any computer and any standard spreadsheet software, but the workshop will use only Macintosh computers and the Microsoft *Excel* spreadsheet software. Each participant will receive a complimentary set of student and instructor spreadsheet materials.

Participants are not required to have previous spreadsheet experience. Some knowledge of how a Macintosh operates with a mouse and button clicks will be helpful.

1. (New York: Wiley, 1988).

Computer Tutorials in Physics

Tutorials on Motion

C. Frank Griffin
Department of Physics, University of Akron, Akron, OH 44325

Two companion computer tutorials, *A Tutorial on Motion* and *A Tutorial on Acceleration*,[1] have been written for the Apple II series of microcomputers. *A Tutorial on Motion* was on the AAPT courseware editor's list, "Favorite Courseware Programs for 1988."[2] All other programs on the AAPT list were experiment-interface packages.

Learning Difficulties in Kinematics

The difficulties of learning kinematics are probably the most thoroughly researched of any learning problem in introductory physics. In an investigation of student understanding of velocity in one dimension, Trowbridge and McDermott find that the failure to make proper comparisons of velocities for two simultaneous motions can be attributed to use of a position criterion to determine relative velocity.[3] Peters finds that students in an honors calculus physics course draw virtually indistinguishable position-time and velocity-time graphs of a classroom demonstration.[4] Similar graphing errors, especially with negative velocity, have been presented at meetings by Goldberg and Anderson.[5] Arons discusses the importance of graphs in helping students develop conceptual understanding,[6] and McDermott, Rosenquist, and van Zee have published a comprehensive study of the difficulties of graphing and kinematics.[7] This research as well as the considerable teaching experience of myself and Lou Turner, my coauthor at Western Reserve Academy, indicates that kinematics is not only a very difficult subject for many students to learn, but it is also an area full of misconceptions.

The most common problem is that students confuse position and velocity. Given a position-versus-time graph, students often draw a velocity-versus-time graph that is indistinguishable from the position-versus-time graph even after having completed an introductory college physics course. In addition, they consider the concept of negative velocity to be alien. They typically believe "You go or you don't. How can you have a negative velocity?"[8] They have difficulty translating a written description of a motion into an accurate graph of this motion and vice versa. Such student difficulties cannot be overcome in the amount of classroom time typically allotted to the topic. In fact, I began writing these tutorial programs from pure frustration when I was attempting to teach velocity and acceleration in the conceptual style of Arons.[9] Even when I devoted one week of lecture time to acceleration, the students still had trouble. These tutorials give the students additional guided learning in kinematics.

Description of Software

The ideas of kinematics build one upon the other. We prefer the tutorial format for the comprehensive development of kinematic concepts because it allows a structured, sequential development of ideas. The software controls the content and order of presentation of the material. The student controls the pace and can repeat a step at any time.

Our tutorials supplement the standard lecture and laboratory approaches to teaching one-dimensional kinematics to high school and college students. They use several descriptive representations of motion, including verbal descriptions, strobe records of a simulated moving automobile, and graphs of position versus time, velocity versus time, and acceleration versus time.

A Tutorial on Motion is a series of programs on position and velocity. Each program introduces an idea with a tutorial and then tests and reinforces the idea with a quiz. The front side of the disk is devoted to position and the back side to velocity. The major topics of "Position" are position, position change, strobe records, and graphing of position versus time. A Tutorial on Motion introduces position and velocity in a very concrete fashion. In "Position," a measuring scale is placed beside a road with a car on it. An arrow points from the front of the car to the scale to mark the position of the car. "Position Change" has the car move and arrows show its displacement along with numerical values. Both of these programs are followed by quizzes.

Graphing creates a strobe record paralleling the S-axis of the position-versus-time graph. The car moves, leaving the strobe record simultaneously with the plotting of a point on the S-T graph, giving the student an opportunity to integrate and compare the motion and both modes of representing motion. A particular goal of this program is to demonstrate that a graph is a strobe record with each strobe track displaced horizontally in equal increments to represent the time. The difficulties students have with motion in a negative direction is overcome here and elsewhere by constantly comparing motion in both positive and negative directions. Differences between graphs of cars speeding up and slowing down are also shown.

Velocity is introduced in terms of the familiar speedometer rather than as displacement divided-by-time interval. The velocity tutorials simulate the motion of the car along the track while showing the corresponding speedometer readings. The speedometer displays negative values when the car backs up. Initially, the scale is shown horizontally just below the horizontal road. Next the speedometer scale is rotated vertically so that it matches the velocity scale of an S-T graph shown just to the right of the speedometer scale. The student observes simultaneously the motion of the car, the speedometer reading, and the corresponding points plotted on the S-T graph. Since the V values on the S-T graphs match the strobe record of the speedometer needle, the S-T plot is convincingly shown to be a plot of speedometer readings. The last tutorial relates the slope of the S-T graph to velocity. In this step-by-step fashion the complexities of the simulated motions, strobe records, and S-T plots are gradually developed and tied together. The quizzes ask the student to match car motions with S-T graphs, strobe records with S-T graphs, and, finally, S-T against S-T graphs.

After going through the programs on *A Tutorial on Motion*, the student should be able to translate from one of the following ways of representing motion to another: verbal, strobe records, and graphs of *S-T*. That is, given any one of these ways of representing motion (strobe track for example) the student should be able to produce or recognize the corresponding version of any of the other representations, including a simulation of the motion itself.

Students need a conceptual understanding of acceleration before they can grasp Newton's laws. Although students intuitively relate force and velocity, they do not intuitively relate force and acceleration. This failure of intuition may happen because they observe position and displacement directly and must infer or calculate velocity and acceleration. Moreover, the direction of motion is the same as the direction of velocity but it may or may not be the same as the direction of acceleration.

The typical textbook begins instruction on acceleration with the statement that acceleration equals change in velocity divided by the time interval. We believe this approach is too abstract. Our belief is borne out by the research of Trowbridge and McDermott, which shows that many students cannot interpret *S-T* graphs, tell velocity from acceleration, nor even velocity from position.[10] Our philosophy in writing these programs is that learning occurs best when the idea being taught builds on what is already known.

The approach in *A Tutorial on Acceleration* is similar to that in *A Tutorial on Motion*. Items are familiar—a moving car, a speedometer, and its moving needle—and there is simultaneous plotting and strobe recording of their motions. Since acceleration is a complicated concept, the direction of acceleration and its magnitude are taught separately. To teach the direction of acceleration, the program compares velocity and acceleration in different ways. First the position change of a car along a scale is simulated and calculated using $\Delta S = S_2 - S_1$. An arrow is added to show the direction of the position change. In parallel fashion a simulated motion of a car on a road is shown and movement of the speedometer needle along its scale is shown. The velocity change is calculated using $\Delta V = V_2 - V_1$, and an arrow is added to show the direction of the velocity change. This makes concrete the definition of the direction of acceleration. Second, simulated motions that produce straight-line *S-T* graphs are compared with those that produce *S-T* graphs. The standard definitions of velocity and acceleration in terms of slopes are presented and correlated to the respective graph. The relationship of the sign and direction of the kinematic quantities to the slopes is also demonstrated with simulations.

The relationship of magnitude of acceleration to the steepness of the slope of the *S-T* graph is illustrated by showing simultaneously, the motions of the car and speedometer needle. The speedometer needle moves along the V-scale of an *S-T* graph while the graph is plotted. The car speeds up or slows down at different but constant rates and the slopes of the *S-T* graph are compared. Each of the ideas presented in tutorial format is followed by a quiz to reinforce the idea and test the student's comprehension.

The second side of the acceleration disk relates strobe records and position-time graphs to acceleration. We think it is particularly important for the student to make the connection between these representations of motion and acceleration to understand Newton's second law. Positions and displacements are observed, but acceleration, which must be inferred from these observations, is related to force. Through simultaneous car-motion simulations with plotting of graphs followed by discussion of change in slope, the program leads the student to see the relationship between the curvature of a plot and acceleration.

"Strobe Record and Acceleration" explains how to use the strobe record of an object to establish whether the object's acceleration is a constant. If the acceleration is constant, the difference between strobe steps is a constant. The program describes how to find the value of the constant acceleration from the strobe steps.

After studying *A Tutorial on Acceleration*, the student should be able to (a) observe the straight-line motion of an object and correctly indicate the direction of acceleration of the object, (b) determine both the direction and relative magnitude of the acceleration of the object from either an *S-T* graph or an *S-T* graph of the object's motion, and (c) study both strobe records and *S-T* graphs and identify cases of constant acceleration and the value of this constant acceleration.

The 24 page manual that accompanies *A Tutorial on Motion* has seven example problems. The manual that accompanies the acceleration disk includes 24 pages of worksheets that may be photocopied for student use. Twenty- one sample test questions are also included.

Instructional Value of Computers

The computer is an ideal vehicle for teaching motion. Piaget has taught that everyone goes through a concrete stage of reasoning before maturing into a student capable of using formal or abstract reasoning. Even a student who is already a formal thinker may go through a brief stage of concrete reasoning when encountering an unfamiliar topic. The concrete thinker studying motion benefits from minimization of the time between seeing a motion and the representation of that motion. The computer can not only show the student a simulated motion (for example, a car decreasing its speed while going in the positive direction), but also simultaneously show (a) a strobe record of the motion, (b) the position and movement of the speedometer needle along its scale, (c) a position-time graph, or (d) a velocity-time graph of the motion. If desired, the student can see the motion and the construction of its representation repeated until he/she understands both. Seeing a real motion and its almost immediate position-time graph has just recently become possible with the advent of the sonic ranger. Laboratory activities or class discussion including use of a sonic ranger followed by use of these disks, provide the student with a rich variety of experiences leading to an improved understanding of motion.

These tutorials provide students with individual tutoring time, which is rarely available from a teacher. This software is patient and nonjudgmental. Students can work through it at their own pace, repeating any tutorial as often as they like. They can also review any section and skip sections previously mastered. However, students always have the option of repeating sections if they need additional work. Each exercise on the disk has been written so that when students believe the material has been mastered, they may exit and move on to the next exercise, allowing more knowledgeable students to spend less time on an exercise. Struggling students can repeat a motion and the simultaneous creation of a record of it until they understand the relationship between them. They can take quizzes more than once because the order of the questions is randomly generated by the computer.

Classroom Management

Making computer tutorials available to students on an individual basis means students must have access to computers. At the University of Akron, six Apple computers are available to students for 25 hours each week. This has proved adequate for up to 110 students. At Western Reserve Academy, students are allowed to use the software in the school library, and they have access to this library at night. If a high school physics teacher does not have microcomputers for use in his personal class, the library is an excellent place to provide such computers. Pressure should be brought to bear on administrators to provide such facilities.

Significant Instructional Features

As already mentioned, the use of graphics to show a car's motion and its various motion representations simultaneously is the most important feature of this disk. There are several additional instructional techniques that make the disk unique.

1. The use of a speedometer has a negative as well as a positive scale. In the world outside of the physics classroom, no one ever talks about positive or negative velocity. Yet when students enter a physics classroom, they suddenly learn that positive and negative signs are the accepted way of indicating direction. This speedometer is an important teaching device of both disks.

2. The speedometer determines the direction of the acceleration. The direction of displacement of the needle on the speedometer scale is compared with the direction of displacement of the car to introduce the direction and sign of acceleration.

3. The relationship between the direction of acceleration and the curvature of the S-T graph is introduced.

4. The motion of the car or speedometer needle close beside and parallel to the graph being plotted, reinforces the meaning of the graph.

Quizzes

Most of the quizzes use one of two techniques to make each repetition a different experience. Several of the quizzes on the disk have infinite loops. The computer will provide randomly generated problems for the student to analyze for as long as the student wants to continue. In other quizzes the answers are presented in a random order in a sequential multiple-choice format. The answers appear in a different order each time the quiz is used. Thus the correct choice can appear first, second, or fifth in the list, and different incorrect choices are likely to be presented before the correct choice.

Motivational reinforcement is provided with correct responses. Most students get a kick out of receiving these compliments. There are 23 of these compliments ("I am impressed," "I am proud of you," "Keep up the good work") built into the programs, and they are chosen randomly. In about 50 percent of the quizzes a tally of student responses is kept, and when the student is finished the number of correct and incorrect responses is displayed along with a percent score.

Professional Quality

A number of the public domain programs and some commercial ones show a number of annoying qualities. Several [Return] may be stored in the keyboard buffer, resulting in an unexpected flipping forward of several screens. Numeric input when string input is expected might cause the program to crash. Several [Return] when input is expected, can cause a menu to scroll off the top of the screen.

None of these difficulties occur with these tutorials. The only known way to crash a program is by pressing [Ctrl] [Return]. Student input is always compared against an answer set presented at the time a response is requested. Any keyboard entry not from this set will be ignored and a bell will ring.

The [Esc] key almost always exits the student from a program when a response is requested. In *A Tutorial on Acceleration*, the student may flip pages backward to review or flip pages forward to skip over material already covered. These two features give the student a great deal of flexibility in going through a tutorial.

Evaluation

There have been no controlled studies of the impact of this disk on student learning, but a comparison can be made due to an unscheduled remodeling problem. After this program had been used for three years at the University of Akron in the algebra-based physics, the computer room was remodeled during the fall of 1987 and these tutorials were not available to the students. The lectures were no different and the weekly tests on kinematics were approximately equivalent in complexity to previous years. The average grade dropped nine percent for the class that did not have access to the computer tutorials.

Student responses at Western Reserve Academy and the University of Akron have been overwhelmingly favorable. Written evaluations indicate that students think the explanations are clear, the graphics helpful and interesting, the ideas clarified, and they say that their overall understanding of motion has been improved. At the University of Akron, where computer study is optional, our records show that 75 percent of the students view the programs on the two disks an average of five hours. Students gather in groups to study, sharing questions and explanations with each other as they work through the programs. We think this group work also enhances learning.[11] Students seem to enjoy the programs, the puzzling questions, and the sometimes surprising answers. We have heard several times, "Now I understand what you were talking about in class."

The most dramatic demonstration of the value of this software is the dismay expressed by students when they learn that we have no other computer tutorials available to them after kinematics.

1. C. Frank Griffin and Louis C. Turner, *A Tutorial on Motion*, and *A Tutorial on Acceleration* (COMPress, P.O. Box 102, Wentworth, NH 03282, 800-221-0419.)
2. J. S. Risley, "My Favorite Physics Courseware Programs for 1988," AAPT Announcer **17**, 94 (1987).
3. D. E. Trowbridge and L. C. McDermott, "Investigation of Student Understanding of the Concept of Velocity in One Dimension," Am. J. Phys. **48**, 1020 (1980).
4. P. C. Peters, "Even Honors Students Have Conceptual Difficulties with Physics," Am. J. Phys. **50**, 501 (1982).
5. F. Goldberg and J. Anderson, "Student Difficulties with Representations of Rectilinear Motion," AAPT Announcer **14**, 78 (1984).
6. A. B. Arons, "Student Patterns of Thinking and Reasoning, Part Three," Phys. Teach. **21**, 88 (1984).
7. L. C. McDermott, M. L. Rosenquist, and E. H. van Zee, "Student Difficulties in Connecting Graphs and Physics: Examples from Kinematics," Am. J. Phys., **55**, 503 (1987).
8. Goldberg and Anderson, "Student Difficulties."
9. A. Arons, *The Various Language* (New York: Oxford University Press, 1977).
10. Trowbridge and McDermott, "Investigation of Student Understanding."
11. E. R. Carnes, J. S. Lindbeck, and C. F. Griffin, "Effects of Group Size and Advance Organizers on Learning Parameters when Using Microcomputer Tutorials in Kinematics," J. Res. Sci. Teach. **24**, 781 (1987).

CAI: Supplementing the Noncalculus Introductory Course

Lisa Grable-Wallace, Joyce Mahoney, and Prabha Ramakrishnan
Department of Physics, North Carolina State University, Raleigh, NC 27695-8202

Teacher input and supervision is an integral part of adding computers to the classroom environment. Instructors must choose computer courseware that will reinforce and enhance concepts presented to the students by other means.

At North Carolina State University we have launched an ambitious project to incorporate microcomputer lessons into a one-semester, noncalculus introductory physics course that covers mechanics, heat, wave motion, and sound. The students typically enrolled in this course are reluctant to use microcomputers, lack confidence in their ability to succeed in physics, and fail to see the relevance of physics to their own experience. Our aim is to create a schedule of computer assignments to accompany the material presented in the course and textbook. We have written workbook exercises to better involve students in the computer activities and to help increase their retention of the material. These exercises consist of instructions for using the courseware, questions to answer while working with the program, and problems to solve after completing the lesson. The completed worksheets are graded as homework assignments.

After reviewing 61 programs, we chose 19 courseware packages for the computer assignments. The majority of the programs used in the assignments are tutorials that present a text treatment of the topic, teach a problem-solving method, or deliver drill-and-practice problem solving. Our most important consideration in selection was the accuracy and clarity of the physics. Other criteria included the amount of user interaction, the quality of the graphics, and how much time an average student needed to complete the program.

Workshop participants will receive hands-on experience with courseware lessons from different portions of the course. The workshop will allow participants to see for themselves the strengths and weaknesses of various commercial software packages.

CAI Geometric Optics for Pre-College

Betty P. Preece
Melbourne High School, Melbourne, FL 32901

This will be a demonstration of the use of software to teach optics in the high school physics classroom. I use such software to demonstrate properties and the associated equations of mirrors and lenses. Students carry out lab activities and are then asked to summarize what they have learned and to compare their summaries with the situations shown in the software. All students are required to provide solutions for various situations shown in the software. I offer extra credit for additional solutions. Students can scan the software both during and outside regular class times.

I also use the software for final visual review of the properties of mirrors or lenses. Students usually indicate that this final viewing of how images are formed enables them to really understand geometric optics.

I will present the syllabi for a unit on geometric optics for regular and honors secondary physics at Melbourne High School. They include labs, observations by students, demonstrations, problems, tests, and software. I will also show examples of how I have integrated commercial software programs into my courses with teacher and student interaction. Although my syllabi are keyed to Holt's *Modern Physics* by Williams, Trinklein, and Metcalf and to Allyn and Bacon's *Physics—Its Methods and Meanings* by Taffel, they can easily be adapted to any texts at the precollege or introductory-college levels. Copies of the syllabi will be available.

Microcomputer-Delivered Homework

Rolf C. Enger and Gary L. Lorenzen
Department of Physics, U.S. Air Force Academy, Colorado Springs, CO 80840-5701

Arguing that students must practice solving problems if they are truly to learn physics, members of the physics department at the U.S. Air Force Academy[1] have written a computer program designed to deliver numerical physics problems with randomly chosen data. Pedagogically, the computer adds little to the exercise. The computer encourages students to practice, but the problems themselves are no different than end-of-chapter problems in most introductory physics texts.

The use of the computer does, however, have advantages. First, the random-number generation makes it difficult for students to bypass problem solving. Second, the computer reduces the administrative burden associated with grading homework. Last fall, for example, approximately 39,000 problems were successfully completed by our students. The actual number of problems worked and graded was probably several times this figure, since most students required several attempts on a given problem.

The program was first put into place in 1984. Computerized physics problems were delivered on several networked computers placed in a classroom. Students came to the classroom both to receive and answer the problems. They liked the idea of receiving points for correctly solving problems, but they didn't like coming to class to solve them or standing in line to gain access to limited computer resources. We were concerned about the access problem, but were so encouraged by the improved test performance that typically followed a computer-delivered problem session that we continued to operate the system.

The access problem has recently been solved. Beginning with the class of 1990, all USAF Academy cadets purchase a Zenith Z-248 microcomputer (IBM AT clone). In August of 1987 we transferred our in-class problem-delivery system to the dormitories and gave it a new name, the *Microcomputer-Delivered Problem* (MDP) system.

The MDP system is programmed in Borland's *Turbo Pascal*, which can be compiled for MS-DOS and several other operating systems. Each physics problem contained in the MDP system is programmed as a separate subroutine, which permits the value of variables used in the problem solution to be randomly generated. A special editor, *MDPAUTHR.PAS*, automatically writes the *Turbo Pascal* code. The system comes with a worksheet that helps an author of new problems determine the parameters required by the MDP system. Further instructions are provided in "README" files.

Operation of the MDP program is very simple. The student simply enters "MDP" at the operating system prompt. The rest of the program is menu driven. When a student takes a test (one of the menu options), the program automatically creates a personal database (called PHYS110.DAT) for that student. Only one database may be present on each floppy disk.

When students successfully complete an MDP problem, a score is awarded and recorded in the student's own personal database. That database is periodically submitted to the student's instructor so that the student's score can be recorded. Identifying information is encoded in the database in order to prevent students from submitting someone else's database as their own. The system also records information about time-on-task and number of tries. This allows us to determine which questions are the most difficult. We hope eventually to offer hints and tutorial support for the more difficult problems.

Student response to the MDP system has been extremely favorable. Last fall, 74 percent of the students said they thought the MDP system was a good idea. Instructor response has also been favorable. The MDP system gives instructors a way of grading homework without hiring teaching assistants. Our faculty is much too small to manually grade a volume of more than 39,000 problems successfully completed and several times that many attempts.

Though the MDP system works well, we are planning a number of improvements, including partial-credit grading and tutorial support.

Administration of the MDP system is as important as the program itself. We have been disturbed that many students wait until the last minute to solve the MDP problems. Last fall, we had just two deadlines, one at midterm and the other at the end of the course. An informal analysis done during summer school 1987 suggested that students who waited until just before the final exam to solve most problems benefited little from the MDP experience. Thus, in the spring of 1988 we experimented with two strategies designed to encourage a more consistent pacing throughout the semester. Results are not yet available as of this writing.

The MDP system allows us to take a problem-solving pedagogical approach to physics teaching without the burden of manual grading. Students like the immediate feedback and reward (points). The system is only a first step. It clearly could be the core of a much larger and more sophisticated system. We think it is a first step well worth taking, and encourage others to take this first step with us.

The software program *Microcomputer-Delivered Problem System* is part of the collection *Computers in Physics Instruction: Software*, which can be ordered by using the form at the end of this book.

1. Lee W. Schrock and Mark V. Tollefson.

Electronic Textbook: A Model CAI System

Duli C. Jain and Frank R. Pomilla
Department of Physics, York College, Jamaica, NY 11451

Electronic Textbook is an innovative concept combining the traditional textbook approach to physics instruction with modern microcomputer technology. *Electronic Textbook* allows a logical development of physics subject matter. All definitions and physical concepts are clearly presented and illustrated with graphics, sound, and animation. The system allows small quizzes to be given to test the student's understanding of fundamental concepts, and gives helpful hints if the student asks for them.

Electronic Textbook contains a large number of solved examples and a variety of randomly generated practice problems of different types. The types of problems are organized in order of increasing difficulty.

The student may review the basic concepts and/or solved examples while solving the practice problem at hand, and is asked to solve one part of a quiz or a problem at a time. The student can use the microcomputer in immediate mode to perform the necessary calculations. After grading student's performance, the system moves on to the next part of the problem or to the next problem. The student's grade record can be printed.

The unit on waves and sound is intended for freshman college physics. It consists of three floppy disks that run on an APPLE IIe microcomputer with a single disk drive. A printer is optional. *Waves and Sound* includes three chapters: "Basic Concepts," "Sound Waves," and "Characteristics of Sound and Doppler Effect."

The software is independent of the textbook adopted for the course, and the student need not have any computer expertise to use the system. Each diskette contains directions for the use of the system and clear and concise instructions at every stage.

Electronic Textbook has been expanded to cover the following topics in the preparatory course, Natural Science 100: molecular formulas, molecular weights, and percent composition; scientific notation; conversion of units; logarithms and exponential functions; introduction to vectors; and graphs.

A sample software program from chapter 1 of *Electronic Textbook: Waves and Sound* is part of the collection *Computers in Physics Instruction: Software*, which can be ordered by using the form at the end of this book.

This project is supported by grant G008200675, from the Minority Institutions Science Improvement Program, U.S. Department of Education.

Tutorials on Motion

C. Frank Griffin
Department of Physics, University of Akron, Akron, OH 44325

Louis C. Turner
Department of Physics, Western Reserve Academy, Hudson, OH 44236

Even after college-level physics work requiring graphical analysis of experiments with dynamic carts or air tracks and photogate timers, many students still have difficulty interpreting graphical representations of motion.[1] In response to this difficulty, we have designed tutorials to supplement the standard lecture and laboratory approaches to teaching one-dimensional kinematics to high school and introductory college students. The tutorials use several descriptive representations of motion, including verbal descriptions, strobe records of a simulated moving automobile, graphs of position vs. time, velocity vs. time, and acceleration vs. time.

Our workshop is a hands-on demonstration of two companion computer tutorials, *Tutorial on Motion* and *Tutorial on Acceleration*.[2] Participants will be able to study these programs under the guidance of the authors. *Tutorial on Motion* was on the AAPT list of "Favorite Courseware Programs for 1988."[3] All other programs on the AAPT list were experiment-interfacing packages.

Tutorial on Motion

Tutorial on Motion is a series of programs on position (filling the front of the disk), and velocity (filling the back of the disk). Each program introduces an idea with a tutorial, and tests and reinforces the idea with a quiz. *Tutorial on Motion* introduces position and velocity in a very concrete fashion. Position of a car is identified not as some vague reference to a coordinate system, but as the number on a scale by the road. Position on a position vs. time graph is shown by having the car move along a track while simultaneously leaving a strobe record. The car's movement parallels the S-axis of the graph and a point is plotted on the S–T graph, giving the student an opportunity to compare and integrate all three modes representing motion. The multiple system clears up many misconceptions.

Velocity is introduced in terms of a familiar speedometer instead of as displacement divided by time interval. The velocity tutorials simulate the motion of the car along the track with corresponding speedometer readings. The speedometer displays negative values when the car backs up. Initially, the speedometer scale is shown horizontally just below the horizontal road. Next, the speedometer scale is rotated vertically so that it just matches the velocity scale of a V–T graph shown to the right of the speedometer scale. The student simultaneously observes the motion of the car, the speedometer reading, and the corresponding points plotted on the V–T graph. Since the V values on the V–T graphs just match the strobe record of the speedometer needle, the V–T plot is convincingly shown to be a plot of speedometer readings. The quizzes ask the student to match car motions with V–T graphs, strobe records with V–T graphs, and, finally, S–T with V–T graphs.

Students need a conceptual understanding of acceleration before they can grasp Newton's laws. Although students intuitively relate force and velocity, they do not intuitively relate force and acceleration. This failure of intuition may be because students observe position and displacement directly, but must infer or calculate velocity and acceleration. Second, the direction of motion is the same as the direction of velocity, but not necessarily the same as the direction of acceleration.

Tutorial on Acceleration

The approach of a *Tutorial on Acceleration* is similar to that of *Tutorial on Motion*. Items are familiar—a moving car, a speedometer and its moving needle—and there is simultaneous plotting and strobe recording of their motions. The direction of acceleration and its magnitude are introduced separately. Velocity and acceleration are compared and contrasted in several different ways: the direction of displacement of the car is compared with the direction of displacement of the needle on the speedometer scale; and simulated motions that produce straight-line S–T graphs are compared with those that produce straight-line V–T graphs.

After using the programs on this disk, the student should be able to: observe the straight-line motion of an object and correctly indicate the direction of acceleration of the object; determine both the direction and relative magnitude of the acceleration of the object from either an S–T graph or a V–T graph of the object's motion; and identify cases of constant acceleration and the value of this constant acceleration from either strobe records or V–T graphs.

1. L. C. McDermott, M. L. Rosenquist, and E. H. van Zee, Am. J. Phys. **55**, 503 (1987).
2. Both programs are published by COMPress, PO Box 102, Wentworth, NH 03282, 800-221-0419.
3. J. S. Risley, "My Favorite Physics Courseware Programs for 1988," AAPT Announcer **17**, 94 (1987).

Density of Orbitals: A Tutorial

John W. Gardner
Department of Physics, Eastern Illinois University, Charleston, IL 61920

Density of Orbitals is a software tutorial for the Macintosh that develops the density-of-orbitals function for a particle in a one-, two-, or three-dimensional box. It is appropriate for upper-class physics majors taking quantum mechanics, thermal physics, or solid-state physics, or for first-year physics graduate students needing a review. The tutorial uses a paging window of text and one or more windows of graphs and other visual aids to teach its subject. After studying the tutorial, the student will understand what density of orbitals means and will know and be able to derive the density-of-orbitals function for a particle in a box.

The tutorial first reviews the quantum energy spectrum for a particle in a one-, two-, or three-dimensional box and enables the student to inspect the degenerate orbitals belonging to any energy level. The tutorial then presents an argument for the $N(E)$ functional form ($N(E)$ = the number of orbitals with energy less than or equal to (E) and then calculates the $N(E)$ function exactly by counting degeneracies so that the student can compare it graphically with the argued functional form. Finally, the tutorial presents the density-of-orbitals function, $D(E)$, as the derivative of the $N(E)$ function.

Density of Orbitals runs on a Macintosh 512KE, Plus, SE, and II, and requires 256K of memory. *Density of Orbitals* makes full use of the Macintosh user interface and permits the student to access data, rescale graphs, and consult review and help sheets solely by use of the mouse.

The operating instructions for *Density of Orbitals* are contained in the tutorial. A fully detailed description of all the features of the program is included as a separate *MacWrite* file. The tutorial has been crafted for ease of use.

The software program *Density of Orbitals* is part of the collection *Computers in Physics Instruction: Software*, which can be ordered by using the form at the end of this book.

Computer-Based Instruction for University Physics

Joseph Priest
Department of Physics, Miami University, Oxford, OH 45056

Computer-based instruction can be used to enhance and enrich topics that are difficult to present in the conventional classroom. Such activity includes discussing motion, plotting functions and changing the parameters, giving the students flexibility in selecting numerical parameters that interest them, drilling in certain types of problems, using the computer to discover new features of the physics, and capturing the fancy of the computer to motivate students to learn physics.

This hands-on workshop introduces participants to the intent and uses of the instructional software produced for two university-level physics textbooks. The programs are written in BASIC and are available for the Apple II and IBM PC series of computers. There are three disks titled *Mechanics, Thermodynamics and Waves*, and *Electricity and Magnetism*. Each disk covers about ten chapters of the text and has about 20 programs. The following descriptions of a program from each of the three disks conveys the flavor and character of the programs.

Mechanics

Physics texts usually have a one-dimensional motion problem phrased something like this: "The driver of a car traveling on a flat, straight stretch of interstate highway is exceeding the speed limit. A patrolman waiting at the edge of the highway pursues the car at constant acceleration. What is the minimum acceleration of the patrol car in order to catch the speeder before she crosses the state line?"

Mechanics programs the equations and uses animated graphics to show the speeder and the patrol car. By experimenting, the student finds the acceleration needed by the patrol car to catch the speeder just as she reaches the state line.

The algebra is then discussed to confirm the minimum acceleration that the student has already discovered. The student is given the flexibility of allowing the speeder to accelerate. Plots of distance versus time and speed versus time for both the speeder and patrolman are displayed along with the animated motion of the cars.

Thermodynamics and Waves

A thermodynamics program concerns the Carnot cycle and the second law of thermodynamics.

The efficiency of a Carnot heat engine is

$$1 - T_L / T_H$$

where T_L and T_H are the Kelvin temperatures of the cool and warm reservoirs. Based on the second law of thermodynamics, no heat engine operating between these two temperatures can have a larger efficiency.

Suppose that the two isothermal processes are retained and different thermodynamic paths are chosen for proceeding from T_H to T_L, and then from T_L to T_H. If the temperatures stay within the bounds of T_L and T_H, then the efficiency will always be less than the Carnot efficiency.

Starting with the pressure and volume at the end of the isothermal expansion, the student chooses straight-line thermodynamic paths leading to the beginning of the adiabatic compression. The process continues by starting at the pressure and volume at the end of the adiabatic compression and proceeding to the initial pressure and volume. The computer calculates the work done and the heat exchanged in each process. When the cycle is completed, the computer calculates the net work, determines the net heat taken into the system, and calculates the efficiency. The student is invited to find a thermodynamic path for which the efficiency exceeds the Carnot efficiency.

Electricity and Magnetism

An electricity and magnetism program involves AC circuits. The student chooses the voltage amplitude and frequency of an AC source and any series combination of an inductor, a resistor, and a capacitor. The program provides plots of all the voltages and the current in the circuit. The student can examine the phases of the voltages and currents relative to the phase of the source, and can check the plots to see that Kirchhoff's voltage rule is satisfied at all times. The computer provides calculations of the rms voltage for each component and the rms current.

Programs like these can enrich student understanding of physics. In many cases they help students understand concepts that are difficult to understand in a conventional classroom.

Sample chapters from *Computed-Based Instruction for University Physics* are part of the collection *Computers in Physics Instruction: Software*, which can be ordered by using the form at the end of this book.

Intelligent Computer Simulations in the Classroom: A Report from the Exploring Systems Earth Consortium

Annette Rappleyea
Physics Department, City College of San Francisco, San Francisco, CA 94112

The learning process involves assimilating new information and comparing and contrasting it with what the learner already knows. In order for learning to take place at all, learners must feel good about themselves and must be able to think about the subject matter. At best, traditional teaching is informative and entertaining; more often it is an oppressive waste of time.

Several of us in the Exploring System Earth (ESE) Consortium[1] are trying to improve classroom learning by designing and building interactive computer simulations that are linked to an "intelligent," friendly tutor.[2] We believe that students learn more, faster, and build positive feelings about a subject by experimenting in a domain-wise environment, instead of simply listening to lectures and reading textbooks. Our simulations are designed to be used in addition to traditional classroom teaching methods in much the same way that laboratory sessions are now used.

The main focus of the project is to couple manipulable computer simulations with an intelligent tutor that will oversee and sometimes guide the student's interaction with the simulation. We call the modules "intelligent tutorial learning environments" (ITLEs). Such simulations are currently operating in kinematics, statics, and dynamics, and new modules are planned in physics, biology, and chemistry.

The simulations are highly interactive. For example, the linear kinematics simulation allows the user to "drive" a small car, controlling the car's position, velocity, or acceleration. Meters and graphs are available for analyzing and monitoring the car's progress. The car can be stopped and restarted at any time. In another mode, the user can set the car's initial values of position, velocity, and acceleration. The car then operates for a few seconds, allowing the user to see how the car moves under the initial conditions.

The tutor is interactive on a different level. It keeps track of the student's progress, and "knows" as many teaching rules as we can put into it. For example, the tutor knows that if a student attempts to drive a square course without stopping at the corners, the student might be confusing velocity and acceleration. The tutor can choose from many examples and exercises designed to overcome that misconception and can present an intermediate example to the student. The tutor operates under its own set of rules that are derived from the best teaching available to us in the subject area.

1. There are many people working on this project. There are faculty members from City College of San Francisco, San Francisco State University, University of California at Berkeley, San Francisco Unified School District, University of Massachusetts, and University of Hawaii. In addition there are experts in related areas of interest (such as expert systems, screen design, etc.) who are available for consultation to the ESE Consortium.
2. Edwin L. Duckworth, James Kelley, Stephen Wilson, "AI Goes to School," Academic Computing **6** (November 1987).

Modeling of Physics Systems: Examples Using the Physics and Automobile Collisions and Physics and Sports Videodiscs

Dean Zollman
Department of Physics, Kansas State University, Manhattan, KS 66506

Physicists often use simplified models as tools to understanding complex systems. Yet we seldom teach modeling until we begin discussing abstract concepts. The videodisc provides a means by which students can build models of concrete situations. Modeling concrete systems prepares students to tackle models of abstract systems such as atoms and particles. To teach concepts of modeling, I use video material from two videodiscs, *Physics and Automobile Collisions* and *Physics and Sports*.[1] The videodisc system allows students to build simple models of a complex system then test those models.

For example, one way to treat the motion of an automobile striking a barrier is to assume that the entire motion is described by a point mass located at the center of mass. Using this model, students can see how the center of mass moves as the car strikes the barrier. By simultaneously watching the motion of their model and the car, students can see the value and limitation of the model. Students can evaluate more complex models, such as series of masses connected by stiff springs, in a similar manner.

Models for the motion of athletes are frequently stick figures. The complexity of the model depends on the number of segments involved. Students can compare models with actual motion by using acetates attached to the video screen. They draw their models on the acetate as they use the videodisc to step through the pictures of the event. With more sophisticated hardware, students can draw the model and perform calculations using computer graphics under the control of a mouse. We are developing and testing both methods.

Supported by the National Science Foundation under grant number MDR-8550145.

1. (New York: Wiley, 1984).

Section 7

Physics Lecture Demonstrations Using Computers

Computer Demonstrations for Optics

Ulrich E. Kruse

Department of Physics, University of Illinois at Urbana-Champaign, Urbana, IL 61801

We have developed computer programs for use as lecture demonstrations to teach optics. Two of the programs are used in demonstration experiments and measure diffraction patterns and compare the results with theory. The other programs are used to explain and illustrate important optical phenomena. The programs run on IBM PC AT computers with EGA 16-color displays. In lectures we use projectors to display the results to students.

The two demonstration experiments measure Fraunhofer diffraction from multi-slit sources and Fresnel diffraction in the shadow of a wire or a half-plane. In the experiments the lecturer uses a laser to produce the pattern on a screen that is viewed by a television camera. The image from the camera is then recorded by an image-grabber board plugged into the computer. The diffraction pattern is displayed on the computer screen and the lecturer selects a rectangular region for analysis. The computer produces the corresponding intensity graph. With a mouse, the lecturer selects four points on the intensity graph to establish a baseline, pattern center, and x and y scales. With this information the computer program scales the corresponding theoretical distribution and makes a direct comparison with the experimental results. The lecture is paced by the verbal explanation because the data acquisition, displays, and calculations are completed in less than two minutes. In a typical lecture several patterns are measured. We have used the Fraunhofer demonstration in the general introductory course offered to sophomores, and the Fresnel demonstration in the optics course for juniors and seniors.

The other demonstrations use computer graphics to explain experimental phenomena or theoretical models. The computer allows the lecturer to calculate accurate patterns and curves quickly so that the input parameters (e.g., the number of slits) can be quickly changed in answer to student suggestions. At this time we have completed programs to illustrate the following: superposition of two waves to form a two-slit diffraction pattern, phasor model to explain phasor representation of a sum of *sine* waves, phasor model of Fraunhofer diffraction from multiple slits, Cornu spiral model for Fresnel diffraction, physical explanation of the rainbow.

For example, in the program that explains Fraunhofer diffraction, the lecturer begins by selecting the parameters for the demonstration. These include the number of slits, the ratio of slit width to slit spacing, the angular range for the pattern to be shown, and the rate at which the demonstration progresses. This facility allows the lecturer to cover several examples and to speed up after the first example. He can also stop the display to point out special features like minima and maxima in the pattern.

In the actual presentation, a diagram of the experiment is shown first and then the diffraction pattern is simulated. The program provides the appearance of continuous variation of intensity by randomly mixing dots of the three intensities available from the EGA card. Then the phasor configuration corresponding to successive angles from the forward direction for the diffraction pattern are diplayed, together with the resulting phasor and a plot of the intensity. Students can follow the effect of the phase difference between the waves from the individual slits through the change of angle between the individual phasors and the change in magnitude of the resultant. The envelope corresponding to the variation of the single slit phasor is included in the plot of the intensity.

For a more advanced discussion, the lecturer can include an explanation of the magnitude of the phasor representing the individual slit. The superposition of "miniphasors" corresponding to the waves coming from parts of the individual slit is shown and displayed with the phasor for the individual slit. The variation of the phasor amplitude from the individual slit then accounts for the fall off of the intensity of the maxima away from the forward direction. By changing the ratio of the slit width to slit spacing, the modulation of the multislit pattern by the single slit intensity can be illustrated.

We use the experimental demonstrations and the model explanations during lectures. We also make the programs available to students so that they can reexamine them for details after the lecture. For the student use, we insert "pages" with an explanation corresponding to the verbal discussion given by the instructor during the lecture.

The demonstrations are easy to modify for related applications. We have made small changes in the labeling and used the demonstration of the superpositon of two waves to illustrate the sound pattern from two loudspeakers. This version is used in a course for nonscientists. We have also adapted the demonstration of the phasor model for the sum of sine waves to illustrate voltages in alternating current circuits. We simply relabeled the plots and changed the amplitudes and phases to correspond to different voltages in a series AC circuit. This demonstration will be used in the circuits part of the general introductory course.

The software program *Optics Demonstrations* is part of the collection *Computers in Physics Instruction: Software*, which can be ordered by using the form at the end of this book.

The Animated Chalkboard (Updated)

Richard W. Tarara
Department of Chemistry and Physics, Saint Mary's College, Notre Dame, IN 46556

The computer has not been as widely used as a lecture aid as it has been for data collection and analysis and individual instruction. During the past four years, I have supplemented my classroom presentation on various topics with short computer-generated animated sequences. Almost any kind of "overhead" material can be presented on the computer (with appropriate graphics software), but its particular strength is in adding animation and bringing pictures to life. This audiovisual capability, the ability to produce customized "filmstrips" at very low cost, makes the computer an important tool for classroom use. The technique is valuable at all educational levels.

The subject matter of my current 100-title library of animations, covers the entire spectrum of the normal physics curriculum. The largest number of programs covers the early study of motion. A common technique is the simultaneous viewing and plotting of a motion. A "Shoot the Monkey" animation, for example, allows the students to pick the speed of the projectile—though this is of little help to the monkey. Animations can also be invaluable in describing relativity. You can show time dilation "clearly" with a mirror and light clock, which can actually be moved on the screen while the path of the light is plotted. Some of the more sophisticated animations that I've done have been associated with a Physics of Technology course where topics as current as nuclear winter, SDI, and cruise missiles, and as common as the operation of a flush toilet provide ample material for animations.

To better illustrate the technique, I will describe one of the more than 100 animations currently in use. Instructors continually struggle to convince students that an object thrown vertically upward always accelerates downward. You can throw a ball straight up and down in front of the class and carefully describe the motion, but the basic principle still eludes many students. In an animation on the computer, you can go beyond simple demonstration. On the computer, you can slow the motion so that the student can really see the gradual change in velocity. The ball thrown up rises quickly at first, then slows until it stops rising and begins to fall—slowly at first, then faster and faster. After the demonstration comes the real trick. Show the motion again, but this time plot the velocity of the ball as a function of time. Do this on the same screen as the animated motion and have the plot appear in synch with the observed motion. After listening to you talk through the motion, viewing the graph, and seeing the obvious fact that the slope is constant and negative, the student really starts to believe that the ball does indeed accelerate downward.

To use an animated chalkboard effectively you have to consider the system configuration in relation to the class size. A single 25-inch monitor can suffice for class sizes in the 50-to-70–student range if you remember that any text or detail has to be large enough to be seen by those in the back of the room. Larger classes will require a projection display or multiple monitors. A definite advantage of this use of the computer is that only one hardware setup is needed, although portability (at least by cart) is desirable. The computer itself can be an inexpensive one—in fact, the Commodore 64 is very well suited to this use at a system price (excluding monitor) of under $300.

The real value of animated chalkboards is in customizing your own animations. This does demand some programming ability, but with the various commercial animation and graphics packages available today, your programming background need not be extensive. With the use of such aids, most of the titles now in use with my classes were prepared in less than two hours per animation. It is important to recognize that it is not necessary to rival arcade-quality graphics to produce effective pedagogical tools.

Student reaction to this technique has been very positive with favorable (and unsolicited) comments appearing often on class evaluations. There is also subjective evidence that those topics amplified through the animations are better understood than in prior years when the same subject matter was presented in more standard ways. The use of computer animations in the classroom and especially the preparation of the software does not necessarily make teaching easier for the instructor, but it does make the instructor a better teacher.

The software program *Physics Demonstrations* is part of the collection *Computers in Physics Instruction: Software,* which can be ordered by using the form at the end of this book.

Software for Physics Demonstrations in an Introductory Physics Class

John S. Risley
Department of Physics, North Carolina State University, Raleigh, NC 27695-8202

Demonstrations provide concrete examples for students learning physics. Software programs are now available that can augment a series of traditional demonstrations by simulating physical phenomena. During a computer demonstration, a physics teacher can clarify concepts by, for instance, stopping the action, displaying vectors that change in time, and experimenting with a situation by trial and error.

The computer display is an important aspect of a computer demonstration. Liquid-crystal displays used with an overhead projector are satisfactory, but they lack high contrast. And as the liquid-crystal display heats up, the contrast is reduced still further. In quick moving animations, the liquid-crystal display is likely to produce ghosting images. The best display option for a class of 50 to 90 students is a 45-inch rear-projection TV monitor, because the size and contrast make for easy viewing.

I have used a number of courseware packages as part of my introductory physics course in mechanics. Some of the best examples are described below. These examples are only a sample of the many possible software packages that can be used to demonstrate physics concepts.[1]

Motion: A Microcomputer-Based Lab.[2] The sonic ranger in this software is perfect for introducing kinematics. The real-time data display feature allows an instructor to move his hand or walk in front of the detector while the class sees a plot being made of the distance from his hand or body to the detector vs. time. The software makes it easy to manipulate the data and make a plot of velocity vs. time or to display a split screen of velocity vs. time and distance vs. time.

For more quantitative discussions, you can use a battery-operated toy bulldozer (for constant velocity motion), or a cart pulled by a string that is fed over a pulley with a weighted end (for constant accelerated motion). Unfortunately, mechanical toys can give out sonic waves that can be detected by the motion detector. These waves appear as spurious data—an effect that troubles students. This is also a problem with air tracks. The high velocity of the air moving out through the pin-holes in the air track apparently produces sonic sound waves that cause false readings for the sonic ranger.

Computer-Based Instructions for College (or University) Physics[3] and *Instructional Software for University Physics.*[4] To demonstrate the concept of simultaneity, I use a program that simulates a police car attempting to overtake a speeder before the speeder crosses the border into the next state. This one-dimensional kinematics problem becomes more interesting when I tell the students that the speeder is an NCSU student returning from a spring break spent at Myrtle Beach, South Carolina, and he is trying to get to classes on time in Raleigh, North Carolina.

The program allows you to choose the acceleration rate for the police car and for the speeder. To help clarify the principles, the student can plot a graph of distance vs. time and velocity vs. time as the cars are shown speeding across the screen. After a trial-and-error experimentation with various accelerations, the student recognizes that solving the problem exactly is going to require a little algebra.

Harmonic Motion Workshop.[5] Simple harmonic motion is easily demonstrated with a pendulum or a mass and spring. Using an excellent animated graphic display of an object in SHM, this software package illustrates the concepts of the temporal variation of the position, velocity, and especially the acceleration. You can illustrate the velocity and acceleration vectors varying in time and in position. Two objects in SHM can be used to illustrate how the motion differs for different amplitudes or phases. The software also allows you to show the relationship between uniform circular motion and SHM, and the traditional sinusoidal plots of position, velocity, or acceleration vs. time.

Animation Demonstration.[6] Illustrating the concept of superposition of waves is difficult in a real-time demonstration. The computer simulation in this software allows you to show how a pulse travels down a string and is reflected at the end, both for a fixed end for which the pulse is inverted and for a loose one for which it remains unchanged. Two and three pulses can be shown. When the two counter-propagating waves overlap, the simulation shows how the two waves add up to give the total wave structure. You can also show how two waves traveling in opposite directions add up to give standing waves. This software package has many other programs that animate numerous subjects in physics. A public domain version of this program is available from the department of physics at NCSU.

1. See M. Gjertsen and J. S. Risley, Phys.Teach., 25, 301.1 (1987) for a listing of over 1,000 software packages along with the addresses and telephone numbers of 100 software publishers.
2. (Pleasantville, NY: HRM Software, 1987).
3. (New York: Addison-Wesley Publishing Company, 1984).
4. (Orlando, FL: Academic Press, Inc., 1984).
5. (Oklahoma City, OK: High Technology Software Products, Inc., 1982).
6. (Iowa City, IA: Conduit).

Section 8

Authoring Tools and Programming Languages

The cT Language and Its Uses: A Modern Programming Tool

Bruce Arne Sherwood
Center for Design of Educational Computing and Department of Physics

Judith N. Sherwood
Center for Design of Educational Computing, Carnegie Mellon University, Pittsburgh, PA 15213

Modern computer applications emphasize windows, mouse interactions, attractive multifont text, and high-quality graphics. End users find such applications directly appealing and easy to use. But the programmers who create these appealing applications face a bewilderingly, complex and daunting programming task, far more difficult than writing programs for earlier computing environments. Not only is the task vastly more complex, but production-level programs often require lengthy batch compilations, making the programmer's environment much less interactive than that of the end user. These factors have created a situation where only exceptionally skilled programmers can produce modern interactive programs.

Another serious problem is the difficulty of moving a modern application from one computer to another. Years ago, it was relatively easy and straightforward to write a FORTRAN calculation that printed a few numerical results on a line printer, and such a program would run on any of the many machines which had a FORTRAN compiler. Now, however, a program with a complex user interface is typically tied to the particular machine on which it was created, because of differences in screen size and interaction aspects such as mouse clicks and menus.

To address these two major issues, the cT programming environment (formerly known as CMU Tutor) has been developed in the Center for Design of Educational Computing at Carnegie Mellon University. The cT programming environment makes it much easier to exploit the power of modern computing environments. It enables those with limited programming experience to develop interactive applications, including educational applications, without having to depend completely on the assistance of professional programmers. Programs written in the cT language can be compiled and run without change on many different kinds of machines. Supported machines include the Apple Macintosh and IBM PS/2 family of personal computers and several brands of professional work stations: IBM RT PC, Sun, and DEC Microvax. The name cT is a trademark of Carnegie Mellon University.

Major features of the cT language include interactive graphics in windowed environments, instant portability across diverse computers, automatic rescaling of text and graphics to fit the window, multifont text, menus, mouse and keyset inputs, analysis of words and sentences, analysis of numbers and algebraic expressions, rich sequencing options, standard calculational capabilities, and numeric and text files.

Major features of the cT programming environment include an integrated editing and execution environment, incremental compilation for fast revision and execution, online reference manual with executable examples, a graphics editor that generates cT graphics statements, and accurate and informative error diagnostics.

In the remainder of the paper these features of the language and its environment will be described in detail.

Description of the cT Language and Environment

Programs written in cT are incrementally compiled. That is, only those program segments that have been changed are automatically recompiled. As a result, the developer sees the effect immediately after making a change in the source code. Since the source code is compiled rather than interpreted, execution speed is very fast. When a user enters a program, those procedures needed immediately are compiled first. Whenever there is a wait for user input, more of the program is compiled in the background, interrupted if necessary for on-demand compilation when the user requires execution of a section not already compiled. This overlapping of execution and compilation ensures that there is hardly any observable delay because of compilation. When necessary, a binary version of the program can be produced.

In cT, compile-time and execution-time error diagnostics are very specific. When an error is detected, a detailed error message is displayed. The source code is automatically scrolled to display the statement containing the error, and the editing caret is placed in the statement at the character location where the error was detected.

Modern multifont text (with italics, bold, large, small, centered, etc.) appears as such in the source code and is manipulated with mouse and menus as in a modern word processor. Execution-time position and margin controls determine a rectangle within which a text statement is displayed. Despite the importance of modern text, there are hardly any other programming languages that handle multifont text directly in the source code.

Graphics capabilities include labeled graphs, rotatable displays, pattern-filled polygons and disks, and animation of icons. Almost all the major graphics abilities of modern machines are accessible through the cT language. Automatic scaling to arbitrary window dimensions is supported with options to scale x, scale y, preserve aspect ratio, and scale text. Graphics scaling factors are constrained by the available font sizes in order to preserve the correct relationship between text and graphics. This automatic scaling makes a big contribution to portability of programs from one machine to another despite differences in screen sizes.

A novel graphics editor directly generates cT source code. Figure 1 illustrates the automatic generation of cT display statements. The source code is at the left and the corresponding display is at the right. At the bottom left is a list of cT commands (only some basic display commands are visible), and these commands can

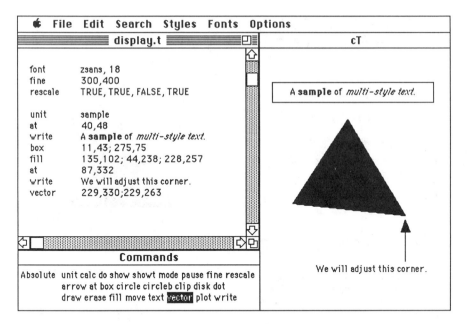

Figure 1.   Automatic generation of cT program statements.

be entered into the source code by pointing at them with the mouse. Clicking the mouse in the display area causes an x-y coordinate pair to be entered into the source code, followed by recompilation and execution of the current unit, which is very fast. The source code for the graphics shown in the figure was generated without touching the keyboard. Note the styled text that was typed into the source code.

Figure 2 shows how existing display routines can be easily modified. The mouse is used to select an x-y coordinate pair in the source code (part of the "fill" statement in the example), as though one were preparing to cut it out or change it to italics. This selected coordinate pair can be visually adjusted by clicking the mouse in the display area, which changes the selected coordinate pair and triggers recompilation and execution to create the altered display. The "fine" command in the example defines a virtual fine-grid coordinate system, and the arguments of the "rescale" command specify that x and y should rescale to fill the window, without preserving aspect ratio, but with scaling of text. The graphics editor supports the device-independent coordinate scheme by unscaling pixel-based mouse clicks into appropriate fine-grid positions.

The cT language handles input from both keyset and mouse. Convenient analysis routines are provided for word, sentence, numerical, and algebraic inputs. Sentence analysis includes checks for spelling and word order. Answer "markup" is supported: unanticipated words and miscapitalized letters are shown in boldface, misspelled words are shown in italics, etc. Mathematical input can be evaluated at high speed, making it easy to create function plotter programs. Implied multiplication (e.g. "3y") is permitted, so input expressions can be written using normal

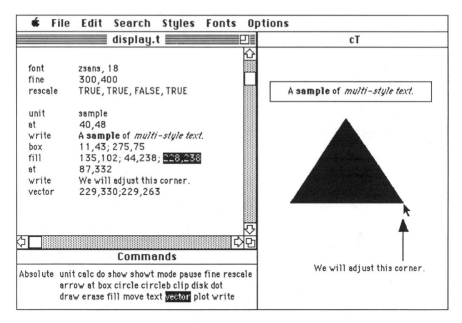

Figure 2.   An adjustment to cT display routines.

mathematical notation. Both mouse clicks and mouse movements can be processed. The cT menu command creates equivalent pull-down or pop-up menus, depending on the type of menus native to a particular machine. This is an example of how the generic nature of cT commands contributes to portability of applications.

The usual calculational structures are provided, including if-else, case, and loop-reloop-outloop (which provides a unified form for iterative, while, and until loops). Case instances need not be constants. Subroutine arguments can be passed by value or by address. Local variables permit recursion. The logical expression "$A < B < C$" is handled correctly (something few languages do). In the expression "$n > 0$ & $K(n) = 3$", the array reference $K(n)$ is not attempted if n is not greater than 0 (this conditional evaluation of expressions is like that in the C language).

Data types include byte, integer, floating-point, file, and markers on character strings, and multidimensional arrays of such data, with full bounds checking and optional offsets. There are bit-manipulation operators. File operations include sequential reads and writes of ASCII numbers, of strings, and of bytes.

String manipulation capabilities are very unusual. A marker variable delimits a portion of text, and insertions earlier in the text cause the marker to move in such a way as to continue to mark the original characters. Marker functions give access to the first or last character of a marked region, the next character after a marked region, or the previous character before a marked region. Text bracketed by a marker variable can be replaced or appended to. The contents of markers can be compared using logical operators (dictionary order governs inequality tests).

Marked text representing an expression can be compiled to machine code and then repeatedly evaluated at high speed.

Library files containing cT routines can have their own global variables and their own initialization routines, which are triggered automatically when the main program is started.

To facilitate the creation of highly interactive programs, there is a richer set of sequencing operations than is typically found in a programming language, For example, the notion of next and previous display is built into the language.

A common syntax exists for conditional commands. The arguments of any command (other than those defining control structures such as if) can be selected by the value of an expression. This provides a compact one-line case statement.

An important component of the cT programming environment is its powerful online reference manual. Language features can be accessed either through hierarchical indices or through names of commands in the language. In either case selections are made simply by pointing with a mouse. Language features are illustrated in context by sample routines. An unusual feature is the ability to execute these samples immediately by using the mouse to copy them into the programming window. These sample routines can act as nuclei for further elaboration by the programmer. Figure 3 shows an example of copying out a sample routine dealing with loops and running it to produce some vertical lines. The upper-left panel of the help window contains a hierarchical list of topics. The upper-right panel contains a list of cT commands and keywords. The bottom panel contains details on the selected feature, including sample routines. The programmer used the mouse to

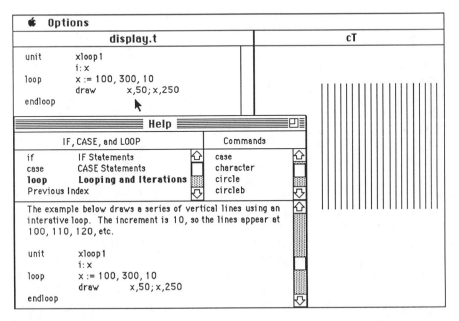

Figure 3.   An online reference manual with sample routines.

copy the loop example into the source code window (just above the help window), then executed the routine, which produced the vertical lines on the right. There is also a printed version in the reference manual.[1]

Although it is already being used for many purposes, cT is not complete. At present there is no support for saving and restoring portions of the screen. Scrolling-text objects are in the planning stage. Data types will be extended to include record structures and pointer variables for dynamic storage allocation. Also planned is the ability to call routines written in C or other languages. cT has already been used for the user interface of large LISP and FORTRAN programs on multiprocessing work stations, with communications between the two programs through files or "sockets."

An important aspect of the cT programming environment is its evolutionary development path. Most new language features are of course additions to the language that do not invalidate existing programs. However, occasionally it has been necessary for further growth to make incompatible changes that would cause existing programs to stop working were it not for the fact that the cT compiler automatically converts old programs to the new format. A syntax-level indicator is kept in the program file to indicate that the appropriate conversions have been made. As a result, despite major growth and change in the language, the oldest programs from 1985 still work. This compatibility from year to year is just as important as the compatibility that cT offers from machine to machine.

Uses of cT

The InterUniversity Consortium for Educational Computing (ICEC) offered one-week workshops on cT to faculty and staff from the ICEC colleges and universities.[2] The prerequisites were that participants had written at least one program in some language, but generally had no experience with modern windowed environments. In less than a week, most participants wrote significant graphics-oriented programs. Later, Vassar College and California State University at Northridge ran cT workshops themselves.

Quantum Well by Bruce Sherwood is an example of an educational program written in cT. The mouse is used to specify a potential well in which an electron is trapped. In Figure 4 the mouse is used to choose a trial energy level, and the quantum-mechanical equations are evaluated left-to-right to plot the corresponding wave function. *Quantum* depends on the display, input analysis, and calculational strengths of cT. Creating the program was greatly facilitated by the ability to test changes rapidly, thanks to incremental compilation.

Other examples of educational programs can be found in the workshop guide and in several of David Trowbridge's conference papers.[2] For an assessment of the suitability of cT for educational programming, see M. Resmer's article and B. A. Sherwood and J. H. Larkin's paper in the proceedings of the 1987 IBM ACIS University Conference.[3] General issues affecting the design of educational programs are treated in a forthcoming article.[4] For a discussion of the difficulties of

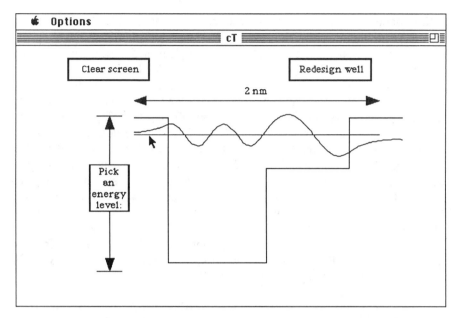

Figure 4.   An example of an educational physics application written in the cT language.

writing modern programs for research purposes, with some assessment of the advantages of cT, see C. Lewis and G. M. Olson.[5]

History

cT is based on the MicroTutor language developed at the Computer-based Education Research Laboratory at the University of Illinois (the PLATO project). It incorporates MicroTutor's important constructs for interactive educational programming.[6] MicroTutor is basically an updated and machine-independent descendent of the TUTOR language, whose development was initiated by Paul Tenczar in 1967.[7] The first version of MicroTutor was created by David Andersen in 1977. Similar languages include Digital Equipment Corporation's DAL (Digital Author Language) and Tenczar's own TenCore for the IBM PC. In large part, cT is compatible with MicroTutor, but it differs from MicroTutor and other Tutorlike languages in its extensions for the modern environments, where windowing, mice, and menus impose additional requirements that the earlier languages did not have to meet. The genesis of cT within the Andrew environment has been carefully detailed.[8] Moreover, we have written a textbook on cT.[9]

Design and implementation of the basic architecture of cT was done by Bruce A. Sherwood, who is associate director of the Center for Design of Educational Computing (CDEC) and professor of physics. Major contributions, including the online reference manual and many of the graphics capabilities were made by Judith

N. Sherwood, who is senior development consultant in CDEC. David Andersen, system programmer with CDEC, has made large contributions, including full machine-code compilation of expressions and implementation of the novel string-handling machinery. CDEC programmer Kevin Whitley has built the text-editing and text-management routines and has continuing responsibility for the Macintosh version. Additional contributions to the development of cT have been made by CDEC staff members Tom Neuendorffer, Chris Koenigsberg, David Trowbridge, and Ruth Chabay. Jill Larkin, director of CDEC during the period of its earliest development, encouraged the project and helped debug cT. Information Technology Center staff members Fred Hansen, Tom Peters, and Andrew Appel contributed much valuable advice. Gregg Malkary, a graduate student of Andy van Dam at Brown University, helped initiate an experimental version for the Macintosh. Special thanks are due Bill Arms, Preston Covey, Michael LoBue, Jim Morris, and Carol Scheftic.

The cT programming environment is currently being distributed by Carnegie Mellon University. For information, write to cT Distribution, CDEC Bldg. B, Carnegie Mellon University, Pittsburgh, PA 15213.

1. J. N. Sherwood, *The cT Reference Manual* (Champaign, IL: Stipes Publishing Company, 1988). This is a printed version of the online reference manual.
2. M. Critchfield, "ICEC-Carnegie Summer Workshop Guide," Carnegie Mellon University internal report (1986). This is a guide for instructors of such workshops. D. Trowbridge, "A Sampler of Educational Software at CMU," *Proceedings of the National Educational Computing Conference*, San Diego, CA, 135 (June 1986); D. Trowbridge, "Quick Generation of Lecture Demonstrations and Student Exercises," *Proceedings of the IBM Academic Computing Information Systems University Conference*, Discipline Symposia–Physics, Boston, MA, 2 (June 1987).
3. M. Resmer, "New Strategies for the Development of Educational Software," Academic Computing **2**, 22 (December–January 1988); B. A. Sherwood and J.H. Larkin, "New Tools for Courseware Production," NERComp Journal (Spring 1989).
4. R. W. Chabay and B. A. Sherwood, "A Practical Guide for the Creation of Educational Software," *Computer Assisted Instruction and Intelligent Tutoring Systems: Shared Issues and Complementary Approaches*, ed. by J. H. Larkin and R. W. Chabay. (Hillsdale, NJ: Erlbaum, in press).
5. C. Lewis and G.M. Olson, "Can Principles of Cognition Lower the Barriers to Programming?" *Empirical Studies of Programmers*, Vol. 2, ed. G.M. Olson, E. Soloway, and S. Sheppard, (Norwood, NJ: Ablex 1987).
6. J. Stifle, S. G. Smith, and D. M. Andersen, "Microprocessor Delivery of PLATO Courseware," *Proceedings of the Association for the Development of Computer-based Instructional Systems*, San Diego, CA, 1027 (1979); B. A. Sherwood and J. N. Sherwood, *The MicroTutor Language* (Champaign, IL: Stipes Publishing Company, 1985).
7. B. A. Sherwood, *The TUTOR Language* (Minneapolis, MN: Control Data Education Co. 1977).
8. J. H. Morris, M. Satyanarayanan, M. H. Conner, J. H. Howard, D. S. H. Rosenthal, and Smith, "Andrew: a Distributed Personal Computing Environment," Communications of the ACM **29**, 184 (March 1986); W. J. Hansen, "The Andrew Environment for Development of Educational Computing," Computers in Education **12**, 231 (1988); B. A. Sherwood, "Work Stations at Carnegie Mellon," *Proceedings of the IEEE-ACM Fall Joint Computer Conference*, Dallas, TX, 15 (November 1986); B. A. Sherwood,

"An Integrated Authoring Environment," *Proceedings of the IBM Academic Information Systems University Advanced Education Projects Conference*, Alexandria, VA, 29 (June 1985).

9. B. A. Sherwood and J. N. Sherwood, *The cT Language* (Champaign, IL: Stipes Publishing Company, 1988).

Learning to Use the cT Language

Bruce Arne Sherwood and Judith N. Sherwood
Center for Design of Educational Computing, Carnegie Mellon University, Pittsburgh, PA 15213

Modern computer applications emphasize windows, mouse interactions, attractive multifont text, and high-quality graphics. End users find such applications directly appealing and easy to use. But these applications are not so easy for the programmers who create them. Creating such programs is a bewilderingly complex and daunting programming task, far more difficult than writing programs for earlier computing environments. Because production-level programs often require lengthy batch compilations, the programmer's environment is much less interactive than that of the end user. As a result, only exceptionally skilled programmers have been able to produce modern interactive programs.

cT (formerly called CMU Tutor) is a programming language developed in the Center for Design of Educational Computing at Carnegie Mellon University. It is especially suitable for rapid creation of interactive programs for modern graphics-oriented environments.[1] cT makes it feasible for a much larger group of people to write modern programs.

Not only is it surprisingly easy to create modern applications with cT, but such applications can be run without change on a wide range of computers, including Macintosh, Mac II, IBM PCs, and PS/2s, and Unix-based work stations (MicroVax, IBM RT, and Sun). Such compatibility is almost without precedent. In the past, any program that had an attractive user interface was tied to one particular computer. This has had a number of bad consequences, perhaps the most serious of which is the split of the educational software market into machine-dependent submarkets. This capability problem has inhibited faculty developers.

cT possesses many features that set it apart from other programming languages for developing and executing interactive applications.

- Rich text and graphics facilities in an easy-to-use form;

- Incremental compilation, combining the revision speed of interpreted languages with the execution speed of compiled languages;

- A display editor, which generates cT source code in a direct and transparent manner;

- Automatic scaling of graphics and text to fit the current window;

- Powerful tools to analyze key-set and mouse inputs;

- Generation of pull-down or pop-up menus for different machines with the same cT menu command;

- Calculational and file-handling features similar to those available in other algorithmic languages;

- Detailed error diagnostics;

- An unusual online reference manual offering instant execution of documented language features.

Despite cT's appeal to novice programmers, it can support sophisticated programming tasks. This makes cT of interest to experienced programmers, and it also provides a growth path for newcomers, who can start simply but progress in easy stages to higher competence. cT is already in use at a number of universities for research computing tasks and creating educational software.

1. B. A. Sherwood and J. N. Sherwood, "CMU Tutor: An Integrated Programming Environment for Advanced-Function Work Stations," *Proceedings of the IBM Academic Information Systems University AEP Conference*, San Diego, April 1986 (San Diego: IBM Academic Information Systems, 1986), IV:29–37; M. Resmer, "New Strategies for the Development of Educational Software," Acad. Comp., 22 (December-January 1988); D. Trowbridge, "Quick Generation of Lecture Demonstrations and Student Exercises," *Proceedings of the IBM ACIS University Conference, Discipline Symposia: Physics*, Boston, June 1987 (Boston: IBM Academic Information Systems, 1987), pp. 2–7.

ILS/2: An Interactive Learning System

James M. Tanner
School of Physics, Georgia Institute of Technology, Atlanta, GA 30332

This session will present *ILS/2*, a software system specifically designed for writing (mode "Author") and presenting (mode "Presenter") interactive courseware on microcomputers. We will demonstrate courseware that has been written using this system.

ILS/2 groups learning units into library modules. Each library module is a learning or training package whose length, content, and design is controlled by the author. A library module could be a single interactive homework problem that provides the assistance to the learner on an as-needed basis. It could be a single-subject tutorial that includes a statement of definition of the subject, a review of important prerequisite material, and drill using, interpreting, or applying the new concept. Or it could be an entire lesson that includes an introduction to new material, a brief review of previous lessons, applications of the new material, conceptual drills on the new material, etc.

"Author" is a menu-driven software mode that allows the user to construct library modules of interactive courseware. When preparing library modules in this mode, an *ILS/2* author designs "screens" that spatially and temporally intersperse text, special characters, symbols, and graphics. On these screens, the author can also plot functions and create simple animations. The system calls for externally compiled programs, learner responses, and branching decisions. A library module consists of a collection of these screens, which, when used, are multilinked by branching decisions based on learner input.

When working in mode "Author," the author designs each screen while working "live" on the computer display. No computer programming experience or knowledge is necessary to work in this mode; there is no programming language to be learned. All of the available authoring options are selected from pop-up menus. Thus, the author's attention is focused entirely on the design and structure of the courseware.

Of particular importance to physicists is the ability to microposition anywhere on the computer display any text character, any special character or symbol, any graphic operation, or any graphic figure that has been previously drawn and saved as a library file. Among the special characters and symbols presently available are all Greek characters, most mathematical symbols, and many characters peculiar to physics. Special symbols or characters may be added by authors. The author can create simple animations by selecting any rectangular portion of a screen as an image and then specifying a screen trajectory for that image.

A learner uses ISL/2 in its "Presenter" mode. In "Presenter," the user moves naturally through a series of screens without needing to know how to manipulate the computer. "Presenter" operates in the background, following the encoded instructions of the author. The learner's path through any module or, for that matter, any sequence of modules is dictated by his responses. Responses may be in the form of menu choices, numeric values, alphanumeric strings, mixtures of numeric values and alphanumeric strings, or even simple graphic input.

Each library module is an ASCII file whose structure is independent of computer model, screen resolution, and graphic mode. After the author gets *ILS/2* to run on a particular generic type of microcomputer (IBM, Macintosh, etc.), he can introduce faster computer models, different computer manufacturers, or high-resolution screens via relatively simple updates of the "Author" and "Presenter" software. That is, the library is immune to hardware developments and will not require modification with the advent of improved/altered computer systems. This, of course, is very important. If authors have spent the time necessary to prepare high-quality library modules, it is imperative that any authoring/presentation system not require that each library module be "repaired" to fit each new microprocessor or graphics system.

The initial version of *ILS/2* executes on IBM PC–compatible systems with CGA, EGA, MCGA, or VGA graphics. A version of *ILS/2* for Apple/Macintosh systems is feasible and its development is under consideration.

Graphics and Screen-Input Tools for Physics Students

Jack M. Wilson and W. MacDonald
Department of Physics and Astronomy, The University of Maryland, College Park, MD 20742

The Maryland University Project in Physics and Educational Technology (M.U.P.P.E.T.) wanted students to use powerful computing tools to solve problems and model phenomena without spending an inordinate amount of time learning computer programming. We satisfied these seemingly contradictory requirements by adopting computer tools that were both powerful and easy to learn.

Nonmajors use an electronic spreadsheet. This widely available tool, which is of great use in many areas outside physics, provides students with the computational power and graphics output necessary for their work in physics classes.

Science majors need more than a spreadsheet; they need a high-level language that they can build on throughout their careers. After considering various languages, M.U.P.P.E.T. adopted *Turbo Pascal* for our physics-major courses. Because it is inexpensive, widely available, and powerful, Pascal is often the first language taught in university computer-science departments. Furthermore, it is easy to move from Pascal to other modern structured high-level languages.

In order to reduce programming overhead, M.U.P.P.E.T. developed a set of input/output and graphics tools that make it possible for students to do professional program development with a minimum of programming effort. These tools work by reducing each program to three short blocks: "Accept(Data)," "Physics," and "PlotData(Data)."

The first block, "Accept(Data)," lets a student develop an input screen in the "Turbo Editor" in WYSIWYG fashion. Each field on the screen is nonmodifiable, numeric modifiable, or alpha-numeric modifiable. Modifiable fields can be given default values so that the user is never completely stumped for values on an input screen. Numeric fields are validated upon input, and cursor positions and function keys pressed are passed back to the user. The second block, "Physics," is the portion of the program containing the physics content.

The third block, "PlotData(Data)," can be as short as a single statement or as elaborate as opening multiple windows and plotting different graphs in each window. Students' first efforts usually consist of the single statement. The routines called by this statement use a default window of the entire screen and automatically scale the data and label the axes. As students become more familiar with the system they learn to define their own windows and scale their own data. All of this can be accomplished without having to have detailed knowledge of the computer system and its graphic devices. Windows are scaled as percentages of the screen and scales are set in user-data units. Students are not required to do detailed mappings to a particular display's pixels.

The M.U.P.P.E.T. tools are available in both *Turbo Pascal 3.0* and *4.0* and have been placed in the public domain. They may be used as "include" files in source-code programs or may be separately compiled and used as extensions to Pascal. The tools are available in files "MUPGRAPH.INC" and "MUPSCRN.INC" on the AAPT/University of Maryland Bulletin Board (301-454-2086). In our presentation we will demonstrate examples of student programs. These are also available from the bulletin board or on disk. The examples include pendulum motion, projectile motion, data analysis with least square, and wave functions in potential wells.

The example programs were originally developed for use in an introductory physics laboratory where students spend one week in the laboratory working with the apparatus and the following week in the computer lab developing computer models and programs for analysis. Students are also required to complete an independent project as part of the laboratory. The tools have now been used for two years in both class and laboratory. We feel that they have been successful in allowing students to develop sophisticated projects without spending much class time learning programming.

HyperCard and Physics

Robert G. Fuller
Department of Physics, U.S. Air Force Academy, Colorado Springs, CO 80840, on leave from the Department of Physics and Astronomy, University of Nebraska-Lincoln, Lincoln, NE 68588-0111

Carl R. Nave
Department of Physics and Astronomy, Georgia State University, Atlanta, GA 30303

David M. Winch
Department of Physics, U.S. Air Force Academy, Colorado Springs, CO 80840, on leave from the Department of Physics, Kalamazoo College, Kalamazoo, MI 49007

The participants in this workshop will work with specific *HyperCard* stacks developed for use by physicists. Specific examples such as physical-constant look-up tables, collections of homework and exam questions, and simple physics animations will be available. Novice users can explore existing physics stacks. More experienced HyperCardians can modify various scripts for their own uses. We will provide copies of existing and modified *HyperCard* physics stacks for the participants to take home. We will also discuss forming a national network of *HyperCard* physics special-interest groups.

Workshop participants should have some previous experience with *HyperCard* (even if only in an Apple exhibit booth) and some experience with a Macintosh. Participants should bring a blank, double-sided 3.5-inch diskette to take stackware home.

Developing CAI Programs with Apple SuperPilot

Robert J. Boye
Physics Department, Morgan State University, Baltimore, MD 21239

Most commercial software has been produced for physics courses at the introductory level. In 1982 the physics department at Morgan State University developed a calculus-based computer assisted instruction (CAI) program for our upper-level courses in mechanics and electricity and magnetism.

The physics department at Morgan State University has 13 Apple microcomputers that were obtained by a MISIP grant from the U.S. Department of Education. We developed our upper-division CAI programs using Apple *SuperPilot*, which had its origin in common Pilot.

The original Pilot language was developed in the early 1970s to enable subject specialists to write CAI programs even if they know little about computer programming. Pilot has only about eight instructions, so time spent learning the language is minimal. The *CAI Sourcebook* estimates that it is about six times faster to create CAI lessons using a language like Pilot than by using a more general purpose language like BASIC.[1]

Apple *SuperPilot* makes use of Apple's graphic and sound capabilities to improve the original Pilot language. The Apple *SuperPilot* program uses the following editors to develop lessons: "Lesson Text Editor," "Character Set Editor," "Graphics Editor," "Sound Effects Editor." These editors have on-screen menus and additional help screens. Student workers can often enter programs and draw graphics and character sets without using the manual.

The "Lesson Text Editor" is used to type instructions for the lesson program. These are stored for student use on a lesson diskette. The instructions tell the computer when to display text, graphic images, sound effects, and character sets. They also accept answers from the student and can branch to different parts of the program depending on the answers.

The "Character Set Editor" is used to create a new set of designs or characters in place of the standard ASCII character set. This can be used to produce Greek letters or mathematical symbols.

The "Graphics Editor" is used to draw diagrams or pictures. This is done by moving a screen cursor using the computer keyboard or game paddles. Using the "Graphics Editor," the user can draw circles, ellipses, and rectangles.

The "Sound Effects Editor" can be used to create sound effects during the lesson. I have not used sound effects because of the disquieting effect when a number of students are simultaneously running lessons.

The following are examples of the most commonly used *SuperPilot* instructions. The type instruction (T:) tells the computer to print anything written after the colon on the screen. The accept instruction (A:) tells the computer to "listen" while the student types something on the keyboard. The match instruction (M:) compares the word or words following the match instruction to the student's response. If the computer finds the match word in the student's response, it stores the answer "Yes;" otherwise it stores the answer "No." This answer can then be used as a condition for executing further instructions.

The jump instruction (J:) tells the lesson program to skip over one or more instructions and branch to a different part of the program. By adding a yes or no conditioner to the Jump instruction (JY: or JN:), you can tell the computer to branch to a different part of the program depending on whether a "Yes" or "No" answer was given to the previous match instruction. The wait instruction (W:n) inserts a delay of n seconds into the program. The student can shorten this delay by pressing a key. Other special instructions control color, animation, and the size or thickness of text on the screen.

At the conference, I will present an example of a *SuperPilot* CAI program in electricity, along with a listing of CAI programs in mechanics and electricity and magnetism.

The software program *Electricity Instruction* is part of the collection *Computers in Physics Instruction: Software*, which can be ordered by using the form at the end of this book.

1. Robert L. Burke, *CAI Sourcebook* (Englewood Cliffs, NJ: Prentice-Hall, 1982), p. 20.

Physics Majors as Research Scientists: Introductory Course Strategy

William M. MacDonald
Department of Physics and Astronomy, The University of Maryland, College Park, MD 20742

The world of the research scientist is open to freshman physics students able to translate the "power tools" of physics into computer programs. Newton's second law is an example of a power tool of classical mechanics that can easily be solved by simple numerical procedures and used directly to explore a variety of physical systems, including nonlinear and chaotic systems for which there exist no mathematically explicit solutions. With the help of computer power tools, students can invent and study physical systems that interest them instead of being confined to "closed-form" exercises. Moreover, writing a computer program usually gives a student a much deeper understanding of physics than he learns from solving set problems. In our experience, a student who experiences the thrill of writing his own programs to solve physics problems begins to read physics equations like a working scientist and his imagination begins to suggest all sorts of variations on the physical problems he tackles.

Using the computer in introductory physics courses requires a carefully planned strategy for integrating the physics with programming and simple numerical methods. Experience gained from courses for physics majors developed in the M.U.P.P.E.T. program has shown that this strategy must include a general-purpose programming language; graded programs; "stub" procedures; libraries; project packages; and simulation packages.

A general-purpose programming language is an important tool for scientists. Such a program should include "strong typing," which requires that every variable used in the program be declared as a real number, an integer, an array, etc. Global variables appear at the beginning; local variables can be introduced in procedures. Strong typing makes for easily readable programs. More important, it practically eliminates frustrating "debugging" for the student (and the instructor) because the compiler checks that each variable is properly used in assigning values, or as arguments to procedures. The programming language should also be well structured. A structured language encourages the student to approach complex problems with an organized, "divide-and-conquer" strategy and to program "top down" by writing outlines in "pseudocode." For example, a student writing a program to calculate projectile motion with air resistance might start with an outline:

```
INPUT x0, y0, v0, angle, drag_coefficient, dt;
REPEAT
STEP tn, xn, yn, vn;
PRINT tn, xn, yn, vn;
UNTIL yn <= 0;
INTERPOLATE for impact_time, range, impact_velocity;
PRINT impact_time, range, impact_velocity;
```

This outline is then fleshed out with the necessary procedures.

Pascal is increasingly regarded as the introductory programming language of choice for high school and college students, but this was the first time most of our students had seen it. We chose *Turbo Pascal* by Borland International because it is an integrated editor, compiler, and run-time module that requires minimal knowledge of the operating system. Two or three one-hour microlabs in the first weeks of the course are sufficient for students to begin writing simple programs that sum series, solve equations, and integrate trajectories. A graded set of programs that introduce programming elements one at a time are used by students as training exercises and as building blocks for their own programs.

"Stub" procedures allow the student to tackle really major problems without spending much time writing inessential parts of a program; the student can focus his attention on the part of a program that contains the essential physics, for example, the force law. Stub procedures are possible because the logical correctness of a Pascal program can be checked even when essential procedures have been left "empty." For example, the following program to add two numbers will compile and run even though it produces no result.

```
PROGRAM Sum;
VAR a,b: real;
PROCEDURE Sum(a,b: real): real;
begin {Fill in statements to Sum two numbers}
end;
BEGIN
WRITE('Enter a and b - '); READLN(a,b);
WRITE('The sum is ', Sum(a,b))
END
```

The student completes this program by replacing the comment with a statement (Sum: = $a + b$).

Just as in research, our students use libraries of programs to reduce the labor of writing programs. Libraries developed in the M.U.P.P.E.T. program provide procedures for graphing results (*MUPgraph*), evaluating functions, and writing input-output screens.

The project package includes the source code for a program that students use in a series of related projects. In the first semester, students use the package *KEPLER*, which contains the source code for a program that integrates the equations for particle motion in a central force and outlines for projects on geosatellites, Halley's comet, the effect of deviations from the inverse-square law, quantized Bohr-Sommerfeld elliptic orbits in hydrogen, Rutherford scattering, and scattering by a shielded Coulomb potential. Each section discusses the appropriate units to use for time, distance, and energy together with the equations of motion in these units.

Students begin with the project on geosatellites, for which the computer program is written. They do all the other projects by modifying this program and using *MUPgraph* to plot the orbit and any other quantities that interest them, e.g., the energy as a function of time. The tabular results and graphs for the first two or three projects are obtained in the first microlab, which runs two or three hours. The remaining projects are done by the students on their own, during free hours of the microlab or in one of the clusters of computers around campus, over a period of about six weeks. Using a second project package (*OSCILLATOR*) in the same way, students explore the behavior of a pendulum in both the linear and nonlinear regime, driven and undriven, and are introduced to the phase plane representation of the motion.

Students use a simulation package as an intuition-training device to explore a variety of related systems. The heart of the package is a sophisticated computer program that allows the student to select, or design, different systems, choose values for the physical constants, and select different graphical presentations of the motion.

The simulation package *ORBITS*, developed by J. Harold and K. Hennacy, allows the student to explore planetary systems with one or two massive bodies and up to five light planets. The motion can be shown in the center of mass or in the rest frame of one of the bodies. For example, when the student views a system consisting of our sun, the earth, and saturn in the earth frame, he sees the sun and saturn executing Ptolemaic epicycles. The program demonstrates the importance of the Copernican choice of the rest frame of the sun in a way that cannot be explained in a textbook.

The Maryland University Project in Physics Educational Technology is supported by the Fund for the Improvement of Post Secondary Education and by the IBM/ACIS program.

Intuition-Building Tool Kits for Physical Systems

Charles A. Whitney and Philip Sadler
Harvard College Observatory, Cambridge, MA 02138

To acquire knowledge, students must confront and clear away misconceptions. One way to do this is to explore and build intuitions about systems. Such heuristic understanding allows students to replace misconceptions and to internalize the essence of basic physical laws.

Computer simulations permit the creation of microworlds that were formerly found only in expensive laboratories or science-fiction novels. Simulations permit an intimate and interactive view of the models that form the basis for much of modern science. They can provide a powerful method for testing preconceptions and for building the qualitative intuitions that precede quantitative reasoning. If properly designed, simulations can also provide teachers with the opportunity to develop their own insights and then use the simulations for further exploration.

We are developing computer tool kits for performing insight-generating experiments. These tool kits will provide open-ended experiences for students and teachers, permitting them to explore their own questions and to set up experiments of their own designs.

Many topics lend themselves to simulation. Our selection of topics will be based on the extent to which the tool kit promises to do the following:

- Raise questions of interest to scientists, teachers, and students.

- Be transparent, so the user can see the process unfolding.

- Be flexible, so the user can design a personalized system, within necessary constraints, and explore preconceptions.

- Be dynamic, so the user can make predictions and compare the behavior of the system with expected behavior and look for discrepancies. This will allow the user to relate new insights to prior knowledge.

- Be coherent, in the sense that the topic can be described verbally, so quantitative variables can be defined explicitly and rules of behavior (laws of nature) can be articulated. This property provides the basis for an analytical approach to the system and places it within the context of school curricula.

We will demonstrate two of the tool kits that appear promising on the basis of preliminary studies. Our heat-engines-and-refrigerators tool kit simulates a classical gas of particles in a box with moving walls and valves. Our star-builder tool kit simulates the interior structure of stars by showing a self-gravitating system of material particles that collide and exchange energy with a photon gas.

Interactive Physics Problems for the Microcomputer

Chris E. O'Connor, Gary M. Fechter, and James M. Tanner
Department of Physics, U.S. Military Academy, West Point, NY 10996

The department of physics of the U.S. Military Academy has developed and is currently using a microcomputer-based software system that facilitates authoring and presenting interactive physics courseware. The system has been used to develop a library of problems that are used by 1,000 students as assigned homework problems in our calculus-based introductory physics sequence. This library, which currently includes over 200 problems and tutorials, is textbook independent and, in principle, could be used by students in any similar introductory sequence. The programs run on IBM PC–compatible computers.

This system consists of three parts: (1) authoring software that assists an instructor in preparing and modifying files containing text, graphics, and branching commands; (2) a library of files that form a set of interactive problems and tutorials; (3) presentation software that interprets the library files and manages the progress of the learner through the problems or tutorials.

The authoring software is a simplified language that "hides" the usual specifics of computer programming, thus permitting the author to focus on the design of courseware. Lessons are written by creating text files containing commands that are interpreted by the authoring and presenting software. Authors can prepare library materials using a word processor or text editor of their own choice. The authoring software can also edit the library materials.

Commands are available to display text, special characters, and symbols on the screen. The current set of special characters includes Greek letters and mathematical symbols, and additional characters can be added to the program as required. Superscript or subscript characters can be displayed, and vector quantities can be indicated with a vector symbol over the character. The spacing between lines can be adjusted to display the information exactly as the author wishes.

The author can create graphical displays using commands that include lines, circles, arcs, dashed lines, filled regions, and vectors. Using these graphical commands, the author can create illustrations and display them intermixed with text anywhere on the presentation screen. Once a picture has been drawn on the screen, a section of the drawing can be copied to memory and repeatedly drawn on the screen, which results in simple animation. An object on the screen can be moved in any direction with constant or accelerated motion, which allows realistic representation of dynamic physical situations.

The author can use a plot command to display polynomials, damped or undamped sinusoids, and exponential functions on a rectangular coordinate system. Such displays allow the author to show the learner how the graphs relate to the problem. Commands for drawing coordinate axes with a grid superimposed are also included.

A series of commands control the progress of the learner through the library. The author can require the student to calculate a numerical answer and then enter the answer into the computer. If the numerical answer is within a specified range of the correct answer and the correct units are included, the program branches to a new file. Incorrect answers branch to a file that leads the learner through the problem step by step. Another answer command provides multiple-choice question processing. This command is also used to branch to different sections of a problem based on responses to menu choices. An exact answer command is also provided. It presents a fill-in-the-blank question that the learner must answer exactly as required before proceeding. This command allows for simple questions that ensure that the student is following the problem presentation.

Each student is provided with a copy of the presentation software that is used throughout the course. The presentation software is a subset of the authoring software. It can interpret commands but not edit them. The lessons are distributed as files containing the commands that are processed by the presentation software. A student simply types the program name to start the presentation software and selects problems from a menu.

Although the library described in this paper is specific to the introductory physics course, the system may be used for writing and presenting material at any level in any discipline.

In my demonstration, I will present sample interactive problems and tutorials from the existing library.

Interactive Inquiry with *TK Solver Plus* Software in the Graduate School Classroom

Kirk A. Mathews

Department of Engineering Physics, Air Force Institute of Technology, AFIT/ENP, Wright-Patterson AFB, OH 45433

I teach an introductory course in physics with engineering applications in an interdisciplinary program at the graduate level. In some respects, this course is comparable to undergraduate level physics for physics majors. In this paper I discuss how I have integrated *TK Solver Plus* software into my classroom.

TK has recently been made available in a college edition at a cost comparable to that of a textbook. I use this edition of *TK* for a course that meets in a computer classroom. *TK* requires a PC or PC-AT compatible computer with PC-DOS or MS-DOS 2.0 or higher and 384K ram. Each of my students has a Z-248 computer; my Z-248 has an overhead projection display.

TK is an effective tool for addressing areas in which students traditionally have difficulty. These include problem solving, visualizing functions, and exploring functional dependencies that are implicit (hence, hidden) in systems of equations.

Problem Solving

The structure of *TK* encourages a properly disciplined approach to problem solving. It uses separate sheets (windows) for variables, equations, unit conversions, lists of values (for tabulation or graphing), and so on. The variable sheet summarizes the input or output value of each variable, its name, its units, and leaves space for a remark on its use. The rule sheet contains equations relating the variables. Filling in these sheets in appropriate order in class, forces the student to approach problem solving in a logical way: what do I know, what do I need to find out, how will I represent these things, what set of units will I use, what equations relate them, what unit conversions do I need? The structure of *TK* maintains this discipline outside of class. *TK* function sheets define functional relationships using table mappings, simultaneous equations, or von Neumann–style programming. They support a top-down analysis to more complicated problems. The seminar-style presentation that *TK* allows gives students hands-on coaching in problem formulation and analysis. The time-consuming algebra (frequently a stumbling block for many students) is eliminated by *TK*'s automated numerical solution. By freeing students from tedious computation, *TK* allows them to concentrate on the problem.

Visualizing Functions

Students have difficulty in visualizing functions and expressions. With *TK* running on each desk, I have the students plot unfamiliar functions as they arise. This gives them both familiarity and confidence. Considerable mathematical sophistication is required to visualize even modestly complicated expressions or compositions of functions, and I find *TK* useful in graphing them, both in the computer classroom and in preparing viewgraphs.

Exploring Functional Dependencies

TK also helps students explore the relationships that are implicit within systems of equations. *TK* solves for lists of values of input and output variables. Since any list can be plotted (linear, semi-log, log-log) against any other list, such relationships are easily investigated without explicit analytical solutions and computer programming. This makes time to do many such graphs. By adjusting other variables and solving again, students can explore parametric variations.

For example, students can explore the influence of various values of mass, spring constant, and damping by plotting the response of an impulse-loaded damped harmonic oscillator. Plotting peak displacement as a function of damping then motivates an analytic solution for such a function.

Another aspect of physics that *TK* addresses is the use of experimental data to confirm a theory or model. Further exploration of the overdamped case revealed that Z, the peak displacement (scaled by $I/(\alpha\ m)$), log-log plotted against $y = (\gamma/\alpha) - 1$ is an elegant curve, asymptotically straight for small y and for large y. This is the sort of prediction that could then be validated by experiment. An analytic solution for this can be obtained (with a few pages of algebra) leading to

$$z=\frac{\left(\frac{y}{y+2}\right)^{\left(\frac{y-1}{2}\right)}}{(y+2)^2}$$

Although this parametric form makes the asymptotic behavior accessible, giving the slopes of $-1/2$ and -2 for the graph described above, students are unlikely to discover this algebraically, nor to find such a proof convincing.

Pascal Programming Templates

John P. McIntyre, Jr. and Edward H. Carlson
Department of Physics and Astronomy, Michigan State University, East Lansing, MI 48824

A program skeleton template[1] is a valuable tool for organizing a programming task and improving program quality. The programmer can use the template to avoid repetitive construction of "bells and whistles"—user aids such as modules, headers, and comments. The template saves programming time and provides a uniform look and feel to a library of programs. The resulting programs are usually more complete and clearly structured than those that result from informal program construction.

The template is the central element in a program-design methodology for small and medium-sized programming tasks (up to about 200 lines of code), it has the property of "code-to ability."[2] You design and construct your program at the keyboard by typing comments in pseudo-code (English masquerading as a programming language) right into the template modules. After the design is complete, you convert these to Pascal. The template is also valuable for coding larger programs, but it is advisable to use a more formal design methodology. Carlson describes many advantages of templates in the teaching of computing for physics.[3]

Pascal is beginning to overshadow FORTRAN for engineering and physics programming. Pascal forces the programmer to employ up-front declaration of variables and encourages modular forms, but still leaves plenty of opportunities to produce muddled, unreliable Pascal code. Templates provide examples of clearly organized code and foster good programming discipline in students and experienced programmers alike.

We present a set of templates written for *Turbo Pascal 4.0.*, which can easily be modified for other versions of Pascal. *Turbo Pascal 4.0* includes a powerful menu-driven set of tools (editor, compiler, linker, and file handler) that ease the task of programming. It also supports "units," which are separately compilable sections of code to be called from the main program or each other.

We distinguish three types of templates: (1) mean, lean ones for production use by programmers familiar both with templates and with programming in the given language, (2) verbose ones to guide a programmer using a template for the first time, and (3) "samplers" to guide an experienced (template-using) programmer in a new language. This third kind of template contains sample constructions so the syntax and logic of the new language can be absorbed without undue page flipping of the reference manual.

Our template set has four pieces: a main program template, a unit template, a set of small utility templates in files for pasting into the main or unit templates, and a utility unit called "MSUutil."

The main template is a runable (but empty) program that has these basic features: header information (author's name and address, machine type and peripherals, file name and version number, date, etc.), modules, empty declaration statements, forward declaration of procedures, and a skeleton menu. Forward declaration allows the user to call the subprograms and functions in any order, and even to order them alphabetically if there are a lot of them.

The unit template is similar to the main template, but contains the structures unique to units.

Turbo Pascal 4.0 allows you to paste files into the current edit window at any spot you want. It pays to build up many small files of useful constructions (in template form) that are likely to reoccur in your programs. Examples include: open files, close files, graphics-screen setup, "default and change variable" menu, and perhaps your favorite sort routine.

The "MSUutil" unit hides several larger utility routines that will probably be used with no change from program to program in your corpus. They are called with a "uses" declaration in your main program. These utilities include a header (for printed output), restore video mode, pause, time, stopwatch, and mouse support, either in this version or a later one.

1. David Marca, *Applying Software Engineering Principles* (Boston: Little, Brown and Company, 1984), p. 52.
2. Roger S. Pressman, *Software Engineering: A Practitioner's Approach* (New York: McGraw-Hill, 1982), p. 262.
3. Edward H. Carlson, "A Template for Writing Programs," Computers in Physics **1**, 65 (1987).

FORTH in the Laboratory

Gary Karshner
Physics Department, Gettysburg College, Gettysburg, PA 17325

FORTH language has been used to bridge the gap between assembly languages and high-level languages in a physics-oriented microcomputer class. It is especially adaptable to interfacing projects in which students are computerizing laboratory experiments.

At Gettysburg College, the microcomputer course has inherited many of the surplus microcomputers from other departments. This collection includes Apples, IBM PCs, and even some old S-100 machines. FORTH language has proven to be an almost machine-independent language that allows students to work on projects using any of these machines and follow the work of other students who have used a completely different architecture. In addition, FORTH proves very useful in allowing students to develop relatively sophisticated interfacing projects in minimal time.

There are distinct advantages to using any high-level language in interfacing experiments to computers: the availability of high-level functions (both mathematical and logical), easy access to disk storage, and efficient debugging. The high-level language FORTH has these advantages and more. It is extendible, it has a built-in editor and assembler, and it can run in two modes: interpreter and compiler.

The extendibility of FORTH is its most powerful asset. It is an extensible language, in which each word is treated on the same level as any other. If the user writes a specific word, say, to read a voltmeter, then this word becomes part of the language. It can be included in any other programs (words) the user writes, or it can be executed in interpreter mode, simply by typing it, to see the current status of the voltmeter. This feature of extendibility suits FORTH uniquely to a laboratory setting.

As an example of this, we have taken two separate classical experimental procedures (measuring resistivity, and monitoring temperature) and have used FORTH to combine them to plot change in resistivity with temperature of a superconducter. In our presentation, we will explicitly detail the implementation of the FORTH code necessary to carry out this experiment.

The use of FORTH in our microcomputer course has drastically improved the quality and complexity of the students' projects in the course.

Logo in the Physics Classroom

Tom Lough
Department of Physics, Piedmont Virginia Community College, Charlottesville, VA 22901, on leave from Department of Physics, University of Virginia, Charlottesville, VA 22901

Since its release to the public in the early 1980s, the Logo computer language has been characterized as a computer language only for children. This could not be further from the truth. Developers Seymour Papert and others at the Massachusetts Institute of Technology originally designed Logo as a computer-based "Mathland," and it has a close kinship to physics.

When Papert introduced Logo to the public in his book, *Mindstorms*,[1] he devoted an entire chapter to the use of Logo in the study of physics. It was not accidental that one of the Logo development group members was an MIT physicist, Andrea diSessa.

Logo, like LISP, performs in ways that enhance physics learning. It is interactive, procedural, extensible, and quickly learned. For example, after I take about

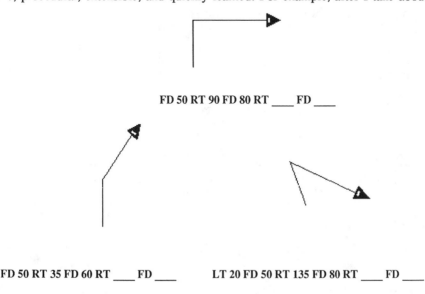

Figure 1. Vector problem.

ten minutes to explain the fundamental concept of turtle geometry (a self-referencing geometry) and a few Logo commands to my students, they immediately plunge into explorations of vectors, displacement, and motion.[2]

The graphics cursor is a self-referencing entity called a turtle. Instead of using a remote coordinate-system origin as a reference point, the turtle remains aware of its heading. Prior to moving, the turtle can turn to the left or the right a specified number of degrees. Then, like its living counterpart, the turtle can move forward in response to the appropriate command.

Using the commands FORWARD, LEFT, and RIGHT, students can make up and solve their own vector-related problems. For example, in the following set, the problem is to figure out what RIGHT (RT) turn and what FORWARD (FD) movement would return the turtle to its starting position (see Figure 1).

Logo allows instructors to write tool procedures for specialized operations, such as calculating the strength of the electric vector in an electrostatic field. Since

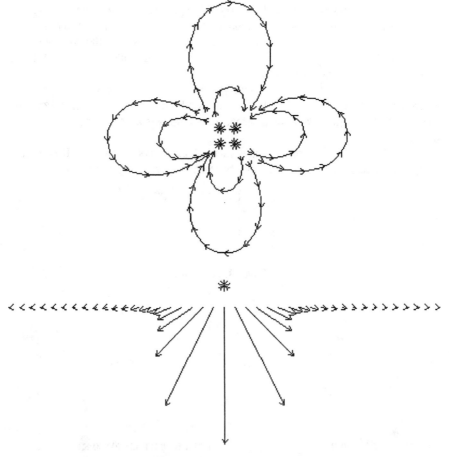

Figure 2. Electric field plots.

the language is extensible, this means that the names of such tool procedures become specialized commands as far as the students are concerned. Yet they can be used in conjunction with any other commands or inside other procedures. Here are some examples of student explorations in electrostatics (see Figure 2). In each case, the locations, signs, number, and sizes of the charges are completely controlled by the students.

LEGO TC logo is an exciting new Logo product recently brought to market. It is a combination of Logo and a specially assembled set of Lego bricks, electric motors, and sensors. This product enables students to build and operate their own laboratory instruments and then to use them in physics experiments by controlling them with Logo. The potential is enormous.

My students and I have found Logo to be a very effective adjunct tool for learning about physics and about the process of experimentation. I heartily recommend it to all physics teachers who are seeking to free their students to learn.

1. Seymour Papert, *Mindstorms* (New York: Basic Books, 1980).
2. Tom Lough, "Logo in Physics," The Physics Teacher **24**, 13 (1986); "Logo in the Physics Class," Collegiate Microcomputer **4**, 353 (1986).

Instructional Scripts with SMP

Russell J. Dubisch
Department of Physics, Siena College, Loudonville, NY 12211

Many computer algebra systems allow the use of scripts that have syntax and capabilities similar to those of compilable programming languages. Computer system–based algebra software has many advantages. It allows the user readily to examine and diagnose student input in algebraic form; it gives the user ready access to numerous operational and diagnostic capabilities of the computer algebra system; and it is easy to program.

A development team consisting of faculty and students at Siena College has been making use of a rapid prototyping system to generate scripts in the computer algebra system *SMP*. The scripting system uses a set of nested predicate lists having the structure of a tree whose elements are "predicates" (*SMP*-scripted functions that return a true or false value when evaluated) and "consequents," (*SMP*-scripted functions that are evaluated or not, depending on their positions [true or false] within the predicate list). For example,

$$[\text{Pred}, (\text{Conseq1}, \text{Conseq2})]$$

results in the evaluation of Conseq1 if Pred evaluates to true, or in the evaluation of Conseq2 if Pred evaluates to false.

The prototyping system is fully recursive and supports hierarchical menuing features. It is presently being used on a VAX network and on the Sun 3/260 work station to develop computer-assisted instructional software for the general physics and first-year calculus courses. We are currently focusing our development efforts on two packages: (1) a system to make SMP easily usable for students having no computer prerequisites or special training in computer algebra systems, and (2) an interactive problem-solving database directed at identifying errors, difficulties, and dead ends in student problem-solving efforts. We will integrate these packages into our physics and mathematics course curricula.

Computer Utilities for Teaching Physics

Curvefit: An Interactive Graphing and Data-Analysis Program for the Macintosh

Gyula J. Lorincz

University of Toronto, Scarborough Campus, Scarborough, Ontario M1C 1A4, Canada

Most physics experiments involve finding the line or curve that best fits the data. *Curvefit* is a Macintosh program that allows students to quickly and easily enter, graph, and fit a polynomial to data. It was developed for use by students in the undergraduate physics labs at Scarborough campus of the University of Toronto. The program was written in MacFORTRAN and runs on a 512K Mac, a Macintosh Plus, or a Macintosh SE.

Graphs drawn by hand are limited to finding the best line through the data; finding the uncertainties in the slope and intercept is difficult. *Curvefit* allows students to do a much more detailed analysis of their data in the time available. It allows higher-order polynomials to be fit, and it gives the standard deviations and correlations of the coefficients. In addition, it plots the deviations of the points from the curve, from which students can evaluate the fit and look for nonlinearity or systematic errors.

Curvefit fits a polynomial of degree less than or equal to 8 to the data points. Either the x or y values may be transformed before the fit; the errors are also appropriately transformed. One of three types of fit may be performed: a standard least-squares fit (equal weights); a fit weighted with y errors only; or a generalized least-squares fit (both x and y have errors).[1]

Because the program has the standard Macintosh interface, (menus, windows, and dialogues), it is easy to teach students to use. There are five menus, and the "Apple," "File," and "Edit" menus are the same as in any Macintosh program. There is also an "Options" menu to control the number of terms, the weighting of the fit, the transformations, and so on; and a "Windows" menu. Since common operations such as opening and saving files, printing, and editing are done with standard Macintosh techniques, the experience gained using *Curvefit* reduces the time students need to learn other programs.

Curvefit is based on four windows: "Data," "Graph," "Results," and "Scatter Plot." The "Data" window contains a four-column array, similar to a spreadsheet; which is used to enter and edit the values (with errors) to be fit. The "Results" window displays the coefficients, with their standard deviations and correlations. The data points and the fitted curve are plotted in the "Graph" window, and the deviations are plotted in the "Scatter Plot" window. The data and graphs can be printed on either the Imagewriter or Laserwriter, or saved to disk.

1. M. Lybanon, "A Better Least-Squares Method When Both Variables have Uncertainties," Am. J. Phys. **52**, 22 (1984).

Dynamic Analyzer

Roger F. Sipson

Department of Physics, Moorhead State University, Moorhead, MN 56560

Dynamic Analyzer is an IBM PC program that allows investigation, within a consistent interface, of a variety of dynamic phenomena that are determined by systems of ordinary differential equations. Examples of currently installed studies include projectile motion, simple harmonic motion, nonlinear oscillations, and motion of a charged particle in an electromagnetic field. Each study allows output of a number of variables, including energies. The program maintains two concurrent runs. The initial release of the program will include at least 13 different studies. Future releases, in the form of *Dynamic Analyzer* volumes, will include other studies, based on the suggestions of users or possible collaborators.

The program is a major redevelopment of the *Analyzer* program.[1] A pull-down, pop-up, dialogue-box interface, which is similar to that which students see in major applications, is used throughout the program. This allows transfer of student computer literacy both to and from the program. An easy-to-use text and graphics windowing system facilitates comparisons between runs, or of various aspects of the same run.

Powerful graphing options include everything from simple, full-screen single graphs to simultaneous animated graphs in multiple windows. Graphics systems supported are CGA, Hercules, MCGA (IBM PS/2 model 30), EGA (monochrome and color), and VGA. The graph mode is set at run time so student disks can be used in whatever machine is available to them. Built-in printer support (IBM Graphics–compatible printers) allows graph printouts of the full screen or any of the available windows.

Numerical "browsers" allow the user to scroll back and forth through the output. The user selects the variables included within a browser and its screen window. The availability of nine browsers allows easy comparison of different variables, different runs, and different parts of runs. This encourages the students to use numerical inspection of run results as well as graphical inspection and to see that each of these modes have particular strengths. The fast and convenient on-screen numerical support discourages wasteful printer output.

Setup files can be used to save and restore run, numerical, and graphical setups. These files are like macro files in that they save and restore conditions that have been established using the menu system. Run files establish all run conditions: which studies, all parameter values (masses, spring constants etc.), initial conditions, computation intervals and any special stop conditions. Graph setup files save and restore full graphing setups, which may include up to four setups of four graphs each. These setup files do not contain actual graphs, but instead include instructions to the program as to which graphs to draw how and where. For example, a setup in a setup file for simple harmonic motion might include an animated view of the motion in the upper left and simultaneously develop X-T, V-T, and A-T graphs in the other three quadrants. This setup will work for any SHM study and is independent of the graphics hardware.

The program includes a built-in clipboardlike work file and editor, into which numerical results and run conditions can be pasted directly. Students can paste parameters, initial conditions and other run information into a work file, then cut and paste significant blocks of numerical results into the file, possibly individual rows or even individual values, and simultaneously use the editor to write their reports around these values. Learning to use cut-and-paste operations is an important part of developing student computer literacy.

At least as important as student use of work files is that instructor-generated work files can also be used as tutorials. In a typical session, the instructor's file guides the students to read in one or more run, browser, or graphics setup files that have been designed to develop certain concepts, and then indicates what to look for. The student might be told to edit certain parameters or initial conditions and then view the results. The instructor file can even include blanks into which the student types a response. The file, or a printout, can then be turned in.

A significant strength of this program is that it permits simultaneous study of two different phenomena. For example, one study can be pendulum motion and the other ordinary SHM. Useful graph setups might include an animated view of the oscillator motion in one window, an animated view of the x-y motion of the pendulum in another window and graphs of X vs. time and θ vs. time. Other graphs might show energy relationships. Students can then adjust the initial conditions for the pendulum to be small amplitude, medium amplitude, and even over-the-top motion and they can then see similarities and differences.

The program *Dynamic Analyzer* is part of the collection *Computers in Physics Instruction: Software,* which can be ordered by using the form at the end of the book.

1. R. F. Sipson, "Analyzer, an Attempt at a Physics Processor," AAPT Announcer **16**, 123 (1986).

The Use of Equation Solvers in Physics Instruction and Research

Robert Rundel
Department of Physics and Astronomy, Mississippi State University, Mississippi State, MS 39762

A new basic category of software has recently appeared that is of particular interest to physicists—the equation solver. A true equation solver is not just a computerized calculator. An equation solver can do iterative solving, find roots of polynomial and transcendental equations, calculate simultaneous equations subject to constraints, and do nonlinear least-squares curve fitting. It can also do fast Fourier transforms and their inverses. The results of these calculations can easily be plotted in a variety of formats for display.

Because they require no programming and their capabilities are easily accessible to those with little or no programming experience, equation solvers are very helpful in all areas of physics research. Because they work in an interactive manner, they are also very useful in classroom instruction. When input to calculations is changed, the results and their graphs change instantly. Equation solvers are to scientists what spreadsheets are to businessmen.

In my presentation, I will review the characteristics, capabilities, and user interface of a number of commercial equations solvers, including *MathCAD*, *Eureka*, *DaDisp*, *Point Five*, and *The Wheel*.

For most physics tasks, the equation solver of choice is *MathCAD*. *MathCAD* combines equation solving with a word processor, a text formatter, and an excellent graph plotter. *MathCAD* includes a truly impressive array of built-in functions and capabilities. It has the normal trig, hyperbolic trig, log, exponential, and power functions, and a random-number generator. It can also do numerical differentiation and integration, find the roots of a polynomial (you must supply an initial guess), do cubic spline interpolation (very useful for connecting experimental data points with a smooth curve), and calculate fast Fourier transforms. It can handle arrays, including finding the largest or smallest member of the array, and it can use complex numbers as well as real ones. If it doesn't have your favorite function built in, you can define your own functions. These functions can be nested, that is, the argument of a function can be itself a function. *MathCAD* is also particularly good at handling arrays of data. It can read from, write to, and append to data files, which contain numbers separated by commas, spaces, or other nonnumeric delimiters. The data may be shown in the worksheet in the form of a nicely formatted table.

As you type an equation, *MathCAD* automatically formats it for you. The formatting is not particularly good, since all text characters are shown the same size, and subscripts or exponents are moved up or down one line. Brackets, radicals, and integrals do expand to match whatever is inside them. However, it is substantially easier to read than, for example, a line of BASIC code expressing the same equation. When you type an opening parenthesis (strangely enough, by typing the apostrophe key), *MathCAD* automatically supplies the closing parenthesis. When you delete one part of a set of parentheses, *MathCAD* automatically deletes the other part as well.

You can graph any function or variable against any other. Typing the @ key inserts a plot region into the worksheet. A small open square appears, with three smaller filled squares along the bottom and left side. These are place holders, in which you enter the function or variable to be plotted on that axis, and the minimum and maximum values of the scale for that axis. You can plot several functions in the same plot, just by entering more than one function name in the place holder. The plot region can be stretched to any desired size, logarithmic axes may be used, and a grid may be superimposed on the plot. If data points are being plotted, you may choose from a variety of plotting symbols, decide whether the points should be connected by lines, and show error bars. Whenever a worksheet is recalculated, the graph is redrawn to reflect the current data.

I find it particularly useful that *MathCAD* understands units. It treats a unit simply as a multiplier in an equation, and keeps track of compounded units by analyzing the dimensionality in terms of mass, length, and time. If you supply a table of units conversions somewhere in the worksheet, *MathCAD* does units conversions by allowing you to edit the unit of calculated quantity. Once the edit is complete, the number is recalculated for the new unit.

The *pièce de résistance* of the *MathCAD* is its math capabilities, vector and matrix operations, and iterative solving. *MathCAD* understands vectors and matrices completely. They can be created and manipulated in a whole variety of ways. You can calculate dot and cross-products, invert matrices, and calculate their trace or determinant. Many math texts use the language of matrices, for example in curve-fitting techniques and it's quite easy just to type the matrix equations into *MathCAD* straight out of the text.

MathCAD also uses the idea of a vector for another very useful purpose. Instead of defining a subscripted array, and then looping through a calculation on each member of the array, you can define a vector, containing the entire array, and then use that vector as if it were a single variable in an equation. The calculation is carried out more rapidly, and the equations are much easier to read.

MathCAD has two forms of iterative solving. You can use the root function, which finds a root of any equation $f(x) = 0$ starting from some guess for x. Although you have to supply the guess, and the root found may depend on the guess, it's so easy to plot an equation in *MathCAD* to see approximately where roots may lie that that's really no problem.

For more complex iterative solving involving constraints, *MathCAD* uses a solve block, which consists of an area in a worksheet starting with the key word "Given," followed by any number of equations or constraints, and ending with an equation that includes the find function. Once you set up this solve block and supply initial guesses for the variables, just hit the "calc" key (F9) and *MathCAD* will iterate whatever is in the solve block until it reaches a solution. You can control the accuracy by setting the internal tolerance variable.

To test *MathCAD*'s iterative solving against *Eureka*'s, I used the "Ladders" example problem supplied with *Eureka* (set up a 45-foot ladder and a 35-foot ladder in an alley between two buildings such that they cross each other 10 feet above the ground). This problem involves solving four simultaneous equations in four unknowns with the constraint that all variables have values greater than zero. Starting from the initial guesses given in the *Eureka* file, *Eureka* produced a nonsensical answer (one of the variables was less than zero), while *MathCAD* properly reported that it could not find a solution. Given a better guess, both programs found good solutions. *Eureka* was faster, by about a factor of two, but *MathCAD* was more accurate.

It's clearly a whole new world out there now. Why, oh why, wasn't this available when I was a graduate student?

MathCAD in the Modern Physics Course

Don M. Sparlin

Physics Department, University of Missouri–Rolla, Rolla, MO 65401

In 1988 I find myself at the dusk of time-consuming programming and poor graphics resolution, and at the dawn of outstanding support for the teaching of physics through interactive computer computational software. The available prepared coursework materials are excellent, but restricting at the upper-division level. I am delighted to discover more and more incredible computational software packages that include user-friendly menus, low time and monetary cost, adequate graphics, complex math, Fourier analysis, and revelant special functions. Used with a math coprocessor, these packages are even fast enough on a 5 MHz machine. EGA is necessary for color graphics presentation, but Hercules resolution is just fine otherwise.

I have recently become acquainted with the Student Edition of *MathCAD*.[1] My students find this product very acceptable for everyday homework preparation and especially for bonus points. I find it useful for preparing tests, and for preparing "living" exercises. The main use of *MathCAD* has been for bonus points. The students work exercises using *MathCAD* to earn a maximum of 150 bonus points above the 1,000 points available through the usual mechanism of tests and homework. The motivational factor of hedging against poor test performance while learning to use software that applies to other courses as well, has proved to be irresistible.

I will present several of the bonus assignments along with the "living exercises." I will also discuss student reactions, focusing on the motivational value of this procedure. Finally, I will address the difficult problem of illegal copies of the software circulating throughout the class.

1. *MathCAD, the Student Edition* (Reading, MA: Addison-Wesley).

Ode: A Numerical Simulation of Ordinary Differential Equations

Nicholas B. Tufillaro

Department of Physics, Bryn Mawr College, Bryn Mawr, PA 19101, and Graham A. Ross, Portland, OR 97215

Ode solves the initial-value problem for a family of first-order differential equations. When provided with an explicit expression for each equation, *Ode* parses a set of equations, initial conditions, and control statements, and then provides an efficient numerical solution.

Ode makes the initial-value problem easy to express. For example, the ode program,

```
# an ode to Euler
y = 1
y' = y
print y from 1
step 0, 1
```

prints 2.718282.

The *Ode User's Manual* contains a guide to applying the program and a discussion of its design and implementation. *Ode* provides a simple problem-oriented user interface, a table-driven grammar, simplifying extensions and changes to the language, a structure designed to ease the introduction of new numerical methods, and (considering that it is an interpretive system) remarkable execution speed and capacity for large problems.

Ode currently runs under the UNIX operating system on machines ranging from micros to mainframes. *Ode* is in the public domain and is in use at numerous educational and industrial sites. The source code, documentation, and copious examples are available from the first author. *Ode* has been used in many research problems and incorporated into coursework.

Computer Desk-Top Management for the Physics Teacher

Carl R. Nave
Department of Physics and Astronomy, Georgia State University, Atlanta, GA 30303

The personal computer is a powerful resource for helping with the strategy and the logistics of teaching. This paper explores the use of a Macintosh computer as a desk organizer and handler to help with the routine tasks of physics instruction. It considers use of several commercial software packages on a stand-alone basis, and then explores their integration with the *HyperCard* environment. Specific attention is given to developing documents with both graphics and text, storing and retrieving data, solving algebraic problems, and searching bibliographies. The overall goal is to show how the machine can serve as a teacher's aide to help with the time-consuming tasks of preparing materials for daily classroom use.

If a physics teacher is to use the computer for preparing exams and developing labs, handouts, and other instructional material, he must integrate word processing with graphics. This was a major problem as recently as five years ago, but there are now numerous alternatives for incorporating graphics quickly and conveniently. I discuss the use of desk accessories and the applications *Switcher* and *Multifinder*. Desk accessories like *Mac∑qn* allow the inclusion of equations.

Next, I discuss how to build a searchable body of teaching information with a word processor. Word processing constitutes the major part of the university professor's use of a computer. In a 1985 survey, our faculty respondents reported that about 60 percent of their computer time was spent in word processing and that about 75 percent of their computer use was with microcomputers rather than the university mainframe. Thus, developing the word processor as a tool for the teacher is of utmost importance.

Algebraic equation solver programs like the commercial software package *TK Solver* are a great help in developing examples and graphics in numerically tedious problems like three-lens zoom telephoto design, loading of a transformer, and motion in a viscous medium. *TK Solver* is also an invaluable tool for solving simple dynamics problems such as the braking distance of an automobile. For more elaborate graphics, an integrated spreadsheet and graphics program like Microsoft *Excel* is an amazing tool for creating models for handouts or slides. For example, using the sensitivity curve of the eye, you can factor the radiation curve from a candle flame to get a graphical representation of the color perception of a candle by the human eye. By patching the *Excel* graphics over to a paint program to create labeling and descriptive text, the instructor can create a one-page handout in one short session.

Constructing a searchable bibliography as a current reading log can be handled well by some of the simpler database programs. These are quickly learned and easy to use. I will discuss developing an organization, developing a quick-entry form, searching, etc.

I also discuss some specific examples of the use of *HyperCard* in physics teaching. *HyperCard* is a major development for enhancing the Macintosh as a teacher's tool. It helps create information files of all types, which can be linked in any way you wish to form an interconnected web of information that can be followed along a logical search path by merely clicking "buttons" that you create on the screen. An example is a periodic table of the elements on which each element symbol is a button that takes you to more extensive data about that element, and other buttons that take you off to the values of fundamental constants or physical data. You access this data bank from within a word-processing program or spreadsheet, and then patch in the needed information. *HyperCard* takes you a long way toward the goal of an affordable "scholar work station" in which information and working tools are almost instantly accessible.

Computer Networking and Workshops

Applications of New Technology for Large-Scale Computer-Based Physics Instruction

Donald L. Bitzer
Computer-based Education Research Laboratory, University of Illinois, Urbana, IL 61801

Dennis J. Kane
Department of Physics, University of Illinois, Urbana, IL 61801

The PLATO system was invented at the University of Illinois in 1960. In the intervening years this system has evolved through several major stages, the last of which has culminated in the new NovaNET system currently in operation. During these stages of development, approximately 16,000 hours of lesson material from 150 different subject areas have been created. Although a few subjects are taught completely by computer, most of these courses use the computer for 30 percent to 50 percent of the instruction.

The PLATO system presents a flexible approach to computer-based education. The system software provides an easy-to-use authoring language, TUTOR, to produce new courseware. Also available is an instructor package that permits a teacher to form a curriculum by selecting and arranging lessons, to enroll students in the course, and to provide tools for measuring and tracking students' performance. The student terminal has a 512×512 pixel graphic display with a keyset and screen-touch input. Some of the terminals also provide for superposition of slides on the graphic display by rear projection.

The physics department at the University of Illinois has used PLATO in four areas: for direct instruction in a complete computer-based introductory classical mechanics course; for administering weekly quizzes in an elementary electricity, magnetism, and thermodynamics course; for isolated instructional exercises of various kinds in other courses from the elementary to the graduate level; and for administrative uses such as enrolling and dropping students in large multisection courses, as well as recording and displaying scores on examinations, homework, and laboratory work for both the students and instructors. Some of these uses will be briefly discussed in this paper. More detailed descriptions are given in the references listed. Finally, a description of the new NovaNET system will be given to show how these PLATO applications can now be expanded across the United States at a very low cost.

The Use of PLATO in Physics

Because over 2,000 students enter the introductory physics sequence each year, the PLATO applications in the physics department have been developed to be usable and manageable for large numbers of students. As the number of students who require individualized attention increases, management problems escalate rapidly.

A computer-based education network consists of a complex human-and-machine system. There are many human players at each level in the system. A few players are highly skilled computer programmers. A larger number consists of those who create lesson material and instructors who mold the lessons into a curriculum. The end users or students, constitutes the largest group, and they are the most naive users in the system.

Problems with the system or courseware are usually detected by the end users. Corrections must be made by the other experts in the system. Consequently, communication is needed between users at all levels for reporting problems and implementing the appropriate corrective steps. In order to be effective, the detection and correction of problems must take place quickly—not in days or months. This process can be implemented either by superimposing a human management layer or by providing a well-thought-out, computer-communication system to accomplish the same task. Education does not escape entropy!

Thus, choosing the option of a central-system approach to solve management problems is an important, time-saving concept. Changing or correcting lesson material instantly for all users at all locations and collecting and distributing student data such as performance records, test scores, class standing, and comments on lessons are some examples of management problems that are solved by the central system approach. Although these data are protected from unauthorized access, they can be viewed by both students and instructors at any location and at any time.

A PLATO-Based Elementary Mechanics Course

The early development of PLATO physics course material began in 1970 with the production of a few lessons in classical mechanics. By 1975, with the help of a National Science Foundation grant, 30 terminals in a single classroom began delivering instruction in classical mechanics to 200 to 500 students per semester. Various kinds of lessons are used in this course, with the major emphasis on the use of tutorial presentations and homework exercises. The flexibility of the PLATO system is important in both of these applications. New principles can be introduced by using graphics to illustrate the concept dynamically. Questions concerning the new principle can be interspersed with the demonstrations. Students can use algebraic responses that include superscripts, subscripts, and dimensional units, with the computer giving meaningful feedback to such responses.

The tutorial lessons help the student to acquire an understanding of fundamental concepts and then to use those concepts in simple applications. The student is introduced to new concepts and terminology while being quizzed on the new material. Throughout the lesson there are short quizzes consisting of problems in random order with random parameters. If the student cannot answer the question, asks for help, or gives an incorrect answer, that question reappears later in the drill with different parameters. On a mastery quiz, which is presented at the end of the lesson, the student must answer six out of eight questions, but no help is provided and answers must be correct on the first try. The mastery quiz can be taken at any time and as many times as the student wishes. However, the questions always appear in a different order and with different parameters.

Graphics and simulations are used extensively to illustrate concepts and applications. The simulations and proof are followed by application examples.

Online homework problems follow the tutorial lesson for each topic. Students buy a booklet containing pictures of these problems and are encouraged to solve the problems outside of the computer classroom, although some opt to work on the terminal exclusively. This online homework provides the student with immediate feedback. Answers that are incorrect or have improper form, draw an appropriate comment on the error. If the response is incorrect but has proper form and correct units, the student is told of the error and is offered help. If the "help" key is pressed, a full discussion of the problem is given.

In the standard course there are two one-hour large lecture sessions, a two-hour small discussion session and a two-hour laboratory session each week. In the PLATO version of the course, the discussion meeting is replaced by a two-hour scheduled PLATO period. The students are free to use more PLATO time if needed for homework and review. Typically, another two hours per week are used this way. Completion of the PLATO lessons and online homework counts as 20 percent of the student's final grade.

The integration of the computer-based version of the physics course with the standard aspects of the course have provided advantages in student management. Since the online materials are so detailed, the lecturer can concentrate more on demonstrations of physical principles and less on elementary exposition of topics. The weekly PLATO session is a time for students to work on the terminals with instructors present to provide additional individual help. Using the online gradebook, instructors can immediately discern what material each student has completed and in what areas each student may need assistance. In addition, students can easily determine their relative standing in homework, quizzes, and examinations, which may provide incentive for them to seek help in particular areas.

Since the online gradebook instantly records students' performance on homework and quizzes, the instructors are not only freed from collecting homework and recording grades, but also have immediate access to these. They can, after reviewing the data, tailor their approaches to fit students' needs and provide extra help in problem areas.

Grades from exams and labs are recorded online as well. Since all the scores are on the computer and can be combined in any grading algorithm, the task of assigning a grade at the end of the semester is simplified.

An electronic mail feature, a computer bulletin board, and an open-forum notes file allow communication between students and teaching staff, as well as among members of teaching staff (who can number as many as 30 in some of the large courses). Using their terminals at the end of the semester, students are able anonymously to communicate suggestions, complaints, and evaluations of the course.

Because of the large enrollment, it has been possible to operate both the traditional and the computer-based versions of the course in parallel. We have therefore been able to study some effects of introducing the computer into the course. The computer-based version is run with one-third fewer teaching assistants. However, results from both the students' scores on tests and longitudinal studies of performance in future physics courses show no significant difference between the students in the computer-based version and those in the traditional course. Each semester the students in the computer section of the course are asked which method of instruction they would prefer for their next physics course. Typically 60 to 80 percent respond in favor of a PLATO-taught course.

PLATO Quizzes in the Elementary Electricity, Magnetism, and Thermodynamics Course

The course following classical mechanics in the elementary physics sequence teaches electricity, magnetism, and thermodynamics. Over the past three years, we have developed a system for administering the weekly quizzes in this course on PLATO. Thus, it allows more faculty control over this important but time- consuming and often unevenly administered component of the course. Students do not have a weekly two-hour, small-group discussion meeting with a teaching assistant as was done previously. Instead they attend an additional one-hour large lecture that stresses problem solving and is taught by a senior faculty member. Students are also scheduled to take a quiz covering the topic for the week on PLATO.

We have developed a set of 12 quiz problems for each of the 14 weekly topics with each problem ranging from four to ten questions. The faculty member in charge of the course selects eight problems to be used out of each set. These eight are available to the students for advance practice prior to the quiz. During the student's scheduled quiz time, PLATO chooses one of these problems at random, changes the numerical values in the problem statement (to make them different from the practice version), and administers it as the student's actual quiz for that week. The student has up to 17 minutes to work on the problem initially. At that time (or before, if the student requests it) the student's answers are graded. In the case of an incorrect answer, the student still receives points for correct units or for an answer calculated correctly but based on an incorrect answer to an earlier question. The student then has up to eight minutes to continue working on the quiz before it is graded again for the final time. This hybrid of quiz and homework has turned out to be popular with the students and is perceived by most of them as an efficient use of their study and class time.

The centralized and secure aspects of the PLATO network are essential for administering this quiz system to as many as 1,100 students who may be enrolled in the course in a given semester. PLATO stores every answer by each student throughout the semester (allowing for later review); assigns, stores, and renormalizes (if necessary) each student's scores; and can present up-to-the-minute, coursewide results instantaneously.

The New NovaNET System

Although the PLATO system has been successful in delivering instruction for many years, the cost of using the system has prevented a large-scale expansion at many levels of education. Now a new system has been designed and tested to deliver the same high-quality computer service at a much lower cost. One important criterion of the new system design, is retaining the important computer-managed features made possible by a central computer approach. This requires efficient central computing along with a much more cost-efficient communication network.

The attempt to reduce total system costs was divided into three components: the student terminal, the central computer, and the communication network.

Low-Cost Terminals

Two approaches were used to provide large quantities of low-cost terminals. First, a new low cost ($600–800) monochrome terminal was designed and produced. It has a resolution of 640×512 pixels and supports all of the central TUTOR programs. This terminal is ideal for those who are interested in using a display for TUTOR programs only and are willing to use a light pen or mouse instead of a touch panel. The second approach was to provide access disks, which convert standard microcomputers such as Macintosh, IBM, Atari, and Zenith into terminals that are compatible with the TUTOR programs. This permits owners of microcomputers to have access to the NovaNET system. However, the quality of the resulting graphic display and keyboard interface, depends on the quality of the microcomputer used. More recent microcomputers, such as those in the IBM PS/2 series, make excellent PLATO terminals.

New Low-Cost Network

In an environment where naive computer users are sharing application packages developed by experts and managed by a computer, central processing makes a low-cost communication network possible. In a computer-based education application, most computing can be done centrally with only the display code transmitted to the user terminal. Data from over eighteen million terminal contact hours have consistently shown that, when averaged over the variety of users and applications, the data rates from the users to the computer are significantly lower than the data rates from the computer to the users.

A more detailed analysis shows that even though a peak data rate of 1,200–9,600 bits/second may be delivered to the user over a brief period, the average data rate is approximately 240 bits/second. The data from the end user to the computer are typically key stroke information. The keystroke rate sent by the user may be as high as 10/second, but the average over time turns out to be 1 key/2 second. This asymmetrical data rate requirement, which occurs with central processing, provides a unique opportunity for low-cost networking.

In the new network discussed here, the communication from the central computer to the terminal is provided by a satellite T-1 carrier. Since this channel is used in a broadcast mode, only one uplink is needed. This is fortunate since uplinks and channel space are relatively expensive. Each distant receiving site needs an inexpensive downlink for receiving the data and a telephone line (9,600 bits/second channel) to return keystroke information to the central computers. Each T-1 channel has the capacity to communicate with up to 4,000 terminals simultaneously, but since not all terminals use the system at the same time, the T-1 channel can support as many as 8,000 connected terminals.

The phone-line return channel has several advantages. First, the phone line can support over 300 terminals because of the low data rate per terminal. Second, the time delay incurred by synchronous orbit satellites is avoided in the return path. This allows the entire communication path to have reasonable human response times for the interactive use. Finally, cost of the phone line from the remote sites to the central computer is very low per terminal.

A network driven by a central computer does not and should not preclude the use of processing at other places in the network, such as at the terminal. There are important applications, like text editing, where very rapid display changes may require local processing. The combination of processing at the terminal and central-system processing takes advantage of the best of both situations.

When the communication system is loaded (4,000 terminals in the T-1 channel and 300 on the return phone-line channel) the estimated cost for round-trip communication between the satellite downlink and the computer totals approximately $10 per month per terminal for full-time access. The network has a round-trip response time of approximately one third of a second. In addition to the remote connection costs, the cost of local redistribution must be added.

High-Data Background for Other Applications

Each T-1 carrier has the capacity to communicate with approximately 4,000 terminals simultaneously. To prevent long data queues when fully loaded with the 4,000 users, there must be sufficient capacity remaining in the communications channel to provide for the statistical variation in data flow. This extra capacity of approximately 20 percent or 300 kb/second average can be used for a background transmission of data if this data is interrupted when the channel space is needed for the regular users. This additional channel space is available for a variety of uses.

1. Downloading programs to microcomputers. There are applications where a program is best run in a microcomputer and there is no need for network interaction. The ability to transmit these programs rapidly to the end user provides an economical method of updating and distributing such programs.

2. High data-rate transmission of output from the supercomputer. Many supercomputer programs generate sophisticated diagrams and color pictures as their output. Obtaining such output in a reasonable period of time usually requires users

to be located near the computer site. This wide-band transmission capability will make possible for many remote users the same interactive capability they would have if they were located at the computer site.

3. Downloading of prestored audio and visual information. One of the new and important directions in computer-based education has been the introduction of prestored video and audio information on the student screen. In addition to the problems of generating this information, the ability to modify, add, and distribute video and audio information has become a barrier to its wide use. Use of the high data-rate network, particularly in lightly used hours, offers an opportunity to send corrections and additions of prestored audio and pictures to remote sites.

New Central Computer

The existing PLATO software consists of approximately one million lines of tested Cyber software code that currently manages the system and executes the lessons. Because of this large amount of operational software, it is important that the new computer be low in cost, powerful enough to support thousands of terminals, and able to execute in exactly the same manner as the present Cyber machine. Such a machine was not available. Therefore, we decided to design and construct such a computer at the Computer-based Education Research Laboratory. Using our existing computer system, which was programmed with the characteristics of the latest ECL chips, we were able to generate a wiring list for the new computer. However, obtaining a wiring list consisting of tens of thousands of connections solves only a part of the problem. Because of the large number of connections, it is necessary to use an automatic wiring machine. Using this procedure, we have produced two prototype computers. Thus, the seemingly impossible task of producing a new large computer consisting of tens of thousands of connections has been reduced to a reasonable task by the use of automatic computer design and wiring. Terminals are currently connected to this new computer system at four experimental downlink sites located in Tucson, Arizona; Des Moines, Iowa; Orono, Maine, and a special supercomputer test site at the University of Massachusetts. This type of special computer implementation works in reality, not just in principle.

Conclusions

The use of a computer for instruction and administration has been successfully integrated into the large introductory physics courses at the University of Illinois. Results show that students in the PLATO version of the course do as well on tests and in future physics courses as the students in the standard course. The use of the computer has reduced the staff load for teaching these courses and the instructors feel that the time that they do spend with the students is more productive.

The new NovaNET system makes it possible for these physics materials and administrative applications to be used at many institutions across the United States at a low cost. This capability might help solve some of the articulation problems of entering students from high school and transfer students from other institutions.

The PLATO system and the NovaNET system are developments of the University of Illinois.

1. S. Smith and B. A. Sherwood, *Proceedings of a Conference on Innovation and Productivity in Higher Education* (Pittsburgh, PA: Carnegie-Mellon University, 1976).
2. B. A. Sherwood, C. Bennett, C. Tenczar, and J. Mitchell, *Proceedings of a Conference on Computers in the Undergraduate Curricula* (Hanover, NH: Dartmouth College, 1971), p. 463.
3. C. D. Bennett, *Proceedings of a Conference on Computers in the Undergraduate Curricula* (Atlanta, GA: Southern Regional Education Board, 1972), p. 369.
4. S. B. Peterson, T. R. Lemberger, and J. H. Smith, *Proceedings of a Conference on Computers in the Undergraduate Curricula* (Pulman, WA: Washington State University, 1974).
5. D. Kane, and B. A. Sherwood, "A Computer-Based Course in Classical Mechanics," Comput. & Educ. **4**, 15 (1980).
6. L. M. Jones, D. Kane, B. A. Sherwood, and R. A. Avner, "A Final-Exam Comparison involving Computer-Based Instruction," Am. J. Phys. **51**, 533 (June 1983).
7. D. Alpert, and D. L. Bitzer, "Advances in Computer-Based Education," Science **167**, 1582 (March 1970).
8. D. L. Bitzer, "A New Network for Computer-Based Education" (Paper presented to the International Conference on Computer Assisted Learning, Calgary, Alberta, Canada, May 1987).

Computer Conferencing: A New Delivery System for College Coursework

Richard C. Smith
Department of Physics, The University of West Florida, Pensacola, FL 32514

The physics teaching community has recently been challenged to reassess its reliance on problem-solving skills as a way of teaching physics. Rigdon puts the matter succinctly when he says that "students can solve problems without understanding the physical concepts involved in them" and suggests strongly the need to allow students to describe physics concepts in words of their own choosing.[1] Recent work in commonsense misconceptions also addresses this problem.[2]

Computer conferencing is one medium for exchanging ideas expressed in words; in addition, it delivers learning experiences to students who are widely dispersed in location and time. An unexpected benefit of computer conferencing is that it challenges traditional assumptions about how physics instructors use scheduled time with their students.

What Is a Computer Conference?

A computer conference is a dynamic collection of comments, questions, responses, or other text-based materials on a common subject of discussion.[3] The specialized conferencing software used to support this activity offers text-editing facilities, easy ways to reference past material, transfer of material between related conferences, and private messaging facilities. The participant charged with leading the discussion is responsible for keeping the discussion focused on the subject.

A computer conference greatly resembles that low-technology means of exchanging information, the hollow tree. Participants visit the hollow tree at times of their own choosing and leave messages that are later retrieved and read by all other participants in the conference. While there, participants can pick up all the notes that have been deposited there by previous visitors and make whatever responses seem appropriate. Today, the hollow tree has become a host computer accessed either by wired terminals or by modem-based dial-up facilities.

The key element is the fact that participants can engage in conversation at times and places of their own choosing. In terms of college course work, it is no longer necessary for the class to meet in room 230 at noon Mondays, Wednesdays, and Fridays. Rather, the dialogue occurs continually in a text-based delivery mode. To offer a course in this mode is roughly equivalent to listing a conventional course as consisting of multiple sections with one student per section, at times and places to be announced.

Requirements for a Successful Course

Not all physics courses are suitable candidates for delivery in this text-based system. A course must meet three requirements if it is to be considered for a successful conference mode application:

1. The information content of the course must already exist in some accessible form. This can take the form of a textbook, a collection of newspaper articles distributed to students, or even text files retrieved from a computer-based file server. The computer conference itself should be reserved for the exchange of reaction to the material or discussion of its main points instead of primary dispersal of information.

2. The material of the course must allow for, and indeed invite, discussion. Courses in which rote memory and fact accumulation play a large role are not well suited for this medium. This requirement does not exclude so many physics courses as it might seem at first, but rather indicates that physics instructors must bring different skills to bear on a course than has been customary in the past.

3. The course must not depend on extensive symbol manipulation. Electronics courses, for example, would be very difficult to teach without the use of electri-

cal device symbols. Geometrical optics would be impossible to teach without ray diagrams, although there is certainly much room for qualitative discussion of the results of geometrical optics.

PHY 3936: Special Relativity

A course that meets these three requirements is the University of West Florida course PHY 3936: Special Relativity, a one semester-hour seminar taught recently in the computer-conference mode. The textbook for the course was *Spacetime Physics* by Taylor and Wheeler.[4] This text was specified because of the large number of word problems and the authors' disposition to resort to calculation only after thorough discussion had paved the way. The computer conference was hosted by The Source, a well-known public-computer facility, and it used the specialized 13 during which time eight students and I contributed 449 messages, an average of 50 messages per participant for the entire course, or three contributions per participant per week.

All students were on campus regularly and could have attended a conventional lecture course. The class met formally during the first week, when I explained the rules and procedures of the course. All agreed that course communication should take place exclusively on *Participate*, and that there should be no hall talk on discussion items. One of the departmental Apple IIs was equipped with modem and terminal software, and students were invited to use these at their own convenience. Several students elected to use their own home computers, and two purchased a modem specifically for use with this course.

The course was structured on *Participate* as a tree-shaped conference, shown in Figure 1. The 13 weeks of the course are shown horizontally, with the branching conferences shown between the first and last week of their existence. Each conference is named at the rightmost end of its existence. The height of each week's representation in each conference is proportional to the number of contributions made during that week, and is shown by the scale in the figure. For example, six separate conferences were active during Week 3, with the "Proper Time" conference eliciting the most (28) contributions.

The root conference, entitled "PHY3936," ran from start to finish of the course and served as take-off point for all the topical sub-conferences. No discussion occurred in "PHY3936" itself. A second special purpose conference was "Procedures," which discussed ways to use *Participate* effectively. This conference was used for general notices unconnected with special relativity topics. Examples included: how to minimize connect time; advance notices of service interruption; comparison of *Participate* with other conference systems. "Procedures" was used almost every week. A third special purpose conference, "Evaluation," was used to gather comments at the end of the course.

Aside from these procedural conferences, 12 topical conferences developed throughout the term: "Events," "Intervals," "Proper Time," "Spacetime," "Lorentz Transformation," "Galilean Transformation," "Velocity Parameter," "Time

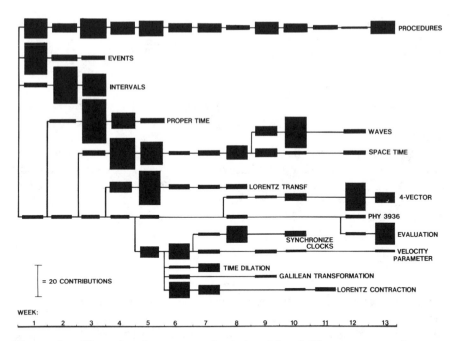

Figure 1. Time development and relationship of 15 separate conferences used in the special relativity course.

Dilation," "Lorentz Contraction ," "Synchronize Clocks," "4Vector," and "Waves." Their relationship to each other and the relative activity levels are depicted in Figure 1.

Typically I would introduce a new topic by posting a textbook reading assignment, specific key ideas to look for, perhaps a small amount of discussion directed toward one of the illustrations in the text, and finally a question designed to elicit student response and interpretation. Often I would assign one or more of the students to make the primary response, the equivalent of calling on that person in class.

From time to time, when I could anticipate the need for a particular graphic explanation, I would produce the graphic and mail it to the students, so that we could all make reference to the illustration as part of our discussion. The details of this particular course have been described in detail elsewhere.[5]

Advantages

There are several pedological advantages to offering college course work in this text discussion mode. A primary advantage is the fact that this mode puts responsibility for course success on the shoulders of the learners, where it belongs, and engages each student as a potential tutor for others in the discussion. If the instructor can learn to use students in a collegial fashion in working through a common body of material, then the experience and knowledge already in the student's minds can be put to use for the benefit of the entire group.

A second advantage is that personal interaction among all participants in the conference is enhanced. This finding often comes as a surprise to those who view computer-mediated communication as being cold and impersonal, but conferencing permits the instructor to notice who is not participating and to encourage reluctant students to become an active part of the group, resulting in better interpersonal relationships.

A final advantage, one that I have not yet succeeded in accomplishing, is the possibility of involving the textbook author or a well-known expert as part of the course itself. This possibility is open because there are no time constraints. One intriguing possibility is to mask the expert's identity until close to the end of the course, so that students would come to perceive this person as a true colleague in the learning of new material.

Disadvantages

There are several significant disadvantages to offering a course in the conference mode. Perhaps the biggest problem is the difficulty in keeping the open-ended discussion from wandering off on tangents. An instructor who finds the discussion leading away from his/her own agenda must use considerable skill to redirect the discussion back to the topic.

The absence of graphics capabilities in text-based discussion is also a major disadvantage. Transmission of graphics is difficult on a general-purpose computer, not only because of the high information content contained in graphic images, but because the mechanism for producing graphics on a terminal is machine specific, so that graphics designed for display on a Macintosh, for example, would be meaningless to a student participating with an MS DOS machine. To be sure, graphics-encoding techniques exist, the principal one being the North American Presentation Level Protocol Syntex (naplps),[6] but the availability of software for encoding and decoding such images is not widespread and often cannot be handled on the general-purpose host computer.[7]

A final disadvantage is that participants are cut off from visual demonstrations of the physical effects that they may be trying to discuss. One way to respond to this problem is to mail out demonstrations on video cassette to each participant, a technique that has been used successfully in other home-delivery course projects.

Dis/advantages

Several features of a computer conference course can be regarded either as an advantage or a disadvantage depending on one's point of view. First, it is difficult to write equations. This can prove frustrating to those instructors who think solely in terms of equations, but this constraint may open up the possibility of understanding more physics through the use of nonmathematical text, precisely the point made by Rigdon.[8]

A second characteristic is the open-endedness of the discussion. An instructor who insists on discussing only the material at hand and not allowing the topic to grow in whatever direction it chooses will be frustrated at keeping students on task. One (perhaps extreme) example of this occurred in the special relativity course, when our stated topic of finite signal velocity led to an unexpected discussion of dispersion, thence to chromatic aberration, and finally to rainbows and glories, admittedly strange topics for a course in special relativity. Whether or not such excursions into new topics is desirable or undesirable in a physics course is left for the reader to decide.

Course Organization

The continuing record of contributions to a computer conference looks much like a typed transcript of a seminar with many participants. One person will make a comment that will be answered by a second participant, who will be responded to by a third, and so on. One can imagine a lively discussion around a common table. The only difference in the two examples is the time scale of interaction. Whereas a live discussion can have several remarks and responses within a few minutes, the same interchange may take a day or two in a computer-conference mode. This means that it may take several weeks of conference proceedings to cover the same material as in an hour of live discussion. A compensating feature of computer conferencing, however, is the fact that several topics can be under discussion at any one time. For example, in the special relativity course, we often had as many as six or seven separate topics active at the same time. If parallel discussions are scheduled in this fashion, then the time expansion noted earlier is neutralized so that the several weeks of computer conference can now be used to work through six separate topics simultaneously, and calendar time is used much more efficiently.

The possibility of parallel topics clearly works only when none of them have a prerequisite relationship to any of the others. Thus if we were scheduling an introductory physics course to be offered in the conference mode, then perhaps we would begin discussion in mechanics, optics, thermodynamics, and electricity and magnetism, all on day one, and use the entire year to develop each of those topics.

New Challenges

The successful conduct of a computer conference course poses new challenges for the physics instructors, who will find that they must develop a new set of skills. The first new skill is clearly that of leading an open discussion, an experience in which few of us are practiced. An important part of leading such a discussion is to learn how to engage other students in the conversation and how to deflect individual requests for information. Our experience in the special relativity course shows that students would often address individual questions to the instructor, whose job it then was to elicit discussion from the rest of the participant group, thus drawing all students into the discussion. The instructor must realize that he cannot form a one-to-one tutorial relationship with each of the students in the conference course.

Instructors must also redefine the traditional "top-dog/bottom-dog" relationship that often exists between instructor and student. The instructor's job is to guide the discussion and assist the learning process, not to dictate responses or appeal to instructor authority, both of which actions defeat the idea of a cooperative discussion.

Instructors must also learn to remain silent during a discussion so as not to wrest control from the rest of the group. I developed the rule in our special relativity course that I would not comment on a new posting by a student until at least two days had passed, thereby giving other students a chance to offer their own comments.

Implications for Conventional Offerings

One of the strongest pedagogical results of offering a computer conference course is a reexamination of some of the traditional ways we offer courses. First, we must consider carefully the way in which we customarily use scheduled time with students in a classroom. Traditionally, the instructor stands at the front acting as authority figure, and students assume passive roles. If we reexamine that relationship, then many possible alternatives for use of that time begin to surface, among them combined lecture/laboratory experiences,[9] student peer conferences, demonstrations, and many other possibilities.

Second, we must rethink the way we schedule conventional courses, particularly in accelerated sessions such as short summer terms. If a physics course is to be offered in half the time, then perhaps it makes good sense to schedule parallel sessions, so that, for example, morning classes could be devoted to one topic and afternoon classes a different one.

Third, we must recognize that alternative communication modes exist and can be mixed in any combination. Thus, when we realize that a course can be offered in a computer-conference mode, we might choose to supplement a conventional class with a computer conference element, perhaps for the purpose of forming student discussion groups or planning joint projects. At least one other physics course has used a computerized bulletin-board system to supplement the workings of a conventional class.[10]

Conclusions

One should use computer conferences only when it makes sense in the context of the student population, the subject of the course, and the computer hardware available. At the very least, experience in computer conferencing opens another channel of communication between instructor and students, the effective use of which can only enhance the teaching and learning of physics.

1. J. R. Rigdon, "Editorial: Problem-Solving Skill: What Does it Mean?" Am. J. Phys. 55, 877 (October 1987).

2. I. A. Halloun and D. Hestenes, "Common Sense Concepts about Motion," Am. J. Phys. 53, 1056 (1985).
3. B. N. Meeks, "An Overview of Conferencing Systems," Byte 10, 169 (December 1985).
4. E. F. Taylor and J. A. Wheeler, Spacetime Physics (San Francisco, CA: Freeman, 1986), p. 60.
5. R. C. Smith, "Teaching Special Relativity through a Computer Conference," Am. J. Phys. 56, 142 (1988).
6. J. Fleming and W. Frezza, "naplps: A New Standard for Text and Graphics, Part 1: Introduction, History, and Structure," Byte 8, 203 (February 1983).
7. D. Zollman, "Delivering a Modern Physics Course to In-Service Teachers via Videocassette and Telephone Conference," aapt Announcer 18, 90 (May 1988).
8. J. R. Rigdon, "Editorial."
9. P. Laws, "Workshop Physics—Replacing Lectures with Real Experience," in this Proceedings.
10. D. Zollman, "Using a Computer Bulletin Board in a Physics Class for Non-Science Students," aapt Announcer 18, 90 (May 1988).

The Personal Computer Impact on Physics Education at the U.S. Air Force Academy

George B. Hept
Department of Physics, U.S. Air Force Academy, Colorado Springs, CO 80840-5701

Robert G. Fuller
Department of Physics, U.S. Air Force Academy, Colorado Springs, CO 80840-5701, on leave from the Department of Physics and Astronomy, University of Nebraska–Lincoln, Lincoln, NE 68588-0111

In the summer of 1986 the personal computer revolution arrived at the U.S. Air Force Academy. When all the new cadets were issued a Zenith 248 computer and a basic set of software, the physics department was faced with the challenge of how to take advantage of this sudden increase in computing power.

During the first year of our experience with "computer cadets," we did not feel the need, nor did we feel we had the experience, to introduce the computer into every course. We did use the PCs to deliver and grade homework problems and act as an electronic messenger for course material and electronic mail.

Freshmen were enrolled only in our large general physics courses covering mechanics and electromagnetism. The PCs were an integral part of the honors versions of these basic courses. With the help of the computer, students could solve harder problems because they were not limited by their math ability.

The honors problems usually introduced a nonlinear term to problems that were done in the basic courses. For example, in a kinematics problem, the honors version introduced a velocity-dependent drag term; in the calculation of the output from a water tower, the head pressure varied as a function of the output; in determining the output of a lightbulb, the filament's resistance changed as a result of Joule heating. Since most of the students had not yet taken a differential equations course, these problems had to be solved algebraically. Students looked at small time intervals and used the answers from a preceding time interval as the new initial conditions for the current time interval. These simple numerical techniques allowed them to solve some basic differential equations.

We were very concerned about adding programming tasks to our student's workload but were determined not to give them "black box" canned software that would give them answers without improving the students' understanding of the solution process. We settled on an electronic spreadsheet program for two reasons: (1) the column format of the electronic spreadsheet and its copy function are ideally suited to the repetitive calculations required for the numerical solution techniques we used; (2) the spreadsheets require minimal programming skills.

All our students had the student edition of Lotus 1-2-3. We met our basic goals with this program, but did encounter some problems. First, the cadets did not have their own graphics-capable printers and the graphics package of the student edition of Lotus 1-2-3 is not fully integrated with the spreadsheet. This made correcting graphs time consuming and cumbersome. Second, the problem statements were deliberately left open-ended so that the students could fully exploit the flexibility of the spreadsheet. This meant that each student's solution was unique, which made debugging and grading student spreadsheets problematic.

Spreadsheet formulas can be difficult to translate into human terms, and students sometimes made subtle errors that were difficult to detect. This could be minimized by making the problem statements highly structured (e.g., how to set up column formats, what cells to put variables in, what values and formulae to use, etc.), but such structuring would compromise the goal of having the students use the spreadsheet as a creative tool to investigate nature. We concluded that spreadsheets would be impractical in our basic courses, where many more mistakes would be encountered and much more extra instruction would be required.

New features are available in the newest generation of spreadsheets, and we are currently reexamining this problem. The most important new feature is its ability to "name" variables so they can be put into formulas in terms humans can readily understand. For instance, the forces on a falling object encountering air drag can now be written as:

$$(mass*gravity) - (drag*velocity^2)$$

instead of something like:

$$(\$a\$4*\$b\$6) - (\$c\$5*b8^2)$$

We feel that this new feature will be a great aid to the student in preparing spreadsheet solutions and an invaluable aid to the instructor in helping the student and grading the results. We plan to present some of the spreadsheet problems we used and introduce a set of spreadsheet problems to be published soon.

The next logical step is to use computers in our majors program, but here we will have different goals. We need not concern ourselves with basic math skills;

our physics majors have these. Instead, we are concerned with broadening our curriculum. Our majors courses in classical mechanics, electomagnetism, and quantum mechanics are now confined to the solution of symmetric or simplified problems that can be solved analytically. We would like to expand our classes to match the growing trend in research laboratories toward computational physics.

In our modern physics course, the first course for our physics majors, we now use a spreadsheet to examine the black-body behavior of a lightbulb. Students solve a set of simultaneous differential equations for the equilibrium condition and integrate Planck's law numerically. The final portion of this course will concentrate specifically on computational physics to give our students a basic grounding in numerical solution techniques that they can take to the advanced physics courses. We also plan to familiarize the cadets with other software they have access to, including MathCAD, Eureka and the Numerical Methods Toolbox used with Borland's Turbo Pascal. At the conference, we will present the lesson plans for our projected six-lesson introduction to computational physics as well as the students' performance and reactions.

Spreadsheets by G. B. Hept for Lotus 1-2-3, the student edition are part of the collection Computers in Physics Instruction: Software, which can be ordered by using the form at the end of this book.

Physics Computer Learning Center at the University of Missouri–Rolla

Ronald J. Bieniek and Edward B. Hale
Physics Department, University of Missouri–Rolla, Rolla, MO 65401

In late 1985 the physics department at the University of Missouri–Rolla was awarded an internal grant of $100,000 to establish a physics computer learning center (PCLC). We converted a classroom and purchased several interacting computers: ten IBM PCs, ten IBM XTs, and seven Zenith ATs along with 12 IBM token-ring network cards.

All the ATs and several of the other computers are connected to form an IBM token-ring network. One AT is employed as a master file server and the other computers as work stations. Each PC and XT has two Tandon floppy drives; a Princeton monochrome monitor, which accepts color input signals; and a Everex graphics card, which supports the Hercules standard. The ATs have 20-megabyte hard drives, two floppy drives, EGA graphics and monitors, and math coprocessors. Every computer has its own local printer. The network has access to several peripherals, including an HP-compatible color plotter, Quadjet color printer, XL IBM Proprinter, Genoa tape-drive unit, and Quadlaser printer.

We purchased a large variety of software for the network. Some of the shareware programs were almost free, and other programs, such as Professional FOR-

TRAN, were moderately expensive. DOS and GW BASIC were purchased for each machine on the network.

A shareware program called Automenu from Magee Enterprises is used as a menu-driven interface between the user and the network. Automenu allows the user to explore and conveniently utilize much of the network software. It give the user access to the power of the network without knowing how the network works. In addition, it allows the use of hidden passwords, by which we can control access to copyrighted network programs.

Automenu's main menu offers the following submenus: "Word Processing," "Data Processing," "Programming Languages," "Scientific Calculators," "Graphics," "Peripherals," "Tutorials and Help," or "Exit to DOS." Each submenu enables the user to select several programs or network commands. Use of replaceable parameters in batch files executed by Automenu loads versions of selected programs appropriate for the particular hardware configuration of the local work station (e.g., type of station, monitor, location, etc.), identified by environmental variables set at power-up. In many cases, it also permits each user to tailor options in each program before loading. While a chosen program is loading, messages appear to inform the user of any special problems or key sequences that the first time user should know. A user can go to the front of the room and read the detailed instruction manual for any program.

We found that software problems were much more numerous and troublesome than hardware problems. The network manuals supplied by IBM were not particularly user-friendly and many problems occurred. For example, protective locks on network server files prevent an application program from opening a file on the network. We solved such problems, but spent a great deal of time and effort in the process. Our token-ring network has been working for almost two years and virtually no problems remain.

Probably the most successful aspect of the network is the ability to share the several network peripherals. Someone is usually plotting or using the network printers. Students favor the relatively fast ATs with their color displays; in retrospect, we should have purchased more ATs. The most popular programs are the user-friendly word processor Volkswriter and the powerful scientific calculator MathCAD.

There are different responses to the network. Some students jump right in and want to explore every program, directory, and feature of the network, and even try to decipher the protection scheme. Other students seem afraid of the system. They still use the PCLC, but they bring their own programs and run them on the off-network computers.

We thought initially that the computer network in the PCLC might be used in the classroom, for example, to work problems or to demonstrate various principles. This now seems unlikely. Students mostly use the PCLC for data processing of lab data or word processing to prepare reports. The PCLC is also used for homework assignments. One faculty member assigns extra-credit homework problems using MathCAD, and several students now work their normally assigned homework problems on the network. It seems likely that more applications of this type will become popular as the faculty develops more expertise with the network. We may also place supplemental commercially available physics tutorials on the network.

The PCLC is continually evolving. New software is added almost weekly. We plan to add considerably more software so that students and faculty can determine which software they like best. We have also improved our hardware. The original 20-megabyte hard drive on the file server was soon filled, so we installed an 80-megabyte hard drive. We are discussing connecting the local network to the campus mainframe and to a campuswide network. To encourage the faculty to utilize the PCLC as an instructional resource, to avoid software duplication costs, and to make internal electronic mail possible, we plan to attach all of the physics faculty to the network.

Computers in Introductory College Physics

Carlos I. Calle

Department of Physics, Sweet Briar College, Sweet Briar, VA 24595

Computers are probably used more often to supplement instruction in physics than in other disciplines.[1] At Sweet Briar College, we have been using both mainframe and microcomputers for physics instruction for over seven years.

The physics department currently has six Apple II microcomputers, one IBM-compatible microcomputer, and two Macintoshes, all with printers. Two of the Apple II computers are equipped with commercial analog-to-digital boards and are connected to equipment. The remaining Apple computers, the Macintoshes, and the IBM are installed on carts so that they can be easily moved from laboratories to classrooms. In addition to physics-department computers, students may use two Macintosh II computers, forty Macintosh Plus computers, and thirty IBM-compatible computers throughout the campus.

The microcomputers are used in introductory physics courses in three ways: for lecture demonstrations, self-paced tutorials, and computations and graph plotting in the laboratory.

Lecture demonstrations are frequently used for introductory physics lectures. Simple, easy-to-set-up lecture demonstrations are readily available for introductory mechanics, electricity and magnetism, optics, and thermal physics, but not for modern physics.

A microcomputer is the ideal tool for developing lecture demonstrations in modern physics. We have developed several computer simulations that make use of graphics and animation on the Apple and the IBM-compatible computers. Our program Rutherford,[2] written for both the Apple and IBM computers, illustrates what will happen in a scattering experiment according to the prediction of

Thomson's "plum pudding" model of the atom. This is followed by an animation of Rutherford's interpretation of the scattering experiment; it shows the nuclear model of the atom, which explains the large scattering angles.

Self-paced tutorials can also benefit from the use of computers. We have developed a tutorial on the introductory concepts of nuclear reactions for the Macintosh computer.[3] The tutorial starts with an animation of an alpha particle interacting with a nitrogen nucleus in the reaction, 14N(a, p)17O, which shows how charge and nucleon number are conserved in nuclear reactions. Two more animations illustrate the reaction 27Al(n, a)24Na, and the subsequent beta decay into 24Mg: 24Na Æ 24Mg + e.

Click buttons allow the user to review the material and view the animations again. Four additional reactions are presented afterward with blank spaces replacing some of the nuclei in the reaction. Using the conservation laws learned in the introductory section of the tutorial, the student can fill in the blank spaces from pull-down menus with selections. By selecting a pull-down menu available at all times, the student can restart or stop the program at any time.

Finally, computers are useful for streamlining calculations and graphing. Nonmajors sometimes find that physics concepts are obscured by the calculations in physics laboratories. One solution to this problem is to accelerate the calculation process so that results can be obtained sooner.

One area where nonscience students can speed up the calculation process is graph plotting. Our program Graphmaker,[4] written for the Apple computer, plots and prints graphs from data input. The program is completely menu driven and therefore requires little training. Some menu selections allow for data modification so that data points can be added, eliminated, or modified. A menu selection directs the computer to plot the graph on the screen. Another menu selection orders a hard copy of the graph. The data are saved to disk in case additional graphs are wanted after the computer is turned off.

The need for repetition of calculations is a major reason for student frustration in physics laboratories. Experiments that require many calculations, like Newton's second law or accelerated motion on an air track, can be frustrating when results do not correspond to expected values and calculations must be redone. Computers also save time when experiments are repeated, new data is collected, and calculations must be done again.

We have developed a series of programs for many common physics laboratory experiments that require significant computations. Each program requests data from the user and calculates the results, which can be viewed on the screen and printed on a line printer. The programs are menu driven and allow for correction of mistakes and reentering data. Students are allowed to use these programs only for checking their own calculations or when they have to redo a experiment.

The software that we have developed for lecture demonstrations, introductory physics tutorials, and introductory laboratories is still in a developmental stage. Student response so far has been positive; students seem to welcome the opportunity of working with the computer. Additional programs are needed in all areas. We are currently developing additional tutorials on the Macintosh in other elementary concepts of physics. Three of these should be available in the near future.

1. A. Bork, Learning with Computers (Bedford: Digital Press, 1981); A. Borghi, et al., "Computers in Physics Education: An Example Dealing With Collision Phenomena," Am. J. Phys. 52, 619 (1984); C.I. Calle and J. A. Roach, "An Interactive Tutorial in Nuclear Physics Using a Microcomputer," Bull. Am. Phys. Soc. 29, 1782 (1985); K. L. Johnston, "Enhancing Engineering Physics with Courseware," Bull. Am. Phys. Soc. 30, 1771 (1985); C. H. Hayn, "Computer Simulation of Non-Uniform Acceleration," Phys. Teach. 25, 293 (1987).
2. C. I. Calle and L. F. Wright, "A Rutherford Scattering Simulation with Microcomputer Graphics," J. Comp. Math. Science Teach. (forthcoming).
3. C. I. Calle and J. A. Roach, "A Nuclear Reactions Primer with Computers," J. Comp. Math. Science Teach. 6, 28 (1987).
4. C. I. Calle and L. F. Wright, "A Program for Plotting Data in the Introductory Physics Laboratory," J. Comp. Math. Science Teach. 5, 44 (1986).

A Summer Institute on Electronics and Microcomputers for High School Science Teachers

Robert B. Muir
Department of Physics and Astronomy, University of North Carolina at Greensboro, Greensboro, NC 27412-5001

One component of Project Archimedes (NSF grant TEI84-70438) was a six-week Electronics and Microcomputer Summer Institute. We conducted the institute twice, in the summers of 1986 and 1987.

We wanted participants to develop confidence in the use of simple hand and shop tools; learn to use modern integrated circuits appropriate to a high school physics course; learn to use electronic measuring instruments, e.g., multimeters and oscilloscope; learn to construct simple instruments; and learn to use a microcomputer for programming and interfacing.

To achieve these goals, participants used shop and hand tools to construct a piece of equipment from scratch; constructed circuits implementing the basic functions of logic ICs; applied logic circuits to physics laboratory experiments, e.g., event counting, timing, etc.; used physical to electronic transducers for temperature, light, sound, position measurement; constructed circuits implementing basic operational amplifier functions; learned about the design and construction of a microcomputer; developed word-processor, spreadsheet, and database management applications relevent to their needs; wrote programs relevent to their needs; performed the fundamentals of data and graphical analysis; interfaced the various transducers to a microcomputer; and learned about peripheral devices, such as printers, disk drives, and monitors, in terms of interfacing and communications standards.

The institute met for five hours a day and five days a week for six weeks (150 contact hours). Such extensive involvement with hands-on activities was very effective for teaching the kind of material dealt with in the institute. Our format allowed participants to leave circuits, experiments, and projects set up from one session to another—a feature seldom found in in-service programs. These advantages notwithstanding, many participants found the schedule a little too exhausting. We therefore recommend that a summer institute or workshop dealing with the same content as the skills institute of Project Archimedes use a schedule within the following ranges:

- three hours a day, five days a week, for five or six weeks (75 to 90 contact hours);

- five hours a day, five days a week, for three to five weeks (60 to 125 contact hours).

A January 1988 evaluation (three teaching semesters after the 1986 institute and one semester after the 1987 institute) validated the content and activities of the institute, but also indicated that more time should have been spent on almost every topic. Based on the evaluation, we recommend that institutes or extensive workshops provided for teachers having backgrounds similar to those involved in Project Archimedes (middle and high school science) deal with two main subjects: use of computers and electronics. Sessions dealing with the use of computers should concentrate on these areas:

- Productivity software (word processor, data-base management, spreadsheet, gradebook, etc.). Spend about 30 hours of hands-on work resulting in something every teacher can take home and use.

- Instructional software. Spend about 40 hours, and include a few good "canned" programs that they can really learn to use effectively. Participants should also write programs themselves. This will require teaching many of them a language. Start by teaching a language that is easy to learn, e.g., BASIC.

- Laboratory applications. Spend about 20 hours on simulations (canned and participant written). Spend another 20 hours on data acquisition (canned and participant written), including analog to digital conversion. Spend a third 20 hours on methods of data analysis (again, canned and participant written), including graphic presentation, statistical analysis, finding maxima and minima, areas under curves, etc.

Sessions on electronics should include:

- Basic DC and AC circuits, properties of electrical signals and wave forms, use of meters and the oscilloscope. This is good review and provides the necessary vocabulary and ability to use instruments for the other electronics topics. Spend about ten hours.

- Transducers, such as the potentiometer for position measurement, thermistor, thermocouple, IC temperature sensors (e.g., Analog Devices AD590), photodiode, phototransistor, CdS cell, Si solar cell, etc. Spend about 15 hours on hands-on work and include lab activities the participants can take back to their classroom.

- Digital ICs. Spend about 30 to 40 hours, and include lab activities the participants can take back to their classrooms. We highly recommend using large breadboarding sockets for constructing circuits.

- Operational amplifier ICs with applications to conditioning the signals from transducers (amplification, voltage to current conversion, offsetting, adding, filtering). Again we suggest constructing circuits on breadboarding sockets. Spend about 20 to 30 hours and include lab activities the participants can take to their classrooms.

- Construction of one or more instruments including a printed circuit board (e.g., with Radio Shack materials). Participants must drill necessary holes in the chassis, solder, assemble, debug, and calibrate. Instruments constructed might include: general-purpose IC-based amplifier, IC-based signal generator, triple power supply ($\pm$ 12V and + 5V), and a four-digit timer.

A Workshop for Secondary School Teachers on Using Computers for Data Collection in Instructional Laboratories

Charles D. Spencer and Peter Seligmann
Department of Physics, Ithaca College, Ithaca, NY 14850

The Ithaca College biology, chemistry, and physics departments received a grant from the National Science Foundation (NSF) to conduct summer workshops for high school science teachers. The workshops have two tracks: one focusing on using computers for data analysis of traditional laboratories, and the other on computer-based data acquisition. The second track, which is described here, is primarily for physical science teachers.

The data–acquisition workshop has been given three times, the first sponsored by an NSF Developments in Science Education grant and the second and third times by Ithaca College. A new grant will support 15 teachers for three weeks beginning in the summer of 1989. Because of the requirement for follow-up visits to participants' home schools, teachers are recruited from New York State and northern Pennsylvania.

The workshop gives teachers a basic knowledge of digital electronics and subsequently the hardware and software needed to make voltage, time, and counting measurements. After developing these skills, participants work on four to six applications chosen from a variety of previously developed experiments in physics, chemistry, and earth science. They return to their home schools with the hardware, software, and skill to incorporate these applications in their courses and subsequently to develop additional experiments.

Several high school teachers who are previous workshop participants assist in the development of materials and new applications.

Participants spend approximately one week on digital electronics, ending with the construction and testing of a general-purpose measurement circuit that works through a computer's game port. During this time, participants also learn to use and alter data acquisition software. Another half-week is spent working with sensors such as a photogate/timer and temperature probe. The remaining week and one-half is spent putting together, testing, and running experiments.

The key to the success of the workshop is the simplicity of the hardware and software. Inexpensive, linear, and accurate voltage-to-frequency converters allow voltage to be determined by measuring frequency. Time is determined by measuring the number of cycles of an accurate clock in an interval. Counting is simply the number of cycles of a signal during a known time interval. This unity of what the computer inputs through the game–port simplifies the electronics and makes the data-collection software the same for the three kinds of measurement. The accurate clock used for testing the system and for time measurement consists of a 1,000 MHz TTL crystal oscillator and six or seven decade counters. A 5-volt DC power supply is required. Batteries are used when necessary for some analog sensors.

Teaching Physics with the Aid of a Local Area Network

Rolf C. Enger, Leonard C. Broline, and Richard E. Swanson
Department of Physics, U.S. Air Force Academy, Colorado Springs, CO 80840-5701

The U.S. Air Force Academy recently installed a 6000-node Local Area Network (LAN) that connects all students, faculty, and administrative offices.[1] To test the hypothesis that a LAN can be a valuable educational tool, we used the LAN extensively in our introductory mechanics course during the past academic year. In this paper we describe our experiences, summarize the lessons we've learned, and discuss the LAN's impact on physics teaching.

We have tried to make maximum use of our LAN. We use the LAN to distribute software and course materials, including the syllabus. Students use the LAN to submit papers and to transfer data files created by course-supplied software. Instructors and students send memos to one another on electronic mail (Email). Some instructors, especially those with home computers and modems, use Email for tutoring. Students who need help posted a message prior to a designated hour in the evening, and instructors check their Email at that time. Email is also used to give and grade quizzes on-line and to solicit end-of-course critiques. In short, if a task could be done with the LAN, we try to do it. As a result of our experiences, we have learned several lessons.

Lesson 1. The introduction of a LAN into an established course requires increased effort on the part of both students and faculty. Time that used to be spent on physics is now diverted to learning how to use the LAN. Students ask faculty members questions both about the LAN and about physics.

Lesson 2. The reaction of students and faculty to a LAN may not be entirely positive. Time is a precious commodity. Some would prefer to spend all available time on physics instead of on learning how to use a new computer system.

Lesson 3. Some information is not effectively communicated with a LAN. Although we distributed the syllabus over the LAN, most students made a paper copy of the syllabus anyway, thus shifting the cost of the printing from the printing plant to more expensive laser printers.

Lesson 4. Every use we made of our LAN involved either file transfer or Email. As of this writing, a bulletin-board feature is just coming online. We believe bulletin boards will ultimately become as valuable as file transfer and Email.

We believe LANs have their greatest impact as communication devices. By freeing the student from the formality of face-to-face contact, a LAN increases the likelihood of communication. This helps compensate for the absence of "body language" and eye contact when using a LAN. In addition, the written communication process required by a LAN forces students to specify and formalize their problems, a process that in itself aids learning.

There are, however, risks associated with LANs. A negative LAN experience, whether because of poor access or LAN malfunction, can rapidly "turn off" students. Students and teachers approach LAN failure from different points of view. Teachers generally discount the specific experience and consider the student's overall ability to access the LAN (which is probably high). Students, however, focus on the repercussions of a specific failure. Finally, typical of any educational use of computers, there is always the risk that students will confuse working on the computer with studying.

Although most schools today do not have local area networks, any school with access to phone lines, computers, modems, and bulletin-board software can establish its own LAN. Since teaching is largely a communication process and a LAN is an alternate communication channel, a LAN has tremendous potential to enhance physics teaching. However, a LAN is a new medium of communication for both students and instructors, and there are bound to be growing pains. It will take time to develop the potential of the local area network as a teaching tool.

1. E. G. Royer, W. E. Ayen, and W. Richardson, "Implementing a Large Local Area Network to Support Undergraduate Education," in Proceedings of the ISMM International Symposium: Mini and Microcomputers and Their Applications—MIMI87, Lugano, Switzerland, 29 June–1 July 1987 (International Society for Mini and Microcomputers, 1987), pp. 10–14.

Online Computer Testing with Networked Apple II Microcomputers

Paul Feldker

Department of Physics, St. Louis Community College at Florissant Valley, St. Louis, MO 63135

At St. Louis Community College at Florissant Valley we have developed an online computer testing system. The system was written primarily in BASIC for a system of networked APPLE II microcomputers. Although the system was developed for a CORVUS network, it could be easily transported to different networks.

This system allows a user to prepare, update, and modify a test item bank; prepare online tests and modify them; administer online tests; keep student records; and prepare grade summaries.

The general setup program allows the user to specify the topics and subtopics for the test-item bank. If a new item bank is being prepared, the user must specify the slot, drive, and volume where these questions are to be stored. Three types of questions may be entered: true-false, multiple choice, and questions that require a single numerical solution. For the last two types of questions, the user may specify the numerical parts of the questions to be randomly generated by the computer.

If the user chooses to ask for random numbers in a question, he must specify a range of acceptable values and enter an algebraic formula for the correct answer. If the question is multiple choice, he must enter formulae for the distractors. If it requires a single numerical result, he must enter a percent of allowable tolerance. Each question in the bank is coded by a four-character access code plus a problem number.

At the beginning of each semester the user sets up a directory of courses and sections and enters the class rosters. To prepare an online test, the user selects the course and sections to be tested and the items for each questions. Each question on the test is selected from up to 15 items from the test bank. The user also selects the amount of credit for a student's second guess, the amount of credit to be given if the test is taken late, and the full-credit time period.

To access the test, the student enters his name, roster number, social security number, course, and section. The test questions are then randomly selected from the items selected by the instructor for each question and the order of answers is randomly arranged. Immediately after selecting an answer, the student is informed about the correctness of his response. After the student completes the test, he is given a score, which is stored in his file. The computer then updates each question that was on the test, recording the number of times it was asked and the number of times it was answered correctly on the first attempt. This information allows a difficulty level to be determined.

Other student grades are entered manually. The software allows the instructor to print a typical gradebook page, print grade records for any or all individual students, or print a class rank report. The instructor may drop low scores and weight score categories.

Some of the unique features of this system are randomization of numbers within questions, assignment of partial credit for items answered correctly on the second attempt, and determination of a cumulative difficulty level that is updated each time a question is asked.

This system was developed as a purely evaluative technique and has been successfully tested at Florissant Valley for the past three years.

Using a Computer Bulletin Board

Dean Zollman
Department of Physics, Kansas State University, Manhattan, KS 66506

I have developed a computer bulletin board to enhance communication among students and between students and the instructor in a physics course. The bulletin board, patterned after some of the commercially available bulletin boards, enables students to perform these tasks: view a week's class schedule and assignments; view the syllabus for the semester; view the learning objectives for recent classes; communicate electronically with other students in the class; read messages from other students in the class; look at a list of interesting readings, which are contributed by students; participate in an "open line" discussion; view the instructor's office hours and class; view their grades for the course; and view solutions to selected homework problems.

The bulletin board is used in Contemporary Physics, a modern physics course for non–science majors. Because it is maintained on the university's mainframe computer, students have access to it from any terminal on campus or through telephone communication. The use of the mainframe enables us to incorporate some of the utilities available on this system without extensive programming. For example, all electronic mail is prepared and transmitted by using BITMAIL. For entering and reading various text files the students use XEDIT (an IBM product) with some macros prepared especially for the bulletin board. The program that controls access to these utilities and presents the menu options is written in EXEC2.

The bulletin board has proven useful for communication between instructor and student, for providing up-to-date course information, and (to a lesser extent) for communication among students. Participation in open discussions on the system and in contributing to the list of interesting reading has been disappointing, perhaps because these students meet in class twice per week.

The bulletin board is now being prepared for use by biology teachers who are located throughout Kansas. This new effort should bring about a very different use pattern.

The use of the bulletin board for in-service teachers is supported by the National Science Foundation under grant number TEI-8751332.

Section 11
Publishing Physics Software

Software Preparation Workshop

Eric T. Lane
Physics Department, University of Tennessee at Chattanooga, Chattanooga, TN 37403

Do you have software that you would like to distribute? Are you considering writing software for physics instruction? Do you dread suffering through the final stages of developing the human interface for your software?

We will critique and revise your software on the spot. We will help you develop your software to a level that will pass technical review for public presentation and possible publication. (Don't worry, we won't critique the pedagogical content.) We will help you develop the recommended introductory screen, the presentation of variables (give range, suggested value), key detection, help screens (F1 on the IBM, ? on the Apple, etc.), error handling, and the end statement (Esc Key). We will also discuss publication possibilities and suggest some other creative ways to use your software.

We will have Apple, Commodore 64, IBM, and Macintosh microcomputers available. Bring your programs, language system, editors, and whatever else you normally use to develop your software.

Practical Advice for the Creator of Educational Software

Ruth W. Chabay and Bruce Arne Sherwood
Center for Design of Educational Computing, Carnegie Mellon University, Pittsburgh, PA 15213

Because the computer is a two-way medium it differs radically from earlier, passive media such as books, lectures, and films. The interaction of the computer medium leads to esthetic new programs.

Based on our extensive experience creating and using educational software, we present some practical suggestions for the would-be creator of educational software. In many cases our advice consists of assertions based only on empirical experience, in the absence of principled theory. Some of what we say may seem utterly obvious and banal, yet we have often seen these "obvious" points missed by intelligent but inexperienced authors, including people who are very skilled with computers but are new to using them for educational purposes. Diligent, committed authors will eventually come to many of the same conclusions we have come to. By offering guidelines and suggestions, we hope to shorten the time required to discover what works and what doesn't.

Our suggestions deal with four crucial components: the layout and interactive management of displays, heightening of interactivity, input analysis, and user control.[1]

Displays: A Visual Medium

The computer is inherently a visual medium. Because of its strong graphics capabilities, and because natural language interaction with a computer is so difficult, communication relies heavily on pictures and diagrams. In this medium even text must be treated as graphics. The considerations involved in creating interactive displays are very different from those that guide the use of text or graphics in books, magazines, or films. However, the dynamic nature of the medium offers new possibilities for presenting and making use of text and graphics.

Interactivity: Intuitive, Transparent Interactions with Displays

Designing interactive programs is surprisingly difficult; it doesn't come naturally. The process is very different from designing a book or a lecture, where there is little or no interaction. It is much easier to tell a student something than to ask him, especially because it is difficult to understand the student's answer or intentions. The uniquely interactive, two-way nature of the computer medium has a striking impact even on seemingly noninteractive components such as static displays. Not only can users interact with the program, they *must* interact for anything to happen at all. How do users know what to do? Sad experience proves that even experienced users, including ourselves, often do not read or attend to such instructions on the screen. A major challenge is to produce graphics and text displays that make intuitively obvious what next moves are possible. This contributes to a subtly different esthetic for what makes a good display.

Input Analysis: Communication Problems

Although modern computers offer exciting capabilities, they do not offer the breadth of interaction of a human. Through tone of voice, facial expression, and other body language, a human teacher provides extremely important extra information, both cognitive and affective, that a computer cannot express. Then too, the computer has only limited ability to guess a student's intentions. Overcoming the constrained communications between computer and student constitutes a challenge to the author of educational software. It is crucial to make text and instructions terse and clear, and to avoid cuteness, insults, or responses that the user might misinterpret in the absence of inflection and gesture. Finding ways to replace extended verbal interaction with direct manipulation of objects in the display can help greatly overcome communication difficulties between computer and student.

Control: Many Paths

Because it depends on user inputs, the behavior of an interactive program is not entirely predictable. In extreme cases, every user of the same program might have very different experiences: different paths through the material, different kinds of messages in response to inputs, different types of exercises, etc. The program must provide sensible context and responses for any path the user may follow; and users must not be allowed to follow paths that the program is not prepared to deal with. The program should offer flexibility. It should be easy to skip forward, go back to review, find out where one is within the program, modify parameters, etc.

Interactive Development: Test Early and Often

Just as the finished program should be interactive, so should the process of developing the program.[2] It is important to test the program with actual students early in the development process, and to revise materials as you observe how users interact with the program.

1. R. W. Chabay and B. A. Sherwood, "A Practical Guide for the Creation of Educational Software," in *Proceedings of the Workshop on CAI and ITS*, edited by Ruth Chabay, Jill Larkin, and Carol Scheftic (Hillsdale, NJ: Erlbaum, in press).
2. B. A. Sherwood and J. H. Larkin, "New Tools for Courseware Production," in *Proceedings of the IBM ACIS University Conference, Discipline Symposia: Computer Science,* Boston, June 1987 (Boston: IBM Academic Information Systems), pp. 13–25.

The Evaluation of Educational Software: The EDUCOM/NCRIP-TAL Higher Education Software Awards Program

Robert B. Kozma
NCRIPTAL, University of Michigan, Ann Arbor, MI 48109

In 1987, EDUCOM and the National Center for Research to Improve Postsecondary Teaching and Learning (NCRIPTAL) established a higher education software awards program with these goals: to identify outstanding higher education software and teaching innovations that use computers, to reward the designers of outstanding software and innovations, and to improve the quality of higher education software and its use.

NCRIPTAL brought together teams of faculty members from academic associations, instructional psychologists, and software experts to review applications from the undergraduate liberal arts and to identify award winners. There are two divisions in the program: the product division for original software developed for higher education, and the curriculum division for professors who use the computer to solve important, local instructional problems.

During 1987, the first year of the program, 139 applications were reviewed; in 1988, nearly 190 are being reviewed. This paper discusses the criteria used to evaluate the applications. Design criteria are used for all applications, and additional, specialized criteria are used for software products and curriculum innovations. In my presentation, I will show video excerpts of the 1987 award winners.

Design Criteria

The reviewers are asked to think of the software developer or curriculum innovator as a "designer," one who creates solutions for problems. To this end, applicants are asked a series of questions that elicit the design rationale implicit in the software or innovations. They are also asked to present data that demonstrate its impact on learning. The following issues are addressed:

- *Problem.* The applicant should have a clear notion of the instructional problem or learning difficulty to be solved. For example, a professor of physics may find that the traditional wet lab does not allow students sufficient opportunity to vary the parameters in the experiment; or an English professor may find few opportunities to work with students on the revision of their compositions.

- *Content.* Various tasks are learned differently, require the use of different prerequisite skills or knowledge, and are facilitated by different instructional experiences. The designer must have an understanding of the content to be addressed, the knowledge or skills to be learned, and how these might be structured and organized by the learner.

- *Students.* Students exhibit a range of previous preparation, motives, and study skills that play an important role in learning and instruction. It would be very difficult for a single piece of software to address the needs of all students taking introductory physics or all those enrolled in English composition. The designer should have a clear notion of the specific set or range of students addressed by the software or curriculum innovation.

- *Instructional Method and Media.* Computers can employ a range of images and sounds; they can also perform calculations, transform numbers into graphs, and even make inferences. Instruction may vary in the amount of structure it provides, in the pace and sequence of presentations and queries, and in the extent of learner control, among other things. The designer should have a clear notion of the the technological capabilities that are used and how they are coupled with instructional strategies to facilitate learning and increase motivation.

- *Coherence and Impact.* The factors described above are all interrelated in the design of software and curriculum innovations. The designer should have a clear sense of the instructional methods most appropriate for the instructional problems, the content, and the students and how all of these considerations fit together. Also, designers should have information, quantitative or qualitative, on the impact of the software or curriculum innovation on learning and motivation.

Specialized Criteria

In addition to the criteria above, which are common to all applications, there are special criteria for software products and curriculum innovations:

- *Software Products.* The software packages are reviewed for their operation (quality of the interface, the treatment of operational errors, and the speed of execution), content (accuracy, breadth of treatment, and importance), and for instructional features (such as presentation mode, learner control, feedback, etc.). These last criteria are customized for type of software (tutorial, simulation, tool).

- *Curriculum Innovation.* A curriculum innovation may involve the educational use of general productivity software or the application of educational software developed by others. In these cases, curriculum innovations are assessed by the added contribution made by the applicant. Also, to endure, curriculum innovations must have support within the institution. Applications are assessed by the amount and kind of support provided by key administrators. Finally, though a curriculum innovation must meet a need that is important in the local context, it is also assessed by how it might help other faculty members, other institutions, or those in other disciplines.

Physics Academic Software: A New AIP Project to Publish Educational Software in Physics

John S. Risley
Department of Physics, North Carolina State University, Raleigh, NC 27695-8202

The American Institute of Physics (AIP), in cooperation with the American Physical Society (APS) and the American Association of Physics Teachers (AAPT), is inaugurating a project, *Physics Academic Software*, to review, select, and publish high-quality software suitable for use in undergraduate or graduate training in physics. Submitted software will be peer-reviewed for excellence in pedagogical or research value.

Physics Academic Software is soliciting authors of software for teaching, laboratory, or research activities in physics to submit their programs and documentation for review. Only finished works will be considered. Although only MS-DOS software can be considered at this time, we intend to add other operating systems to the scope of the project in the future. A royalty will be paid for published software. A licensing agreement with the AIP is required.

Acceptance of software for publication depends on the quality of the program, completeness of the documentation, ease of use, clarity of screen displays, strength of instructional design, logic of presentation, accuracy of numerical algorithms, and effectiveness for education or research.

Submissions are reviewed by the editor, his staff, external peer referees through blind reviews, and the AIP-APS-AAPT Advisory Committee. Authors receive critical reports commenting on the instructional design and pedagogical value of their software. If necessary, they are encouraged to improve it for publication.

Educational software should be structured for inherently intuitive operation. It should guide the user with prompt messages, option menus, pop-up windows, on-screen lists of commands, and context sensitive help screens. The software should protect the user from inadvertent loss of data, and recover from drive and printer errors. The program should use accurate numerical techniques and correctly model the physics concepts.

The documentation should follow the style of previous manuals from *Physics Academic Software*. The front matter should contain the exact title of the program, the name and institutional affiliation of the author, acknowledgments, exact copyright information, and a list of the minimum and recommended hardware and auxiliary software. A preface should describe the software and the contents of the documentation.

The main part of the documentation should begin with a short overview of the program, followed by an orientation tutorial, a brief description of the physics involved and how the program models it, and instructions for operating the program.

The appendices should include a "key order" description of the commands (if necessary), a quick reference sheet, information for DOS users, an annotated list of files, pedagogical advice for a teacher, and sample exercises.

Additional advice, styles, and conventions can be obtained from the editor. To submit software, prepare a transmittal letter containing the name, address, and phone number of the author who will serve as the contact person; a statement designating, in compliance with any rules of the author's institution, to whom royalties should be paid; and, if applicable, a list of similar existing software packages and an explanation of how the submitted program differs from them. Also prepare a set of printouts showing an example of every screen display, and a listing of all on-screen prompts, messages, labels, etc.

Send the above along with four complete copies of the software package to Prof. John S. Risley, Editor. Published software can be ordered from Physics Academic Software, American Institute of Physics, 335 East 45th Street, New York, NY 10117, Tel. (800) AIP-PHYS.

Videodiscs and Visualization for Physics

LVs, CDs, and New Possibilities

Robert G. Fuller

*Department of Physics, United States Air Force Academy, Colorado Springs, CO 80840
and Department of Physics, University of Nebraska-Lincoln, Lincoln, NE 68588-0111*

"In the beginning..." I remember going to a videodisc conference about a decade ago in Pajaro Dunes, California, organized by Alfred Bork and others. I started to write my first proposal to obtain funds to produce a videodisc that same year. Consider that. If you think of a generation of computers lasting about two and a half years, then a decade is four computer generations ago. In human terms that's equivalent to about 288 years ago! 1700! In human terms the videodisc player arrived on the scene in 1700. Where have we all been?

For most of us, the demands of our jobs and the rewards of our profession are such that the videodisc and the use of optical techniques for the storage of data has, I think, largely passed us by. How many instructors have videodisc players in their classrooms that they can regularly use for teaching physics? How many use videodiscs interactively with small groups of students? How many have videodisc players in a space where students interact directly with the videodisc's images to collect data or learn physics?

If you were transported back in time to 1700, you could hardly have any human experience whatsoever that you would not recognize instantly as different from your life today. On the other hand, if you were transported all the way back to a 1978 physics course, in lecture, recitation, and perhaps even in laboratory work, you might be hard put to tell whether it was then or now.

This, then is the first observation of my talk—optical storage technology has been very slow to be brought into physics instruction.

Let me offer a short examination of the capabilities of videodisc technology itself. Videodiscs, which became readily available commercially in 1978 and hit the home entertainment market in 1979, now have the commercial trade name of LaserVision (LV). A 30-cm LV disc can contain up to 54,000 analog video pictures with two audio tracks. Each picture, or frame, on the disc is numbered and the player enables the user to access frames on the disc in almost any sequence chosen. Many players offer slow motion forward and reverse, three times normal speed forward and reverse, single-frame step forward and step back, and rapid random access. At normal speed (30 frames per second) a LV disc will play for 30 minutes on each side. (The small 12-cm LV disc is known as CDV for compact disc video.) A typical LV disc ranges in price from about $20 for popular entertainment discs to about $400 for professional photographic data discs. A player requires a television monitor and a stereo system for stereo sound. LV players cost from $200 to $1,600 each, depending on age and quality. Most commercial/education players have a standard computer-interface port. By now, there is a pretty good list of titles, in the tens, of videodiscs that can be used to teach physics and astronomy.

When one talks about interactive video, one usually means a system with a computer controlling a videodisc player. The education potential of an interactive videodisc system is almost beyond our wildest dreams. It offers color overlay graphics, computer-based lessons, tutorials and simulations in conjunction with high-quality, real-world video to which can be added high-fidelity stereo sound.

From the point of view of 1700, one can ask why has it taken so long for this technology to work its way into learning physics. But a decade in social reform is hardly any time at all. Simply put, the social structure of the physics community as it applies to teaching and learning physics is not able to keep up with changing technologies. Here we have a technology that will let us bring real-world audio/visual events into the physics classroom for ready access to individuals or small groups of students. The responsibilities for learning can be turned over to the learner and the physics professor can assume the role of facilitator, encouraging students to use the conceptual tools of the discipline to increase their understandings of nature. What will it take to break loose the log jam that seems to be preventing the wide distribution of this technology into physics classrooms?

Physics educators seem especially resistant to change. We are not prepared in our institutions to cope with a new generation of technology every two and a half years. We still use voltmeters built in the 1940s. Physicists and educational institutions do not easily part with capital equipment. Once it has been useful, we hate to let it go. We do not like the risks involved in bringing in new technology to do the same old job that we have been doing quite well (we think) without.

In the United States our markets are generally demand driven. It is difficult to get money to undertake educational projects when the maximum potential market extends only to a small part of the physics community. We need a much larger base of videodisc players installed in schools, colleges, and universities if we hope to see a large number of superior videodiscs produced for use in physics classrooms.

Make note of this: one of the best things you can do for this technology is to insist that your school or department buy a videodisc player. New models capable of being hooked directly to a computer cost less than $1,000, and if you are willing to buy last year's model, you can find them for as little as $175.

You and your physics-teaching colleagues must get into the era of optical data storage. The place to start is with the analog storage of 54,000 photographs on a single side of a videodisc. As soon as you get back home, talk to whoever controls the funds for your institution: your chairman, the media-center director, the film librarian, or the software-control director. I'm serious. The availability of good software in a largely capitalistic economy is fundamentally driven by perceived demand. Until you make a demand for physics discs look real, we will continue to have difficulty getting new videodiscs produced for our use.

We reluctantly admit computers are here to stay, but we secretly think this video stuff is just a passing fancy. And as if to prove the point, more and more video technologies seem to be coming along every day, each new one not compatible with the old. Let's just wait and see what will happen. And of course what did happen was the compact disc revolution. Also before we could blink our eyes, laser-storage technology leaped from analog video signals to the digital audio signals of the compact disc; the economies of the music industry drove the CD market faster than we could imagine. In the past few years compact discs have become the darling of the home entertainment industry.

Compact disc audio, widely known as CD, became available in about 1983. A compact disc will play about 70 minutes of digital stereo audio. CDs cost from about $8 to $20. They require a CD player that costs over $100 and earphones or a stereo system. While they are currently used mostly for entertainment, there is one CD with demonstration sounds available for teaching physics. It was developed by Tom Rossing and is being distributed by the Acoustical Society of America.

Perhaps because the teaching of the physics of sound is not a high priority for every physics department, there have been few articles or discussions about the use of CDs in teaching physics. But that may soon change as new ways of storing data digitally on compact discs become available. There are now five additional types of compact disc storage systems in use. Let's briefly review the capabilities of each one and then let me close with a physics scenario for the CD ROM.

Compact Disc Read Only Memory (CD ROM) is a publishing medium. A CD ROM disc is a permanent record of digital data. It can be used to store up to 600 megabytes of digital data, about equivalent to 250,000 pages of text or about 700 floppy disks. This is equal to about 2,000 high-resolution images. It can hold combinations of text, graphics, and sounds in digital format. The use of a CD ROM requires a computer and a CD ROM disc drive. At the present time the disc drives to work with MS DOS system vary in price from $600 to about $1,500 each and are available from most of the main disc companies such as Sony and Hitachi. There is a CD ROM drive available for the Macintosh family from Apple for about $1,200. Unfortunately, there is a difference in the format of the discs that can be read by the MS DOS drives and the Mac drives. It is to be hoped that the High Sierra format of the MS DOS systems will eventually be readable on the Mac system drives. At this time there are about 200 disc titles available for the MS-DOS system. The one most likely to be useful to physicists is the *Microsoft Bookshelf* collection of dictionary, thesaurus, and style manual. The *Science Helper* available from PC-Sig, Inc., for about $200 contains kindergarten through eighth grade science lessons developed under grants from the NSF. The *Science and Technology Encyclopaedia* from McGraw Hill is available for about $300 and the *Electronic Encyclopaedia* from Grolier is about $300. I see this as the most immediately promising of the new CD technologies and I want to return to discuss it later, but now let me go on to talk briefly about other CDs.

Write Once–Read Many Times (WORM) which became available in 1988, serves as an archival computer digital optical storage system. A WORM will hold about 400 megabytes per side. A double-sided disk costs about $200 and the drives run about $4,000. They are available for both MS DOS and Macintosh systems. They are excellent storage systems for such things as student records and research data that you never want to lose or erase.

Magneto-Optical Erasable Compact Discs have been announced by Sony, Philips, and Tandy. They will hold about 300 megabytes per side and the drives will cost from $2,000 to $10,000. At this time no such drives are commercially available. They have been promised by the end of 1988. This technology and the WORMs may be destined to take on the job of storing vast amounts of data for computer users. I find it difficult to imagine their having a direct impact on the physics classroom.

The digital storage of video images has long been discussed. However, the problem is formidable: a good-quality color video image requires from 600K to 1 Mb to be of acceptable resolution. Normal television plays 30 such images each second. Real-time video uses up computer memory at the rate of nearly 30 Mb per second! However, new products using various forms of data compression have been demonstrated. The first to come on the scene was compact disc interactive.

Compact Disc-Interactive (CD-I) has been promoted by Sony and Philips as the wave of the future. Of course neither one of these corporations has always been able to predict correctly what the market would support. For example, the Sony Beta format for videotape lost out to the inferior VHS format. The new CD-I system uses a Philips proprietary hardware and operating system and will be available in fall 1988 for about $1,200. This system will play specially encoded compact discs and will offer 72 minutes of partial-screen motion video. A disc will be able to hold up to 5,000 natural pictures with a 384 x 280 resolution, coupled with sound-over-stills capability. Philips and Sony are talking about a big home entertainment market with education and self-improvement as secondary.

Digital Video Interactive (DVI) was unveiled two years ago by RCA at the compact disc conference hosted by Microsoft, just at the time they were planning to sweep everyone off their feet with the power of CD-I. But of course the exciting possibilities of DVI wowed people at the CD ROM conference. This process uses a VAX minicomputer to compress images for storage on a compact disc: then a specially designed set of chips that can be put into the slots of a personal computer decompresses the images in real time. This would offer 72 minutes of full-screen motion video at 30 frames a second with a resolution of 256×240, or up to 40,000 low resolution stills (256×240) with sound-over-still capability. RCA is talking primarily about industrial/educational products first with growth into the consumer market down the road a piece. The special add-in boards will cost about $4,000 and fit into a standard MS DOS computer, to which you will need to attach a standard CD ROM drive.

So interactive digital video data for classroom use is just around the corner. As a group of educators who primarily use chalk and a blackboard, we're not ready.

Now let me return to discuss the CD ROM for a few minutes. I believe that this technology is so compelling that it will change our educational system, perhaps sooner than we are ready. A single CD ROM is practically a library by itself. Think with me about a possible physics CD ROM. Let's play out a physics scenario. Let's take a conservative estimate of 150,000 pages of text and about 1,500 line drawings (graphs or apparatus). That's about 150 books of text and data. The CD ROM disc will be accompanied by a search-and-retrieval diskette of software to control computer access to all of the data on the disc. It would have find-and-sort software as well as cut-and-paste and insert functions. It should be able to access the data of the CD ROM data from your favorite applications software either the word processor or data base or spreadsheet program that you usually use. You will be able to paste data, graphs, or mathematical functions from the CD ROM text to the document on which you are currently working.

What will we have on such a physics CD ROM Disc? Reference data—about 20 books worth of physical constants, mathematical functions, the periodic table and data on the elements. Biographical data—about 16 books worth of the lives of famous physicists, AAPT membership, AIP memberships, NSTA, the AIP directory of physics departments, and the names, telephone numbers, and addresses of the participants at this conference. Publishers and equipment supply companies—about ten books worth of addresses, catalogs and prices. Textbooks—about 70 books worth of physics books that can range from beginning-level conceptual books up through the standard graduate-level textbooks or monographs. Valuable research articles—about 34 books worth of Nobel prize winners' articles, review articles, history of physics articles, etc.

Think of this: when you sit down at your computer, you will have this whole data set at your finger tips. You can bring on your computer screen in a few seconds any of the information stored on this CD with only a few key strokes. If you are an expert writing an advanced physics research article, you can use the reference data section. If you are a novice trying to learn physics, you can be guided through the levels of materials on the disc by tutorial software. If you wanted to scan a whole range of physics-related materials, as you might in a library, you can sample a broad range of physics materials without ever having to leave your computer keyboard.

This technology is here today, waiting for us to master the first physics compact disc. Here, on the eve of the 550th anniversary of the Gutenberg printing press, we stand on the brink of a new era in publishing and in education.

What is possible? We are limited only by our vision of what ought to be possible.

Beyond TV—Interactive and Digital Video in Physics Teaching

Dean Zollman
Department of Physics, Kansas State University, Manhattan, KS 66506

Encouraging physics students to interact with visual images is not new. Long before videotape or videodiscs were available, students looked at the filmed image of Don Ivy and said, "He's upside down, isn't he?" Later, developers of the Project Physics Curriculum built interaction into a number of Super-8 film cartridges. Students used these cartridges by stopping the film and collecting data from individual images. While these types of interactions were limited by the available technology, they were early attempts to provide students with the ability to interact with visual images and to learn physics from such interactions.

In the mid 1970s technological advances provided a new means for high levels of interactions between students and video images. The optical videodisc (commonly called the LaserDisc) offers random access to any one of 54,000 individual pictures (frames). With no more effort than that required to enter numbers with a keypad, students can get access to any one of these pictures. The access time ranges to a maximum of approximately two seconds on modern videodisc players and 15 seconds on very early videodisc players. Once students have reached the picture they have requested, they have the options of watching or analyzing that picture for any length of time, playing forward or backward at normal speed, playing a video sequence in slow motion, or playing at several speeds which exceed the normal projection rate. Thus with the interactive videodisc students can control the way they view instructional images.

Even greater flexibility can be obtained by connecting the videodisc player to a computer. By programming a computer to control the display of videodisc images and to interact with students, a physics instructor can develop sophisticated lessons which engage students in materials which help them understand physics and how it is applied. In principle, any type of student interaction with a computer can be augmented by video scenes. In addition to being able to create interactive computer materials with the addition of video, the instructor can create materials which involve video as the primary medium of instruction. For example, applying physics to automobile collisions is most useful when students are able to analyze the collisions of real cars. The videodisc *Physics and Automobile Collisions*[1] provides that option. Or, an astronomy teacher may wish to have students look at a large number of images of certain types of stars. The large number of still frames on *Space Disc 6: Astronomy*,[2] when coupled with an appropriate data base, provide students with the opportunity to search out a particular type and to compare various aspects of the images. The microcomputer coupled to a videodisc player provides these opportunities in a way that would be difficult and extremely tedious in any other format.

In this paper I present a short history of the hardware and software available for using interactive video in physics teaching, and then describe a few examples of this type of instruction. Finally, I will look at the most recent attempts to use the sophisticated systems such as IBM InfoWindow and Macintosh *HyperCard*. I conclude with a brief look at the future, which includes a new generation of digital video on compact disc, a method by which students and instructors will have complete control over the image and changes in individual video images.

History

The development of videodiscs for teaching physics began in the late 1970s. The first commercially available videodisc for physics instruction was *The Puzzle of the Tacoma Narrows Bridge Collapse*.[3] This videodisc operated on a stand-alone videodisc player which had an internal microprocessor. The control program, which can operate today on the Pioneer 6000 series, enabled students to control the videodisc images, to complete qualitative experiments on the interaction between a model bridge and the wind, and to complete a quantitative experiment on the standing waves in the vibrations of a string. The nature of the interaction was such that the video, which required 30 minutes to view as a linear program, usually took students three or more hours to complete as an interactive lesson.

Following this initial effort, several more interactive videodiscs for physics teaching have been produced. In addition, some of the classic films for physics teaching, such as *Frames of Reference*[4] and *Powers of Ten*,[5] have been transferred to videodisc so that individual teachers may prepare interactive lessons using them. Because the space sciences and physical sciences series produced by the Optical Data Corporation[6] are also quite useful for teaching many aspects of physics, the number of videodiscs available for physics teaching grew rather rapidly. When *The Mechanical Universe* series was placed on videodisc in the spring of 1989, the total number of the available physics videodiscs more than doubled.

Almost all types of interaction and computer interfacing have been demonstrated in completing physics lessons. *The Puzzle of the Tacoma Narrows Bridge Collapse* is the only one which made extensive use of the small microprocessing capabilities of some videodisc players. However, others such as *Studies in Motion*[7] and *Energy Transformations Featuring the Bicycle*[8] demonstrated the use of computer graphics overlays with video images. At the other end of the spectrum in terms of hardware sophistication, videodiscs such as *Physics and Automobile Collisions* and *Physics of Sports*[9] were developed for use on stand-alone videodisc players. However, as described below, these videodiscs still involve a high level of interaction between the student and the video images.

Hardware

At present, all interactive videodiscs available in the United States play on machines which use optical reflective technology. The laser playback assures rapid access to all of the available frames and minimum deterioration in the quality of the image.[10]

The level of interaction has been defined in a standard way as shown in Table 1. With the exception of level 4, use with artificial intelligence programming, interactive video lessons in physics have been prepared for all levels.

Table 1.

Levels of Interactive Video

Level 0: Linear play only. Random access and still frame not available.
Level 1: Stand-alone player. Random access and still frame available through keypads.
Level 2: Stand-alone player. Program control available with the internal microprocessor of the videodisc player.
Level 3: Player connected to an external computer. Program controlled by the computer operating in a standard language environment.
Level 4: Player connected to an external computer operating in an artificial intelligence environment.

A particular difficulty arises when one wants to mix interactive graphics generated by a computer with the videodisc signal. Videodisc players produce standard television signals (in the United States NTSC video). Essentially all computers produce signals that are not entirely compatible with this standard. While many of the early computers could be played on standard television, their signals were still not entirely compatible and could not be easily mixed with the NTSC video. Many modern computers have resolutions that are much higher than standard television and thus also run into difficulties when the mixture is attempted. However, circuitry is available to make the two signals compatible. Peripherals or boards may be connected to many computers to create a situation in which the two signals can in fact be placed on top of each other. These boards, available for a few hundred dollars for some computers and several thousands for others, provide a level 3 interaction, which includes the possibility of computer-generated graphics placed over a video image. Because of the complexity and expense of this type of hardware, many early interactions involved two-screen systems. One screen would be used for the computer output while a different screen would be used for the videodisc images. This type of arrangement still is used heavily in machines such as the Macintosh SE, which has a video signal that is very difficult to combine with NTSC video.

In general, physics teachers have not attempted to produce complete, self-standing lessons using interactive video. Instead, most lessons have been made part of a course in physics and have been used to supplement existing materials to replace classroom discussions and to act as laboratory exercises. Recently, we have attempted to develop lessons which could be used on a stand-alone player. If the lesson is available on a stand-alone player, we can provide it to many physics teachers without the need for each individual school or university having the same software and hardware. The basic technique involves a videodisc player, its keypad, and a transparent sheet of acetate. Students use the keypad to find an appropriate scene on the videodisc. We then recommend that they play the scene once so they have a feel for what is happening. Then they return to the beginning of the scene (again using the search capabilities of the videodisc player) and begin collection of data. The data collection involves making marks on the transparent acetate. For an analysis of a point particle or the center of mass of an extended particle such as an automobile, the students make individual marks for the location of the center of mass on consecutive frames. Because the rate at which the film was recorded is known, the time between consecutive images is also known. Students may remove the acetate, place it on a piece of graph paper and enter the coordinates into a productivity program such as a spreadsheet. Very quickly then students may obtain velocities, accelerations, and other kinematic information.

A more complex situation involves the collection of data for an object which changes its shape. We have studied objects such as automobiles during collisions and athletes performing high jumps, long jumps and pole vaults. For these situations we have followed the same techniques used by researchers in biomechanics and kinesiology. Students are asked to develop a model of the object or athlete and use the stick figure models to learn more about the motion. The example models can be drawn on the acetate for each of the video images in the scene. Again, the acetate is removed, the coordinates are read into a computer program, and the analysis is completed. This system allows students to begin to understand the nature of a physical model and to apply physics to more complex situations than are usually treated in introductory physics courses. (This system is also beginning to be used in kinesiology and biomechanics classes to help those students understand the value of this technique.)

While these techniques can be very useful, the time involved in drawing the models on the acetate and entering the data can be extremely long. We have found that the analysis of a high jumper using a five-segment model can take students many hours of tedious work. To circumvent that difficulty and still enable students to understand the power of this type of modeling, we have developed a program using interactive graphics overlayed on top of the video screen. Our first attempt at this uses an Amiga computer with a Genlock peripheral. The peripheral provides the mixing of the computer graphics with the video signal while a computer program enables students to locate the end of each segment of the model by moving a pointer with a mouse. As the students click the mouse on each end of each segment, the coordinates of that segment appear on the screen and are simultaneously stored in a file. The analysis can be completed for each of the video frames and the end result is a file containing all of the data concerning the model. This data collection technique is clearly much quicker than that of the acetate overlay.

We believe that it is wise for students to learn to use existing productivity software, and, of course, we wish to minimize our own programming efforts. Thus we have not written an analysis program for these data. Instead, we suggest that students import the data into a spreadsheet program and analyze them in that way. For some students a more appropriate technique may be to write their own programs to analyze the data.

This technique is quite useful. However, it suffers from a lack of hardware standardization. For each type of computer, one would need to produce a different driver and set of materials for both the videodisc control and the graphics overlay. Thus at this point we have only produced a demonstration lesson.

Interactive video materials can be created quite easily with existing videodiscs. The large number of available videodiscs enables any instructor who wishes to create materials to do so. An instructor can choose to use a level 1 system with paper and pencil or acetate sheets to prepare a videodisc lesson. A number of examples of this type of material are available.[11] An instructor with computer programming knowledge or with students who have computer programming knowledge can develop lessons and then drive them with a small computer. Instructors who wish to use authoring systems have a number of options available.[12]

The production of video for a videodisc has until recently been limited to those individuals who have access to professional production studios. However, recent advances in the quality of recording equipment and desk-top video software for the Amiga and Macintosh II have opened up new possibilities.[13] Videodiscs have been produced in which all of the title frames were created using desk-top video on the Amiga; some video materials have been recorded on standard home video equipment and converted to videodisc as well. Thus the capability of preparing video for a videodisc is within the reach of many physics faculty at this point.

Videodiscs that will not be commercially distributed and will only be used for local purposes can be obtained inexpensively at this time. Companies such as Crawford Video in Atlanta and Spectral Images in Burbank can produce a single copy of a videodisc for $300. While the quality of this videodisc is not as high as that of the discs produced by companies such as 3M, Pioneer and Dupont-Philips, it is quite adequate for lessons on an individual campus. Thus one can create one's own video and obtain a videodisc for much lower cost than was possible as recently as one to two years ago.

The Future

Two developments will have great impact on the future use of interactive video in teaching. One of those, hypertext, is available now on some systems. The other, fully digitized video on compact discs, should become available in the very near future.

The hypertext format of software provides a new way for teachers to use some aspects of interactive video relatively easily. Full multimedia presentations using the Macintosh *HyperCard* software have been developed in a number of areas.[14] These programs are particularly useful when one wishes to have students explore a visual data base. For example, *HyperCard* stacks for the National Gallery of Art videodisc provide students of art with a large number of options.[15] They may search through the textual data base associated with the videodisc for any one of a number of descriptors listed on the data base. Thus if a student wished to view all of Van Gogh's treatments of flowers, he or she could find those descriptors in the data base and see each individual picture on a video screen as the description was being displayed on the computer screen. Similar types of *HyperCard* stacks have been developed for a number of topics in astronomy[16] and are under development for some of the physics videodiscs.

A much bigger departure from the present analog video signals will take place when video is available in digital form on a compact disc. Two separate systems are under development at this time. One is Digital Video Interactive (DVI) being developed at the David Sarnoff laboratories.[17] A second system, compact disc interactive (CD-I) is a different format being developed by Philips and Sony.

Because both of the systems provide video in a digital format, the images can be treated as complete computer information. The result will be that future video images can be manipulated in a variety of ways that are only available now on very expensive digital effects generators in broadcast-level TV studios. Images will be able to be changed in size and shape. Even the motion of an object will be able to be changed by processing the image with an appropriate computer graphics program. The difficulties of mixing video signals and computer signals will go away completely because in fact there will be only one signal.

One can begin to understand the power of the CD-I and DVI systems by digitizing individual pictures using today's technology. When these individual pictures are imported into paint programs, one can then cut part of one picture out of an image and paste it onto another one. Or, one can use archival film and video in which the camera moves by manipulating the image to obtain a common reference point and isolating small areas of a video image to use in other graphics presentations. These and other capabilities will enable physics teachers to open up new possibilities for their students.

Conclusion

Interactive video during the past ten years has greatly changed the way in which visual images are used to teach physics. The possibilities of easily collecting data from scenes of real events has enabled us to bring quantitative analysis of the world outside the classroom and into our everyday laboratories. The capabilities of stopping some of the classic physics teaching films at critical points and allowing students to think about and interact with the visual images as well as branch to different places, has provided both students and instructors ways to structure these materials to be more appropriate for individual students or groups of students. New software capabilities and the ever-increasing number of materials available on videodisc mean that many teachers can develop appropriate lessons for their classroom situations.

The advent of digital video will provide another large change in the capabilities of various systems. The digital video systems will enable us to manipulate change and present images in ways that we can only think about at the moment. The future for interactive video is bright and will continue to get brighter as the capabilities of the hardware and the creativity of the physics instructor expand.

Introductory Physics Videodisc Lesson Guide

David M. Winch
Physics Department, U.S. Air Force Academy, Colorado Springs, CO 80840, on leave from the Physics Department, Kalamazoo College, Kalamazoo, MI 49007

The type of interactive education that is crucial for teaching physics is now possible with computer/videodisc technology. An interactive computer/videodisc system combines the logical control of the computer software and the flexibility of the instructor with the suberb audiovisual characteristics of the laser videodisc. The laser videodisc can store 54,000 individual, digitally numbered, color-television images accompanied by two independent audio tracks.

Unlike other forms of audiovisual media, the videodisc does not need to be presented in the same way to every audience. Its random-access capability allows an educator to tailor its use to the needs of a particular class. Videodisc lessons may be used with a computer-controlled videodisc player, a videodisc player remote-control unit, or a combination of a computer with an instructor/student control by a remote-control unit.

Maximum flexibility is obtained by using a combination of computer control and remote control. The instructor has the power to program the lesson for classroom or individual student use and to override the computer with the remote-control unit. This allows the instructor to have the lesson prepared before the presentation and to react to situations that might require stopping a videodisc sequence for discussion, or replay a sequence for better comprehension.

Videodisc systems are widely available. There are two authoring programs (*LaserWrite* and *LaserWork*) with connecting videodisc cables for the Apple IIe; *HyperCard* software products that drive videodisc players run on the Macintosh; the Info Window computer/videodisc system runs on IBM systems, and the U.S. Air Force Academy Physics department has developed a Zenith-based system.[1]

To accompany the U.S. Air Force Academy system, we are developing a set of instructor lesson guides. Each lesson guide includes a title, the lesson objective, the title of the videodisc, frame numbers of videodisc scenes, a description of videodisc scenes, suggested activities, and sample discussion questions.

An average lesson requires approximately ten minutes of class time and includes a number of learning methods (motivation, discovery, problem solving, and discussion). The lesson guides present the required videodisc information for the lessons. The instructor chooses the most suitable hardware and software to drive the videodisc player.

1. T. Gist and G. Lorenzen, "Hardware/Software Implementation of an Integrated In-Class Instructor Work Station," TITE **87**, 19 (1987).
2. R. G. Fuller, D. Zollman, and T. C. Campbell, *The Puzzle of the Tacoma Narrows Bridge Collapse* (New York: John Wiley and Sons, 1982); R. G. Fuller and D. Zollman, *Energy Transformations Featuring the Bicycle* (Lincoln, NE: Great Plains Media Center, University of Nebraska–Lincoln, 1984); D. Zollman and R. G. Fuller, *Studies in Motion* (Lincoln, NE: Great Plains Media Center, University of Nebraska–Lincoln, 1984); D. Zollman, *Physics and Automobile Collisions* (New York: Wiley, 1984); P. Morrison and P. Morrison, *The World of Charles and Ray Eames Videodisc, Powers of Ten* (Lexington, KY: Ztek Co. 1986); P. Hume and D. Ivey, *Frames of Reference* (Princeton, NJ: Educational Services Incorporated, 1960).

Using Videodiscs Interactively with *HyperCard*

David M. Winch
Department of Physics, U.S. Air Force Academy, Colorado Springs, CO 80840, on leave from the Department of Physics, Kalamazoo College, Kalamazoo, MI 49007

Robert G. Fuller
Department of Physics, U.S. Air Force Academy, Colorado Springs, CO 80840, on leave from the Department of Physics and Astronomy, University of Nebraska-Lincoln, Lincoln, NE 68588-0111

The laser videodisc stores 54,000 individual, digitally numbered, color-television images accompanied by two independent audio tracks. There are many physics videodiscs available (*The Puzzle of the Tacoma Narrows Collapse, Energy Transformations Featuring the Bicycle, Studies in Motion, Physics and Automobile Collisions, The World of Charles and Ray Eames* (*The Powers of Ten*), *Skylab Physics, Frames of References, and the Mechanical Universe Series*).

At the simplest level, the instructor may use the videodisc remote unit as a lecture-demonstration device to access an image, play a motion sequence, pause, search, etc. This can be a simple and effective method to bring real-world events into the classroom and promote discussion. This method is, however, cumbersome to use when the presentation requires recalling many visual images.

An instructor can design more complicated interactive lessons by coupling a computer to the videodisc player. These lessons may be used in the lecture, in the laboratory, or in an individual learning environment. It is possible to use authoring languages to control videodisc players by computer, but these systems are hard to use. Now, however, videodiscs can be controlled with the easy-to-use *HyperCard* environment. *HyperCard* is a software erector set—a group of tools you can use to create your own Macintosh software applications using Macintosh icons and mouse style. The basic tools consist of buttons, fields, graphics, and cards. The developer can add *HyperTalk* commands to these basic tools to expand the range of applications. *HyperCard* stacks are now available to drive videodisc players, and have increased the ease of faculty lesson development and student use.

Hardware and software requirements for such a computer/videodisc system include a Macintosh Plus, SE, or Mac II with a hard disk drive; a videodisc player with an RS-232 interface (Pioneer LVP4200, Pioneer LDV6000, Sony1500, HitachiVideo, etc.); an appropriate cable connector (Macintosh to videodisc player); *HyperCard* software; and an appropriate videodisc driver stack.

The "Mechanical Universe" Videodisc Project

David Campbell
Department of Physics, Saddleback Community College, Mission Viejo, CA 92675

Don Delson
Department of Physics, California Institute of Technology, Pasadena, CA 91125

Dean Zollman
Department of Physics, Kansas State University, Manhattan, KS 66506

The television series "The Mechanical Universe and Beyond" consists of 52 programs that span the topics in an introductory physics course. The programs can provide instruction in broadcast or cassette form, but such a format does not allow classroom or independent-study random access. To provide random access the series is being placed on videodisc. Accompanying software for the IBM InfoWindow interactive video system will offer teachers and students the ability to select segments from the discs and to use the materials in an interactive mode. I will demonstrate examples of progress to date.

The "Mechanical Universe and Beyond" television series received major funding from the Annenberg/CPB Project. The videodisc development is being supported by IBM.

Interactive Laser Videodisc

John A. Rogers
Science Center, Educational Service Unit #3, Omaha, NE 68137

The problem of providing quality instruction has reached crisis proportions in the general education community. This crisis has been caused by the decrease in the amount of real money available, the increase in the diversity of our population (non–English speaking, etc.), the decrease in the number of qualified teachers (especially in the physical sciences), and the increase in pressure to better prepare students in fundamentally important skills, concepts, and thinking/learning processes.

A variety of technologies can help the educational community meet many of the instructional requirements facing us today and in the near and far futures. If "technologies" are properly designed and implemented, they can capture the essence of our best instructors' teaching methods so that our students can achieve their potentials.

We are a visually oriented society; we seek high-quality images. One way to enhance the quality of education is to use high-fidelity color images of both real (concrete) and abstract objects and concepts. Both still images and real-time motion sequences can be of educational value. The best method we presently have available to deliver these images to students is the laser videodisc (LaserVision).[1]

This talk will be a brief introduction to interactive video technology (LaserVision disc). Conference participants will have opportunity for hands-on viewing time and working with a variety of laser videodiscs, players, and micro-computer systems.

Some laser videodiscs are available through general distribution. Each disc is designed for a specific level of control—some level 1, some level 2 and some level 3—but all have educational value and can be used in a variety of systems and teaching methodologies. Prices range from $40 to $400 per disc. Table 1 lists specific discs that relate to physics.

Teachers can use authoring programs to create their own interactive videodisc lessons. Authoring programs make designing, scripting, and programming lessons easier. The current versions of most programs control many of the decisions that must be made while authoring by menu. Menu choices include frame/chapter searches, play sequences, types of participant responses, branching information, page style, computer-generated graphic image recalled from floppy disk, and the way participants' responses/choices are recorded.

In choosing an authoring program, the following considerations are of primary concern: (1) ease of use in both software programming and videodisc player-computer interfacing; (2) broad use in education; (3) low cost; (4) features and power.

Complexity is more often a major stumbling block than cost for a neophyte author. Some programs offer more features and instructional power, but they require more costly equipment and more time to learn. Although simpler programs are somewhat limited in their features, they have lower learning curves.

Authoring is a very complex and time consuming process and it is made still more complex and time consuming by hard-to-use programs. My philosophy is to keep it simple when beginning and grow as needed. As a teacher's need for more features and power increases, growth into more complex programs is easy. Table 2 shows the authoring programs that I support and work with.

The interface is the physical/electrical connection between the microcomputer and the laser videodisc player. The general trend is to make the interfacing simple. As with the software, I believe that the system should start simple and grow as funds and needs increase. All of the single-cable interface systems require two screen presentations. Prices range from $25 to $1000 depending on the author's requirements for features of the system. Table 3 lists interfaces that I support and use.

Scientists and engineers, through industry, can now provide educators with computers that are low in cost, and have high operating speeds, larger RAM, and improved graphics. Mass information storage such as CD-ROM and 40M+ RAM magnetic hard drives, and powerful, flexible, and user-friendly software are now readily available. In addition, there are increasingly better methods of visual technologies, e.g., laservision, DVI, etc. The educational community is benefiting from better intercommunications, which will enable instructors to share information and ideas. These technological and communication advances will allow us to see the development of better instructional delivery tools and systems. The advanced technology will free teachers to concentrate on what they can do best—interact with students.

Table 1

Companies Distributing Videodiscs Related to Physics

Great Plains National Video, Tel. (800) 228-4630

Studies in Motion	Uses swimmers, gymnasts, and a dancer to illustrate and study linear and rotational motion. Computer program available.
Energy Transformation	Uses several bicycles to study energy systems. Computer program available.

John Wiley, Tel. (212) 850-6418

Puzzle of Tacoma Narrows Bridge Collapse	Originally a Level 2 disc, useful Level 1 and 3. Has several "experiments" on it.
Physics and Automobile Collisions	Straight-line and 2D collisions–from side and above.

American Association of Physics Teachers, Tel. (301) 345-4200

Skylab Physics	Includes print materials for background and student activities.

Zytek, Tel. (800) 252-7276

Powers of Ten	Contains other video materials from Eames, included is the book *Powers of Ten*.
National Air and Space Museum, Vols. 1, 2, and 3	The three discs contain approximately 100,000 still images each of aircraft, etc.
Space Watch: Home Planetarium	

Central Scientific Co., Tel. (800) 262-3626

Frames of Reference	The original PSSC film can be still-framed.
Photons and Interference of Photons	Two PSSC films, both on one disc.

Optical Data Corp., Tel. (800) 524-2481

Physical Science	Disc 1: *Matter, force and motion.*
	Disc 2: *Waves, E and M and reactions.* Good physics.
Space Flight	Space Archive Vol. 1: *Space Shuttle.*
	Space Archive Vol. 2: *Apollo 17.*
	Space Archive Vol. 3: *Mars and Beyond* and *Landings on Mars.* Some images are 3D.

Space Archive Vol 4: *Shuttle Downlink*. Includes repair of Solar Max satellite and free walk in space.

Space Archive Vol. 5: *Greeting from Earth*. High altitude and satellite images of Earth.

Space Archive Vol. 6: *Encounters*. Contains Halley's comet information and Skylab "filmloops."

Pulling G's	Motion sequences of a variety of aircraft. A fun disc.
Astronomy	Very good for general astronomy classes.
The Sun	First side: many images of the sun; second side: the movie *Universe*.

Image Premastering, Tel. (612) 454-9622

Image Premastering Service-Shared Disc	Approximately 200 AAPT slides and a variety of many other images.

Rose-Hulman Institute of Technology, Tel. (812) 877-1511

Enlivening Physics	Project between Rose-Hulman and area high schools. Limited distribution.

National Technical Information Service, Tel. (703) 487-4650

Planetary Image Disc	Two volumes.

Additional companies distributing videodiscs related to physics:

Teaching Technologies, Tel. (805) 541-3100

Sargent Welch, Tel. (312) 677-0600

VideoDiscovery, Tel. (800) 548-3472

Minnesota Educational Computing Corporation, Tel. (800) 228-3504

Arbor Scientific, Tel. (313) 663-3733

In preproduction:

Mechanical Universe	College Version
Physics of Sport	

Table 2
Authoring Programs

Program Name	Producer/Distributor	Telephone
Apple II (+, e, c, gs) family of microcomputers		
LaserWorks	Teaching Technologies	(805) 541-3100
	Ztek.	(800) 247-1603
	Video Discovery.	(900) 548-3472
LaserWrite/Lesson Maker	Optical Data Corp.	(800) 524-2481
Macintosh (Plus, SE and Mac II microcomputers)		
HyperCard	Apple Computer Corp.	Local Apple Dealer
Voyager Video	StackVoyager Company	(213) 474-0032
MS-DOS (IBM PC-DOS and compatible microcomputers)		
VidKit	VideoDiscovery	(800) 548-3472

Table 3
Computer-Videodisc Player Interface

Program Name	Producer/Distributor	Telephone
Apple II (+,e,c,gs) family of microcomputers		
Laser Talk	Optical Data Corp	(800) 524-2481

Single cable between Apple II (+, e and gs); two-screen system only; no video switching. Cables for Pioneer LD-V2000 (and similar players) and LD-V4200.

LDI	Teaching Techniques	(805) 541-3100

Two versions, both for the Apple II (+, e, c, and gs) family of microcomputers; (1) single cable, two-screen-only system, no video switching, and (2) video switching interface that allows either one screen, video-switching systems, or two-screen systems without video switching. Pioneer LD-V2000 (and similar players) and LD-V4200 players supported.

MS-DOS (IBM PC-DOS)		
Visual Database Systems Cables	Visual Database Systems	(408) 438-8396
Cables	VideoDiscovery	(800) 548-3472

Macintosh (*HyperCard*)

Cables for the Macintosh microcomputer to a variety of players are available from the companies given above. An interface between Macintosh microcomputers and Pioneer LD-V2000 players will be available during the summer of 1988. Contact the above companies for more information.

1. Ed Schwartz, The Educator's Handbook to Interactive Videodisc (Washington, DC: Association for Educational Communications and Technology, 1987); Should Schools Use Videodisc? (Alexandria, VA: Institute for the Transfer of Technology to Education, National School Boards Association); Videodiscs in Education (Chelmsford, MA: Merrimack Education Center); The Videodisc Monitor (a monthly videodisc industry publication).

Smooth Animation on the IBM PC for Lecture Demonstrations Using Page Flipping

Randall S. Jones
Physics Department, Loyola College, Baltimore, MD 21210

There are two main problems with using computer graphics to demonstrate physics concepts in the lecture room. First, the cost of projection systems is high. Second, a lecturer who uses an IBM color graphics adapter (CGA) can't do page flipping.

Page flipping is the graphics technique of drawing on an invisible screen in memory while a second screen is displayed, and then flipping the two pages. Since none of the line drawing is visible, page flipping produces smooth animation.

The first of these problems has been solved by the recent arrival of liquid crystal display (LCD) projection systems[1] for use on overhead projectors. The second problem can be solved by using an extended graphics adapter (EGA) standard, which allows page flipping for smooth animation in several of its graphics modes. Unfortunately, the current crop of LCD projection displays do not support the EGA standard.

There are, however, several inexpensive EGA graphics cards, for example, EGA Wonder by ATI Technologies Inc., that will connect to CGA monitors. These cards may be used to drive LCD projection displays. Thus, smooth animation is now available even for those of us on tight budgets.

We will show the effectiveness of page flipping in generating smooth graphics effects by demonstrating several simple programs that we have written in BASICA to simulate wave interference effects. Although such programs are easy to produce, they are effective instructional aids.

1. Frank Bican, "Real-Time Overhead Displays for the Big Screen," PC Magazine 7, (5) 161 (1988).

Trademarks

MacFORTRAN™ is a trademark of Absoft Corporation. Analog Devices® is a registered trademark of Analog Devices, Inc. Apple®, Apple SuperPILOT®, Applesoft BASIC®, Laserwriter®, MacPaint®, and MacWrite® are registered trademarks of Apple Computer, Inc. Apple DOS 3.3™, HyperCard™, HyperTalk™, MacPascal™, Multifinder™, and Switcher™ are trademarks of Apple Computer, Inc. Macintosh™ is a trademark licensed to Apple Computer, Inc. ASYSTANT+2™ is a trademark of ASYST Software Technologies, Inc. UNIX® is a registered trademark of AT&T Bell Labs. Beagle Bros© is copyrighted by Beagle Bros Micro Software, Inc. Eureka: The Solver™ and Turbo Pascal Numerical Methods Toolbox™ are trademarks of Borland International. Turbo BASIC® and Turbo Pascal® are registered trademarks of Borland International. The PLATO® system is a registered trademark of Control Data Corporation. Commodore 64®, Commodore 128® and VIC-20™ are registered trademarks of Commodore Electronics, Ltd. Amiga® is a registered trademark of Commodore-Amiga, Inc. AmigaDOS™ is a trademark of Commodore-Amiga, Inc. Compaq™ is a trademark of Compaq Computer Co. SuperCalc™ is a trademark of Computer Associates. Cricket Graph™ is a trademark of Cricket Software. Digital PRO™, MicroVax™, and VAX™ are trademarks of Digital Equipment Corporation. DADiSP™ is a trademark of the DSP Development Corporation. EPSON™ is a trademark of Epson America, Inc. Exxon® is a registered trademark of Exxon Company, U.S.A. Great Plains™ is a trademark of Great Plains Publishing Co. Hercules® is a registered trademark of Hercules Computer Technology Corp. Hewlett-Packard® and HP® are registered trademarks of Hewlett-Packard Corp. Hitachi™ is a trademark of the Hitachi Corp. HRM™ is a trademark of HRM Software. IBM PC®, IBM XT™, IBM AT®, IBM RT®, IBM PS/2®, and PC-DOS® are registered trademarks of International Business Machines, Inc. IBM Color Graphics Adaptor™, IBM Enhanced Graphics adaptor™, IBM CGA™, IBM EGA™, IBM VGA™, IBM MCGA™, IBM InfoWindow™, IBM Proprinter™, and LAN™ are trademarks of International Business Machines, Inc. Intel™ is a trademark of the Intel Corp. Keithley® is a registered trademark of Keithley Instruments, Inc. Volkswriter™ is a trademark of Lifetree Software. Lotus 1-2-3® is a registered trademark of Lotus Development Corporation. MathCAD™ is a trademark of MathSoft, Inc. BenchTop™ is a trademark of Metaresearch, Inc. GW BASIC™ and QuickBASIC™ are trademarks of Microsoft Corp. Excel®, Microsoft®, MS-DOS®, and Works® are registered trademarks of Microsoft Corporation. LaserTalk™ is a trademark of Optical Data Corp. Point Five™ is a trademark of Pacific Crest Software, Inc. MacViz™ and ProViz™ are trademarks of Pixelogic, Inc. MacΣqn© is copyrighted by Software for Recognition Technologies. Palette™ is a trademark of Software Wizardry, Inc. Sony® is a registered trademark of Sony Corporation. Sun™ is a trademark of Sun Microcomputers, Inc. Radio Shack® is a registered trademark of Tandy Corp. LaserWork™ is a trademark of Teaching Technologies, Inc. TrueBASIC™ is a trademark of True Basic, Inc. TK Solver™ is a trademark of Universal Technical Systems, Inc. The NovaNET™ system is a registered trademark of University Communications, Inc. Zenith® is a registered trademark of the Zenith Corp.

Author Index

Index

Software Contents

For the Macintosh (800K diskettes)

Disk 1
*Spacetime and
Collision*
E. F. Taylor

MacScope
E. R. Huggins

Disk 2
Density of Orbitals
J. W. Gardner

Bouncing Ball
N. B. Tufillaro and
T. A. Abbott

*Fixed and Free Reflection,
Friction, Car & Truck*, and
Hunter & Monkey
I. A. Miller

For the Amiga

Disk 3
*Scattering of a Wave
Packet from a Potential*
J. E. Lewis

For the IBM-PC

Disk 4
*Sample Chapters from
Computer-Based
Instruction for
University Physics*
J. R. Priest

Rigid Pendulum Simulation
J. A. Goodman

Disk 5
Optics Demonstrations
U. E. Kruse

Disk 6
*Microcomputer-Delivered
Problem System*
R. C. Enger and
G. L. Lorenzen

Disk 7
*Sample Simulations from
Fields and
Electromagnetism*
P. B. Visscher

Disk 8
Spacetime
E. F. Taylor

Disk 9
Collision
E. F. Taylor

Disk 10
Dynamic Analyzer
R. F. Simpson

Disk 11
Simulation Lab
M. H. Krieger and
R. N. Martin

*Two-Dimensional Ising
Model Simulation*
J. R. Fox

*Billiard in a
Gravitational Field*
B. N. Miller and
H. Lehtihet

*Solar Neighborhood
Cluster Dynamics
Simulator*
J. J. Dykla

Disks 12 and 13
*The Ohio University
Chaotic Dynamics
Workbench*
R. W. Rollins

Disk 14
*Physics Spreadsheets
for Lotus 1-2-3*
R. Novak

*Physics Spreadsheets
for Student Edition of
Lotus 1-2-3*
G. Hept

Computers in Physics Instruction: Software

A collection of software programs selected from the Conference on Computers in Physics Instruction is available for $90.00 plus $4.00 shipping and handling per set. Each set contains 21 diskettes (for five computers) and an instruction booklet describing the 34 software programs. Diskettes are not available individually.

Send check, money order, credit card number, or purchase order to the following address and allow two to four weeks for delivery:

> CPI Conference
> Department of Physics
> North Carolina State University
> Raleigh, NC 27695-8202
> Tel. (919) 737-7059

ORDER FORM

(Cut out or photocopy form and mail to address above.)

Please send _____ sets of Computers in Physics Instruction: Software for $90.00 each, plus $4.00 shipping and handling for each set.
    Enclosed is $_____ .

Check (make payable to North Carolina State University)
    _____ Money Order
    _____ Purchase Order No._____
    _____ Charge Card: _____ MasterCard _____ VISA

        Card No. _____ Exp. Date_____

        Authorized Signature _____

Ship to:

_____

_____

_____

The Addison-Wesley **Advanced Book Program** would like to offer you the opportunity to learn about our new physics and scientific computing titles in advance. To be placed on our mailing list and receive pre-publication notices and special offers, just **fill out this card completely** and return to us, postage paid. Thank you.

Title and Author of this book: **Date purchased:**

Name _____

Title _____

School/Company _____

Department _____

Street Address _____

City _____ State _____ Zip _____

Telephone/s(   ) _____ (   ) _____

Where did you buy/obtain this book?

☐ Bookstore   ☐ Mail Order   ☐ School (Required for Class)
☐ Campus Bookstore   ☐ Toll Free # to Publisher   ☐ Professional Meeting
☐ Other _____ ☐ Publisher's Representative

What professional scientific and engineering associations are you an active member of?

☐ AAPT (Amer Assoc of Physics Teachers)   ☐ APS (Amer Physical Society)   ☐ SPS (Society of Physics Students)
☐ AIP (Amer Institute of Physics)   ☐ Sigma Pi Sigma   ☐ AAAS (Amer Assoc for the Advancement of Science)
☐ Other _____

Check your areas of interest.

⑩ ✔**Physics**

11 ☐ Quantum Mechanics	18 ☐ Materials Science	25 ☐ Geophysics
12 ☐ Particle/Astro Physics	19 ☐ Biological Physics	26 ☐ Medical Physics
13 ☐ Condensed Matter	20 ☐ High Polymer Physics	27 ☐ Optics
14 ☐ Mathematical Physics	21 ☐ Chemical Physics	28 ☐ Vacuum Physics
15 ☐ Nuclear Physics	22 ☐ Fluid Dynamics	
16 ☐ Electron/Atomic Physics	23 ☐ History of Physics	
17 ☐ Plasma/Fusion Physics	24 ☐ Statistical Physics	
29 ☐ Other _____		

Are you more interested in: ☐ theory ☐ experimentation?

Are you currently writing, or planning to write a textbook, research monograph, reference work, or create software in any of the above areas?
       ☐ Yes   ☐ No
       Area: _____

(If Yes) **Are you interested in discussing your project with us?**
       ☐ Yes   ☐ No

Physics

fold and staple

||.|...|..|||....|.|.|.|.|.|..|.|....||||.|...|.||

||| ||| |

No Postage
Necessary
if Mailed in the
United States

BUSINESS REPLY MAIL
FIRST CLASS PERMIT NO. 828   REDWOOD CITY, CA   94065

Postage will be paid by Addressee:

ADDISON-WESLEY
PUBLISHING COMPANY, INC.®

Advanced Book Program
390 Bridge Parkway, Suite 202
Redwood City, CA 94065-1522